Lecture Notes in Business Information Processing

547

Series Editors

LNBIP reports state-of-the-art results in areas related to business information systems and industrial application software development – timely, at a high level, and in both printed and electronic form.

The type of material published includes

- Proceedings (published in time for the respective event)
- Postproceedings (consisting of thoroughly revised and/or extended final papers)
- Other edited monographs (such as, for example, project reports or invited volumes)
- Tutorials (coherently integrated collections of lectures given at advanced courses, seminars, schools, etc.)
- Award-winning or exceptional theses

LNBIP is abstracted/indexed in DBLP, EI and Scopus. LNBIP volumes are also submitted for the inclusion in ISI Proceedings.

Jānis Grabis · Tanja E. J. Vos ·
Maria José Escalona · Oscar Pastor
Editors

Research Challenges in Information Science

19th International Conference, RCIS 2025
Seville, Spain, May 20–23, 2025
Proceedings, Part I

Springer

Editors
Jānis Grabis
Riga Technical University
Riga, Latvia

Tanja E. J. Vos
Open Universiteit
Heerlen, The Netherlands

Maria José Escalona
Universidad de Sevilla
Seville, Spain

Oscar Pastor
Universidad Politécnica de Valencia
Valencia, Valencia, Spain

ISSN 1865-1348 ISSN 1865-1356 (electronic)
Lecture Notes in Business Information Processing
ISBN 978-3-031-92473-6 ISBN 978-3-031-92474-3 (eBook)
https://doi.org/10.1007/978-3-031-92474-3

This Springer imprint is published by the registered company Springer Nature Switzerland AG
The registered company address is: Gewerbestrasse 11, 6330 Cham, Switzerland

Preface

Volumes 547 and 548 of the Lecture Notes in Business Information Processing series contain the proceedings of the 19th International Conference on Research Challenges in Information Science, RCIS 2025, held in Seville, the capital of Andalusia in Spain, during 20-23 May, 2025. Seville is famous for its stunning Moorish and Gothic landmarks, including the Alcázar of Seville, a UNESCO World Heritage Site. Also, the city is the birthplace of the flamenco dance and it hosts the world-famous *Feria de Abril*, a lively festival filled with colourful dresses, horse parades, and Andalusian culture.

The scope of RCIS covers a broad range of thematic areas, including information systems and their engineering, user-oriented approaches, data and information management, enterprise management and engineering, domain-specific information systems engineering, data science, information infrastructures, and reflective research and practice. RCIS 2025 focused on the special theme "Advancing Information Science and Information Systems Quality in the Era of Complexity."

We received a total of 103 full paper submissions that were carefully reviewed and selected. Among these, five submissions were desk rejected for being outside the scope of the conference. The remaining papers underwent a rigorous single-blind review process, each evaluated by at least three Program Committee (PC) members. Following this, the review process included a discussion period moderated by Program Board (PB) members, culminating in a PB meeting held in Paris to finalize the selection of papers for the conference program. Eventually 33 papers were accepted, meaning the acceptance rate for the main conference was 34%. Authors were provided with meta-reviews and recommendations to incorporate in their camera-ready versions.

The proceedings also include 10 short papers accepted for presentation at the Forum, which aims to showcase new research ideas and prototypes in information science, and three short papers accepted for discussion at the Doctoral Consortium.

The conference program commenced with workshops, followed by the main conference featuring keynotes, research paper sessions, tutorials, the Forum, the Doctoral Consortium, research projects, and journal-first presentations. RCIS 2025 hosted three distinguished keynote presentations:

Tijs van der Storm, from the CWI, The Netherlands about *The next 700 Low-Code Platforms.*

Rebeca Gutierrez from the University of Seville, Spain, about *Challenges in Forming European Consortia for Ethical and Human-Centered AI: Overcoming Barriers to Collaborative Innovation.*

Giancarlo Guizzardi from the University of Twente, The Netherlands on *Explanation, Semantics, Ontology & Trustworthiness.*

Additionally, three accepted tutorials provided deep dives into timely topics central to the RCIS community.

Tutorial 1. *Problematising and Ideating for Design Science Research*, delivered by John R. Venable, Curtin University, Australia.

Tutorial 2. *Taking control of our privacy when using mobile apps: checks and tools*, delivered by M. Mercedes Martínez-González, Alejandro Pérez-Fuente, Amador Aparicio, and Pablo-Abel Criado-Lozano, Privacy Engineering Research Group, Universidad de Valladolid, Spain.

Tutorial 3. *Introduction to Fractal Enterprise Model (FEM) and FEM toolkit*, delivered by Ilia Bider, Stockholm University, Sweden

We extend our sincere gratitude to all authors who submitted their work to RCIS 2025, as well as to the Program Committee and Program Board members for their dedication in reviewing and discussing the submitted papers. We also wish to thank all members of the Organization Committee and the student volunteers for their invaluable contributions to making this conference a success.

Jānis Grabis
Tanja E. J. Vos

Organization

Conference Chairs

General Chairs

Oscar Pastor	Universitat Politècnica de València, Spain
María José Escalona	Universidad de Sevilla, Spain

Program Chairs

Jānis Grabis	Riga Technical University, Latvia
Tanja E.J. Vos	Open Universiteit, The Netherlands and Universitat Politècnica de València, Spain

Steering Committee

Saïd Assar	Institut Mines-Télécom Business School, France
Marko Bajec	University of Ljubljana, Slovenia
Xavier Franch	Universitat Politècnica de Catalunya, Spain
Renata Guizzardi	University of Twente, The Netherlands
Haralambos Mouratidis	University of Essex, UK
Pericles Loucopoulos	Institute of Digital Innovation and Research, Ireland
Selmin Nurcan	Université Paris 1 Panthéon-Sorbonne, France
Oscar Pastor	Universitat Politècnica de València, Spain
Jolita Ralyté	University of Geneva, Switzerland
Colette Rolland	Paris1 Panthéon Sorbonne, France
Maribel Yasmina Santos	University of Minho, Portugal
Jelena Zdravkovic	Stockholm University, Sweden

Local Organizing Team

Lola de Acuña	University of Seville, Spain
José Luis Alonso-Rocha	University of Seville, Spain
Irene Barba	University of Seville, Spain
Elena Enamorado	University of Seville, Spain
Julián García-García	University of Seville, Spain
Manuel García Romero	University of Seville, Spain
José Gonzalez-Enríquez	University of Seville, Spain
Javier Gutiérrez	University of Seville, Spain
Andrés Jimenez	University of Seville, Spain
Antonio Martinez-Rojas	University of Seville, Spain
Ana Jaleh Mehrabi de Acuña	University of Seville, Spain
Leticia Morales	University of Seville, Spain
Antonio Rodríguez-Ruiz	University of Seville, Spain
Nicolás Sánchez	University of Seville, Spain
Tomasz Wojdynski	University of Seville, Spain

Doctoral Consortium Chairs

João Araújo	Universidade NOVA de Lisboa, Portugal
Nora Koch	Universidad de Sevilla, Spain

Forum Chairs

Greta Adamo	Universitat Politècnica de València, Spain
Beatriz Marín	Universitat Politècnica de València, Spain

Workshop Chairs

Raian Ali	Hamad Bin Khalifa University, Qatar
Sergio España	Utrecht University, The Netherlands
Sameha Alshakhsi	Hamad Bin Khalifa University, Qatar

Tutorial Chairs

Rebecca Deneckere	Centre de Recherche en Informatique, France
Mohamad Gharib	University of Tartu, Estonia

Research Projects Chairs

Isabel Brito	Polytechnic Institute of Beja, Portugal
Irina Rychkova	Université Paris 1 Panthéon-Sorbonne, France

Journal-First Chairs

Saïd Assar	Institut Mines-Télécom Business School, France
Iris Reinhartz-Berger	University of Haifa, Israel

Publicity Chairs

Istvan David	McMaster University, Canada
Ben Roelens	Open Universiteit, The Netherlands
Pedro Guimaraes	Universidade do Minho, Portugal
Glenda Amaral	University of Twente, The Netherlands

Student Internationalization Chair

Selmin Nurcan	Université Paris 1 Panthéon - Sorbonne, France

Proceedings Chairs

Elizabete Citskovska	Riga Technical University, Latvia

Program Board

Saïd Assar	Institut Mines-Télécom Business School, France
Marko Bajec	University of Ljubljana, Slovenia
Dominik Bork	TU Wien, Austria

Xavier Franch	Universitat Politècnica de Catalunya, Spain
Renata Guizzardi	University of Twente, The Netherlands
Haralambos Mouratidis	University of Essex, UK
Pericles Loucopoulos	Institute of Digital Innovation and Research, Ireland
Selmin Nurcan	Université Paris 1 Panthéon-Sorbonne, France
Oscar Pastor	Universitat Politècnica de València, Spain
Jolita Ralyté	University of Geneva, Switzerland
Manfred Reichert	University of Ulm, Germany
Colette Rolland	Paris1 Panthéon Sorbonne, France
Maribel Yasmina Santos	University of Minho, Portugal
Jelena Zdravkovic	Stockholm University, Sweden

Program Committee

Ademar Aguiar	University of Porto, Portugal
Raian Ali	Hamad Bin Khalifa University, Qatar
Carina Alves	Universidade Federal de Pernambuco, Brazil
Vasco Amaral	NOVA University Lisbon, Portugal
João Araújo	Universidade NOVA de Lisboa, Portugal
Fatma Başak Aydemir	Utrecht University, The Netherlands
Clara Ayora	Universidad de Castilla-La Mancha, Spain
Anna Bernasconi	Politecnico di Milano, Italy
Isabel Sofia Brito	Polytechnic Institute of Beja, Portugal
Jean-Michel Bruel	IRIT, France
Faiza Bukhsh	University of Twente, The Netehrlands
Nelly Condori Fernández	Universidad de Santiago de Compostela, Spain
Mario Cortes-Cornax	Université Grenoble Alpes, France
Fabiano Dalpiaz	Utrecht University, The Netherlands
Maya Daneva	University of Twente, The Netherlands
Istvan David	McMaster University, Canada
Andrea Delgado	Universidad de la República, Uruguay
Rebecca Deneckere	Centre de Recherche en Informatique, France
Chiara Di Francescomarino	University of Trento, Italy
Rodrigo Falcão	Fraunhofer IESE, Germany
Hans-Georg Fill	University of Fribourg, Germany
Andrew Fish	University of Brighton, UK
Agnès Front	Université Grenoble Alpes, France
Ignacio García Rodríguez de Guzmán	University of Castilla-La Mancha, Spain
Sepideh Ghanavati	University of Maine, USA

Mohamad Gharib	University of Tartu, Estonia
Ana-Maria Ghiran	Babeş-Bolyai University of Cluj-Napoca, Romania
Giovanni Giachetti	Universitat Politècnica de València, Spain
Cesar Gonzalez-Perez	Incipit CSIC, Spain
Jaap Gordijn	Vrije Universiteit Amsterdam, The Netherlands
Miguel Goulão	Universidade Nova de Lisboa, Portugal
Martin Henkel	Stockholm University, Sweden
Jennifer Horkoff	University of Gothenburg and Chalmers University of Technology, Sweden
Felix Härer	FHNW University of Applied Sciences and Arts, Switzerland
Mirjana Ivanovic	University of Novi Sad, Serbia
Christos Kalloniatis	University of the Aegean, Greece
Oliver Karras	Leibniz Information Centre for Science and Technology, Germany
Evangelia Kavakli	University of the Aegean, Greece
Manuele Kirsch Pinheiro	Paris 1 Panthéon-Sorbonne University, France
Elena Kornyshova	CNAM, France
Tong Li	Beijing University of Technology, China
Lidia Lopez	Universitat Politècnica de Catalunya, Spain
Beatriz Marin	Universitat Politècnica de València, Spain
Andrea Marrella	Sapienza University of Rome, Italy
Raimundas Matulevicius	University of Tartu, Estonia
Giovanni Meroni	Technical University of Denmark, Denmark
Fredrik Milani	University of Tartu, Estonia
Denisse Muñante	ENSIIE & Télécom SudParis, France
Elena Navarro	University of Castilla-La Mancha, Spain
Kathia Oliveira	Université Polytechnique Hauts-de-France, Spain
Ana Paiva	University of Porto, Portugal
Jose Ignacio Panach Navarrete	Universitat de València, Spain
Oscar Pastor	Universitat Politècnica de València, Spain
Rūta Pirta	Riga Technical University, Latvia
Thomas Polacsek	ONERA, France
Henderik A. Proper	TU Wien, Austria
Ben Roelens	Open Universiteit, The Netherlands
Patricia Rogetzer	University of Twente, The Netherlands
Marcela Ruiz	Zurich University of Applied Sciences, Switzerland
Irina Rychkova	Université Paris 1 Pantheon-Sorbonne, Paris
Rainer Schmidt	Munich University of Applied Sciences, Germany
Florence Sedes	IRIT, Université Toulouse III Paul Sabatier, France

Denis Silveira	Universidade Federal de Pernambunco, Brasil
Anthony Simonofski	Université de Namur, Belgium
Samira Si-Said Cherfi	Conservatoire National des Arts et Métiers, France
Monique Snoeck	KU Leuven, Belgium
Pnina Soffer	University of Haifa, Israel
Eric-Oluf Svee	Stockholm University, Sweden
Daniel Stanley Tan	Open Universiteit, The Netherlands
Ernest Teniente	Universitat Politècnica de Catalunya, Spain
Olivier Teste	IRIT, France
Porfirio Tramontana	University of Naples Federico II, Italy
Nicolas Travers	Léonard de Vinci Pôle Universitaire, France
Gianluigi Viscusi	Linköping University, Sweden
Yves Wautelet	Katholieke Universiteit Leuven, Belgium
Hans Weigand	Tilburg University, The Netherlands

Program Committee, Forum

Judith Barrios Albornoz	University of the Andes, Venezuela
Victoria Döller	University of Vienna, Austria
Riccardo Coppola	Politecnico di Torino, Italy
Georgios Koutsopoulos	Stockholm University, Sweden
Istvan David	McMaster University, Canada
Clara Ayora	Universidad de Castilla-La Mancha, Spain
Michalis Pavlidis	University of Brighton, UK
Abdelaziz Khadraoui	University of Geneva, Switzerland
Christoforos Ntantogian	Ionian University, Greece
Francisca Pérez	Universitat Politècnica de València, Spain
Ben Roelens	Open Universiteit, The Netherlands
Mark Mulder	TEEC2, The Netherlands
Giovanni Meroni	Technical University of Denmark, Denmark
Patricia Martin-Rodilla	Spanish National Research Council, Spain
Dominik Bork	TU Wien, Austria
Elena Kornyshova	CNAM, France
Cinzia Cappiello	Politecnico di Milano, Italy
Sotirios Liaskos	York University, Canada
Giovanni Giachetti	Universitat Politècnica de València, Spain
Yves Wautelet	Katholieke Universiteit Leuven, Belgium
Gianluigi Viscusi	Linköping University, Sweden
Pedro Valderas	Universitat Politècnica de València, Spain
Jose Luis de la Vara	University of Castilla-La Mancha, Spain

Program Committee, Doctoral Consortium

Carina Alves	Universidade Federal de Pernambuco, Brazil
Cristina Cachero	Universidad de Alicante, Spain
Emilio Insfran	Universitat Politècnica de València, Spain
Gustavo Rossi	Universidad Nacional de La Plata, Argentina
Jean-Michel Bruel	University of Toulouse, France
Jolita Ralyté	University of Geneva, Switzerland
Jose Luis de la Vara	University of Castilla-La Mancha, Spain
Maribel Yasmina Santos	University of Minho, Portugal
Miguel Goulão	Universidade Nova de Lisboa, Portugal
Nelly Condori Fernández	Universidad de Santiago de Compostela, Spain
Renata Guizzardi	University of Twente, The Netherlands
Saïd Assar	Institut Mines-Télécom Business School, France
Selmin Nurcan	Université Paris 1 Panthéon - Sorbonne, France
Sergio Firmenich	Universidad of Loyola Andalucía, Spain
Xavier Franch	Universitat Politècnica de Catalunya, Spain

Additional Reviewers

Acitelli, Giacomo
Adiba, Eudes
Baldwin, Wilder
Bono Rossello, Nicolas
Borcard, Daniel
Bromuri, Stefano
Bulmer, Dylan
Casciani, Angelo
Cavalcanti Ribeiro, Ana Beatriz
Chan, Anouck
Curty, Simon
Danthine, Antoine
Gallego, Víctor
Gaspar Vilallba, Alberto
Georgiou, George
Giachetti, Giovanni
Hilvert, Adir
Jokste, Lauma
Jonathan, Gideon Mekonnen
LaChance, Clark
Lektauers, Arnis
Linkevics, Gusts
Macías, Aurora
Matisons, Ralfs
Milani, Fredrik
Mosāns, Guntis
Muff, Fabian
Parshutin, Sergei
Pecerska, Jelena
Pourghasemi Fatideh, Ali
Pretel, Elena
Romanovs, Andrejs
Roponena, Evita
Schinckus, Malik
Zangogianni, Paraskevi-Chrysovalantou

Keynote Talks

The Next 700 Low-Code Platforms

Tijs van der Storm[1,2]

[1] Centrum Wiskunde & Informatica (CWI), Amsterdam, Netherlands
storm@cwi.nln
[2] University of Groningen, Groningen, Netherlands

Abstract. There's too much code in the world and it's growing at staggering exponential rates. No approach in software engineering or programming paradigm has made a dent in this trend. What should we do? Buy a low-code platform? Will AI bring us salvation? The key to successful software engineering is maintaining domain knowledge and architectural knowledge, the "what" and "how". Why not maintain them as code? I argue that the in-house development of domain-specific languages (DSLs), modeling different aspects of a software system, offers a potential way out. State-of-the art language workbenches make this approach feasible, which I demonstrate with a functional low-code platform, developed in under 1000 lines of code. Language engineering for all!

Keywords. Software engineering · domain-specific languages (DSLs) · language engineering · language workbenches · knowledge management

Programming is a form of encrypting requirements in code. Software projects end up as large-scale knitted castles, tangled webs of implementation detail, boilerplate, API appeasement, and through it all, one might find remnants of actual features, scattered and dispersed over thousands if not millions of lines of code. As the code base grows, quality inevitably deteriorates. The ripple effect of changes becomes harder and harder to predict or understand. Too afraid to break stuff in distant parts of the system, developers give up on refactoring. At a certain point in time the only fixes we dare to commit are workarounds...

Complexity is the bane of software development. It's an oft-cited fact that for every 25 percent increase in problem complexity, there is a 100 percent increase in solution complexity. Whether this is literally true or not, I leave to the reader to ponder. What we do know is that software needs to evolve to be alive, and that increase in size hinders maintenance. A recent estimate states that a 2020s hybrid car already contains 150 MLOC, and projects that fully self-driving cars will run on 500 MLOC. There's too much code in the world, and it's increasing at staggering rates.

No programming language or software technology has managed to make a dent in these trends. Yes, we have better, safer, nicer, or faster programming languages, but the economy of code size has largely remained the same. If you squint, all programming languages are the same: different flavors of procedures, expressions, conditionals, and loops. The problem is: general-purpose languages cannot cut corners in terms of expressivity; they have to serve as many clients as possible. The result is that they are mediocre (but functional) at everything, but they excel at nothing.

No software engineering methodology has had any significant effect on programmer productivity or code reduction. Components, frameworks, libraries, patterns, objects, functional programming, or AI—while each of these things has its merits from some point of view, the fact remains that code is all around us, it's not going away, and it's growing. In fact, one could argue that the current hype around AI makes things even worse: AI for programming makes it *much* too easy to produce code *faster*. I will not be surprised if a plot of the global code footprint over time will show a spike when generative AI took the stage...

Another trend is offered by so-called low-code or no-code platforms (e.g., Mendix, Outsystems, Bettyblocks, etc.): often graphical environments which allow citizens to define applications with little or no coding. Indeed, such platforms do increase productivity in certain domains (most prominently business apps), but nevertheless, just like the 4GLs of old, organizations become dependent on such platforms or tools. When a vendor stops support or collapses, what then? The fact that 4GLs were *bought but not owned* caused many companies and organizations to end up with large code bases of unmaintainable legacy code, written in unsupported and obscure languages, with no influx of new generations of programmers. In a sense, one might expect that low-code platforms are paving the way for a 4GL crisis of the future.

What if we could build *our own* low-code platform? So that we own it, so that we understand it fully, so that we can change it to our ever-evolving requirements, wishes, and circumstances? Would it be feasible for ordinary organizations to maintain a set of little languages to define their applications in? In this keynote I argue that maybe this is a way out of the tar pit of complexity, by presenting a radical design experiment: building a low-code platform from scratch, in fewer than 1000 lines of code.

I present the argument that if a single person, using the right tools, can build an operational low-code platform with so little effort (gratuitously taking lines of code as a proxy for effort), wouldn't it be feasible to scale it up, say, a hundred times, to develop a platform suitable for production use? If so, businesses could reap the benefits of increased productivity, without depending on third parties, without paying for features they don't use, without dealing with the constraints of the offered product. In other words, they would own a bespoke low-code platform.

Software engineering where DSLs are the central software development artifact, not designed as a *product*, but as a way to organize the software development process and the knowledge about the problem domain in a traceable, executable manner. Language engineering and design becomes the central activity. Don't buy into a low-code platform—build your own!

Challenges in Forming European Consortia for Ethical and Human-Centered AI: Overcoming Barriers to Collaborative Innovation

Rebeca Gutierrez

Ayesa Advanced Technologies S.A., Seville, Spain
http://rgutierrezs.com/

Abstract. In the context of the increasing complexity of the information systems that surround us and the intensified competition in the generation of innovative ideas, collaboration around European values and distinctive approaches emerges as a key catalyst to achieve significant advances in artificial intelligence, overcoming obstacles and cultivating a high-impact innovation ecosystem for the transformation of companies, entities and where high-impact actions are obtained in the citizenship. This talk will analyse strategies that strengthen cooperation between academia, industry and the public sector, with a particular focus on flagship initiatives such as Horizon Europe, CINEA, InnoFund or the recently launched European Network of AI Centres of Excellence and AI Factories. The discussion will focus on how these collaborations facilitate access to critical resources, foster efficient knowledge transfer and promote agile adoption and development of AI technologies. In parallel, the conference will address the ethical and regulatory challenges inherent in collaborative AI innovation by exploring regulatory frameworks that encourage the development of responsible and trustworthy AI, without compromising creative and innovative potential. It will also examine open innovation models designed to boost European competitiveness in the global AI arena, balancing the need for technological advancement with ethical and societal imperatives.

Keywords. Artificial Intelligence · Innovation · European Collaboration · AI Ethics · Human-Centered AI

References

1. European Commission Homepage. https://commission.europa.eu/
2. AI Factories. https://digital-strategy.ec.europa.eu/en/policies/ai-factories
3. AI Europe Collaborations. https://digital-strategy.ec.europa.eu/en/news/second-wave-ai-factories-set-drive-eu-wide-innovation
4. AI Literacy. https://digital-strategy.ec.europa.eu/en/events/third-ai-pact-webinar-ai-literacy
5. AI Act. https://digital-strategy.ec.europa.eu/en/policies/regulatory-framework-ai

Explanation, Semantics, Ontology and Trustworthiness

Giancarlo Guizzardi

Semantics, Cybersecurity & Services (SCS), University of Twente, The Netherlands
g.guizzardi@utwente.nl

Abstract. Cyber-human systems are formed by the coordinated interaction of human and computational components. The latter are justified to the extent that they are meaningful to humans, in both senses of 'meaning', that is, in the sense of *semantics* and in the sense of *purpose* (or *significance*). On the one hand, the data these components manipulate only acquire meaning when mapped to shared human conceptualizations of the world; on the other hand, they can only be justified if ethically designed. Cyber-human systems are trustworthy if the interoperation of their components is meaning preserving, i.e., if we can semantically interoperate these components; and transparently demonstrate (i.e., explain) how their interoperation positively contributes to human values and goals. In this talk, I will present a notion of explanation termed *Ontological Unpacking*, which aims at explaining symbolic domain descriptions (e.g., conceptual models, knowledge graphs, logical specifications). I show that it is this explanatory nature that is required for semantic interoperability, and hence trustworthiness. Finally, I will argue that the current trend in XAI (Explainable AI) in which 'to explain is to produce a symbolic artifact' is an incomplete project, as these artifacts are not 'inherently interpretable', and that they should be taken as the beginning of the road to explanation, not the end. The talk is strongly based on [1, 2, 3, 4].

Keywords. Real-world semantics · Ontology explanation · Ontological unpacking · Semantic interoperability · Trustworthiness

References

1. Confalonieri, R., Guizzardi, G.: Explanation, semantics, and ontology. Data Knowl. Eng. **153** (2024)
2. Guizzardi, G., Guarino, N.: On the multiple roles of ontologies in explanations for neuro-symbolic AI. Neurosymbolic Artif. Intell. **1** (2025)
3. Bernasconi, A., Guizzardi, G., Pastor, O., Storey, V.C.: Semantic interoperability: ontological unpacking of a viral conceptual model. BMC Bioinform. **23**(1) (2022)
4. García, A., Bernasconi, A., Guizzardi, G., Pastor, O., Storey, V. C., Panach, I.: Assessing the value of ontologically unpacking a conceptual model for human genomics. Inf. Syst. **118** (2023)

Contents – Part I

Information Systems Quality

TRACE4PM: Trace Related Analysis and ClustEring for Process Modeling of Users' Interactions in Information Systems 3
Marwa Trabelsi, Noura Joudieh, Amira Ania Dahache, Cyrille Suire, and Ronan Champagnat

Challenges in Data Quality Management for IoT-Enhanced Event Logs 20
Yannis Bertrand, Alexander Schultheis, Lukas Malburg, Joscha Grüger, Estefanía Serral Asensio, and Ralph Bergmann

Blockchain-Based Trust Management System for Enhancing Security in SIoT 37
Raouf Jmal, Mariam Masmoudi, Ikram Amous, and Florence Sèdes

A Comprehensive Review on Equivalent Mutant Detection Using Machine Learning 52
Gabriel Guerrero-Contreras, Sara Balderas-Díaz, Pedro Delgado-Pérez, and Inmaculada Medina-Bulo

Security, Risk and Strategy

Identifying and Detecting Patterns in Work Organization with Active Window Tracking 71
Mari A. J. Braakman, Iris Beerepoot, Maria Peeters, Eva Knies, and Hajo A. Reijers

Black Swan Theory for Navigating Trust in Mixed-Traffic Environments 87
Hind Bangui, Barbora Buhnova, and Mouzhi Ge

Friend, Foe, or Target? Domain Models as Risk Deterrents, Risk Sources, and Assets at Risk 103
Isadora Valle, Tiago Prince Sales, Eduardo Guerra, Ítalo Oliveira, Renata Guizzardi, Luiz Olavo Bonino da Silva Santos, Henderik Proper, and Giancarlo Guizzardi

Conceptual Modelling and Ontologies

Towards an Ontology of Type-Level Phenomena for System Modeling 121
Rodrigo Fernandes Calhau, João Paulo A. Almeida, Giancarlo Guizzardi, Raquel Hoffmann, Luís Ferreira Pires, James Logan, and João Rebelo Moreira

Low-Code Browser Front-End Automation Using RDF Graphs and a Domain-Specific Language for UX Representation 140
Ştefan Uifălean and Robert Andrei Buchmann

Risk Response: An Adversarial Discourse Game and Group Modelling Tool ... 156
Max Willis, Greta Adamo, and Anna Sperotto

Modeling Methods and Requirements Engineering

Modelling Hierarchies of Organisational Rules 175
Jöran Lindeberg, Martin Henkel, and Katarina Fast Lappalainen

If Complexity Is the Problem, Collaboration Is the Solution: Drivers and Trust-Building Interventions in DSR 192
Xabier Garmendia and Oscar Díaz

Synthline: A Product Line Approach for Synthetic Requirements Engineering Data Generation Using Large Language Models 208
Abdelkarim El-Hajjami and Camille Salinesi

The Role of Ethics in Requirements Engineering for Developing Information Systems 226
Christopher Julian Kern, Karin Hübner, Leo Poss, Stefan Schönig, and Julia Kroenung

Databases and Information Management

Model-Agnostic Evolution Management 245
Pavel Koupil, Jáchym Bártík, and Irena Holubová

A Flexible Framework for Transposition-Aware Querying of a Musical Score Database 262
Adel Aly, Olivier Pivert, and Virginie Thion

Implementing Digital Twin Query Views 279
Emilio Carrión and Pedro Valderas

Framework for the Design and Development of an Automated Virtual Data Integration System for Genomic Data 295
Ana León Palacio

Human Factors in Information Systems

Developing a Design Features Taxonomy of Human-Computer Interaction in Social Media that Affect User Engagement and Addictive Behaviors 313
Maria Fernanda Granda, María-Belén Sarmiento, Ana-Gabriela Nuñez, Ricardo Maldonado, and Otto Parra

User Correction of Misinformation on Social Media: Perceived and Actual Social Norms 331
Selin Gurgun, Emily Arden-Close, Keith Phalp, and Raian Ali

Do Social Media Simultaneously Contribute to Well-Being and Use Disorder? Empirical Evidence and Design Challenges 347
Tourjana Islam Supti, Ala Yankouskaya, Sameha Alshakhsi, Areej Babiker, Dena Al-Thani, and Raian Ali

Business Process Engineering and Management

Multi-perspective Next Event Prediction in PPM via Heterogeneous Graph Neural Networks 365
Sebastiano Dissegna, Chiara Di Francescomarino, and Massimiliano Ronzani

OCPQ: Object-Centric Process Querying and Constraints 383
Aaron Küsters and Wil M. P. van der Aalst

Investigating Tailored Retraining for Online Process Predictions Using Log Features 401
Suhwan Lee, Xixi Lu, and Hajo A. Reijers

Navigating the Challenges of Process Mining Use Case Definition: A Qualitative Study 418
Kerstin Haug, Pol Schumacher, Holger Wittges, and Stefanie Rinderle-Ma

Zero-Shot Approaches for the Extraction of Event Logs from Medical Notes ... 435
Allmin Susaiyah and Natalia Sidorova

Review of Design of Business Process Simulation Models 452
Samira Khraiwesh and Luise Pufahl

Exploring Webcam Eye Tracking Software for Robotic Process Automation: A Pilot Benchmarking Study 470
Manuel García-Romero, Antonio Martínez-Rojas, José González Enríquez, and Andrés Jiménez-Ramírez

Author Index 487

Contents – Part II

Machine-Learning and Generative AI Applications

Can Generative AI Mitigate Strategic Decision-Making Complexity? An Empirical Exploration of RAG-Augmented Decision-Support Systems 3
Jordan Abras, Corentin Burnay, and Stéphane Faulkner

Cross-Lingual Entity Linking Using GPT Models in Radiology Abstracts 20
Mariana Dias and Carla Teixeira Lopes

Modeling Out-of-Vocabulary Words via Grammatical Fusion 38
Dror Mughaz

Early Length of Stay Prediction at Admission in Short-Stay Hospitals 52
Mohamed Gharbi, Christine Verdier, Maria Di Mascolo, and Jean-Marc Babouchkine

Alignment of Schema-Only and Instance-Only Data Sources Using Large Language Models 67
Nour Elhouda Kired, Franck Ravat, Jiefu Song, and Olivier Teste

RCIS Forum

Can Llama 3 Accurately Assess Readability? A Comparative Study Using Lead Sections from Wikipedia 89
José Frederico Rodrigues, Henrique Lopes Cardoso, and Carla Teixeira Lopes

Research Challenges in Routine Optimization for Synthesizing Software Robots 98
J. L. Alonso-Rocha, A. Martínez-Rojas, A. Jiménez-Ramírez, and J. G. Enríquez

Mining for Meaning: Ontology-Aware Process Mining Methods Through Knowledge Patterns 109
Riley Moher and Michael Gruninger

How to Use a FEM Model as a Basis for Strategic-Level Risk Analysis 120
Toomas Saarsen and Ilia Bider

Modelling Neural Network Models 130
Nadia Daoudi, Ivan Alfonso, and Jordi Cabot

Quantifying the Magnitude of Violation: Predictive Compliance Monitoring Approaches 140
Qian Chen, Stefanie Rinderle-Ma, and Lijie Wen

Evaluating Programming Optimization Techniques in C and Python: Impact on Energy Consumption 151
Carlos Pulido, Félix O. García, Mł Ángeles Moraga, Coral Calero, Miguel Baños-González, Jorge Cancho-Casado, and Javier Corral-García

An Analysis of Resilience in Digital Business Ecosystems 162
Beāte Krauze

AI Auditing: Towards a Practicable Model 172
A. B. van Wingerden and H. Weigand

From Acquiring to Suggesting DL Design Choices with Agility: A System Design 183
Gustavo Rodrigues dos Reis, Mario Cortes Cornax, Adrian Mos, and Cyril Labbé

RCIS Doctoral Consortium

Digital Twins for Incident Detection and Response 197
Konstantinos E. Kampourakis

Towards an Enterprise Architecture Based Approach for the Development of Digital Twins for Sustainable Real Estate Management 207
Marianne Schnellmann

A Method for Domain Reference Model Inference Through Knowledge and Data Intelligent Unifiers 218
Pedro Guimarães

Tutorials

Problematising and Ideating for Design Science Research 231
John R. Venable

Taking Control of Our Privacy When Using Mobile Apps: Checks and Tools ... 232
M. Mercedes Martínez-González, Alejandro Pérez-Fuente, Amador Aparicio, and Pablo-Abel Criado-Lozano

Introduction to Fractal Enterprise Model (FEM) and FEM Toolkit 234
Ilia Bider

Author Index 237

Information Systems Quality

TRACE4PM: Trace Related Analysis and ClustEring for Process Modeling of Users' Interactions in Information Systems

Marwa Trabelsi, Noura Joudieh(✉), Amira Ania Dahache, Cyrille Suire, and Ronan Champagnat

L3i Laboratory, La Rochelle University, La Rochelle, France
{marwa.trabelsi,noura.joudieh,amira.dahache,cyrille.suire, ronan.champagnat}@univ-lr.fr

Abstract. The rapid growth in information systems usage has highlighted the importance of modeling user behavior to improve usability, deliver personalized recommendations, and refine system design. User interactions with these systems generate rich datasets of event data crucial for such analyses. Process mining plays a pivotal role in modeling these behaviors, often complemented by trace clustering techniques that help manage the complexity of user interactions. In this context, we introduce TRACE4PM, an automated pipeline encapsulated within an API explicitly designed to model user behaviors in information systems using process mining techniques focusing on trace clustering techniques. The platform offers services, including parsing, tagging, clustering, and process modeling of the event data, all based on state-of-the-art Process mining techniques. This API-based solution promotes ease of integration and usage, enabling developers and researchers to use the API platform without requiring them to delve into technical aspects or requirements. To demonstrate the real-world applicability of TRACE4PM, we developed a Moodle plugin that collects and analyzes Moodle logs to cluster and model student interaction traces, successfully extracting helpful learning models from the educational data. Furthermore, researchers and educators feedback validated TRACE4PM's and the Moodle plugin's efficiency, applicability, and educational utility, highlighting their potential for broader adoption and refinement.

Keywords: API-based platform · Modeling Users Behaviors · Process Mining · Trace Clustering · Moodle · Learning Scenarios

1 Introduction

With the increasing use of information systems, analyzing and modeling user behaviors has become vital for improving system design and user experience.

J. Grabis et al. (Eds.): RCIS 2025, LNBIP 547, pp. 3–19, 2025.
https://doi.org/10.1007/978-3-031-92474-3_1

Accurate modeling of user interactions enables personalized support, recommendations, and proactive measures to anticipate needs and optimize workflows, resulting in more intuitive information systems [29]. These systems involve structured activities aimed at specific objectives, such as the defined steps in purchasing a product on an e-commerce platform. However, not all processes are clearly defined. Although system designers create models to reflect user behaviors during information searches, users often deviate from these paths based on their goals, encountering complexity, failure, or stress, diminishing the system's relevance.

Modeling user behaviors and interactions in information systems has evolved through disciplines like Web Usage Mining (WUM) and Web analytics. WUM aimed to predict and understand user behavior by analyzing interactions to uncover preferences and navigation patterns [23,27]. While insightful, it struggled with the growing complexity of web systems. Web analytics emerged to address this, focusing on defining KPIs, collecting data, and presenting results via dashboards. It involves measuring, gathering, and analysis of user navigation paths to optimize their web experiences [7,20,35]. Recently, **Process Mining** (PM) has emerged as a more comprehensive method, combining data mining and business process management. PM discovers, monitors, and improves processes by extracting knowledge from event data stored in information systems. These event logs, generated through user interactions, reveal operational processes and enable the discovery of models to verify workflows and enhance process understanding [1]. Unlike WUM and Web analytics, which focus on abstract, page-level insights, PM examines complete, end-to-end user behavior. It visualizes entire user journeys using notations like Petri nets and Directly Follows Graphs.

The increasing use of information systems generates vast and complex event data, which often pose challenges for PM techniques. Directly mining entire datasets can result in spaghetti models, a significant challenge addressed through Trace Clustering [26,34]. By grouping similar interaction sequences and uncovering hidden patterns and behaviors, Trace Clustering is typically applied first. This analysis enables organizations to make informed decisions, enhancing system efficiency and user experience across various platforms [6,9,12,19,26,29,34]. Researchers analyzing user interactions typically follow a multi-step process, including data cleaning, optional enrichment, trace clustering, and process discovery [19,29,30]. However, performing these steps manually is time-intensive, repetitive, and requires specialized knowledge. Motivated by this challenge, we sought to automate the entire pipeline from log parsing to PM.

Thus, we introduce **TRACE4PM**, an API that automates the pipeline for modeling user behaviors in information systems using advanced PM techniques, with an emphasis on trace clustering. TRACE4PM offers services such as **trace clustering** and **process modeling** algorithms from the process mining domain. A key feature is the **Parser service**, which prepares event logs from raw data. To enrich logs, the domain-specific **Tagger service** uses configurable rules for annotation. The **Trace Clustering service** includes algorithms tailored for digital libraries [30], education [19], and state-of-the-art techniques [5,6,26,34]. Finally, the **Process Discovery service** models user behaviors, uncovering patterns and enabling stakeholders to optimize user experiences across platforms. TRACE4PM simplifies research by automating repetitive tasks, allowing researchers to focus on innovation. While tools like ProM, Disco, and Celonis [1] support process modeling, TRACE4PM extends the pipeline to include trace clustering to categorize user interactions in information systems. To demonstrate its applicability, we developed a Moodle plugin as a real-world use case, demonstrating how TRACE4PM models learning behaviors and extracts insights from educational data using PM techniques.

In what follows, Sect. 2 details the TRACE4PM API's architecture and services while providing background information to contextualize the work, and highlights its contributions compared to existing tools. Section 3 describes the Moodle plugin and its evaluation process. Section 4 presents feedback from researchers and teachers, evaluating both the API and the Moodle plugin. Finally, Sect. 5 concludes the paper and outlines directions for future research.

2 TRACE4PM: An End-to-End Automated API

TRACE4PM[1] builds on the motivation to automate log analysis and the typical pipeline seen in existing research [1]. The developed services are inspired by established methodologies and integrate recent contributions [19,29]. TRACE4PM has evolved from a non-web application focused on digital library logs [30] into a robust, scalable framework for analyzing complex user interactions and enabling effective PM. FastAPI[2] was chosen for its efficiency, modern architecture, and built-in security, making it ideal for building REST APIs within TRACE4PM. Its architecture supports rapid development and scalability, crucial for tasks like trace clustering and PM. TRACE4PM's microservices-based architecture (Fig. 1) ensures flexibility, scalability, and independence, with FastAPI serving as the API Gateway to manage external requests and secure communication. Data consistency and integrity are maintained using MongoDB[3]. The following sections detail each service.

2.1 Parser Service

The **Parser service** preprocesses raw logs, forming the foundation for the entire pipeline by extracting key elements to create an event log ready for process modeling and analysis. An **event log** $L = \{t_1, t_2, ..., t_k\}$ is a set of k **traces** where each trace t_i $(1 \leq i \leq k)$ is a set of n_i consecutive events $t_i = <e_{i1}, e_{i2}, ...e_{in_i}>$ made by the same CaseId. For instance, consider the following raw log entry from a digital library:

```
2024-07-26 14:35:20 192.168.1.1 GET /index.html 200
```

[1] Available at https://github.com/TRACE4PM.
[2] https://fastapi.tiangolo.com/.
[3] https://www.mongodb.com.

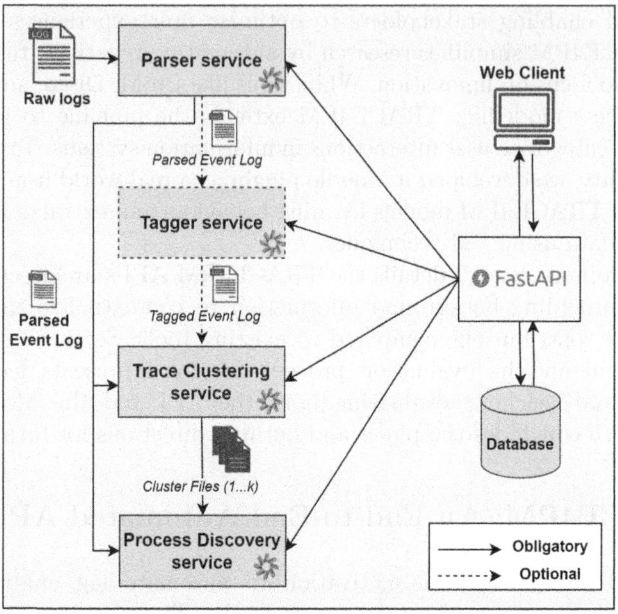

Fig. 1. TRACE4PM Micro-services-based architecture.

The parser extracts key fields such as the timestamp (e.g., **2024-07-26 14:35:20**), user IP address (e.g., **192.168.1.1**), HTTP method (e.g., **GET**), requested URL (e.g., **/index.html**), and HTTP response code (e.g., **200**) to identify the event's details. These elements are organized by trace ID (e.g., IP address), activity, and timestamp, providing a structured output for effective PM.

To support various log formats, the API includes multiple parsers. One handles raw logs from information systems, while another processes Comma Separated Values (CSV) files, such as those from Moodle, transforming structured data for analysis. After processing, the event logs are returned to the API and stored in the database for further use. In summary, the Parser service extracts an event log from raw input logs, enabling subsequent analysis.

2.2 Tagger Service

The **Tagger service** is an optional step that enhances low-level event logs by transforming them into higher-level logs for improved analysis. It is commonly used in research across domains like digital libraries [29], healthcare [3], and business [14]. For example, in education, [33] analyzed interactive behaviors in MOOCs, turning basic browser events into complex behavioral patterns for assessing engagement and progress. In process model analysis, accurate activity labeling is particularly valuable for non-technical stakeholders, such as teachers.

This service is **domain-specific**, requiring customized tagging rules depending on the field. While we describe a tagger for digital libraries, it can be adapted for other domains like educational logs. Raw URLs (e.g., /index.html) offer limited context about user behavior, so they must be transformed into meaningful activity labels. This transformation is vital for making actionable recommendations [22]. However, since tagging rules are domain-specific, this step is optional. Users can still utilize our tool to analyze URLs with limited semantic meaning, gaining an initial understanding of the events executed within their information systems. This allows for a preliminary exploration of generated models, even without fully customized tagging. The tagging process uses a function f to map a requested URL r to a corresponding understandable activity label a, making the data more accessible to designers and users.

In literature [22], the set of activities $A = \{a_1, a_2, \ldots, a_{|A|}\}$ can be defined through two main approaches: (1) **Manual approach** where activities are manually defined. In this case, the function f applies a set of predefined rules:

$$f(r) = \begin{cases} a_1 & \text{if } (Exp_1 \text{ is in } r) \\ \vdots & \\ a_{|A|} & \text{if } (Exp_{|A|} \text{ is in } r) \end{cases} \tag{1}$$

where Exp_i are expressions created by experts, based on the site's structure. For example, an HTTP request containing '/index.html' could be labeled as "Homepage". And (2) an **Automatic Approach** where unsupervised clustering methods can be employed to define A by grouping similar activities based on patterns detected in the log data [24,33]. However, in our case, we opted for the manual approach [29]. This choice is particularly justified in contexts where domain knowledge plays a critical role in interpreting user behaviors. By using expert-defined expressions, the manual approach ensures that the labels are not only accurate but also aligned with the specific objectives of the analyzed system. In summary, the Tagger Service maps low-level interaction data to high-level user actions by assigning labels. These labels provide an abstract perspective on user journeys, offering a clearer reflection of their actions within the system.

2.3 Trace Clustering Service

Trace clustering methods are generally classified into three approaches: (1) **Trace-based** clustering, which compares the structure or content of traces using similarity measures to assess alignment [5,9]. (2) **Feature-based** clustering, where traces are transformed into feature vectors (e.g., event frequencies) and clustering is performed in this feature space [19,26,29]. (3) **Model-based** clustering, which groups traces based on their similarity to discovered process models, with clustering quality dependent on the metrics used to evaluate model accuracy [8,12,15]. Additionally, **hybrid approaches** combine these methods for improved results [11]. In the current version of TRACE4PM, the **Trace Clustering** service addresses two trace clustering categories: the **Trace-based approach** and the **Feature based approach**.

Trace Based Clustering. To perform trace-based clustering, users configure parameters such as the clustering algorithm and distance measure. For algorithms like Hierarchical Agglomerative Clustering (HAC) [28], additional parameters, such as the linkage criterion, are required (Fig. 3). In TRACE4PM, these settings are centralized in the Config File (Fig. 4), which stores all the configurations for each clustering task.

The trace-based workflow shown in Fig. 2 starts by computing a distance matrix using the chosen distance measure, which quantifies the difference between traces by calculating the minimal number of edits needed to transform one trace into another. Based on this matrix, TRACE4PM supports two clustering algorithms: DBSCAN (Density-Based Spatial Clustering of Applications with Noise) [13] and HAC. Users select the algorithm and its parameters, such as *min_samples* and *eps* for DBSCAN, or linkage criteria and the number of clusters for HAC. The clustering quality can be evaluated using the Silhouette score, which measures how well traces fit within their assigned clusters [25]. By adjusting algorithm parameters and assessing results with the Silhouette score, users can optimize the clustering process to achieve meaningful groupings that reflect the data's structure.

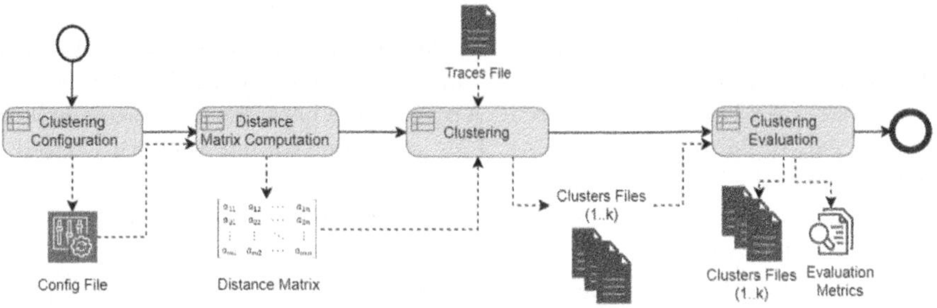

Fig. 2. Trace based clustering workflow

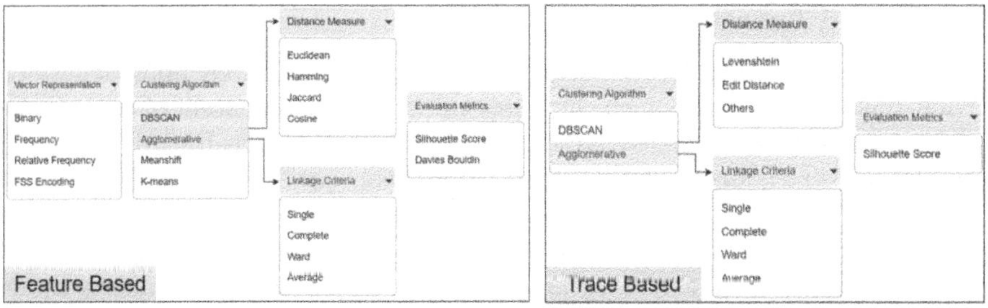

Fig. 3. Clustering choices

Feature Based Clustering. Similar to the trace-based approach, clustering parameters for the feature-based approach are stored in a configuration file. As shown in Fig. 4, users can choose from four techniques to extract features from traces and convert them into vectors for clustering.

- **Binary representation**: this approach uses a binary vector where each element indicates the presence (1) or absence (0) of a specific event within the trace [26]. **Example:** Given the traces:

  $$T_1 : [a, b, c, d], \quad T_2 : [a, c, b, d], \quad T_3 : [a, e, d]$$

  Their corresponding binary vectors are:

  $$T_1 : [1, 1, 1, 1, 0], \quad T_2 : [1, 1, 1, 1, 0], \quad T_3 : [1, 0, 0, 1, 1]$$

- **Frequency and Relative Frequency representations**: these representations, inspired by the Binary representation [5,26], use vectors where each element indicates the frequency or relative frequency (proportion) of a specific event or activity within the trace.
- **Frequent Sub-Sequences (FSS) encoding**: this representation entails a vector that encodes the presence of frequent sub-sequences within the trace. These sub-sequences are recurring patterns identified in the event logs. The process involves extracting features from traces by identifying and encoding these frequent sub-sequences. This encoding method considers various parameters to effectively distinguish between users' traces, enhancing the granularity and accuracy of the analysis [19,29].

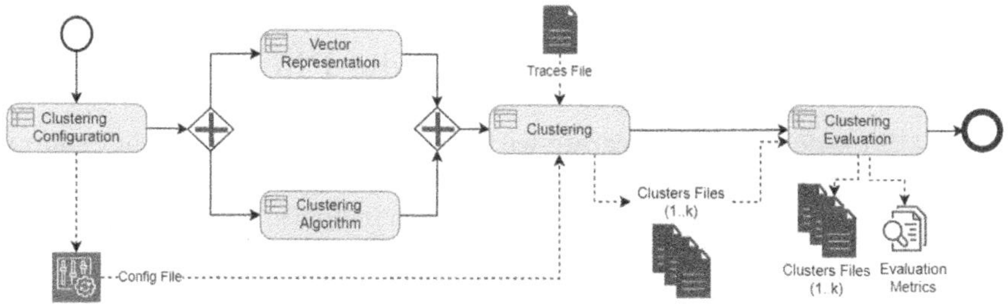

Fig. 4. Feature based clustering workflow

Users can choose to either apply clustering algorithms directly to feature vectors or first compute a distance matrix using metrics like Cosine similarity, Jaccard index, or Hamming distance [26]. With the distance matrix, algorithms like DBSCAN or HAC can be used. This flexibility enables tailored clustering based on dataset characteristics. Finally, the quality of clusters can be evaluated using metrics like the Silhouette Coefficient and Davies-Bouldin Index. The

Silhouette Coefficient measures how similar a point is to its own cluster compared to other clusters. This coefficient ranges from −1 to 1, where values close to 1 indicate well-separated clusters, values near 0 suggest overlapping clusters, and negative values indicate incorrect clustering. Whereas, Davies-Bouldin Index evaluates the average similarity between each cluster and its most similar one. A lower value of this index indicates better clustering.

2.4 Process Discovery

PM analyzes event logs to model business processes using techniques like Directly-Follows Graphs (DFGs) and Petri nets. DFGs highlight direct transitions between activities [21], while Petri nets [2] model more complex behaviors such as concurrency and causality.

The final service in TRACE4PM is the **Process Discovery** module, which automatically generates process models from event logs. This module leverages established process discovery techniques from the literature [1], utilizing the PM4PY library[4] to implement algorithms like Alpha Miner, Heuristic Miner, and Inductive Miner. These algorithms enable the generation of effective process models. The service also evaluates the quality of the discovered models using metrics such as Fitness, Precision, Generalization, and Simplicity [1]. To enhance user understanding, we incorporated an animation API through the BupaR library[5], which creates visual representations of the process models.

TRACE4PM also provides detailed statistics on parsed logs, providing insights into trace durations, action counts, frequencies, and unique trace variants. It also includes targeted analyses for digital library logs, such as identifying peak usage times, most accessed resources, and common user action sequences. These statistics help understand user behavior and optimize system performance.

3 TRACE4PM in Action: A Moodle Plugin Case

TRACE4PM is designed to integrate with any information system that supports logging. This paper focuses on Moodle, a widely used Learning Management System (LMS) for creating and managing online courses. Known for its robust features and flexibility, Moodle is adopted by educational institutions worldwide, from primary schools to universities [10]. Its community-driven development ensures continual improvements, offering a wide range of plugins that support various teaching methods, including fully online and blended learning. With a MySQL database for data storage, Moodle is a popular choice for accessible, high-quality education [4]. Our focus lies on **Moodle logs**, which capture every user action within a course, provide valuable insights into learning processes. These logs are a rich resource for understanding student progress and have been widely studied in the context of process mining, demonstrating their potential for educational research.

[4] https://processintelligence.solutions.

[5] https://bupar.net/.

For example, Hobeck et al. [17] conducted a case study showcasing how student trajectories can be translated into curriculum recommendations, highlighting the potential of PM to support personalized learning paths. Similarly, Hidalgo et al. [16] used event abstraction techniques to analyze behavioral patterns in MOOCs, uncovering dynamics that drive student engagement and progress. Further extending the application of PM, Wagner et al. [32] introduced a methodology that combines PM with rule-based artificial intelligence to analyze and better understand student study paths using data from campus management systems and program models. In another study, Umer et al. [31] integrated machine learning with PM to create early prediction models that help identify at-risk students and enable timely interventions.

This section explores the applicability of TRACE4PM within a Moodle plugin. We begin by presenting the plugin's user interface and explaining the structure of Moodle logs. Next, we demonstrate the plugin's functionality, showcasing the data collection process and the results generated by its services. Finally, we discuss the plugin's effectiveness and reflect on the insights it enables.

3.1 Moodle Plugin Interfaces

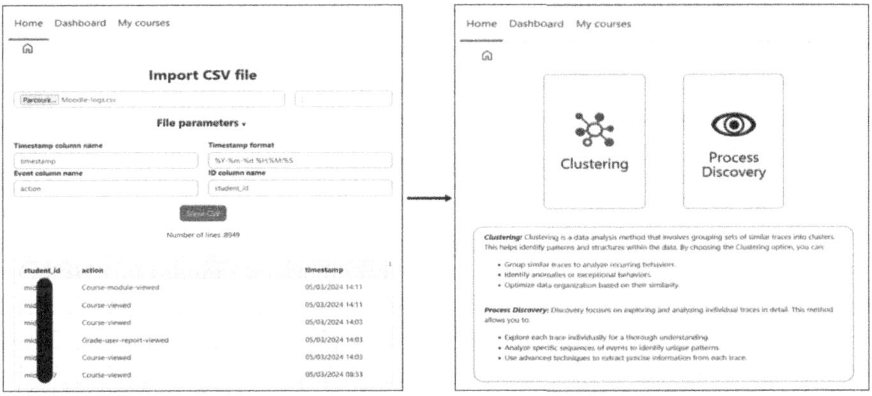

Fig. 5. Log Handling User Interfaces

The Moodle plugin acts as the web client depicted in Fig. 1, managing user requests while ensuring logical consistency. Upon validating requests, the plugin interacts with the API to invoke essential services such as parsing, trace clustering, or process discovery. It features three categories of User Interfaces (UIs), each corresponding to the primary services offered by TRACE4PM:

Log Handling User Interfaces. This set of interfaces manage and process uploaded Moodle logs. As shown in Fig. 5, the first interface allows users to select three key fields: case identifier, activities performed, and timestamp. These fields

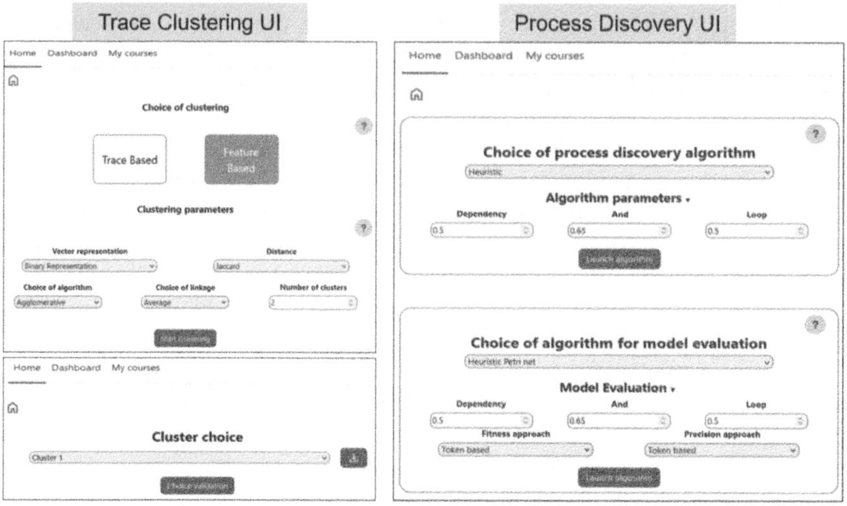

Fig. 6. Clustering and Process Discovery Algorithm Configuration

form the foundation for creating an event log from the uploaded file, enabling process discovery and trace clustering. Once this step is completed, users can move to the second interface, where they can choose either trace clustering or process discovery, offering flexibility based on the analysis goals.

Trace Clustering User Interfaces. When users opt for trace clustering, the Trace Clustering UI shown in Fig. 6 appears. This window facilitates the selection of Trace-Based or Feature-Based clustering (Sect. 2) and enables users to configure clustering settings. Users can specify the algorithm, distance measure, vector representation, and the desired number of clusters. Additional parameters, such as linkage criteria for HAC or epsilon and minimum samples for DBSCAN, are set as needed. After configuring the options, users can initiate the clustering process by clicking "Start Clustering". The interface then displays the clustering results, along with evaluation metrics like the Silhouette Coefficient and Davies-Bouldin Index. Additionally, a "Process Discovery" button allows users to analyze process models generated from the clustered traces.

Process Discovery User Interfaces. After processing the log in Fig. 5, users can proceed directly to the Process Discovery interface shown in Fig. 6. If users wish to discover a process model for a specific cluster, they must first select the desired cluster before accessing this interface. The Process Discovery UI allows users to configure discovery settings in two main blocks. The first focuses on algorithm selection and parameter configuration, which vary based on the chosen algorithm. For instance, selecting Heuristic Miner requires setting thresholds for dependency, AND relations, and loops. The second block is dedicated to configuring evaluation criteria like measuring fitness and precision.

3.2 Moodle Log Data

To transition from the interface to the core, it is crucial to understand Moodle logs, which typically include attributes like **Time** (chronological timestamp), **User full name** (performing user), **Affected user** (action recipient), **Event context** (platform context), **Component** (module type, e.g., Wiki, Quiz), **Event name** (action type, e.g., view, delete), **Description** (event details), and **Origin** (source and IP address). Table 1 highlights various types of event logs captured by Moodle, tracking key user activities. Events like "A submission has been submitted" and "Quiz attempt submitted" document assignment submissions and quiz completions, while "Quiz attempt reviewed" and "Submission updated" reflect review or modification actions by instructors or students. Logs such as "Course module viewed" and "Discussion viewed" track access to course materials and forums, whereas "A file has been uploaded" and "User profile viewed" log file uploads and profile views. Events like "Grade user report viewed" and "Course viewed" capture activities related to grade checks and course navigation. Moodle's descriptive event naming convention provides clear insight into user actions and their contexts.

3.3 Moodle Plugin in Action

This section demonstrates the use of the plugin with a course's Moodle log. First, the data is described, then the results from the trace clustering and process discovery services are elaborated.

Data Collection and Preprocessing: The Parser Service. This study used event logs from 80 students in the "Modeling Databases" course at La Rochelle University over six weeks (February-March 2024). Logs were anonymized and cleaned to retain only student activities, excluding system- and lecturer-generated actions. A unique Moodle ID was assigned to each student as the process instance identifier (student_id). The resulting event log included 80 traces, with 8,730 events spanning 27 unique event classes and trace lengths ranging from 66 to 177 events. The cleaned file was uploaded via the **Log Handling User Interface**, powered by the parser service (Sect. 2). After selecting student_id for the ID column, clustering was performed on the data.

Using Trace Clustering and Process Discovery Services. We used the Moodle plugin to test both trace clustering and process discovery services, focusing on a single example to demonstrate the methodology without overloading the analysis with multiple models. Trace-based clustering was applied using the HAC algorithm, followed by process discovery for one selected cluster. To determine the optimal number of clusters, an external dendrogram analysis was performed, as suggested by [18], since this feature is not integrated into the plugin. Based on the analysis, four clusters were selected. Table 2 presents the plugin's results and evaluation metrics. Cluster 3, which exhibited the best silhouette score, was chosen for process discovery.

Table 1. Sample of student event logs.

CaseId	Timestamp	Event names
1	2018-01-12 10:34:25	Course viewed
2	2018-01-12 10:36:25	Course viewed
1	2018-01-12 10:34:26	Course module viewed
1	2018-01-12 10:34:28	Submission viewed
3	2018-01-12 10:36:26	Course viewed
3	2018-01-12 10:36:27	Submission form viewed
2	2018-01-12 10:36:28	Submission submitted

Table 2. Clustering Evaluation

Metric/Cluster	Silhouette Score	Number of Traces
Silhouette Score	0.2167	
Cluster 0	0.1916	26
Cluster 1	0.2243	43
Cluster 2	0.1748	2
Cluster 3	0.2619	9

Using the "Heuristic Miner" algorithm, we discovered the Directly Follows Graph (DFG) for the selected cluster, visualized in Fig. 7. A DFG [21] is formally defined as a directed graph (N, E), where N comprises a set of activities, including the special nodes *start* and *end*, and E represents transitions between these activities. The structural constraints of the DFG ensure that *start* nodes lack incoming edges and *end* nodes lack outgoing edges, thereby establishing clear boundaries for process flows. The cluster under study represents students' learning behaviors, encapsulating sequences of events such as "Course module viewed", "Quiz attempt started", and "Submission form viewed". These activities are represented as blue rectangles (nodes), and transitions (edges) indicate the order of events, with edge labels denoting transition frequencies. For instance, the self-loop on the "Course module viewed" node, labeled "74", highlights repeated visits to the same module by users. Node labels also include event frequencies (e.g., "Course module viewed (192)" reflects 192 occurrences of the event). Start and end nodes, marked by green circles and orange ovals respectively, demarcate the boundaries of the learning process. This process model highlights key behaviors of the cluster, including frequent interactions with quizzes, submissions, and course materials. Students revisited quizzes, updated submissions, and reviewed progress, while submission actions followed a distinct flow from creation to monitoring. Analyzing this cluster can help educators identify strategies to support similar students, enhancing educational outcomes and success.

4 Evaluation and Discussion

To evaluate TRACE4PM and the Moodle plugin, two questionnaires were designed to assess their effectiveness in modeling user behavior and enhancing teaching strategies. These included both quantitative (using a Likert scale from 1: Strongly Disagree to 5: Strongly Agree) and qualitative open-ended questions.

The first questionnaire (Q1) had nine questions evaluating **TRACE4PM** targeting **researchers**. It focused on three core criteria: efficiency, applicability, and extensibility (outlined in Table 3). Q1 evaluated TRACE4PM's efficiency in modeling user behaviors, its ability to address domain-specific requirements, and

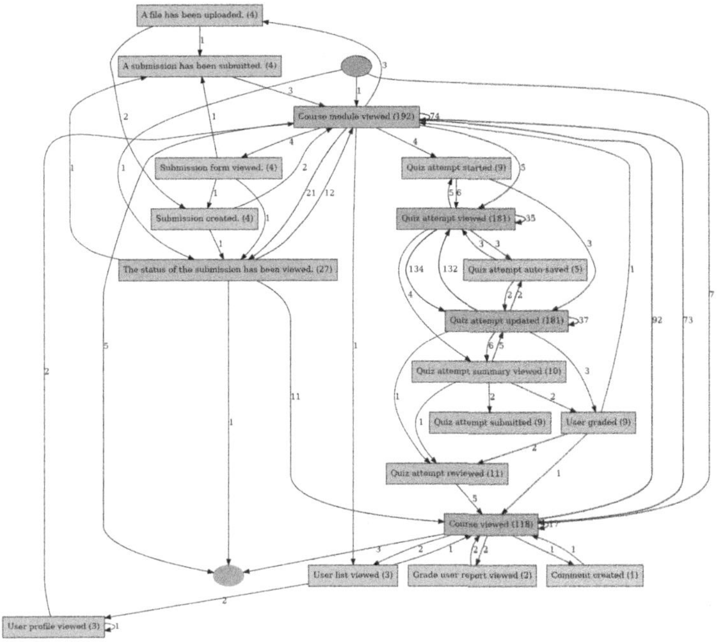

Fig. 7. The process model (DFG) of a student's logs cluster using the trace-based approach and discovered using the heuristic miner algorithm

researchers' interest in adopting or contributing to the API. A total of 10 participants from diverse fields, including computer science, natural language processing, process mining, adaptive interactive systems, and energetics responded. The results revealed several key strengths. Notably, 80% of respondents rated TRACE4PM positively for efficiency, with 87.5% strongly agreeing on its ability to meet domain-specific needs. Researchers' open feedback valued TRACE4PM for streamlining the entire PM pipeline, appreciating its automation, versatility, and modularity. They expressed interest in adopting or contributing to the API, showing its collaborative potential. Suggestions for improvement focused on enhancing the user interface, expanding documentation, and simplifying tagging. Users also proposed extending capabilities to include predictive analytics, anomaly detection, and support for custom algorithms, showcasing its capacity for broader research and practical applications.

The second questionnaire (Q2) on the other hand, evaluated the usefulness of the **Moodle Plugin** with five questions targeted at **educators**. Q2 first assessed the plugin's impact on course design and teaching strategies, as well as its ability to identify students in need of support using the process models generated by the plugin. Secondly, it examined the plugin's utility in informing resource allocation decisions through the provided statistics, and finally, questioned educators' interest in testing the plugin within their courses. The feedback from 8 educators demonstrated robust support for the plugin's capabilities in improv-

Table 3. TRACE4PM API and Moodle Plugin Evaluation

Evaluation Criteria	Strongly Agree (%)	Agree (%)	Neutral (%)	Mean
TRACE4PM API Evaluation (Q1)				
Efficiency	40%	40%	20%	4.2
Addresses Requirements	87.5%	12.5%	0%	4.9
Interest in Using/Contributing to TRACE4PM	100%	0%	0%	5.0
Moodle Plugin Evaluation (Q2)				
Course Design/Teaching Strategies	37.5%	62.5%	0%	4.4
Student Support Identification	62.5%	25%	12.5%	4.5
Resource Allocation Decisions	62.5%	37.5%	0%	4.6
Interest in Testing with Course Data	62.5%	37.5%	0%	4.6

ing teaching practices. Most educators agreed that the models generated by the plugin could enhance their course design and teaching strategies (62.5% agreed, and 37.5% strongly agreed). Many also highlighted its potential to help identify students who might be struggling and need personalized interventions (62.5% strongly agreed). The plugin was also appreciated for its ability to inform better resource and activity allocation decisions (62.5% strongly agreed). Notably, there was significant enthusiasm for testing the plugin on their own course data, with 62.5% strongly agreeing. Despite this, some educators pointed out challenges in interpreting the generated models, citing the need for improvements to the user interface. Suggestions from open feedback questions included making the models more intuitive and explanatory and simplifying workflows to cater to non-expert users. These insights underscore the plugin's potential while highlighting areas for refinement to maximize its impact across diverse educational contexts.

5 Conclusion

In this paper, we introduced **TRACE4PM**, an innovative API-based platform that automates the well-established yet often manually implemented pipeline for analyzing and modeling user behaviors in web applications through navigation traces. Unlike existing tools, TRACE4PM integrates multiple state-of-the-art techniques within a unified framework, reducing the manual effort required for preprocessing, clustering, and process discovery. This automation facilitates more efficient and reproducible research in process mining and user modeling. TRACE4PM comprises four microservices: the Parser, Tagger, Trace Clustering, and Process Discovery services. The **Parser** handles preprocessing of log files, while the **Tagger** assigns labels to logs based on predefined rules. The **Trace Clustering** service groups sequences of actions from parsed or tagged logs using trace-based and feature-based approaches, enabling comprehensive user behavior pattern identification. Finally, the **Process Discovery** service employs advanced algorithms to generate process models, evaluate process quality, and provide a robust mechanism for understanding and improving users' processes.

The practical applicability of TRACE4PM was demonstrated through a Moodle plugin, offering a user-friendly interface for educators and researchers. By analyzing Moodle course logs, the plugin effectively clustered and modeled student interactions, showcasing its potential in e-education for real-world data analysis. Two questionnaires validated the impact of TRACE4PM and the Moodle plugin, highlighting their efficiency, applicability, and educational utility. Researchers appreciated TRACE4PM's domain-specific capabilities and collaborative potential, while educators emphasized the plugin's role in enhancing teaching strategies, supporting interventions, and informing resource allocation, alongside suggestions for intuitive models and streamlined workflows. These findings underscore TRACE4PM's value and guide its future development.

Despite its promising contributions, TRACE4PM has certain limitations. The number of integrated clustering and process discovery techniques remains limited, as the current version includes only a subset of state-of-the-art methods. While designed to be extensible, future work will focus on incorporating a broader range of techniques to improve adaptability. Although TRACE4PM has been evaluated with researchers and educators, a more diverse user base could offer deeper insights, richer feedback, and new suggestions for enhancement. Another key consideration is the potential variability in algorithm outputs, as TRACE4PM automates a pipeline based on preselected techniques, different datasets may yield varied results. While the tool itself is not responsible for these variations, clearer guidelines on algorithm selection and interpretation would improve usability.

Looking ahead, we aim to enhance TRACE4PM by inviting researchers and practitioners to contribute their trace clustering algorithms, tagging methods, and PM techniques. This collaborative approach will expand the API's capabilities and broaden its user base across fields such as education, healthcare, and business analytics. A recommendation system could further assist users in selecting appropriate algorithms based on their dataset characteristics. Additionally, validating the tool with a broader and more diverse audience will provide richer insights into its usability and potential enhancements. Furthermore, elevating explainability in process mining results would make TRACE4PM more accessible to non-expert users. Explainable Artificial Intelligence techniques for example, could provide transparent, human-readable justifications for clustering and process discovery outcomes, allowing decision-makers to better interpret and act on insights. Interactive visualizations and natural language explanations could further bridge the gap between technical complexity and practical usability. By integrating additional algorithms and offering more options, we envision TRACE4PM evolving into a versatile tool that meets diverse user needs, improving personalized recommendations, and providing deeper insights into user behaviors.

References

1. Van der Aalst, W.: Process Mining: Data Science in Action. Springer, Cham (2016)
2. Van der Aalst, W., Weijters, T., Maruster, L.: Workflow mining: discovering process models from event logs. IEEE Trans. Knowl. Data Eng. **16**(9), 1128–1142 (2004)
3. Alharbi, A.M.: Unsupervised abstraction for reducing the complexity of healthcare process models. Ph.D. thesis, University of Leeds (2019)
4. Aulia, D., Waspada, I.: The design of exploratory application and preprocessing of event log data in LMS Moodle-based online learning activities for process mining. Khazanah Informatika: Jurnal Ilmu Komputer dan Informatika **5**, 124–133 (2019)
5. Bose, R.J.C., Van der Aalst, W.M.: Context aware trace clustering: towards improving process mining results. In: Proceedings of the International Conference on Data Mining, pp. 401–412. SIAM (2009)
6. Bose, R.J.C., van der Aalst, W.M.: Trace clustering based on conserved patterns: towards achieving better process models. In: International Conference on Business Process Management, pp. 170–181. Springer, Cham (2009)
7. Boufenneche, W., Hebboul, M., Benabderrahmane, O.: Web analytics tools for e-commerce: an overview and comparative analysis. In: International Conference on Managing Business Through Web Analytics. pp. 51–72. Springer, Cham (2022)
8. Cadez, I., Heckerman, D., Meek, C., Smyth, P., White, S.: Model-based clustering and visualization of navigation patterns on a web site. Data Min. Knowl. Disc. **7**(4), 399–424 (2003)
9. Chatain, T., Carmona, J., Van Dongen, B.: Alignment-based trace clustering. In: International Conference on Conceptual Modeling, pp. 295–308. Springer, Cham (2017)
10. Cole, J., Foster, H.: Using Moodle: Teaching with the Popular Open Source Course Management System. O'Reilly Media, Inc. (2007)
11. De Koninck, P., De Weerdt, J.: Scalable mixed-paradigm trace clustering using super-instances. In: 2019 International Conference on Process Mining, pp. 17–24. IEEE (2019)
12. De Weerdt, J., Vanden Broucke, S., Vanthienen, J., Baesens, B.: Active trace clustering for improved process discovery. IEEE Trans. Knowl. Data Eng. **25**(12), 2708–2720 (2013)
13. Ester, M., Kriegel, H.P., Sander, J., Xu, X., et al.: A density-based algorithm for discovering clusters in large spatial databases with noise. In: KDD, vol. 96, pp. 226–231 (1996)
14. Fazzinga, B., Flesca, S., Furfaro, F., Masciari, E., Pontieri, L.: Efficiently interpreting traces of low level events in business process logs. Inf. Syst. **73**, 1–24 (2018). https://doi.org/10.1016/j.is.2017.11.001
15. Ferreira, D., Zacarias, M., Malheiros, M., Ferreira, P.: Approaching process mining with sequence clustering: experiments and findings. In: International Conference on Business Process Management, pp. 360–374. Springer, Cham (2007)
16. Hidalgo, L., Munoz-Gama, J.: Domain-driven event abstraction framework for learning dynamics in MOOCs sessions. In: International Conference on Process Mining, pp. 552–564. Springer, Cham (2022)
17. Hobeck, R., Pufahl, L., Weber, I.: Process mining on curriculum-based study data: a case study at a German university. In: International Conference on Process Mining, pp. 577–589. Springer, Cham (2022)
18. Joudieh, N., Eteokleous, N., Champagnat, R., Rabah, M., Nowakowski, S.: Employing a process mining approach to recommend personalized adaptive learning paths

in blended-learning environments. In: 12th International Conference in Open and Distance Learning, Athens, Greece (2023)
19. Joudieh, N., Trabelsi, M., Champagnat, R., Rabah, M., Eteokleous, N.: Using trace clustering to group learning scenarios: an adaptation of FSS-encoding to moodle logs use case. In: CSEDU (2), pp. 247–254 (2024)
20. Kumar, L., Singh, H., Kaur, R.: Web analytics and metrics: a survey. In: Proceedings of the International Conference on Advances in Computing, Communications and Informatics, pp. 966–971 (2012)
21. Leemans, S.J., Poppe, E., Wynn, M.T.: Directly follows-based process mining: exploration & a case study. In: 2019 International Conference on Process Mining (ICPM), pp. 25–32. IEEE (2019)
22. Nouvellet, A., Beaudoin, V., D'alché-Buc, F., Prieur, C., Roueff, F.: Analysis of gallica usage traces. Research report, Télécom ParisTech ; Bibliothèque nationale de France (2017). https://hal.archives-ouvertes.fr/hal-01709264
23. Ouatik, F., Erritali, M., Ouatik, F., Jourhmane, M.: Predicting student success using big data and machine learning algorithms. Int. J. Emerg. Technol. Learn. (iJET) **17**(12), 236–251 (2022)
24. Poggi, N., Muthusamy, V., Carrera, D., Khalaf, R.: Business process mining from e-commerce web logs. In: Daniel, F., Wang, J., Weber, B. (eds.) BPM 2013. LNCS, vol. 8094, pp. 65–80. Springer, Heidelberg (2013). https://doi.org/10.1007/978-3-642-40176-3_7
25. Rousseeuw, P.J.: Silhouettes: a graphical aid to the interpretation and validation of cluster analysis. J. Comput. Appl. Math. **20**, 53–65 (1987)
26. Song, M., Günther, C.W., Van der Aalst, W.M.: Trace clustering in process mining. In: International Conference on Business Process Management, pp. 109–120. Springer, Cham (2008)
27. Talakokkula, A.: A survey on web usage mining, applications and tools. Comput. Eng. Intell. Syst. **6**(2), 22–29 (2015)
28. Tokuda, E.K., Comin, C.H., Costa, L.d.F.: Revisiting agglomerative clustering. Physica A: Stat. Mech. Appl. **585**, 126433 (2022)
29. Trabelsi, M., Suire, C., Morcos, J., Champagnat, R.: A new methodology to bring out typical users interactions in digital libraries. In: 2021 ACM/IEEE Joint Conference on Digital Libraries (JCDL), pp. 11–20. IEEE (2021)
30. Trabelsi, M., Suire, C., Morcos, J., Champagnat, R.: User's behavior in digital libraries: process mining exploration. In: International Conference on Theory and Practice of Digital Libraries, pp. 388–392. Springer, Cham (2019)
31. Umer, R., Susnjak, T., Mathrani, A., Suriadi, S.: On predicting academic performance with process mining in learning analytics. J. Res. Innov. Teach. Learn. **10**(2), 160–176 (2017)
32. Wagner, M., et al.: A combined approach of process mining and rule-based AI for study planning and monitoring in higher education. In: International Conference on Process Mining, pp. 513–525. Springer, Cham (2022)
33. Yu, H., Harper, S., Vigo, M.: Modeling micro-interactions in self-regulated learning: a data-driven methodology. Int. J. Hum.-Comput. Stud. **151**, 102625 (2021)
34. Zandkarimi, F., Rehse, J.R., Soudmand, P., Hoehle, H.: A generic framework for trace clustering in process mining. In: 2020 2nd International Conference on Process Mining, pp. 177–184. IEEE (2020)
35. Zheng, G., Peltsverger, S.: Web analytics overview. In: Encyclopedia of Information Science and Technology, 3rd edn., pp. 7674–7683. IGI Global (2015)

Challenges in Data Quality Management for IoT-Enhanced Event Logs

Yannis Bertrand[1(✉)], Alexander Schultheis[2,3], Lukas Malburg[2,3], Joscha Grüger[2,3], Estefanía Serral Asensio[4], and Ralph Bergmann[2,3]

[1] Department of Business Informatics and Operations Management, Faculty of Economics and Business Administration, Ghent University, Ghent, Belgium
yannis.bertrand@ugent.be

[2] German Research Center for Artificial Intelligence (DFKI), Branch Trier University, Behringstraße 21, 54296 Trier, Germany
{alexander.schultheis,lukas.malburg,joscha.grueger, ralph.bergmann}@dfki.de

[3] Artificial Intelligence and Intelligent Information Systems, Trier University, 54296 Trier, Germany
{schultheis,malburgl,grueger,bergmann}@uni-trier.de

[4] Research Centre for Information Systems (LIRIS), KU Leuven, Brussels, Belgium
estefania.serralasensio@kuleuven.be

Abstract. Modern organizations make frequent use of *Internet of Things* (IoT) devices, such as sensors and actuators, to monitor and support their so-called IoT-enhanced *Business Processes* (BPs). These IoT devices collect vast amounts of data which, when processed appropriately, can yield crucial insights into the working of the BPs. However, IoT data, such as sensor data, is notoriously of poor quality, e.g., suffering from noise or having some missing data points. These problems are referred to as *Data Quality Issues* (DQIs), which often interfere with the analysis of IoT data in an industrial context. In this paper, we present a list of challenges that have to be tackled to achieve *Data Quality* (DQ) management in IoT-enhanced event logs. These challenges are derived and refined by leveraging expert knowledge and experience in DQIs within a focus group interview. In addition, we provide directions for solutions to these challenges based on input from the focus group interview and the literature. Finally, we discuss the challenges and their impact on typical DQ management tasks. The insights provided can help guide future research to achieve better DQ in event logs of IoT-enhanced BPs.

Keywords: Business Process Management · Internet of Things · Data Quality Management · Data Quality Issues

1 Introduction

Recently, the field of *Business Process Management* (BPM) is getting more and more attention in the context of *Internet of Things* (IoT) environments [15,19,33] such as smart homes or smart factories. In this context, BPM can be

J. Grabis et al. (Eds.): RCIS 2025, LNBIP 547, pp. 20–36, 2025.
https://doi.org/10.1007/978-3-031-92474-3_2

applied to execute modeled processes with support from IoT devices in smart environments, realizing IoT-enhanced *Business Processes* (BPs). In these processes, IoT sensor data can be used to detect certain emerging situations and react to them [26]. The integration of IoT data and event data generated by BPM systems is not straightforward, and previous research has presented multiple challenges [7,15]. However, the literature has so far not dealt with an essential hurdle to IoT-enhanced BPM, namely *Data Quality* (DQ) [17]. IoT sensor data are prone to *Data Quality Issues* (DQIs), which can have various root causes, such as sensor failures, network problems, or environmental influences [14,37].

DQIs can significantly affect both the quality of the data and the results of subsequent analyzes [8]. Previous research has identified patterns that can characterize recurring DQIs in the data. However, there remain crucial challenges to overcome in DQ management in IoT-enhanced BPM. For example, it remains challenging to address DQIs that only rarely occur, but have a significant impact on the DQ. These infrequent issues often require domain specialists for the detection and handling of these DQIs. Moreover, methods and approaches supporting domain experts in this process are currently still in their infancy and only support certain types of DQIs or have a very high computational complexity (see, e.g., Schultheis et al. [28]), leading to low user satisfaction. In total, this gap limits the ability to handle DQIs, leading to poor analysis and decisions based on the event log.

In general, the goal of this paper is to highlight critical challenges in managing DQIs within IoT-enhanced event logs. This paves the way for the development of intelligent techniques to resolve DQIs and ensure the acquisition of high-quality sensor data, which can be analyzed alongside process data provided by a process-aware information system, or abstracted to detect process events in the absence of such a system. Specifically, this paper's contribution is twofold:

1. It identifies challenges that hamper DQ management in IoT-enhanced event logs, following a mixed-method approach that combines a literature review with a focus group interview; and
2. it examines the relationship between these challenges and the typical DQ management tasks discussed in the literature.

Therefore, this paper is structured as follows: Sect. 2 provides the background necessary to understand the topic of IoT-enhanced BPM and the challenges in managing DQIs, particularly within IoT event logs, and the main phases of DQ management addressed in the literature. Then, Sect. 3 introduces the methodology followed to derive the challenges, involving a focus group interview with IoT BP experts. In Sect. 4, the key challenges associated with addressing DQIs are identified and elaborated. These challenges are positioned within the broader context of DQ management in Sect. 5, illustrating their relevance to established DQ management tasks. Subsequently, in Sect. 6 possibilities based on which the various groups of challenges can be addressed are outlined. Section 7 offers a discussion of the implications of these challenges and how they relate to existing approaches as well as threats to the validity of the results. Finally, Sect. 8

concludes the paper, summarizing the main findings and suggesting directions for future research.

2 Background

This section presents the research background of this work. Section 2.1 presents the state of research in the field of DQ, while Sect. 2.2 shows the IoT-enhanced research in the BPM area. Section 2.3 covers preliminary work that combines both research areas. The management of DQ, which arises in this context, is introduced in Sect. 2.4.

2.1 Data Quality

DQ is a research area that mainly deals with the detection and handling of DQIs in databases and datasets. In general, DQ is considered to determine the extent to which the data meet the requirements of their users [4,36]. Various dimensions are defined to describe and quantify DQ, among them: accuracy, timeliness, precision, completeness, reliability, and error recovery [17]. The importance of each of these dimensions depends on the application scenario and the type of data.

2.2 IoT-Enhanced Business Process Management

IoT-enhanced BPs are characterized by the multitude of IoT devices (i.e., sensors and actuators) that support their execution, for example, by automating tasks and tracking physical process parameters [15]. Recently, there has been increasing awareness of the potential integration of IoT data with BPM techniques in IoT-enhanced BPs in various sectors, including manufacturing, healthcare, logistics, and smart spaces [3,10,19,29].

There, IoT data can be used to contextualize BPs with data describing their physical environment, e.g., sensor data in modern production processes or patient vital signs in healthcare. For example, IoT-enhanced process mining techniques can provide a more profound understanding of the BPs by, e.g., deriving IoT-based decision rules, process variants, or anomalous sensor data patterns. There, typical steps include preprocessing the raw data (i.e., cleaning, formatting), event correlation to retrieve the cases each event belongs to, and event abstraction to derive meaningful process events from sensor data [10,18].

2.3 Data Quality in IoT Business Process Management

IoT data quality is a broad topic, ranging from detecting DQIs to improving DQ through cleaning methods [17,34]. IoT applications often rely on low-cost sensors with limited battery and processing power, often deployed in hostile environments [34]. This leads to sensor issues such as low sensing accuracy, calibration loss, sensor failures, improper device placement, range limit, and loss of data packages. Such sensor faults, in turn, cause various types of errors in the generated data, complicating further analysis. Teh et al. [34] review the sensor DQ literature and list the following error types (in decreasing order of frequency): outliers; missing data; bias; drift; noise; constant value; uncertainty; stuck-at-zero. When left untreated, these errors lead to incorrect data and unreliable analysis, leading to poor decision-making.

Although basically all previous research applying BPM with IoT data had to tackle DQIs, the existing body of literature specifically addressing DQIs in IoT-enhanced BPM is still scarce. Bertrand et al. [8] review the IoT BPM literature to describe the IoT DQIs and the event log DQIs encountered by previous research. Based on this state of affairs, patterns that relate IoT DQIs with resulting event log DQIs are derived. More specifically, these patterns are of the following shape: Sensor fault $\Longrightarrow$ sensor DQI $\Longrightarrow$ event log DQI(s). For example, one of the patterns describes the following: unstable environment $\Longrightarrow$ noisy sensor data $\Longrightarrow$ incorrect case ID. A concrete example of the occurrence of this pattern in the literature is provided by Brzychczy and Trzcionkowska [10], where data from sensors placed on a drilling machine in a mine were used to derive an event log. Unfortunately, the sensors produced noisy data, making it difficult to recognize the start and end activities of the mining process and resulting in some incorrect case IDs in the log derived from the sensor data.

2.4 Data Quality Management

Various steps for addressing DQIs are discussed in the literature (e.g., in [5, 12, 14, 34]). In this section, we introduce some of the most commonly discussed tasks in DQ management. Although so far a widely accepted standardized model has not yet been proposed that describes the process of DQ management from detection to handling of DQIs, we outline the main tasks in Fig. 1.

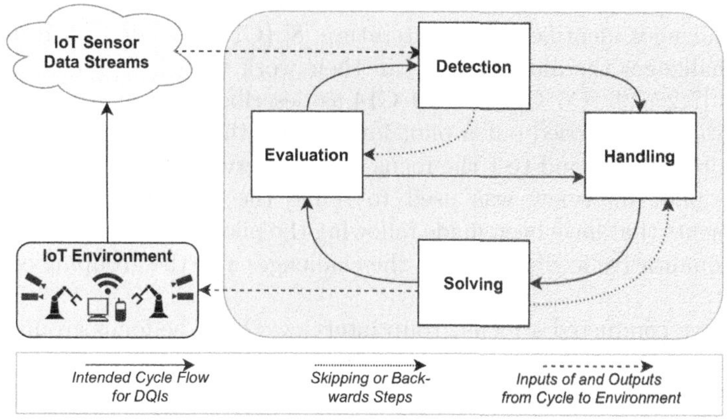

Fig. 1. Visualization of the Relationships Between Common DQ Management Steps (In Accordance With the Plan-Do-Check-Act Cycle [21])

1. Detection: In this task, incoming IoT sensor data are monitored to identify possible DQIs [14,34]. The result of this is the classification of whether DQIs are present and, if so, which ones. If DQIs are detected, the data are passed on to the handling phase.
2. Handling: This task typically builds on the classification of DQIs output by the detection [14]. The aim of handling is to provide recommendations for action to rectify each individual DQI. These may include both the repair of the components that caused the DQI and the cleaning of the event log. As a result, recommendations for action are given for the solving task.
3. Solving: In this task, scheduling is carried out to prioritize the recommended actions from handling [12]. For example, duplicate steps are combined, and actions are checked for interdependencies. The actions are then implemented automatically, semi-automatically or manually [5,34]. If it turns out that the proposed action cannot be executed, the problem is returned to the handling task.
4. Evaluation: The evaluation verifies the correctness of the result, i.e., whether the log is cleaned and/or the cause of the DQIs is solved [12,14]. This task should also check whether new DQIs have been generated as a result of the application of a data cleaning technique. If insufficient quality is detected in the evaluation, it is checked whether the identification and classification of the DQI in the detection phase is correct and whether the recommended and executed actions have really solved this problem. If this is not the case, the detection and handling tasks can be triggered again.

3 Focus Group Methodology

In this section, we introduce the mixed method approach we used to derive the challenges (depicted in Fig. 2).

First, we created an initial list of challenges, derived from two main sources: 1) the challenges identified in the literature [8] (C1–C5 as described in Sect. 4); and 2) challenges the authors faced in their work with IoT-enhanced business processes [7,28,29] (C6–C7, and C9–C14 as described in Sect. 4).

Afterward, we carried out a pilot interview with an expert in IoT-enhanced processes to prepare and test the focus group interview. The feedback obtained from the pilot interview was used to refine the interview process. Notable improvements that have been made following the pilot interview include a greater focus on qualitative feedback about the challenges and the grouping of the challenges.

Then, we conducted a focus group interview [24]. The focus group interview method involves selected participants discussing a specific topic, using group dynamics to produce more in-depth insights than individual interviews [35]. Participants are chosen for their relevance and expertise in the topic and their previous experience in group discussions [25]. We selected four researchers who are specialized in IoT-enhanced BPM, all members of the *Internet of Processes*

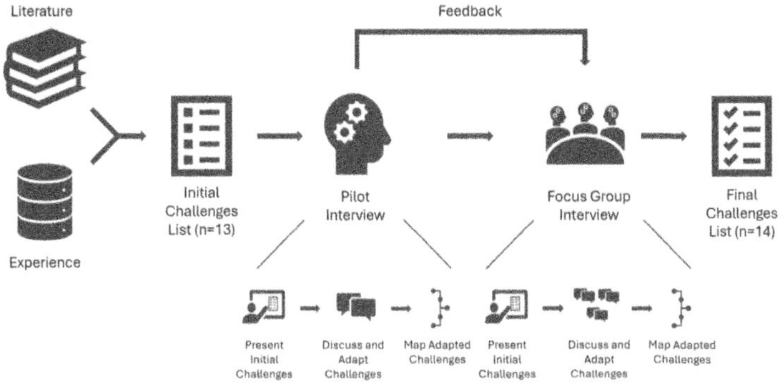

Fig. 2. Overview of the Methodology Followed to Derive the Challenges Presented in the Paper.

and Things (IoPT) initiative[1], a group of researchers pioneering industrial IoT applications. As we were able to consult these members of the IoPT group, we gained detailed insights and challenges in current research works on the topic of IoT-enhanced BPM through the interview. In addition to their research work, this also includes the practice-oriented research projects the members are conducting with industry partners, which go beyond laboratory experiments and, thus, provide more realistic and industrial challenges. The focus group has been designed to facilitate discussion and stimulate interaction between participants. Specifically, the participants were first presented with the initial list of challenges. The participants have then been asked to discuss the challenges, with the discussion revolving around three main questions:

1. Which challenges have they already encountered?
2. Which challenges do they consider irrelevant?
3. Which challenges do they believe are missing?

During the interview, the group evaluated, adapted, and grouped the proposed challenges, and multiple challenges were proposed and debated. Challenges have only been included by the focus group if a consensus has been reached, and the focus group ultimately resulted in the inclusion of one more challenge in the final list (C8). Meanwhile, the authors moderated the exchanges, pointing out some similarities and differences between the proposals made by the participants. The discussion has been recorded and transcribed for analysis[2]. This group interview resulted in a general consensus on the final list of challenges presented in Sect. 4.

Throughout the discussion, suggestions have been made to solve the challenges. These suggestions are summarized and completed in Sect. 6, where, for each group of challenges, we present potential solutions and approaches to mediate the impact of the challenges. Finally, the participants were asked to reflect

[1] https://zenodo.org/communities/iopt/about.
[2] Transcript available at: https://zenodo.org/records/15058151.

upon which task(s) of DQ management presented in Sect. 2.4 each group of challenges impacted most, resulting in the mapping presented in Sect. 5.

4 Challenges for Addressing Data Quality Issues

In this section, we present the list of challenges derived from the focus group interview. In total, 14 challenges were identified, which are clustered into four groups: Data Complexity (see Sect. 4.1), Knowledge Acquisition (see Sect. 4.2), Knowledge Management (see Sect. 4.3), and Conflict of Interests (see Sect. 4.4).

4.1 Group 1: Data Complexity

This group addresses the challenges related to the fundamental characteristics of big data in the context of IoT environments. Specifically, it focuses on issues that arise from the immense volumes of data generated by sensors, the variety of data formats, the velocity at which data must be processed, and the granularity required for effective DQI identification and resolution. The following four challenges are identified for this group:

C1 – **Volume**: This challenge pertains to the big data issue of volume [16], which involves the continuous generation and collection of vast amounts of data. In particular, IoT BPs equipped with numerous sensors, especially in long-running operations, produce an overwhelming volume of data. Analyzing these large data sets is very computationally intensive [28]. Although approaches that simplify or abstract data representation (e.g., [2,20,31]) can reduce computational load, they often result in a trade-off between data compression and the loss of critical information, potentially causing some DQIs to go undetected.
C2 – **Variety**: Variety refers to the big data challenge of managing data produced in multiple formats – structured, unstructured, or semi-structured – from diverse sources [16]. In the context of DQIs, this variety emerges from different types of sensors that generate time series data, which represent various values over time, such as Boolean states or spatial coordinates [2]. Additionally, the complexity increases with dependencies between data sources, requiring careful consideration in data storage and processing.
C3 – **Velocity**: Velocity refers to the challenge of processing data at high speed, as described in the context of big data [16]. In IoT BPs, sensors perform frequent measurements, typically every second. To effectively address DQIs, it is crucial to identify and correct failures in near real time. Delaying this process can result in a trade-off where failures are only analyzed after they have impacted the data, compromising the timeliness of DQI management.
C4 – **Granularity**: Granularity addresses the level at which DQIs are identified and solved. The key is to choose the appropriate method or algorithm for taking the data to the desired level of detail. For instance, it can be advantageous to pre-process or abstract data (see the trade-off in C1) instead of relying on raw sensor data. The granularity level enables different processing places, such as

distributed pre-processing at the edge level [27,32]. However, data abstraction is performed at the risk of overlooking dependencies, such as those between time series, which may only be fully captured at lower levels of granularity.

4.2 Group 2: DQI Knowledge Acquisition

To effectively identify and handle DQIs, it is crucial to acquire knowledge about them. The challenges in this category are associated with the collection and processing of knowledge regarding the DQIs present in the data, which is closely aligned with the research area of knowledge acquisition [23]. This process pertains to the specific methods of extracting knowledge from the data. Therefore, four key challenges are identified.

C5 – New Data Quality Issues: This challenge deals with DQIs that have never been observed before. Hence, there cannot be knowledge about this type of failure, making automatic detection and classification difficult. This raises the problem of how knowledge about such new DQIs can be suitably acquired and generated, e.g., for automated processing in an Artificial Intelligence model. Although there are methods to identify the existence of an unknown issue, such as anomaly detection [11], concrete classification requires further information.
C6 – Rare Data Quality Issues: Rare DQIs present significant challenges due to the scarcity of data on these infrequent occurrences. This scarcity complicates pattern recognition, algorithm training, and strategy development to manage them. Classifying such DQIs and obtaining relevant information through appropriate solutions is particularly difficult. Moreover, there is a heightened risk of, e.g., overfitting when training Machine Learning models to detect these rare DQIs. The core challenge, therefore, lies in effectively addressing rare DQIs throughout the various steps of DQ management described in Sect. 2.4, despite the limited and imbalanced available data.
C7 – Causality Ambiguity: In the context of DQIs, ambiguity means that known patterns cannot be clearly traced back to an issue. For example, noisy data can result from a sensor being incorrectly calibrated or from an issue in the environment, such as someone walking near the sensor. In this case, the DQI patterns left in the data would be very similar, but the cause would be entirely different, creating difficulties to correctly classify and handle the DQI.
C8 – Solution Ambiguity: This challenge involves the difficulty of selecting the appropriate solution for a DQI when ambiguity exists. Even if the cause of a DQI is known, there may be multiple solutions with no clear best choice. This uncertainty creates challenges in evaluating and applying different possible remedies. Implementing a solution might trigger a conflict of business interests (see C13), such as requiring significant financial investment or resources. Furthermore, there might be dependencies between solutions that could introduce new DQIs (see C14). The key issue is effectively balancing and choosing among various solutions, each with its own trade-offs and potential implications.

4.3 Group 3: DQI Knowledge Management

In this section, we explore the challenges related to the management of existing knowledge about DQIs. Effective knowledge management [6, pp. 9–11] is essential for the identification, handling, and resolution of DQIs, especially when dealing with complex and evolving data environments. The challenges discussed here focus on the representation, reuse, and refinement of knowledge. These aspects are crucial to ensure that knowledge about DQIs is not only well-documented and structured, but also effectively leveraged and continuously improved to address both existing and emerging issues in DQ management.

C9 – Knowledge Representation: This challenge addresses the problem of finding a suitable representation of knowledge. It considers the preprocessing of existing knowledge about DQIs (e.g., manuals, workers' experience, patterns, causes, and solutions) into a format interpretable for machines so that it can be processed automatically (so-called externalization [22]). The appropriate knowledge representation is the basic requirement for applying algorithms or more advanced methods to address DQIs.
C10 – Knowledge Reuse and Transfer: Knowledge Reuse and Transfer involves leveraging existing, well-structured knowledge stored in a knowledge base (e.g., a database) to detect, manage, and solve new DQIs. The key challenge is to automatically identify the most relevant knowledge within the database and adapt it to address new DQIs. This process can occur within the current environment (e.g., factory or machine) or across different environments, which may require further generalization to ensure effective application.
C11 – Knowledge Refinement: This challenge refers to the constant evaluation and, if necessary, updating of knowledge. For example, solutions that are proposed for a new DQI as part of knowledge management and applied where necessary must be checked to see whether they actually increase the DQ for this use case. Furthermore, it is possible that the application of solutions may cause new problems (see C14). To ensure that a DQ management system, that is as automated as possible, eliminates causes of DQIs and increases DQ, constant refinement of the knowledge must therefore be carried out. In this context, knowledge might also be acquired again, referring to the challenges of Group 2 (see Sect. 4.4).

4.4 Group 4: Conflict of Interests

This section addresses the challenges that arise when DQ management concerns clash with broader organizational matters. These include issues such as privacy, where data protection requirements can impede the identification and resolution of DQIs; business interest conflicts, where financial considerations may restrict the resources dedicated to DQ management; and cascading effects, where the resolution of DQIs might unintentionally cause adverse consequences in the data processing workflow. Addressing these challenges is essential for implementing effective and sustainable DQ management practices.

C12 – Privacy: This challenge highlights potential conflicts of interest related to data privacy, which can occur in two main situations. First, privacy may be required to protect personal interests or to comply with legal regulations. For instance, these privacy concerns may conflict with the analysis of certain data, such as video recordings, or may entirely prohibit the use of whole data sets and data sources. This means that data that could be used to identify and rectify DQIs may be missing, or their use could be restricted. On the other hand, privacy can also be viewed in terms of trade secrets and confidentiality, where sensitive information is protected from unauthorized access or disclosure. This means that a cross-enterprise exchange of possible DQI knowledge may be restricted to prevent competitors from obtaining information about production (which exacerbates the problem of few failure data, see C6).
C13 – Business Interest Conflicts: This challenge describes financial interests within a company that may stand in the way of DQ management. For example, this DQ management may play a subordinate role within the company and there may be no willingness to spend money on addressing it at all, e.g., by deciding not to replace a faulty sensor. Certain DQIs may also be considered unimportant and contrary to other internal company guidelines. For example, Predictive Maintenance [38] is often carried out within a factory. If it is recognized that a sensor is producing faulty or missing data, it may still be economical from this perspective to keep the sensor running shortly before or even until total failure. On the other hand, DQ management would have an interest in repairing or even replacing the sensor as fast as possible.
C14 – Cascading Effects: This challenge can occur if the semantics of a process can get lost due to the elimination of DQIs. In general, there is an interest in rectifying DQIs if possible. However, this can lead to undesirable side effects, as a result of which the data are still not suitable for analysis, its quality may even deteriorate or non-existent other failures may be searched for. For example, a faulty sensor might disturb the production in a smart industrial process. By repairing the sensor causing the failure and the corresponding DQI, the process can be continued. However, repairing the event log would mean that the semantics that caused production to stop would no longer be retained. As a result, the event log would be searched for the failure that caused the stop and none would be found, leading to an unnecessary failure investigation. In this case, the interest of a clean event log takes precedence over the semantics of the process. Such cases must therefore be identified and handled accordingly, possibly by discarding a part of the log.

5 Mapping of Challenges to Data Quality Management Tasks

The final part of the focus group interview aimed at mapping the challenges with the DQ management tasks discussed in Sect. 2.4. This positions the challenges in the broader context of DQ management. Depending on the addressed step, it is possible to determine which challenges relevant to this step are likely to occur and based on this, define a prioritization for dealing with them. Depending on the step, various solution approaches can be developed, each addressing the most relevant challenges. This prioritization was derived and evaluated through discussions with the experts in the focus group interview. As the participants saw little difference between the challenges within each group, the mapping is given for each group of challenges. The mapping is shown in Table 1.

Table 1. Mapping of Challenges to the Four Main DQ Management Tasks Presented in Sect. 2.4.

Group	Detection	Handling	Solving	Evaluation
Group 1 (Section 4.1)	✗	✗		
Group 2 (Section 4.2)	✗	✗	✗	✗
Group 3 (Section 4.3)	✗	✗	✗	✗
Group 4 (Section 4.4)	✗	✗	✗	

First, for Group 1, it was recognized that challenges C1–C4 mostly impact the detection and the handling of the DQIs, as these tasks usually assume a simplification of the data which negates the effect of the challenges on subsequent tasks. For example, with large volumes of data (C1), the share of data where DQIs are detected is likely to be minimal, which means that the volume of data that will go through the next tasks will typically be more manageable. Then, for Group 2, all tasks are impacted: in C5 and C6, the rarity of the DQI implies difficulties in detection as well as solving and the evaluation of cleaned data, while ambiguity can affect both the detection and evaluation tasks (C7) as well as the handling and solving phases (C8). Next to this, the Group 3 challenges also impact all tasks, as the knowledge about the DQIs covers the techniques to detect them as well as the approaches to solve them and their effects. Finally, Group 4 affects all tasks but the evaluation, as these challenges relate to the availability of data (especially C12) and the prioritization of the solutions that can be applied (C13 and C14).

6 Towards Addressing the Challenges

In this section, we discuss possible solutions to the challenges based on the results of the focus group interview, completed with the literature. We present the solutions along the four groups of DQIs identified in Sect. 4.

6.1 Group 1: Data Complexity

The challenges in this group are deeply interrelated and are best addressed all together. For instance, more fine-granular data (C4) and different types of data (C2) often lead to a higher volume of data (C1), which in turn exacerbates the challenge of velocity (C3). Overall, a trade-off has to be made between them: To tackle the challenge of velocity and solve DQIs faster, one can, e.g., abstract the data to a higher granularity level, thereby also reducing the volume of data, making it easier to process the data in a timely way (albeit at the cost of potentially missing some DQIs).

Two primary approaches were proposed to tackle these challenges holistically:

a) *Expert-Guided Data Prioritization & Filtering*: By focusing on the most relevant bits of data and aggregating data to a semantically meaningful level, the volume of data to process can be reduced, making it easier to analyze various types of data (C2), timely, and at the right granularity level. This technique can be supported by a divide-and-conquer approach to data preprocessing, where an initial layer of data filtering occurs at the edge, streamlining subsequent processing.
b) *Increasing Computational Power*: Expanding processing capacity enables the handling of larger data volumes (C1), faster processing of the same amount of data (C3), analysis at a finer level of granularity (C4), or a combination of these improvements simultaneously.

6.2 Group 2: DQI Knowledge Acquisition

For the challenges in group 2, potential solutions lie in specific data analysis techniques and expert knowledge. To address new DQIs (C5), anomaly detection techniques can be employed to identify new types of DQIs (for an in-depth treatment of anomaly detection in IoT time series, see [11,30]). Next to this, expert knowledge can guide the discovery of new DQI types, in particular when the process is being modified, i.e., there is concept drift. Then, for rare DQIs (C6), data augmentation techniques can be used to generate more cases of emerging DQI types to refine the DQI detection and handling techniques. Next, to address causality ambiguity (C7), root cause analysis based on expert knowledge is the most straightforward option. This way, occurrences of similar DQIs due to different causes could be more easily disentangled. Solution ambiguity (C8), in turn, can be addressed by ranking solutions and trying to apply the best ranked solutions until the DQI is solved. This ranking could be performed based on the likelihood that each solution will solve the issue and its cost.

6.3 Group 3: DQI Knowledge Management

This group of challenges revolves around knowledge representation, and the choice of a suitable knowledge representation, addressing knowledge representation (C9), is a prerequisite to tackling the other challenges. To this end, we expect a suitable knowledge representation to store, for each DQI detected, the data, the cause of the DQI and the solution applied. Knowledge-based approaches (like case-based reasoning [1,6]), ontologies [9] or knowledge graphs [13] could be used to store this type of information. Moreover, they are supported by data retrieval techniques that can be applied to address knowledge reuse and transfer (C10). Knowledge refinement (C11), finally, can also be addressed by the techniques used for new DQIs (C5) and rare DQIs (C6), as knowledge refinement is often required when concept drift occurred or when a DQI is not well-known yet.

6.4 Group 4: Conflict of Interests

The challenges in this group are not as interconnected as the challenges in other groups and require different solutions. Depending on the type of data and the level of confidentiality required, data privacy (C12) can be addressed by anonymizing the data, using federated learning to learn from multiple data sources without directly sharing the data or generating synthetic data. For business interest conflicts (C13), the issue is mainly a management one, which requires prioritizing DQIs based on a cost-benefit analysis, and only solving DQIs which cause more financial damage than the cost of handling them. Regarding cascading effects (C14), one possibility is to keep both the original log (containing DQIs) and the cleaned one, so that root cause analysis is always possible on the original data which caused the issues in the process execution.

7 Discussion

One of the main difficulties encountered in the collection and description of the challenges presented in Sect. 4 is that such challenges are very interdependent. This interdependence is often observed between challenges of the same group. For example, granularity (C4) and volume (C1) impact each other, as being able to abstract the raw sensor data without losing important information can greatly reduce the burden caused by large volumes of data. Moreover, solution ambiguity (C8) can be mildened by accurate detection of DQIs, i.e., resolving causality ambiguity (C7).

Some challenges can also have an effect on other challenges in other groups, e.g., the granularity level at which data are analyzed (see C4) can be impacted by the knowledge representation (C9) chosen. As another example, a suitable procedure for business interest conflicts (C13) can mitigate rare DQI challenges (see C6).

Next to this, some overarching factors that can affect multiple challenges are identified. First, in the handling and solving phases, two types of techniques can be proposed and applied:

1. Approaches to clean the data, e.g., scripts to remove outliers or impute missing data,
2. Actions to repair the data collection process, e.g., recalibrating a sensor or replacing a faulty network component.

These have very different effects and can solve different types of issues, as repairing the data collection process is typically more costly (e.g., regarding hardware costs or downtimes) and takes longer. But some DQIs can only be solved by such interventions. There, business interest conflicts (C13) play a crucial role in balancing the costs and benefits of both types of solutions.

Second, the timeliness of the DQ management (i.e., how fast DQIs can and need to be solved), while having an obvious link with the velocity challenge (C3), also impacts multiple other aspects of DQ management and determines how serious some challenges are. As such, the volume of the data (C1) can be handled more easily if the data can be stored and processed per batch. The challenges of Group 2 (see Sect. 4.2) are also easier to address if there is enough time to have a process/data expert analyzing the problem in depth. On the other hand, being able to address DQIs in (near) real-time makes some challenges much less problematic. For example, cascading effects (C14) can be defused before they have a chance to occur if the initial DQI is immediately handled.

Finally, a last point of discussion in the focus group has been the importance or gravity of each challenge. There, no consensus could be achieved, and it has been suggested that the importance of each challenge is highly dependent on the perspective taken. For example, considering the challenges from a Process Mining perspective, the granularity of the data (C4) has been mentioned as a particularly important challenge.

While these insights provide valuable perspectives on the identified challenges, it is also important to consider potential limitations of our findings. Therefore, we acknowledge the following threats to the validity of our results:

a) *Limited Sample in the Focus Group Interview*: The challenges were validated by a group of four researchers only as well as one additional researcher for the pilot interview. These focus group participants were selected for their expertise in the domain, which allowed the authors to gather high-quality feedback to validate the challenges.
b) *Purely Academic Setting*: All participants in the study (authors and focus group interviewees) are university or research institute members. However, all participants have experience working with industry partners, which makes them familiar with the challenges faced when handling real-life data.
c) *Non-formal Validation Approach*: In this study, we chose the focus group for its ability to generate new ideas through the interaction between participants. However, a focus group is not a systematic validation method, and a controlled survey would allow a more robust validation of the challenges.

8 Conclusion and Future Work

In this paper, we address the issue of DQ in the research topic of IoT meets BPM. Our main contribution lies in the discussion of the main challenges that have to be overcome to achieve DQ management in IoT BP, which were derived from a focus group interview with experts in IoT BPM. Next to this, we propose a mapping of the different challenges groups with the main DQ management tasks discussed in the literature. We put forward the summary of the challenges that we present as a foundation for future research in the domain, and also present initial ideas for addressing these.

In future works, we plan to build on this foundation and develop a framework for semi-automated DQ management in IoT BPs, capable of addressing the challenges we presented, based on the main steps described in Sect. 2.4.

Acknowledgement. This work is supported by the Flemish Fund for Scientific Research (FWO) with grant number G0B6922N. This work is also partially funded by the Federal Ministry for Economic Affairs and Climate Action under grant No. 01MD22002C *EASY* [27].

This research was supported by the Internet of Processes and Things (IoPT) community: https://zenodo.org/communities/iopt/about.

References

1. Aamodt, A., Plaza, E.: Case-based reasoning: foundational issues, methodological variations, and system approaches. AI Commun. **7**(1), 39–59 (1994)
2. Allen, J.F.: Maintaining knowledge about temporal intervals. Commun. ACM **26**(11), 832–843 (1983)
3. Banham, A., Leemans, S.J., Wynn, M.T., Andrews, R., Laupland, K.B., Shinners, L.: xPM: enhancing exogenous data visibility. AI Med. **133**, 102409 (2022)
4. Batini, C., Scannapieco, M.: Data Quality: Concepts, Methodologies and Techniques. DCSA. Springer, Cham (2006)
5. Batini, C., Scannapieco, M.: Data and Information Quality. Data-Cent. Syst. Appl. 63 (2016)
6. Bergmann, R.: Experience Management: Foundations, Development Methodology, and Internet-Based Applications. LNCS, vol. 2432. Springer, Cham (2003)
7. Bertrand, Y., De Weerdt, J., Serral, E.: Assessing the suitability of traditional event log standards for IoT-enhanced event logs. In: BPM 2022, pp. 63–75. Springer, Cham (2022)
8. Bertrand, Y., Van Belle, R., De Weerdt, J., Serral, E.: Defining data quality issues in process mining with IoT data. In: ICPM 2022 Workshops. LNBIP, vol. 468, pp. 422–434. Springer, Cham (2022)
9. Breitman, K.K., Casanova, M.A., Truszkowski, W.: Semantic Web: Concepts, Technologies and Applications. NASA MSSE. Springer, Cham (2007)
10. Brzychczy, E., Trzcionkowska, A.: Creation of an event log from a low-level machinery monitoring system for process mining purposes. In: IDEAL 2021, pp. 54–63. Springer, Cham (2018)
11. Cook, A.A., Mısırlı, G., Fan, Z.: Anomaly detection for IoT time-series data: a survey. IEEE Internet Things J. **7**(7), 6481–6494 (2020)
12. Ehrlinger, L., Wöß, W.: Automated data quality monitoring. In: 22nd ICIQ Proceedings, pp. 15.1–15.9. Little Rock, AR (2017)
13. Fensel, D., et al.: Knowledge Graphs - Methodology, Tools and Selected Use Cases. Springer, Cham (2020)
14. Goknil, A., et al.: A systematic review of data quality in CPS and IoT for industry 4.0. ACM Comput. Surv. **55**(14s), 1–38 (2023)

15. Janiesch, C., et al.: The Internet-of-Things meets business process management. A Manifesto. IEEE Syst. Man Cybern. Mag. **6**(4), 34–44 (2020)
16. Kaisler, S.H., Armour, F., Espinosa, J.A., Money, W.: Big data: issues and challenges moving forward. In: 46th HICSS Proceedings, pp. 995–1004. IEEE Computer Society (2013)
17. Karkouch, A., Mousannif, H., Al Moatassime, H., Noel, T.: Data quality in Internet of Things: a state-of-the-art survey. JNCA **73**, 57–81 (2016)
18. Koschmider, A., Janssen, D., Mannhardt, F.: Framework for process discovery from sensor data. In: EMISA, pp. 32–38 (2020)
19. Leotta, F., Mecella, M., Mendling, J.: Applying process mining to smart spaces: perspectives and research challenges. In: Persson, A., Stirna, J. (eds.) CAiSE 2015. LNBIP, vol. 215, pp. 298–304. Springer, Cham (2015). https://doi.org/10.1007/978-3-319-19243-7_28
20. Malburg, L., Schultheis, A., Bergmann, R.: Modeling and using complex IoT time series data in case-based reasoning: from application scenarios to implementations. In: 31st ICCBR Workshop Proceedings, vol. 3438, pp. 81–96 (2023)
21. Moen, R., Norman, C.: Evolution of the PDCA cycle (2006)
22. Nonaka, I., Takeuchi, H.: The Knowledge-Creating Company: How Japanese Companies Create the Dynamics of Innovation. Oxford University Press, New York (1995)
23. Olson, J.R., Rueter, H.H.: Extracting expertise from experts: methods for knowledge acquisition. Expert. Syst. **4**(3), 152–168 (1987)
24. Rabiee, F.: Focus-group interview and data analysis. PNS **63**(4), 655–660 (2004)
25. Richardson, C.A., Rabiee, F.: A question of access: an exploration of the factors that influence the health of young males aged 15 to 19 living in Corby and their use of health care services. Health Educ. J. **60**(1), 3–16 (2001)
26. Rinderle-Ma, S., Mangler, J.: Process automation and process mining in manufacturing. In: Polyvyanyy, A., Wynn, M.T., Van Looy, A., Reichert, M. (eds.) BPM 2021. LNCS, vol. 12875, pp. 3–14. Springer, Cham (2021). https://doi.org/10.1007/978-3-030-85469-0_1
27. Schultheis, A., et al.: EASY: Energy-Efficient Analysis and Control Processes in the Dynamic Edge-Cloud Continuum for Industrial Manufacturing. KI (2024)
28. Schultheis, A., et al.: Identifying missing sensor values in IoT time series data: a weight-based extension of similarity measures for smart manufacturing. In: 32nd ICCBR Proceedings. LNCS, vol. 14775, pp. 240–257. Springer, Cham (2024)
29. Seiger, R., Malburg, L., Weber, B., Bergmann, R.: Integrating process management and event processing in smart factories: a systems architecture and use cases. J. Manuf. Syst. **63**, 575–592 (2022)
30. Sgueglia, A., Di Sorbo, A., Visaggio, C.A., Canfora, G.: A systematic literature review of IoT time series anomaly detection solutions. Futur. Gener. Comput. Syst. **134**, 170–186 (2022)
31. Shahar, Y.: A framework for knowledge-based temporal abstraction. Artif. Intell. **90**(1–2), 79–133 (1997)
32. Shi, W., Dustdar, S.: The promise of edge computing. Computer **49**(5), 78–81 (2016)
33. Soffer, P., et al.: From event streams to process models and back: challenges and opportunities. Inf. Syst. **81**, 181–200 (2019)
34. Teh, H.Y., Kempa-Liehr, A.W., Wang, K.I.-K.: Sensor data quality: a systematic review. J. Big Data **7**(1), 1–49 (2020). https://doi.org/10.1186/s40537-020-0285-1

35. Thomas, L., MacMillan, J., McColl, E., Hale, C., Bond, S.: Comparison of focus group and individual interview methodology in examining patient satisfaction with nursing care. Soc. Sci. Med. **1**(4), 206–220 (1995)
36. Wang, R.Y., Strong, D.M.: Beyond accuracy: what data quality means to data consumers. JMIS **12**(4), 5–33 (1996)
37. Zhang, L., Jeong, D., Lee, S.: Data quality management in the Internet of Things. Sensors **21**(17), 5834 (2021)
38. Zonta, T., Da Costa, C.A., da Rosa Righi, R., de Lima, M.J., da Trindade, E.S., Li, G.P.: Predictive maintenance in the Industry 4.0: a systematic literature review. Comput. Ind. Eng. **150**, 106889 (2020)

Blockchain-Based Trust Management System for Enhancing Security in SIoT

Raouf Jmal[1(✉)], Mariam Masmoudi[2], Ikram Amous[1], and Florence Sèdes[3]

[1] MIRACL, Enet'Com, Sfax University, Sfax, Tunisia
raouf.jmal14@gmail.com , ikram.amous@enetcom.usf.tn
[2] MIRACL, FSEGS, Sfax University, Sfax, Tunisia
[3] IRIT, Paul Sabatier University, Toulouse, France
Florence.Sedes@irit.fr

Abstract. In the Social Internet of Things (SIoT), where users interact in a distributed manner, attackers exploit system vulnerabilities to manipulate trust. These attackers spread false information and services, build deceptive reputations, and gain user trust to achieve malicious goals. Such strategies, known as trust-related attacks, involve falsified ratings or manipulated feedback to artificially boost the reputation of malicious entities within the network. To counter these attacks, trust management systems play an essential role in identifying and mitigating malicious activity. Blockchain technology has revolutionized decentralized and distributed systems, providing enhanced security through various applications. Although integrating blockchain into trust management poses challenges, it significantly improves trust evaluation and strengthens the overall security framework. This work introduces a blockchain-based secure trust management system that addresses vulnerabilities through an authentication layer powered by zero-knowledge proof technology, ensuring privacy and robust validation. In addition, a trust evaluation model based on federated learning is proposed, designed to manage heterogeneous data from diverse SIoT nodes with constrained computational resources. The proposed approach is designed to detect various types of trust-related attacks, fostering trustworthy interactions within SIoT environments.

Keywords: Trust management · SIoT · Blockchain · Zero-knowledge proof · Federated learning

1 Introduction

The Social Internet of Things (SIoT) extends the capabilities of Internet of Things (IoT) by integrating it with social networks, enabling seamless control, optimization, and communication among IoT devices. These networks facilitate data sharing and service requests between users' IoT devices, enhancing utility and connectivity [49]. This convergence of IoT and social networking gave rise

J. Grabis et al. (Eds.): RCIS 2025, LNBIP 547, pp. 37–51, 2025.
https://doi.org/10.1007/978-3-031-92474-3_3

to the concept of SIoT [4]. While SIoT significantly enhanced IoT functionality, it introduced distinct security challenges not encountered in traditional IoT systems [34]. Among these challenges, trust management emerged as a pivotal concern [24].

Academic literature has commonly defined trust as the belief of one party (the truster) in another (the trustee) to act in a beneficial or expected manner under uncertain or interdependent conditions [42]. In blockchain and SIoT contexts, trustees have often been nodes, devices, or service providers responsible for data handling, computations, or resource management, while trusters have been entities such as users, devices, or systems that delegated trust for tasks such as data storage or network operations.

In SIoT, trust played a fundamental role in fostering peer interactions, analogous to social dynamics where caution was exercised in sharing sensitive information or engaging with unfamiliar entities [41]. However, SIoT often involved interactions among anonymous users, amplifying the complexity of trust assessment. A significant challenge lay in determining trustworthy entities, as malicious nodes could exploit trust-related vulnerabilities to inflate reputations dishonestly or disrupt operations [28,32,33]. A robust trust management system became essential to detect such attacks, evaluate the trustworthiness of node, and protect sensitive trust data.

In blockchain applications, trust management played a vital role in ensuring the credibility of transactions and participants [12]. Studies explored the integration of blockchain technology to enable decentralized trust management [6,25]. The properties of the blockchain, such as immutability, decentralization, and transparency, enhanced the propagation of trust and facilitated the reliable sharing of trust-related information [20,46,48]. In this context, blockchain transactions often included trust metrics that were crucial for calculating trust scores or identifying trust-related attacks.

Despite advances, existing trust management approaches faced challenges in securely storing and updating trust metrics while ensuring user authentication and privacy protection [1,2,21,30,31]. Zero-knowledge proofs (ZKPs) emerged as a pivotal solution to enhance privacy across various systems. ZKPs enabled the validation of statements, transaction authentication, and data integrity verification without exposing sensitive information [11,22,39].

Consensus protocols in trust management systems often relied on third-party entities for trust aggregation and authentication, leading to inefficiencies. Federated Learning (FL), a collaborative machine learning approach, offered a promising alternative. By allowing distributed devices to train a shared model without exchanging raw data, FL preserved privacy while addressing the challenges of machine learning-based consensus protocols [23].

Smart contracts further streamline trust management by automatically executing predefined constraints, eliminating the need for manual intervention or intermediaries [45]. Storing transactions on blockchain improved security through immutability, but incurred high costs for direct storage of sensitive data. The InterPlanetary File System (IPFS), a decentralized protocol for file storage and retrieval, provided a cost-effective alternative. Governed by smart contracts, IPFS ensured secure, efficient, and low-cost storage through a distributed network of nodes [13,19].

The remainder of this paper is structured as follows: Sect. 2 reviews related work, offering essential context and background. Section 3 introduces the proposed system, which integrates zero-knowledge proof techniques and a federated learning model for evaluating trust and authenticating users in SIoT. Section 4 presents the experimental results of trust-based zero-knowledge proofs and the trust evaluation model. Finally, Sect. 5 concludes the paper and outlines future research directions.

2 Related Works

Ensuring trust in the Social Internet of Things (SIoT) required addressing the trade-offs inherent to IoT systems, including the limited data-handling capabilities of nodes, the heterogeneity of devices, and the necessity of protecting privacy while facilitating information sharing across social networks. To address these challenges, [47] proposed a blockchain-based system that integrates feedback from trust evaluations. This system enabled the edge nodes to assess the trustworthiness of the sensor nodes, analyze the accuracy of the trust, and examine the resilience and convergence of the trust computation process. However, this approach did not adequately address critical concerns, such as data privacy and feedback sparsity, nor did it resolve issues related to internal credibility, which could not be managed through trust computation alone.

To bridge these gaps, researchers highlighted the need for a comprehensive system that accounted for feedback sparsity and incorporated a robust credibility management framework. Cryptographic primitives had been used to strengthen trust systems [40], and access control mechanisms were used to grant access exclusively to trusted entities. However, these approaches often relied on centralized architectures, leaving them vulnerable to single-point-of-failure attacks. Although each method had its limitations, the combination the cryptographic primitives with trust management techniques presented a more resilient solution.

Despite its importance, the processes of updating, authenticating, and securing sensitive trust data received limited attention in previous research [3,5,7,18,21,29]. These processes were crucial for maintaining reliable decision-making models by dynamically updating trust-related information and retraining machine learning (ML) models. Failing to address these phases often led to diminished accuracy over time, particularly in dynamic environments like SIoT networks. Such environments were especially vulnerable to sophisticated trust-related attacks, including On-Off Attacks (OOA) [26,44], where attackers alternated between malicious and benign behavior to avoid detection.

The diversity and varying severity of trust-related attacks further highlighted the need for more nuanced patterns and metrics to accurately identify and classify each type. However, a significant gap persisted in the availability of real or open-source datasets capable of detecting a comprehensive range of trust-related attacks, including Bad Mouthing Attacks (BMA), Ballot Stuffing Attacks (BSA), Self-Promoting Attacks (SPA), On-Off Attacks (OOA), On-Off Selective Attacks (OSA), and Discriminatory Attacks (DA).

To address these challenges, [35] proposed the "PoTA" consensus protocol, which utilized Support Vector Machine (SVM) classification to differentiate between various attack types. This innovative protocol improved system security by tailoring defense mechanisms to counter various identified trust-related attacks effectively.

Various consensus protocols had been developed to identify malicious nodes [10,17,35,36,50], shaping the evolution of distributed trust management systems. While these protocols demonstrated promising results, their performance often depended on resource availability, such as processing power, memory, and energy. Furthermore, many protocols were integrated into machine learning or deep learning frameworks to detect malicious activities. However, they frequently overlooked the authentication of registered nodes, overburdening validator nodes with computationally intensive tasks. Consequently, these protocols faced trade-offs that limited their scalability and efficiency [16].

Several methods have been proposed to detect trust-related attacks, including anomaly detection, reputation-based systems, and behavior analysis. However, scalability and reliability remained significant challenges, particularly in consensus protocols requiring high computational complexity. Limited node capabilities exacerbated these challenges, impeding real-time attack detection and processing. In addition, user authentication and the management of trust metrics, such as their storage, updates, and protection, were often neglected in existing studies.

This study aimed to alleviate the computational burden on SIoT nodes by deploying a lightweight machine learning model distributed as a global framework in the blockchain network. The proposed approach also incorporated user authentication through a smart contract that uses zero-knowledge circuits, ensuring the integrity of the trust data stored in IPFS. By proving trust without revealing personal information, this method improved privacy while mitigating vulnerabilities and a broad spectrum of trust-related attacks during the authentication process.

3 Blockchain-Based Trust Management System

Our system is organized into two primary phases, which form the foundation of the consensus mechanism within the trust management framework, as illustrated in Fig. 1. The first phase, user authentication, employs blockchain technology and zero-knowledge proof techniques to preserve user privacy while securely validating trust metrics for each participant. The second phase, trust evaluation, uses a federated learning model to identify trust-related attacks during interaction requests. This phase is specifically designed to accommodate the constrained computational resources of nodes, while effectively managing the diverse data and services these nodes process.

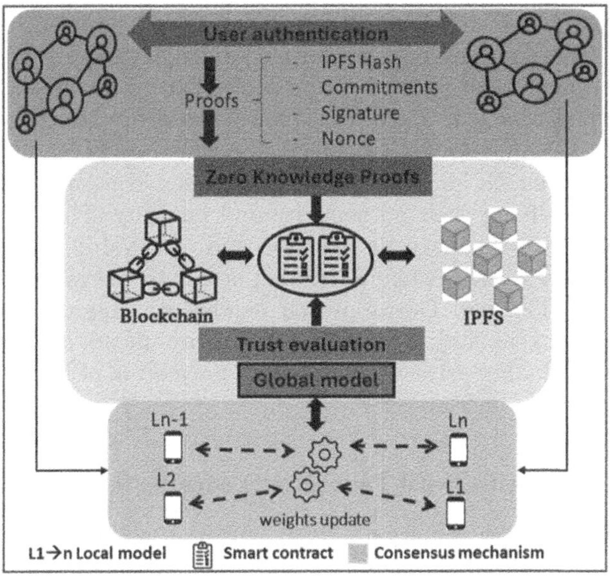

Fig. 1. System architecture for authentication and trust evaluation using blockchain.

3.1 User Authentication with the ZKTP Protocol

We developed a novel zero-knowledge proof protocol, termed Zero Knowledge Trust Proof (ZKTP), which employs advanced cryptographic techniques to ensure the integrity and authenticity of trust metrics in decentralized environments. The protocol facilitates secure interactions between a prover, a user, and a verifier, implemented as a smart contract, by leveraging zero-knowledge proofs and cryptographic commitments.

As illustrated in Fig. 1, the ZKTP process begins with a user's request to connect to the blockchain. The protocol relies on four critical proofs for validation. First, the prover generates a Pedersen commitment to encapsulate their trust metrics, which are securely stored on the blockchain. Second, to mitigate replay attacks, the prover includes a unique nonce associated with each transaction. Third, the prover submits the InterPlanetary File System (IPFS) hash of the trust metrics, ensuring that the actual data can be retrieved and independently verified. Finally, the prover generates a Schnorr signature to authenticate the submission, utilizing their private key and a challenge issued during the protocol. Schnorr signatures, known for their simplicity and efficiency, are particularly suitable for blockchain applications [14].

The verifier, implemented as a smart contract, validates the received commitment, nonce, IPFS hash, and Schnorr signature against the blockchain records. The IPFS hash provided by the prover must match the one stored on the blockchain to pass verification. The commitment proof, generated using the trust metrics and blinding factors, is compared to the stored commitment to confirm that the metrics genuinely belong to the prover. Users consistently update their trust metrics by generating new commitments. The Schnorr signature proof is verified using the prover's public key, ensuring robust authentication, while the nonce proof guarantees that the provided nonce differs from the previously stored one on the blockchain, effectively preventing replay attacks.

By implementing ZKTP, the system ensures that trust metrics remain confidential while verifying their authenticity and integrity. The protocol enables the smart contract to manage the creation and retrieval of commitments and IPFS hashes for each user, while restricting access to the proofs stored on the blockchain to ZKTP only. This approach significantly improves trust management security and reliability in decentralized environments by addressing critical challenges, including privacy preservation, prevention of replay attacks, and robust authentication mechanisms.

3.2 Trust Evaluation with Federated Learning Model

The trust-related attack detection model, as shown in Fig. 1, is integrated into the consensus mechanism and built on a federated learning framework to ensure effective trust evaluation. Malicious nodes can engage in various trust-related attacks, each necessitating the collection of specific trust metrics that vary between nodes. By categorizing each attack type for an individual node or a group of nodes, the model facilitates the precise classification of interactions. Local models are trained on distinct subsets of data, and their weights are aggregated to update a global model that encompasses all attack classes, thus ensuring comprehensive and efficient detection of trust-related attacks.

The consensus mechanism represents an innovative federated learning architecture tailored to address the unique challenges of trust-based systems. Unlike conventional federated learning models, such as Federated Averaging (FedAvg) [43], which assume Independent and Identically Distributed (IID) data between clients, our proposed model is optimized for non-IID environments often characterized by data heterogeneity. This optimization is achieved by incorporating a proximal term in the loss function, inspired by the FedProx algorithm [27], which mitigates the divergence between local and global models. The proximal term ensures a balance between local optimization and global consistency, enabling the global model to maintain robust performance even under highly biased local data distributions, a common challenge in trust-based systems.

Traditional Federated Learning (FL) methods aggregate model updates from clients without explicitly addressing alignment between local and global models. Our proposed FL model addresses this limitation by introducing a fine-tunable parameter, denoted as 'mu', in the proximal term. This parameter penalizes substantial deviations of local weights from global weights, allowing local models to adapt to unique data distributions while ensuring meaningful contributions to the global model during aggregation.

Based on the classification results, the system communicates the results, along with the global model, to other blockchain nodes via a smart contract for validation. Interactions are either accepted if no trust-related attacks are detected or rejected if attacks are identified.

4 Implementation and Evaluation of the Authentication Protocol

The authentication protocol was implemented using zero-knowledge proof circuits, which were subsequently implemented as a smart contract. This section provides a comprehensive overview of the experimental setup and evaluates the protocol's performance, focusing on the proof size and the associated cost of executing the smart contract.

4.1 Implementation of the ZKTP Protocol

The proposed ZKTP protocol was implemented using Circom [8] and the Groth16 verification system [15], which offers improved efficiency compared to the Pinocchio verification scheme [37]. Specifically, Groth16 requires only three elliptic curve points for verification, as opposed to Pinocchio's eight, enhancing its suitability for smart contract integration. The implementation used zk-SNARKs, which further optimized the performance for decentralized applications.

The authentication circuit developed as part of the ZKTP protocol consisted of 5,377 constraints. The size of the proving key was measured at 3.2 MB, while the verification key was significantly more compact at 3.4 KB, illustrating the inherent disparity between the two key sizes. Notably, the proof size was exceptionally small, measuring just 804 bytes, underscoring the protocol's efficiency. For a complete understanding of zero-knowledge circuits, as well as the concepts of proving keys, verification keys, and their respective optimizations, the reader is referred to [9,15,37].

4.2 Analysis of Verification Costs

To assess the efficiency and practicality of our proposed authentication protocol, we conducted experiments to analyze gas overheads, a critical metric for computational costs in blockchain environments, across multiple blockchain networks. As of July 2024[1], the average gas prices for Ethereum, Arbitrum, Optimism, and Polygon are detailed in Table 1.

The experimental results, based on a contract call (verification off-chain) with a gas consumption of 209,130, demonstrate significant variation in costs across the evaluated chains. Ethereum exhibited the highest total cost at $7.79, attributed to its elevated gas prices. In contrast, layer-2 scaling solutions such as Arbitrum, Optimism, and Polygon substantially reduced these costs, showcasing their economic benefits. Among these, Polygon achieved the lowest cost at $0.00614, followed by Optimism and Arbitrum.

[1] CoinMarketCap. https://coinmarketcap.com/ (2024).

These findings in Table 1 directly relate to our objective, where resource constraints were a key consideration in implementing the authentication framework. By evaluating gas consumption across different blockchain environments, we validated that our protocol remains viable for deployment, particularly on cost-effective layer-2 solutions. This aligns with our goal of designing an authentication mechanism that ensures security in SIoT while minimizing computational overhead.

Table 1. Gas costs for off-chain verification across different blockchains (July 2024)

Blockchain	Gas Price (Gwei)	Gas Cost in ETH/MATIC	Current Price (USD)	Total Cost (USD)
Ethereum	20	0.0041826	1,862.65	7.79
Arbitrum	0.02	0.0000041826	1,862.65	0.00779
Optimism	0.016	0.00000334608	1,862.65	0.00623
Polygon	42	0.0000087826	0.6994	0.00614

5 Experiments and Results of the Federated Learning Model

Our experiments and results were conducted using the Sigcomm2009 dataset [38], selected for its rich features and patterns representative of a Social Internet of Things (SIoT) environment. Additionally, we created a dataset with simplified features to enhance its applicability in real-world SIoT scenarios while ensuring efficiency in detecting various types of trust-related attacks.

5.1 Dataset

To evaluate the performance of the federated learning model, we simulated Sigcomm2009 traces to create a dataset that contains up to 26,000 samples. This dataset spans seven distinct classes: Bad Mouthing Attack (BMA), Ballot Stuffing Attack (BSA), On-Off Attack (OOA), Self-Promoting Attack (SPA), On-Off Selective Attack (OSA), Discriminatory Attack (DA), and a none-attack class. The dataset includes six key metrics calculated from user ratings, which are feedback scores exchanged during interactions. These ratings typically follow a predefined scale, with higher scores reflecting positive feedback and lower scores indicating negative feedback. Using these metrics, the dataset provides a robust foundation for evaluating trust-related attacks in decentralized systems. The six features are defined as follows:

1. **Reputation Change Rate (RCR):**

$$RCR = r_n - r_{n-1}$$

   where:

- r_n: The latest rating value for a user.
- r_{n-1}: The previous rating value for the same user.

This feature measures the change in a user's reputation based on their most recent ratings.

2. **Consistency of Ratings (CR):**

$$CR = \frac{1}{n}\sum_{i=1}^{n}(r_i - \bar{r})^2$$

- r_i: Individual rating given by the user.
- $\bar{r}$: Mean of all ratings given by the user.
- n: Total number of ratings.

This feature measures the variance in the user's ratings, with lower values indicating more consistent behavior.

3. **Rating Disparity (RD):**

$$RD = |r_u - \bar{r}_{\text{all}}|$$

- r_u: The average rating of the user.
- $\bar{r}_{\text{all}}$: The average rating across all users.

This feature evaluates how a user's rating differs from the community's average.

4. **Temporal Rating Pattern (TRP):**

$$TRP = \sigma(\Delta\ \text{t})$$

- Δt: Time differences between consecutive ratings.
- σ: Standard deviation.

This feature captures irregularities in the timing of user ratings.

5. **Negative Rating Ratio (NRR):**

$$NRR = \frac{|R_{\text{neg}}|}{|R_{\text{total}}|}$$

- $|R_{\text{neg}}|$: Number of negative ratings (e.g., ratings < 3).
- $|R_{\text{total}}|$: Total number of ratings.

This feature indicates the proportion of negative ratings given by the user.

6. **Proximity-Based Rating Similarity (PBRS):**

$$PBRS = \frac{1}{|U|}\sum_{u\in U}\text{cosine}(r_u, r_{\text{int}})$$

- U: Set of users who interacted with the target user.
- r_u: Ratings given by the target user.
- r_{int}: Ratings given by an interacted user.
- cosine: Cosine similarity function.

This feature measures the similarity between the ratings of a user and those of their interacted users.

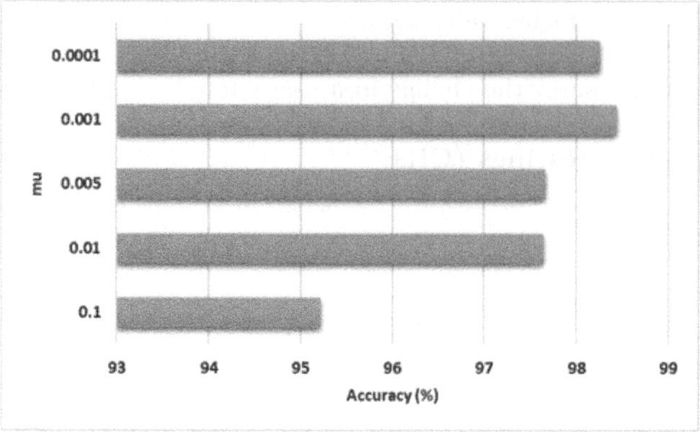

Fig. 2. Impact of the mu parameter on the accuracy of the federated learning model.

5.2 Performance of the Trust Evaluation Model

To evaluate the efficiency of our federated learning model, we conducted experiments incorporating a proximal term and a neural network to enhance classification accuracy. The dataset consisted of 2,000 samples for each attack type, with an additional 14,000 samples classified under the None-attack class. The data was split into 80% for training and 20% for testing.

To test the model's capability to handle heterogeneous data, we utilized two nodes with distinct attack classes. Node 1 was trained on BMA, BSA, DA, and a portion of the None-attack samples, while Node 2 was trained on OSA, SPA, OOA, and the remaining None-attack samples. Initially, we experimented with various neural network architectures to determine the optimal design. Subsequently, we focused on the parameters 'mu' and the number of rounds, as these are critical to the scalability of weight updates. The 'mu' parameter controls the deviation between local and global weights, while the rounds parameter determines the number of updates made during training.

Figure 2 demonstrates the impact of the 'mu' parameter on model accuracy. The best accuracy of 98.44% was achieved with a mu value of 0.001. We observed that smaller values of mu facilitated better model training, with 0.0001 yielding an accuracy of 98.25%, slightly lower than 0.001 but still outperforming larger values such as 0.1, 0.005, and 0.01.

Figures 3 and 4 depict the test accuracy of the FL model across varying numbers of rounds for different mu values. Comparing these figures reveals that the model with a mu value of 0.0001 converged more rapidly than with a mu value of 0.001. Specifically, with mu set to 0.0001, the model reached its maximum test accuracy within approximately 60 rounds, whereas for mu equal to 0.001, convergence began only after 100 rounds.

The model effectively handled subsets with distinct labels, demonstrating its robustness in non-IID environments. Experiments with smaller mu values, such as 0.001 and 0.0001, significantly enhanced the model's adaptability by enabling

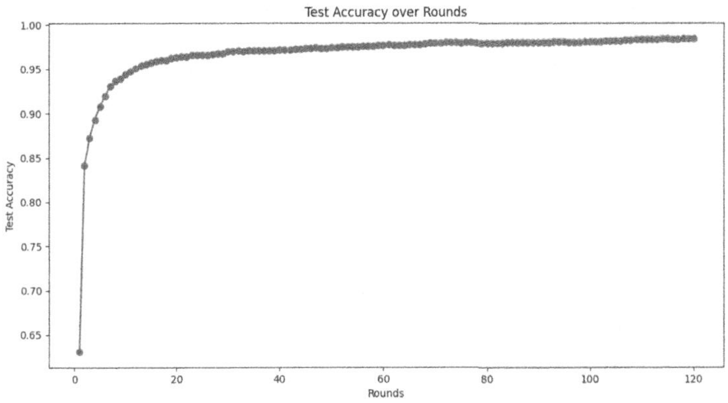

Fig. 3. Accuracy test across rounds with mu = 0.001.

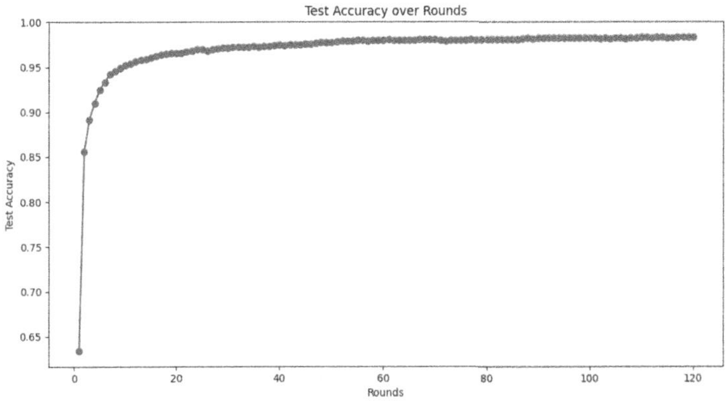

Fig. 4. Accuracy test across rounds with mu = 0.0001.

local models to better adjust to unique data distributions. These results highlight that lower mu values improve performance in handling highly non-IID trust-related data. This adaptability is particularly crucial for trust-based systems, where local behaviors often deviate from global patterns.

Figure 5 presents the standard evaluation metrics (precision, recall, and F1-score) of the federated learning model based on the parameter values from the best accuracy experiment, achieving 98.44%. The evaluation metrics highlight the model's strong performance, with the BSA, None-attack, and OOA labels attaining the highest F1-scores, each exceeding 99%. Similarly, the BMA and DA labels recorded F1-scores above 97%. The lowest F1-scores were observed for the OSA and SPA labels, with values of 94.80% and 95.21%, respectively. These results confirm the model's effectiveness in accurately classifying various trust-related attacks while maintaining robust overall performance.

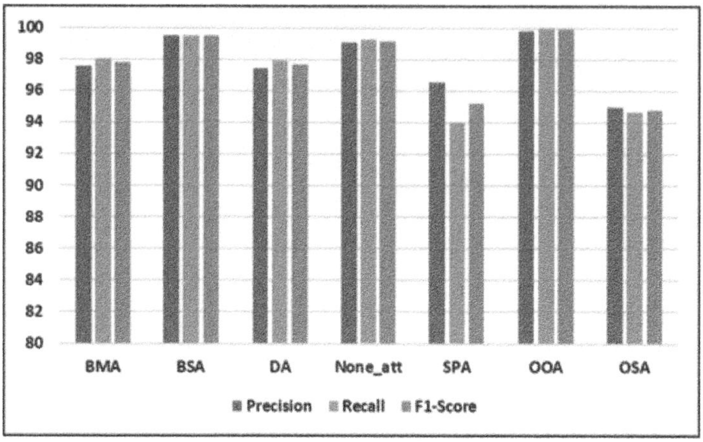

Fig. 5. Evaluation of federated learning model classes (rounds: 120, mu: 0.001, epochs: 140)

Overall, the proposed FL model consistently demonstrated high efficacy in detecting six trust-related attacks, positioning it as a reliable solution for decentralized trust management in SIoT environments.

6 Conclusion

To ensure trust and privacy in SIoT environments, we proposed a blockchain-based trust management system. In this system, user authentication was ensured, followed by trust evaluation using a federated learning model. This process was implemented through a ZKP circuit, which was subsequently converted into a smart contract. The smart contract verified the validity of defined proofs for authenticated users while maintaining low gas costs with certain blockchains.

Additionally, we computed features based on ratings and temporal patterns to effectively detect six types of trust-related attacks. The trust evaluation model demonstrated its ability to handle non-IID data, achieving an accuracy of 98.44%. This scalability enables the model to identify various trust-related attacks across different real-world scenarios. Furthermore, we emphasized the importance of the mu parameter during training, which controlled the weight updates between local models and the global model. This parameter played a pivotal role in enhancing classification accuracy and optimizing the number of rounds required, thereby reducing execution time.

For future work, we aim to introduce additional features and further refine the model to improve accuracy across various types of trust-related attacks. However, ensuring the security of weight updates against vulnerabilities and cyberattacks remains a significant challenge. Addressing this issue will be crucial as the model is adapted for integration with blockchain and Web3 applications.

References

1. Abdelghani, W., Zayani, C.A., Amous, I., Sèdes, F.: Trust evaluation model for attack detection in social internet of things. In: International Conference on Risks and Security of Internet and Systems, pp. 48–64. Springer, Cham (2018)
2. Abderrahim, O.B., Elhdhili, M.H., Saidane, L.: TMCoI-SIOT: a trust management system based on communities of interest for the social internet of things. In: 2017 13th International Wireless Communications and Mobile Computing Conference (IWCMC), pp. 747–752. IEEE (2017)
3. Alghofaili, Y., Rassam, M.A.: A dynamic trust-related attack detection model for IoT devices and services based on the deep long short-term memory technique. Sensors **23**(8), 3814 (2023)
4. Atzori, L., Iera, A., Morabito, G., Nitti, M.: The social internet of things (SIoT)-when social networks meet the internet of things: concept, architecture and network characterization. Comput. Netw. **56**(16), 3594–3608 (2012)
5. Awan, K.A., Din, I.U., Almogren, A., Guizani, M., Altameem, A., Jadoon, S.U.: Robusttrust-a pro-privacy robust distributed trust management mechanism for internet of things. IEEE Access **7**, 62095–62106 (2019)
6. Awan, S., Javaid, N., Ullah, S., Khan, A.U., Qamar, A.M., Choi, J.G.: Blockchain based secure routing and trust management in wireless sensor networks. Sensors **22**(2), 411 (2022)
7. Bahri, L., Girdzijauskas, S.: When trust saves energy: a reference framework for proof of trust (PoT) blockchains. In: Companion Proceedings of the The Web Conference 2018, pp. 1165–1169 (2018)
8. Bellés-Muñoz, M., Isabel, M., Muñoz-Tapia, J.L., Rubio, A., Baylina, J.: Circom: a circuit description language for building zero-knowledge applications. IEEE Trans. Dependable Secure Comput. **20**(6), 4733–4751 (2022)
9. Ben-Sasson, E., Chiesa, A., Tromer, E., Virza, M.: Succinct {Non-Interactive} zero knowledge for a von neumann architecture. In: 23rd USENIX Security Symposium (USENIX Security 2014), pp. 781–796 (2014)
10. Bi, L., Muazu, T., Samuel, O.: IoT: a decentralized trust management system using blockchain-empowered federated learning. Sustainability **15**(1), 374 (2022)
11. Commey, D., Hounsinou, S., Crosby, G.V.: Securing blockchain-based IoT systems with physical unclonable functions and zero-knowledge proofs. arXiv preprint arXiv:2405.12322 (2024)
12. Dai, H.N., Zheng, Z., Zhang, Y.: Blockchain for internet of things: a survey. IEEE Internet Things J. **6**(5), 8076–8094 (2019)
13. Dwivedi, S.K., Amin, R., Vollala, S.: Smart contract and IPFS-based trustworthy secure data storage and device authentication scheme in fog computing environment. Peer-to-Peer Netw. Appl. **16**(1), 1–21 (2023)
14. Fuchsbauer, G., Wolf, M.: Concurrently secure blind Schnorr signatures. In: Annual International Conference on the Theory and Applications of Cryptographic Techniques, pp. 124–160. Springer, Cham (2024)
15. Groth, J.: On the size of pairing-based non-interactive arguments. In: Fischlin, M., Coron, J.-S. (eds.) EUROCRYPT 2016. LNCS, vol. 9666, pp. 305–326. Springer, Heidelberg (2016). https://doi.org/10.1007/978-3-662-49896-5_11
16. Guru, A., Mohanta, B.K., Mohapatra, H., Al-Turjman, F., Altrjman, C., Yadav, A.: A survey on consensus protocols and attacks on blockchain technology. Appl. Sci. **13**(4), 2604 (2023)

17. Hegde, P., Maddikunta, P.: Secure PBFT consensus-based lightweight blockchain for healthcare application. Appl. Sci. **13**(6), 3757 (2023)
18. Huo, C., He, D., Liang, C., Jin, D., Qiu, T., Wu, L.: TrustGNN: graph neural network-based trust evaluation via learnable propagative and composable nature. IEEE Trans. Neural Netw. Learn. Syst. (2023)
19. Jain, A., Jat, D.S.: A review on consensus protocol of blockchain technology. In: Intelligent Sustainable Systems: Selected Papers of WorldS4 2021, vol. 2, pp. 813–829 (2022)
20. Jmal, R., Masmoudi, M., Amous, I., Sèdes, F.: A blockchain and IPFS-enhanced model for attack detection and resource efficiency. In: IFIP International Internet of Things Conference, pp. 163–174. Springer, Cham (2024)
21. Jmal, R., Masmoudi, M., Amous, I., Zayani, C.A., Sèdes, F.: Apache spark based deep learning for social transaction analysis. In: WEBIST, pp. 365–372 (2023)
22. Júnior, T.A.F., Vasconcelos, R.O., Ribeiro, A.d.R.L.: A comparative study between zero-knowledge proof (ZKP) and ring signatures targeting the legal and regulatory implications with the general data protection law (GDPL) and general data protection regulation (GDPR). OBSERVATÓRIO DE LA ECONOMÍA LATINOAMERICANA **22**(2), e3098–e3098 (2024)
23. Katyayani, M., Keshamoni, K., Murthy, A., Rani, K.U., Reddy, S.L., Alapati, Y.K.: Federated learning: advancements, applications, and future directions for collaborative machine learning in distributed environments. J. Electr. Syst. **20**(5s), 165–171 (2024)
24. Khan, W.Z., Hakak, S., Khan, M.K., et al.: Trust management in social internet of things: architectures, recent advancements, and future challenges. IEEE Internet Things J. **8**(10), 7768–7788 (2020)
25. Kudva, S., Badsha, S., Sengupta, S., La, H., Khalil, I., Atiquzzaman, M.: A scalable blockchain based trust management in VANET routing protocol. J. Parallel Distrib. Comput. **152**, 144–156 (2021)
26. Kumari, S., Kumar, S., KR, V.: Trust management in social internet of things: challenges and future directions. Int. J. Comput. Digit. Syst. **14**(1), 1–xx (2023)
27. Li, T., Sahu, A.K., Zaheer, M., Sanjabi, M., Talwalkar, A., Smith, V.: Federated optimization in heterogeneous networks. Proc. Mach. Learn. Syst. **2**, 429–450 (2020)
28. Liu, Y., Wang, J., Yan, Z., Wan, Z., Jäntti, R.: A survey on blockchain-based trust management for internet of things. IEEE Internet Things J. **10**(7), 5898–5922 (2023)
29. Malik, S., Dedeoglu, V., Kanhere, S.S., Jurdak, R.: Trustchain: trust management in blockchain and IoT supported supply chains. In: 2019 IEEE International Conference on Blockchain (Blockchain), pp. 184–193. IEEE (2019)
30. Marche, C., Nitti, M.: Trust-related attacks and their detection: a trust management model for the social IoT. IEEE Trans. Netw. Serv. Manag. **18**(3), 3297–3308 (2020)
31. Masmoudi, M., Abdelghani, W., Amous, I., Sèdes, F.: Deep learning for trust-related attacks detection in social internet of things. In: International Conference on e-Business Engineering, pp. 389–404. Springer, Cham (2019)
32. Masmoudi, M., Amous, I., Zayani, C.A., Sèdes, F.: Real-time mitigation of trust related attacks in social IoT. In: International Conference on Model and Data Engineering, pp. 303–318. Springer, Cham (2023)
33. Masmoudi, M., Amous, I., Zayani, C.A., Sèdes, F.: Real-time prevention of trust-related attacks in social IoT using blockchain and apache spark. Comput. Commun. **225**, 65–82 (2024)

34. Masmoudi, M., Amous, I., Zayani, C.A., Sèdes, F.: Trust attack prevention based on spark-blockchain in social IoT: a survey. Int. J. Inf. Secur. **23**(5), 3179–3198 (2024)
35. Masmoudi, M., Zayani, C.A., Amous, I., Sèdes, F.: A new blockchain-based trust management model. Procedia Comput. Sci. **192**, 1081–1091 (2021)
36. Nawab, F., Sadoghi, M., et al.: Consensus in data management: from distributed commit to blockchain. Found. Trends® Databases **12**(4), 221–364 (2023)
37. Parno, B., Howell, J., Gentry, C., Raykova, M.: Pinocchio: nearly practical verifiable computation. Commun. ACM **59**(2), 103–112 (2016)
38. Pietilainen, A.K., Diot, C.: Crawdad thlab/sigcomm2009 (2022). https://doi.org/10.15783/C70P42
39. Prasad, S., Tiwari, N., Chawla, M., Tomar, D.S.: Zero-knowledge proofs in blockchain-enabled supply chain management. In: Sustainable Security Practices Using Blockchain, Quantum and Post-Quantum Technologies for Real Time Applications, pp. 47–70. Springer, Cham (2024)
40. Samuel, O., Javaid, N., Khalid, A., Imrarn, M., Nasser, N.: A trust management system for multi-agent system in smart grids using blockchain technology. In: 2020 IEEE Global Communications Conference, GLOBECOM 2020, pp. 1–6. IEEE (2020)
41. Schilke, O., Reimann, M., Cook, K.S.: Trust in social relations. Ann. Rev. Sociol. **47**, 239–259 (2021)
42. Simpson, T.W.: Trust: A Philosophical Study. Oxford University Press (2023)
43. Sun, T., Li, D., Wang, B.: Decentralized federated averaging. IEEE Trans. Pattern Anal. Mach. Intell. **45**(4), 4289–4301 (2022)
44. Suryani, V., Sulistyo, S., et al.: The detection of on-off attacks for the internet of things objects. In: 2018 International Conference on Control, Electronics, Renewable Energy and Communications (ICCEREC), pp. 1–5. IEEE (2018)
45. Taherdoost, H.: Smart contracts in blockchain technology: a critical review. Information **14**(2), 117 (2023)
46. Tariq, N., Asim, M., Khan, F.A., Baker, T., Khalid, U., Derhab, A.: A blockchain-based multi-mobile code-driven trust mechanism for detecting internal attacks in internet of things. Sensors **21**(1), 23 (2020)
47. Wu, X., Liang, J.: A blockchain-based trust management method for internet of things. Pervasive Mob. Comput. **72**, 101330 (2021)
48. Yaga, D., Mell, P., Roby, N., Scarfone, K.: Blockchain technology overview. arXiv preprint arXiv:1906.11078 (2019)
49. Yan, Z., Zhang, P., Vasilakos, A.V.: A survey on trust management for internet of things. J. Netw. Comput. Appl. **42**, 120–134 (2014)
50. Zhang, F., Wang, H., Zhou, L., Xu, D., Liu, L.: A blockchain-based security and trust mechanism for AI-enabled IIoT systems. Futur. Gener. Comput. Syst. **146**, 78–85 (2023)

A Comprehensive Review on Equivalent Mutant Detection Using Machine Learning

Gabriel Guerrero-Contreras(✉), Sara Balderas-Díaz, Pedro Delgado-Pérez, and Inmaculada Medina-Bulo

Department of Computer Science and Engineering, University of Cadiz, Av. Universidad de Cádiz, 10 Puerto Real, 11519 Cádiz, Spain
{gabriel.guerrero,sara.balderas,pedro.delgado,inmaculada.medina}@uca.es

Abstract. Mutation testing evaluates test suite effectiveness by introducing small code modifications ("mutants") and checking if tests detect them. A major challenge is equivalent mutants, which are code changes that do not alter program behavior, making detection computationally expensive and undecidable. This review examines Machine Learning approaches, including Abstract Syntax Tree Neural Networks and Tree-Based Convolutional Neural Networks, to address this issue. Key challenges include limited datasets, inconsistent metrics, and reproducibility concerns. The study highlights the need for standardized frameworks and future research directions in mutation testing.

Keywords: Equivalent mutants · mutation testing · machine learning · neural networks · software testing

1 Introduction

Mutation testing is a robust fault-injection-based software testing technique designed to evaluate the effectiveness of test suites by introducing controlled modifications, known as mutants, into the source code of a program [30]. These mutants simulate potential errors that developers might introduce during coding [1]. A test suite's ability to detect and eliminate these mutants is indicative of its strength in capturing faults [28]. The method has garnered attention due to its capability to highlight weaknesses in test cases, aiding developers in improving software reliability and quality [25]. However, mutation testing is computationally expensive and faces significant challenges, particularly in identifying equivalent mutants [9].

Equivalent mutants are syntactically altered versions of a program that, despite their differences, exhibit the same functional behavior as the original program [13]. These mutants cannot be detected by any test case, as they are functionally indistinguishable from the original [3]. Their presence introduces noise into mutation testing metrics by inflating the count of live mutants, leading

J. Grabis et al. (Eds.): RCIS 2025, LNBIP 547, pp. 52–67, 2025.
https://doi.org/10.1007/978-3-031-92474-3_4

to inaccurate assessments of test suite quality [12]. Detecting equivalent mutants is inherently challenging because it is an undecidable problem in the general case. The undecidability stems from the fact that determining whether two programs are functionally equivalent is analogous to solving the Halting Problem, which has been proven to lack a general algorithmic solution [5]. Consequently, no systematic method can guarantee identifying all equivalent mutants for arbitrary programs. This theoretical limitation complicates mutation testing, particularly for large-scale systems where the combinatorial explosion of possible mutants further exacerbates the problem, making manual identification infeasible [2].

Nevertheless, various techniques have been proposed to tackle the equivalent mutant problem, as reported in the following systematic reviews [24,31]. Among them, traditional techniques often utilize symbolic execution [33,36], a method that replaces concrete input values with symbolic variables. This approach systematically explores all possible execution paths of a program by analyzing the logical conditions that guide the program's flow. By comparing the symbolic outputs of the original program and its mutants under all input conditions, symbolic execution can theoretically classify mutants as equivalent if their outputs match for every possible input. Despite its rigor, symbolic execution faces practical limitations such as path explosion [36]. Additionally, manual identification of equivalent mutants remains time-intensive and challenging, often requiring manual inspection of program structures to confirm that the mutants have no semantic effect [33].

To overcome these limitations, different methods have been developed. One such approach is Dependence Guided Symbolic Execution (DGSE), which reduces computational costs by identifying and eliminating redundant paths while preserving fault detection capabilities [36]. Pattern-based approaches, such as those in the MEDIC framework [20], identify equivalent mutants by matching specific data flow patterns in program code. These patterns represent program constructs where mutations do not alter the program's behavior, such as certain loop conditions or assignments that are inherently equivalent. By focusing on these predefined patterns, MEDIC automates the detection of many equivalent mutants, significantly reducing the need for manual analysis or exhaustive symbolic execution. However, the approach is inherently limited to the predefined patterns it recognizes, meaning it cannot detect equivalent mutants arising from complex or unconventional program structures that fall outside these patterns. As a result, MEDIC provides an effective but partial solution to the equivalent mutant detection problem, requiring complementary techniques to achieve broader coverage. One such complementary technique is Trivial Compiler Equivalence (TCE) [21], which leverages compiler optimizations to detect equivalent mutants by comparing the object code of the original program and its mutants. If the compiler optimizes a mutant back into an identical form of the original, it is classified as equivalent. An extension of this approach, TCE+ [17], combines compiler optimizations with abstract syntax tree (AST) analysis by examining structural similarities in the program's syntax. This allows TCE+ to detect equivalences that arise from transformations not optimized by the compiler, such as refactoring changes or variations in control structures. However, the effectiveness of both TCE and TCE+ is constrained by the limitations of compiler optimizations and the complexity of AST analysis. Overall, these techniques have made detecting equivalent mutants more efficient, but complementary techniques remain necessary.

In this regards, machine learning (ML) has emerged as a promising solution to automate the detection of equivalent mutants [27,34]. By leveraging the patterns and relationships within program features, ML models can predict equivalence with a significant degree of accuracy [6]. A wide range of techniques, including supervised learning [27], deep learning (DL) [23], and hybrid models [12], have been explored for this purpose. These approaches often use program representations, such as ASTs [37], Control Flow Graphs (CFGs) [32], or Program Dependency Graphs (PDGs) [4], as feature inputs to ML models. Despite the promise of these methods, their adoption faces several challenges, including data scarcity, feature complexity, and the lack of standardized benchmarks for evaluation.

The body of research on ML-based equivalent mutant detection is diverse, spanning various programming languages, datasets, mutation operators, and modeling approaches. This diversity reflects the field's exploratory nature but also introduces fragmentation, making it difficult to compare results across studies or identify effective practices. Furthermore, the lack of publicly available datasets and benchmarks hinders reproducibility and the development of generalizable solutions. This study aims to bridge this gap by conducting a comprehensive review of the application of ML to equivalent mutant detection.

The rest of this paper is structured as follows: Sect. 2 details the research methodology, including the systematic protocol adopted, research questions, and selection criteria. Section 3 presents the outcomes of the review, addressing the research questions on applied ML techniques, datasets, preprocessing methods, model effectiveness, and challenges in the field. Finally, Sect. 4 summarizes the key insights gained from the review.

2 Research Method

The adopted research method follows a systematic protocol designed according to the PRISMA (Preferred Reporting Items for Systematic Reviews and Meta-Analyses) guidelines [26]. This methodology ensures a transparent, rigorous, and reproducible process for conducting systematic reviews. The key stages of this protocol, including the identification, screening, eligibility, and inclusion of studies, are detailed in the following subsections.

2.1 Research Questions

The primary objective of study is to provide a comprehensive overview of existing research on equivalent mutant detection using machine learning techniques and to present fundamental insights into this field. In alignment with this goal, we formulated the following research questions:

- **RQ1**: *What machine learning techniques have been applied to detect equivalent mutants in mutation testing?* This question identifies the range of methods explored, including supervised, unsupervised, and hybrid approaches.
- **RQ2**: *What datasets or benchmarks are commonly used to train and evaluate machine learning models for equivalent mutant detection?* This question analyzes the datasets' characteristics, availability, and limitations.
- **RQ3**: *What preprocessing techniques are employed to prepare data for machine learning in equivalent mutant detection?* It focuses on strategies such as feature extraction, data augmentation, dimensionality reduction, and handling class imbalances.
- **RQ4**: *How effective are machine learning models in identifying equivalent mutants?* This question evaluates the models' performance, highlighting their advantages and limitations.
- **RQ5**: *What are the current challenges, limitations, and emerging trends in the application of machine learning to equivalent mutant detection?* It examines barriers, methodologies, and future research directions in this area.

2.2 Search String

We constructed the search string using the terms listed in Table 1. These terms were derived from the main concept of "mutation testing" by considering alternative expressions and different spellings to ensure comprehensive coverage of relevant literature.

Table 1. Search terms used for constructing the search string.

Major Terms	Alternative Terms	Different Spelling
Mutation testing	Mutant analysis	Mutants
Equivalence detection	Equivalent mutants	–
Machine learning	Artificial intelligence, ML models, AI techniques	–

We employed logical operators, "OR" and "AND," to link the search terms effectively. After conducting multiple tests to refine the query, we adopted a search string that retrieved the most significant number of relevant articles while maintaining specificity. The final search string is presented below:

("mutation testing" OR "mutant analysis") AND ("equivalence detection" OR "equivalent mutants") AND ("machine learning" OR "artificial intelligence" OR "ML models" OR "AI techniques")

2.3 Selection Criteria

To select the primary studies, we established a set of inclusion and exclusion criteria, as presented in Table 2. In summary, we selected papers written in English, available online, and directly related to the detection of equivalent mutants in software testing using artificial intelligence.

Table 2. Inclusion and exclusion criteria.

Inclusion Criteria	Exclusion Criteria
The paper is related to the software engineering area.	Out of scope: not related to equivalent mutant detection, artificial intelligence, or software testing.
The paper is related to detecting equivalent mutants using artificial intelligence.	Not available online.
The paper is written in English.	Not written in English.
The paper is a journal paper, conference paper, or workshop paper.	Abstracts, posters, reviews, conference reviews, theses, keynotes, doctoral symposiums, and patents.

2.4 Search Process

The search process was conducted in January 2025, as illustrated in Fig. 1, following the PRISMA guidelines [26]. The studies were gathered by querying a total of five databases: IEEE Xplore, Springer, ACM Digital Library, Scopus, and Web of Science. These databases were selected due to their reputation and comprehensive coverage of high-quality publications in software engineering and related fields. We did not consider Google Scholar because it does not guarantee that all papers are peer-reviewed, which may introduce grey literature and less reliable references. The initial search yielded 172 papers. After removing 16 duplicate entries, 156 unique records were screened based on their titles, abstracts, and keywords. At this stage, 41 records were excluded. Subsequently, 115 papers were sought for retrieval. Of these, 2 papers could not be retrieved, leaving 113 papers to undergo a more in-depth assessment of their full content for eligibility. During this detailed evaluation, 89 papers were excluded for being outside the study's scope, as their primary objectives did not align with the focus on detecting equivalent mutants. Many of these studies treated equivalent mutant detection as a secondary outcome of methodologies aimed at tasks like mutant generation or selection. Additionally, 16 papers were excluded for relying on non AI based methods. To ensure the inclusion of all relevant studies, a forward and backward snowballing process was performed on the reference lists and citations of the selected papers. This process identified one additional paper, which was ultimately excluded because the journal did not follow a peer-review process. In the end, 8 studies met all the inclusion criteria and were included in the systematic review. The selected works comprise those by Naeem et al. [27], Brito et al. [6], van Hijfte et al. [16], Chung et al. [8], Tian et al. [34], Kusharki et al. [23], Peacock et al. [29], and Jammalamadaka et al. [19].

No restrictions on publication year were applied, ensuring a comprehensive inclusion of studies across the entire timeline of the field's development. However, gray literature, such as technical reports, white papers, and preprints, was excluded to focus on peer-reviewed and published works. The selection process was independently conducted by two authors to ensure rigor and minimize bias. After that, they held a discussion to decide whether to include or exclude the papers that were in doubt. The search string was applied to the title, abstract, and keywords of each record. Consequently, Fig. 2a illustrates the distribution of studies included in the review, categorized by publication year and source type. The studies span from 2020 to 2024, highlighting the recent growth in research activity related to equivalent mutant detection. The stacked bars reveal a balanced contribution from conferences and journals, with conferences being the predominant source across most years.

A complete list of included and excluded papers, along with the reasons for exclusion, is available in the replication package repository on Zenodo [14].

3 Outcomes

3.1 RQ1 - Applied ML Techniques for Equivalent Mutant Detection

Figure 2b illustrates the distribution of methodological approaches employed in studies focused on equivalent mutant detection. A significant portion of the research, representing 43% of the analyzed studies, utilizes classic supervised learning techniques. For instance, Naeem et al. [27] used classifiers such as Random Forest (RF), Gradient Boosted Trees (GBT), and Support Vector Machines to predict mutant equivalence. Brito et al. [6] expanded this exploration by evaluating seven algorithms, including k-Nearest Neighbors, Decision Trees, Linear Discriminant Analysis, Logistic Regression, and Gaussian Naive Bayes. Symbolic execution has also been used to augment datasets for machine learning models, as demonstrated by Chung et al. [8], where a model classified mutants based on features derived from symbolic execution.

DL techniques account for 29% of the studies, Kusharki et al. [23] introduced a Tree-Based Convolutional Neural Network (TBCNN) that integrates ASTs for hierarchical feature extraction, further enhanced by Word2Vec embeddings. Similarly, Peacock et al. [29] adopted an Abstract Syntax Tree Neural Network (ASTNN) using Bi-directional Gated Recurrent Units (Bi-GRUs) to process semantic and syntactic information from ASTs.

Other studies have employed hybrid models to enhance prediction accuracy. Jammalamadaka et al. [19] proposed a Hybrid Wavelet Convolutional Rain Optimization (HWCRO) framework, combining wavelet-based convolutional neural networks with optimization algorithms to refine model parameters. This technique utilizes features like semantic similarity and information entropy.

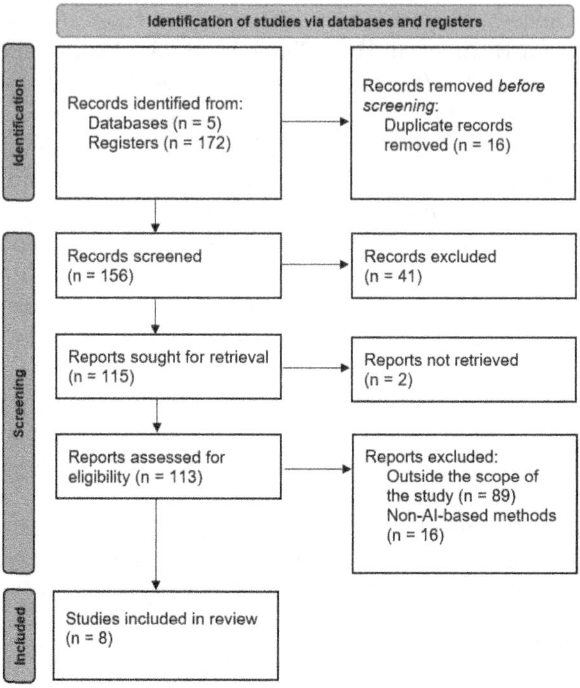

Fig. 1. PRISMA flow diagram depicting the systematic process of identification, screening, and selection of studies included in the review.

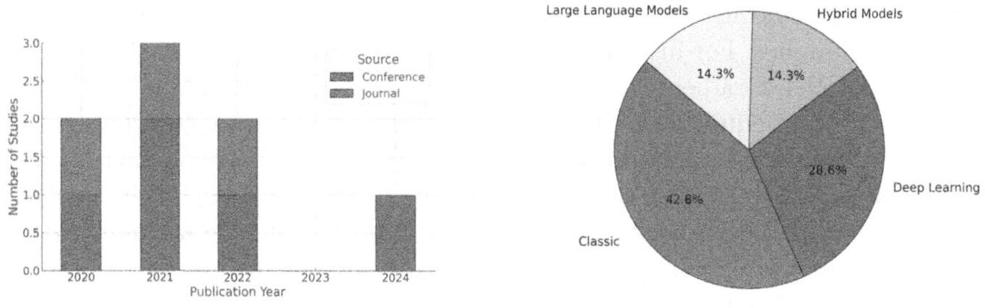

(a) Distribution of studies by publication year and source type.

(b) Distribution of articles by methodological approach.

Fig. 2. Comparison of methodological approaches and publication trends.

Moreover, the emergence of Large Language Models (LLMs) marks a significant advancement in this domain. Tian et al. [34] investigated state-of-the-art LLMs, such as UniXCoder and CodeT5+, fine-tuning them for equivalent mutant detection. These models leverage pre-trained code embeddings and integrate prompting strategies like zero-shot and few-shot learning. UniXCoder, in particular, outperformed traditional machine learning and neural network approaches.

3.2 RQ2 - Datasets and Benchmarks

The datasets utilized in equivalent mutant detection research vary significantly in programming language, mutation tools, and the proportion of equivalent mutants identified, as summarized in Table 3. Naeem et al. [27] employed 10 Java programs, including 5 benchmark programs and 5 open-source projects such as Jtar[1] and Parsii[2], with sizes ranging from 32 to 13,767 lines of code (LOC) and an average of 3,337 LOC. Mutants were generated using the MuJava tool, applying 12 mutation operators, including the Arithmetic Operator Replacement (AOR) and Relational Operator Replacement (ROR). From the 3,024 live mutants analyzed, 46% were identified as equivalent.

Table 3. Summary of datasets for equivalent mutant detection.

Reference	Language	Mutation tool	Projects	Mutants	EM (%)	Public
Naeem et al. [27]	Java	MuJava	10	3,024	46	No
Brito et al. [6]	C	Proteum	27	23,411	8	Yes
van Hijfte et al. [16]	Java/C	-	37	4,400	32	Yes
Chung et al.* [8]	Java	MuJava	19	1,811†	-	No
Tian et al.* [34]	Java	-	19	3,302	15	Yes
Kusharki et al. [23]	C/C++	-	-	52,000	40	No
Peacock et al.‡ [29]	C	MILU	5	582	50	No
Jammalamadaka et al. [19]	Python	-	-	2,500	51	No

* Uses a subset of the MutantBench dataset.
†Number of equivalent mutants; total number of mutants is not specified.
‡Uses a subset of the dataset provided by Yao et al. [38].

In contrast, Brito et al. [6] used a dataset of 27 C programs, totaling 49 functions with LOC ranging from 9 to 142 and an average of 31.78 LOC. They generated 23,411 mutants using the Proteum mutation tool, reporting that 8% were equivalent. This study provides a publicly accessible dataset[3], which enhances its utility for further research and replication.

The introduction of MutantBench [16] marked a significant milestone in standardizing equivalent mutant detection research. This open-source benchmark consolidates datasets from prior studies, including those by Yao et al. [38], Baer et al. [5], Kintis et al. [22], and Houshmand and Paydar [17]. These datasets were processed and validated for integration. To ensure reliable conversion, the AST of mutants before and after conversion were compared using GumTree [11], revealing no changes in any mutant. This meticulous validation step underscores the robustness of the conversion process. The final merged dataset comprises

[1] http://github.com/kamranzafar/jtar.
[2] http://github.com/scireum/parsii.
[3] https://github.com/claudineibjr/ClassifyingMutants.

4,400 mutants from 37 programs (19 in Java and 18 in C), with 32% being equivalent mutants, as determined through manual labeling in the original studies. MutantBench encompasses a diverse set of 25 mutation operators, such as Absolute Operator Insertion (ABSI) and Relational Operator Deletion (ROD). Additionally, MutantBench adheres to the FAIR principles (Findability, Accessibility, Interoperability, and Reusability) [18], making both tools and datasets publicly available[4].

Chung et al. [8] utilized the Java subset of MutantBench dataset, with an average size of 170 LOC, containing 462 labeled equivalent mutants. To address the limitations of small datasets and potential manual labeling errors, the authors proposed two augmentation techniques. The first method, *Generate & Validate* (G&V), relied on symbolic execution, SMT solvers, and EvoSuite to identify equivalent mutants from a pool of live mutants. The second method, *Equivalent Mutant Synthesis*, used static analysis and control flow graphs to directly generate equivalent mutants. These methods augmented the dataset with an additional 1,349 unique equivalent mutants, increasing the original MutantBench dataset size from 462 to 1,811.

Tian et al. [34] also used a subset of MutantBench focusing on Java programs, analyzing 3,302 mutant pairs. The train set contained 250 equivalent mutants and 1,402 non-equivalent mutants, while the test set included 249 equivalent mutants and 1,401 non-equivalent mutants.

Kusharki et al. [23] focused on mutants generated only using the Absolute Value Insertion (ABS) operator from C and C++ programs, resulting in a dataset of 52,000 mutants. This included 20,800 equivalent mutants and 31,200 non-equivalent mutants. Similarly, Peacock et al. [29] explored a subset of five C programs from Yao et al. [38], generating mutants with the MILU mutation tool. Their dataset included 582 mutants split between equivalent and non-equivalent mutants, using ABS and Unary Operator Insertion (UOI) operators.

Jammalamadaka et al. [19] utilized Python programs and manual inspection to analyze 2,500 live mutants, identifying 51% as equivalent. Their approach considered equivalent mutants across five classes based on CBT theory [10], such as reachability and execution errors. However, the study lacks details about the mutation operators used or the specific programs analyzed, and the dataset is not accessible.

The limited availability of datasets in equivalent mutant detection research presents significant challenges to the field. As Table 3 illustrates, while some studies, such as Brito et al. [6] and MutantBench [16], provide publicly accessible datasets, the majority remain unavailable. This lack of accessibility limits reproducibility and restricts the potential for independent validation. For instance, the datasets developed by Naeem et al. [27] and Kusharki et al. [23], remain confined to their respective studies, limiting their broader impact.

[4] https://github.com/MutantBench/MutantBench.

3.3 RQ3 - Preprocessing Techniques

Studies highlight the utility of PDGs, ASTs, and CFGs to derive meaningful features, as summarized in Table 4. For instance, Naeem et al. [27] extracted features such as reachability constraints, semantic distances, and information entropy using PDGs and ASTs, integrating these with metrics such as PageRank and hub/authority values. Similarly, Brito et al. [6] emphasized control flow information, mutation operator types, and complexity measures as key features. Chung et al. [8] also leveraged symbolic execution to extract features such as primitive arc distances and mutation operator scores, further enriching datasets.

Table 4. Comparison of feature extraction techniques and metrics used in mutation equivalence detection.

Feature Type	Naeem et al. [27]	Brito et al. [6]	Chung et al. [8]
Source	PDG, AST, and CBT theory	Control flow, source code, and test execution	Control flow and symbolic execution
Mutation Op.	-	Categorical feature transformed via one-hot encoding	Fault-revealing capacity
Graph	PageRank, hub, authority	CFG distances (min, max, avg)	PageRank, hub, authority, and primitive arcs
Complexity	-	Code complexity at mutation points	Ratio of mutations per node
Fault-Revealing	Semantic distance, reachability, necessity, sufficiency	-	Position score and operator score
Statement	-	Types such as loops, assignments, etc.	Block, While, For, If, Assignment, Return, Function Call
Other	Information entropy	-	Distance metrics (Begin/End Min, Max, Avg)

In contrast, embedding techniques have been applied to enhance feature representation. Kusharki et al. [23] employed Word2Vec to encode program tokens and combined these embeddings with Bi-GRU models to generate semantic representations for TBCNNs. Similarly, Peacock et al. [29] utilized AST embeddings to model program structures dynamically, combining them with neural networks.

Datasets are often dominated by non-equivalent mutants, for that reason, studies have employed resampling techniques to address this imbalance. Naeem et al. [27] implemented SMOTE (Synthetic Minority Over-sampling TEchnique) [7] to upsample equivalent mutants. Conversely, Brito et al. [6] adopted random undersampling of non-equivalent mutants to achieve balanced datasets. Stratified sampling strategies, as applied by Tian et al. [34], involve dividing the dataset into homogeneous subgroups called strata, based on class labels such as equivalent and non-equivalent mutants. Samples are then drawn proportionally from each stratum, ensuring that the class distribution in the original dataset is preserved across both the training and testing subsets. This process prevents overrepresentation or underrepresentation of any class, maintaining a balanced dataset.

Moreover, Information Gain Ratios [15] have been used to determine the most impactful features, such as position scores, operator scores, and reachability constraints [27]. Wavelet decomposition has also been introduced within convolutional layers of neural networks to reduce feature dimensionality while retaining critical information [19].

LLM approaches have introduced preprocessing strategies designed to handle code semantics while complying with token limitations. Tian et al. [34] employed pre-defined regular expressions to remove natural language comments from Java code, significantly reducing input length and aligning it with LLM token constraints. Furthermore, by performing mutant detection at the method level rather than the program level, they further minimized input size while preserving the semantic structure crucial for equivalence detection.

3.4 RQ4 - Effectiveness of Machine Learning Models

The reported performance of the ML models for detecting equivalent mutants is summarized in Table 5. Naeem et al. [27] reported a good performance of GBT model, particularly in the *split-project scenario* (subset of classes within a single program is used as the test set, while the remaining classes of the same program are used to train the classifier). Their model achieved high precision (88.8%) and recall (90.5%) for equivalent mutants, with even better performance for nonequivalent mutants, with precision and recall values of 99.1% and 98.5%, respectively. The overall metrics, such as area under the curve (AUC) and Kappa statistic, both exceeded 88%. These metrics remained consistent in *cross-project scenarios* (one entire program is used as the test set, while the models are trained on data from other programs), although with slight reductions (AUC: 88.5%, Kappa: 88.0%). Brito et al. [6] identified RF as the most effective classifier, reporting an average F1-score of 81.95% for identifying minimal mutants and 80.49% for classifying equivalent mutants.

Neural network-based approaches have also shown significant performance. Peacock et al. [29], using an ASTNN, achieved an F1-score of 90% for equivalent mutants generated by the ABS mutation operator, while performance improved to 94% for the UOI operator. Similarly, Kusharki et al. [23] demonstrated the efficacy of TBCNN, reporting a recall of 96%, precision of 93%, and overall accuracy of 94%.

Chung et al. [8] emphasized the effectiveness of dataset augmentation via symbolic execution. This approach enabled their models to reach a F1-score of 97%, with accuracy and recall values of 96%. Futhermore, Jammalamadaka et al. [19] with their hybrid optimization approach, HWCRO, achieved an F1-score of 85.16%, a recall of 84.89%, and a precision of 85.36%.

Table 5. Comparative performance of machine learning models for equivalent mutant detection.

Reference	Model	Precision (%)	Recall (%)	F1-score (%)	Accuracy (%)
Naeem et al. [27]	GBT	88.8	90.5	-	-
	RF	-	-	80.5	-
Peacock et al. [29]	ASTNN (ABS operator)	86.0	94.0	90.0	-
	ASTNN (UOI operator)	92.0	96.0	94.0	-
Kusharki et al. [23]	TBCNN	93.0	96.0	-	94.0
Chung et al. [8]	Augmented Dataset Model	97.0	96.0	97.0	96.0
Jammalamadaka et al. [19]	HWCRO	85.4	84.9	85.2	-
Tian et al. [34]	UniXCoder (LLM)	94.3	81.8	86.6	-
	TCEjavac (Compiler)	40.6	38.2	39.3	-
	TCEsoot (Compiler)	51.3	51.8	50.8	-

Tian et al. [34] recently highlighted the effectiveness of LLMs in detecting equivalent mutants. Their approach fine-tuned a code embedding strategy using UniXCoder, resulting in remarkable performance metrics: a precision of 94.33%, a recall of 81.81%, and an F1-score of 86.58%. This LLM-based method significantly outperformed traditional compiler-based techniques such as TCEjavac (leveraging static analysis through the Java compiler) and TCEsoot (utilizing advanced static analysis via intermediate code representations with the Soot framework [35]). These conventional methods achieved considerably lower F1-scores of 39.31% and 50.80%, respectively.

3.5 RQ5 - Challenges and Limitations

The high computational cost associated with generating and analyzing large numbers of mutants is a persistent issue. Tools such as MuJava often produce a vast number of mutants, even for small programs, leading to significant resource consumption. Identifying equivalent mutants, which behave identically to the original program despite syntactic differences, further complicates this process due to its undecidable nature. Manual verification remains a common but labor-intensive approach, slowing down scalability [8,27].

A major limitation is the imbalance in datasets, where equivalent mutants are vastly underrepresented compared to non-equivalent ones. This imbalance negatively impacts the training of ML models, leading to suboptimal performance [6]. Despite efforts to address this issue through techniques like data augmentation, the scarcity of diverse and high-quality datasets remains a barrier to progress [8]. Additionally, the lack of standardized datasets complicates cross-study comparisons, emphasizing the need for open benchmarking frameworks like MutantBench, which adhere to FAIR data principles [16].

Feature extraction and representation pose another critical challenge. Models must capture the nuanced semantic and syntactic features of mutants, yet determining which features are most relevant is non-trivial. Approaches using PDGs, semantic distances, or error-propagation rates show promise but require further refinement [29]. DL models, such as TBCNNs or WCNNs, offer advanced feature extraction capabilities but demand substantial computational resources [19,23].

Emerging methodologies, such as integrating symbolic execution with ML, provide promising avenues for improving equivalence validation. However, scalability remains a concern, particularly for industrial-scale programs [8]. LLMs have also demonstrated potential for handling code semantics, yet their effectiveness is constrained by data scarcity and computational cost [34].

Future research should prioritize addressing these challenges by developing hybrid methods that combine static and dynamic analyses, refining feature engineering processes, and expanding the diversity of training datasets. Enhancing benchmarking tools and frameworks to support reproducibility and scalability, as MutantBench, will be essential for advancing ML-based approaches to equivalent mutant detection [16]. Additionally, exploring architectures like Bi-GRU or LSTM, along with tailored prompting strategies for LLMs, could further enhance adaptability and performance in diverse programming environments [23].

4 Concluding Remarks

This paper reviews research on equivalent mutant detection, emphasizing the use of ML techniques. It examines detection methods, datasets, preprocessing strategies, model effectiveness, and challenges and limitations in this domain.

Our analysis identified 8 studies published in diverse venues, predominantly in software engineering and testing domains. While the number of publications is limited, this reflects the scarcity of research specifically addressing equivalent mutant detection using ML techniques. Despite extensive searches in major digital libraries and the application of forward and backward snowballing, no additional studies directly tackling this problem with ML were found. This highlights that the field is still emerging, and while there is a growing interest, the number of contributions remains constrained. The majority of studies leverage supervised learning techniques, such as RF and GBT, or DL models, including TBCNNs and ASTNNs. Also, it is observed a recent trend toward employing LLMs, such as UniXCoder and CodeT5+, which offer promising capabilities in capturing the semantic structure of code.

The datasets used in these studies are varied, covering multiple programming languages (e.g., Java, C, Python) and mutation tools (e.g., MuJava, Proteum). However, a significant limitation is the scarcity of high-quality, publicly available datasets, which hinders reproducibility and generalization. MutantBench is a notable exception, providing a benchmark that adheres to FAIR principles and consolidates diverse datasets. Nevertheless, most datasets are small in size and exhibit a substantial imbalance between equivalent and non-equivalent mutants, which complicates model training and evaluation.

Preprocessing techniques commonly involve feature extraction from program representations such as ASTs, CFGs, and PDGs. These features are often enriched with additional metrics, such as semantic distances, PageRank scores, and mutation operator types. While some studies employ data augmentation and resampling techniques (e.g., SMOTE and stratified sampling) to address data imbalance, these methods remain underexplored in the context of equivalent mutant detection.

ML models' effectiveness varies across studies. Classic supervised models (e.g., GBT) achieve high precision and recall, while DL models (e.g., ASTNNs, TBCNNs) leverage hierarchical and semantic code representations for greater potential. LLMs like UniXCoder further enhance code understanding, surpassing traditional methods. However, challenges persist, including high computational costs, undecidability in equivalence detection, token limits, and limited diverse datasets. Hybrid approaches, such as integrating symbolic execution with ML, show promise but need further validation.

Acknowledgments. This publication is part of the I+D+i grant PID2021-122215NB-C33 funded by MICIU/AEI/10.13039/501100011033 and by ERDF/EU, and by the Spanish Ministry of Science and Innovation under the Excellence Network AI4Software (Red2022-134647-T).

References

1. AbuSalim, S.W.G.: Comparative analysis between selective and non-selective mutation techniques. Int. J. Emerg. Trends Eng. Res. (2020)
2. Amorim, S., et al.: Reducing manual efforts in equivalence analysis in mutation testing. J. Softw. Eng. Res. Dev. **12** (2024)
3. Ayad, A., Marsit, I., Tawfig, S., Loh, J.M., Omri, M.N., Mili, A.: Quantitative analysis of mutant equivalence. In: van Sinderen, M., Maciaszek, L.A. (eds.) Software Technologies, pp. 58–80. Springer, Cham (2020)
4. Baah, G.K., Podgurski, A., Harrold, M.J.: The probabilistic program dependence graph and its application to fault diagnosis. In: Proceedings of the 2008 International Symposium on Software Testing and Analysis, pp. 189–200 (2008)
5. Baer, M., Oster, N., Philippsen, M.: MutantDistiller: using symbolic execution for automatic detection of equivalent mutants and generation of mutant killing tests. In: 2020 IEEE International Conference on Software Testing, Verification and Validation Workshops (ICSTW), pp. 294–303 (2020)
6. Brito, C., Durelli, V.H., Durelli, R.S., de Souza, S.R., Vincenzi, A.M., Delamaro, M.E.: A preliminary investigation into using machine learning algorithms to identify minimal and equivalent mutants. In: 2020 IEEE International Conference on Software Testing, Verification and Validation Workshops (ICSTW), pp. 304–313. IEEE (2020)
7. Chawla, N.V., Bowyer, K.W., Hall, L.O., Kegelmeyer, W.P.: SMOTE: synthetic minority over-sampling technique. J. Artif. Intell. Res. **16**, 321–357 (2002)
8. Chung, S., Yoo, S.: Augmenting equivalent mutant dataset using symbolic execution. In: 2022 IEEE International Conference on Software Testing, Verification and Validation Workshops (ICSTW), pp. 150–159. IEEE (2022)
9. Dang, X., Yao, X., Gong, D., Tian, T., Sun, B.: Multi-task optimization-based test data generation for mutation testing via relevance of mutant branch and input variable. IEEE Access **8**, 144401–144412 (2020)
10. DeMillo, R.A., Offutt, A.J., et al.: Constraint-based automatic test data generation. IEEE Trans. Softw. Eng. **17**(9), 900–910 (1991)

11. Falleri, J.R., Martinez, M.: Fine-grained, accurate and scalable source differencing. In: Proceedings of the IEEE/ACM 46th International Conference on Software Engineering, pp. 1–12 (2024)
12. Ghiduk, A.S., Girgis, M., Shehata, M.H.: Employing dynamic symbolic execution for equivalent mutant detection. IEEE Access **7**, 163767–163777 (2019)
13. Grün, B.J., Schuler, D., Zeller, A.: The impact of equivalent mutants. In: 2009 International Conference on Software Testing, Verification, and Validation Workshops, pp. 192–199. IEEE (2009)
14. Guerrero-Contreras, G., Balderas-Díaz, S., Delgado-Pérez, P., Medina-Bulo, I.: Replication Package for "Equivalent Mutant Detection Using Machine Learning" (RCIS 2025) (2025). https://doi.org/10.5281/zenodo.15037838
15. Harris, E.: Information gain versus gain ratio: a study of split method biases. In: AI&M (2002)
16. van Hijfte, L., Oprescu, A.: MutantBench: an equivalent mutant problem comparison framework. In: 2021 IEEE International Conference on Software Testing, Verification and Validation Workshops (ICSTW), pp. 7–12. IEEE (2021)
17. Houshmand, M., Paydar, S.: TCE+: an extension of the TCE method for detecting equivalent mutants in Java programs. In: Dastani, M., Sirjani, M. (eds.) FSEN 2017. LNCS, vol. 10522, pp. 164–179. Springer, Cham (2017). https://doi.org/10.1007/978-3-319-68972-2_11
18. Jacobsen, A., et al.: FAIR principles: interpretations and implementation considerations (2020)
19. Jammalamadaka, K., Parveen, N.: Equivalent mutant identification using hybrid wavelet convolutional rain optimization. Softw. Pract. Experience **52**(2), 576–593 (2022)
20. Kintis, M., Malevris, N.: MEDIC: a static analysis framework for equivalent mutant identification. Inf. Softw. Technol. **68**, 1–17 (2015)
21. Kintis, M., Papadakis, M., Jia, Y., Malevris, N., Le Traon, Y., Harman, M.: Detecting trivial mutant equivalences via compiler optimisations. IEEE Trans. Softw. Eng. **44**(4), 308–333 (2017)
22. Kintis, M., Papadakis, M., Papadopoulos, A., Valvis, E., Malevris, N.: Analysing and comparing the effectiveness of mutation testing tools: a manual study. In: 2016 IEEE 16th International Working Conference on Source Code Analysis and Manipulation (SCAM), pp. 147–156. IEEE (2016)
23. Kusharki, M.B., Misra, S., Muhammad-Bello, B., Salihu, I.A., Suri, B.: Automatic classification of equivalent mutants in mutation testing of Android applications. Symmetry **14**(4), 820 (2022)
24. Madeyski, L., Orzeszyna, W., Torkar, R., Józala, M.: Overcoming the equivalent mutant problem: a systematic literature review and a comparative experiment of second order mutation. IEEE Trans. Softw. Eng. **40**(1), 23–42 (2014)
25. Mishra, D., Acharya, B., Rath, D., Gerogiannis, V., Kanavos, A.: A novel real coded genetic algorithm for software mutation testing. Symmetry **14**, 1525 (2022)
26. Moher, D., Liberati, A., Tetzlaff, J., Altman, D.G., Group, P., et al.: Preferred reporting items for systematic reviews and meta-analyses: the PRISMA statement. Int. J. Surg. **8**(5), 336–341 (2010)
27. Naeem, M.R., Lin, T., Naeem, H., Liu, H.: A machine learning approach for classification of equivalent mutants. J. Softw. Evol. Process **32**(5), e2238 (2020)
28. Ojdanić, M., Soremekun, E., Degiovanni, R., Papadakis, M., Traon, Y.L.: Mutation testing in evolving systems: studying the relevance of mutants to code evolution. ACM Trans. Softw. Eng. Methodol. **32**, 1–39 (2021)

29. Peacock, S., Deng, L., Dehlinger, J., Chakraborty, S.: Automatic equivalent mutants classification using abstract syntax tree neural networks. In: 2021 IEEE International Conference on Software Testing, Verification and Validation Workshops (ICSTW), pp. 13–18. IEEE (2021)
30. Petrović, G., Ivankovic, M., Fraser, G., Just, R.: Practical mutation testing at scale: a view from Google. IEEE Trans. Softw. Eng. **48**, 3900–3912 (2021)
31. Rani, S., Suri, B.: Equivalent mutant problem and its problem-solving techniques: a retrospective view. In: 2018 8th International Conference on Cloud Computing, Data Science & Engineering (Confluence), pp. 791–797 (2018)
32. Rimsa, A., Nelson Amaral, J., Pereira, F.M.: Practical dynamic reconstruction of control flow graphs. Softw. Pract. Experience **51**(2), 353–384 (2021)
33. Schuler, D., Zeller, A.: Covering and uncovering equivalent mutants. Softw. Testing Verification Reliab. **23**(5), 353–374 (2013)
34. Tian, Z., Shu, H., Wang, D., Cao, X., Kamei, Y., Chen, J.: Large language models for equivalent mutant detection: how far are we? In: Proceedings of the 33rd ACM SIGSOFT International Symposium on Software Testing and Analysis, pp. 1733–1745 (2024)
35. Vallée-Rai, R., Co, P., Gagnon, E., Hendren, L., Lam, P., Sundaresan, V.: Soot: a Java bytecode optimization framework. In: CASCON First Decade High Impact Papers, pp. 214–224 (2010)
36. Wang, H., Liu, T., Guan, X., Shen, C., Zheng, Q., Yang, Z.: Dependence guided symbolic execution. IEEE Trans. Softw. Eng. **43**(3), 252–271 (2016)
37. Wang, Y., Li, H.: Code completion by modeling flattened abstract syntax trees as graphs. In: Proceedings of the AAAI Conference on Artificial Intelligence, vol. 35, no. 16, pp. 14015–14023 (2021)
38. Yao, X., Harman, M., Jia, Y.: A study of equivalent and stubborn mutation operators using human analysis of equivalence. In: Proceedings of the 36th International Conference on Software Engineering, pp. 919–930 (2014)

Security, Risk and Strategy

Identifying and Detecting Patterns in Work Organization with Active Window Tracking

Mari A. J. Braakman(✉), Iris Beerepoot, Maria Peeters, Eva Knies, and Hajo A. Reijers

Utrecht University, Princetonplein 5, 3584 CC Utrecht, The Netherlands
m.a.j.braakman@uu.nl

Abstract. Collecting data on the organization of work by individual employees has become increasingly feasible, with time-tracking applications offering valuable insights into how individuals structure their tasks and projects. Active Window Tracking (AWT) is one such method that captures data from computers on the systems the individual used and windows that were active at a certain point in time. The use of AWT data provides an opportunity to enhance research approaches by offering objective, data-driven insights into work behavior across multiple systems. In semi-structured interviews, we identified four work organization patterns within the context of academic staff: Bundling, Starting, Ending, and Dividing. We show how the work organization patterns can be detected from AWT data of three individuals covering multiple months. Using examples from the data, we demonstrate how work organization patterns can provide insights into work behavior. The findings highlight the potential of AWT data, which can be leveraged to inform strategies to optimize performance and employee well-being.

Keywords: Work Organization Patterns · Employee Behavior · Pattern Detection · Active Window Tracking

1 Introduction

Collecting data about one's work is becoming easier in today's digital world. Many time-tracking apps monitor time spent on a particular website or app, allowing managers and individual workers to estimate the total duration of work on a task or project [1]. Adding temporal information, such as detailed timestamps each time a window is opened and closed, would allow for insights beyond mere time allocation, enabling analysis of the sequencing of tasks. Active Window Tracking (AWT) is one such means of data collection, which provides an extensive log of the system and window title are active when working on a laptop or computer, at a specific point in time [2]. As such, monitoring one's work with AWT goes beyond tracking how long someone works on tasks or general time management [3]. AWT gives the opportunity to investigate the combination of tasks performed in sequence, their durations, and how the individual has organized their work as a whole, instead of looking at the total time spent on individual tasks and projects. Collecting such data about the organization of work provides an opportunity

J. Grabis et al. (Eds.): RCIS 2025, LNBIP 547, pp. 71–86, 2025.
https://doi.org/10.1007/978-3-031-92474-3_5

to investigate how individuals plan and organize their days of work and detect patterns in their behavior. We refer to work organization patterns as recurring ways individuals structure, prioritize, and manage their work tasks over time. These patterns can reveal consistent approaches to daily work planning and execution.

Although underexplored, AWT has been used to investigate work behaviors and work organization, including disruptions during work [4] and the use of digital environments [5]. These studies focus on specific behaviors that may be crucial predictors of work performance and employee well-being. Potentially, AWT might also enable the analysis of daily work behavior by identifying work organization patterns based on how, when, and for how long an individual worked on a particular task. Previous studies on workflow patterns [6] and workflow resource patterns [7] have shown how, from a business process perspective, patterns can be identified to analyze work systems. We propose AWT data can be used to analyze work organization patterns from a data-driven perspective, providing insights into how individuals allocate their time and manage their tasks. AWT data offers the opportunity to complement and extend existing approaches to focus on analyzing individual work organization to improve work system optimization.

In this mixed-method study, we investigate work organization patterns through AWT data by answering the following research questions:

1. *What work organization patterns are applied by employees?*
2. *How can these work organization patterns be detected in AWT data?*

In this study, we answer these questions in the context of academic staff, i.e., the people who teach and research within academic institutes. We chose this setting as academics generally have high autonomy and flexibility when planning their workweek [8], which suggests they have an influence on how they organize their work. We answered the first question by conducting semi-structured interviews where we identified four types of work organization patterns. The second research question is answered using real-life and longitudinal AWT datasets from three academics; we demonstrate how the identified work organization patterns can be detected. In doing so, we propose an objective, data-driven approach for analyzing individual work organization. This study contributes to the relatively underexplored domain of individual work organization by providing an initial definition of work organization patterns and the means to detect these patterns automatically. The perspective on individual work organization allows us to gain further insights into the intricacies of the work system and how it can be optimized for both individual workers and organizations.

2 Theoretical Background

2.1 Important Concepts

To arrive at work organization patterns, it is important to establish a shared understanding of the concepts we will refer to throughout this paper. In Fig. 1, an illustration can be found of how the different concepts relate to each other. Below, we will explain each concept with the help of examples related to the context of our research, namely academic staff.

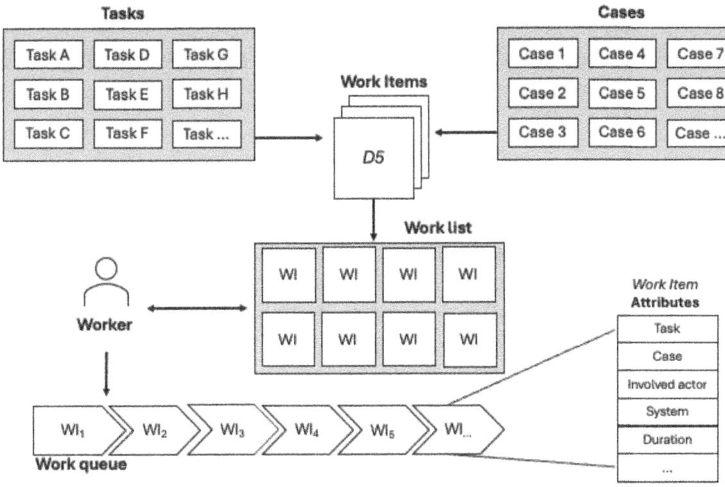

Fig. 1. Overview of the important concepts.

Work is generally described as "a set of coordinated and goal-directed activities that require sustained effort and are conducted in exchange for something else" [9]. A *task* can be defined as "a single unit of work" [7]. Examples of tasks are providing lectures or seminars. Academic staff provide several such lectures and seminars for different courses. A *work item* (WI) can be described as a specific task instance linked to a case [7]. Examples of work items are lectures 1 through 4 of the statistics course in the first period of the academic year 2024/2025. The *case* in this example would be the statistics course in the first period of the academic year 2024/2025. If the lectures are to be repeated next year, these are considered separate work items for another case.

The set of work items that one worker plans to undertake is part of a personal *work list*. Work items can be divided into smaller *sub-work items*, such as creating slides for a lecture, practicing the lecture, giving the lecture, sending the lecture slides to students, and taking notes of possible changes for the next time the lecture is given.

Apart from the work item's link to the case it belongs to and the task it concerns, a work item may carry other *attributes*. The work item attribute *involved actor*, refers to the person performing the activity, similarly to resource in [6, 7]. Within an organization, department, or team, multiple resources are performing work, sometimes collaborating. *System* is an attribute that refers to the medium used to perform the work item. An example is Microsoft's PowerPoint, which can be used to create lecture slides. *Duration* is an attribute that refers to the time taken to complete a work item. More attributes can be found for work items, such as physical location, cost, and deadline, to name a few.

Individuals plan and organize their work items from their personal work list, deciding when to perform the specific work items; this results in their *work queue*: an ordered list of work items [7]. The transformation of the work list into the ordered work queue can be considered *work organization*, i.e., how individuals plan and organize their work. The expectation is that individuals showcase repetitive behavior in how they approach this transformation, i.e., that it is possible to identify *work organization patterns*: generic patterns in which employees plan and organize items from a work list into a work queue. This study aims to identify various work organization patterns that can be applied in individual work organization.

2.2 Related Work

Active Window Tracking (AWT) is a means of automatically collecting data by recording which system and window title of that system is active, at what time [2]. This data collection method results in a dataset consisting of the app's name, the window title, the start time and date, and the end time and date. In Table 1, an illustration of a dataset consisting of AWT data is given. In previous research, AWT has been used to measure individual employee behavior. Beerepoot et al. showed how AWT data can be used to measure interruptibility during work [4]. Vuouri et al. linked AWT data to heart rate variability and a daily survey about the feeling of control and achievement [5]. The advantage of AWT is the possibility of analyzing across systems rather than receiving information from just a single system, which is a challenge in the business process management field [1, 10]. Additionally, AWT can collect data from multiple employees, processes, and cases with little effort. In this research, we explore the detection of several work organization patterns within AWT datasets.

Table 1. Data set with AWT data

App	Title	Begin	End
Microsoft Outlook	Inbox	09:11:21	09:12:00
Google Chrome	New tab	09:12:00	09:12:04
Google Chrome	Springer – Google search	09:12:04	09:13:11
Microsoft PowerPoint	Slides-project1	09:15:38	09:19:12

AWT data shares similarities with event data, commonly used in process mining, which is defined as "extracting valuable, process-related information from event logs" to discover business processes [10]. However, AWT data differs from event data due to the absence of a case notion. Rather than analyzing several process instances of one process, our approach examines work from the perspective of an individual worker, and their involvement in several processes. Workers prioritize tasks based on urgency, importance, or personal preference, often independently of the overarching process structure. This paper will focus on individual employees and how they organize their work.

3 Identifying Work Organization Patterns

The first part of this study aimed to discover general work organization patterns as applied by academics. We conducted a series of interviews to do so. We describe the approach and the resulting patterns below.

3.1 Method

Interview participants were contacted personally or through a supervisor and asked to volunteer for participation in the study. We aimed at arriving at a representative sample from different faculties, ultimately resulting in nine academic employees from various faculties of Utrecht University in the Netherlands having participated in the interviews: the Faculty of Geosciences, Faculty of Humanities, Faculty of Law, Economics, and Governance, and Faculty of Social and Behavioural Sciences, with two or three participants from each faculty. We excluded the faculties of Medicine and Veterinary Medicine as the work performed in those faculties is expected to differ significantly from the typical academic. We excluded the Faculty of Science from the interview study because the AWT data was collected there, and the automatic detection of patterns was applied.

The participants' roles varied from Ph.D. candidate[1] to full professor. An information letter was sent beforehand with information about the goal of the research, information about the interview procedure, and how the interview data would be stored and analyzed. The interviews lasted approximately 45 min and were conducted in the participant's office or online via Microsoft Teams. During the interview, the participants reflected on the previous work week, starting from seven days before the interview. They were asked to describe what tasks they did at what time. All interviews were audio-recorded and transcribed, after which the recordings were deleted.

The interview transcripts were coded using NVivo 14 [11]. Coding took place in two phases. The first phase consisted of open and axial coding by one of the authors, resulting in overarching categories related to the context of the participant's work. These related to various context factors of the performed tasks, such as contact with colleagues or students, time and location, task content (related to research, education, or managerial work), and if the task was directly associated with another performed task (i.e., in preparation of another task or finishing up another task). Through discussion between the authors, work organization patterns were formulated. There were three discussion rounds among the research team in which the patterns were altered and reformulated. In the second phase, the newly developed work organization patterns were used for another round of selective coding to verify whether the definitions had been accurately formulated based on the interviews.

3.2 Analysis

From the interviews with academic workers, we identified four work organization patterns:

1. *Bundling*: Performing work items in succession where, at least for one attribute, the attribute values for the work items are the same.
2. *Starting*: Performing work items at the start of workdays where, at least for one attribute, the attribute values for the work items are the same.

[1] In the Dutch Academic system, most Ph.D candidates, including the Ph.D candidates participating in this research, have a contract as 'employed PhD candidates', meaning they are part of the academic staff and have both teaching and research duties.

3. *Ending*: Performing work items at the end of workdays where, at least for one attribute, the attribute values for the work items are the same.
4. *Dividing*: Splitting one work item into two or more sub-work items performed throughout the week, where the sub-work items are not performed in immediate succession.

Each pattern may appear in different variants, with the Dividing pattern as an exception. For Bundling, Starting, and Ending, variants were found corresponding to all work item attributes: case, task, resource, system, and duration. For example, a worker can perform several work items in succession that are all related to the same case, or all related to the same task. As we approach work from the point of view of one worker, the resource attribute is used in a special way: not to refer to the person performing the work – as that is always the person that is the object of the analysis - but to other people involved in the work. To avoid confusion, we will therefore refer to involved actors, meaning other participants involved in performing a specific work item or someone a work item depends on when collaborating. For example, when two colleagues meet to discuss the content of one of the lectures that will be given next week for the statistics course, those employees are seen as involved actors for that work item. The involved actor does not have to be an employee of the organization. In the context of academic employees, students can also be seen as involved actors. An involved actor can also be more passive when one individual sends an e-mail (active involved actor) to a colleague to whom the e-mail was sent (passive involved actor).

Sometimes, more than one pattern variant can be found in one part of someone's work organization. For example, consider the quote, "*And then [at the end of the workday] I did a few more things, like some emails and small practical tasks, I think from six to seven or so*". This participant bundled work items with a short duration at the end of the workday. As such, it is an example of the *bundling pattern*, specifically the *duration variant*. However, if the participant often ends the day with short-duration tasks, this is also evidence of the *ending pattern*, also the *duration variant*. In Fig. 2, the four work organization patterns are depicted, along with the number of times each pattern occurred in the interviews.

Bundling. The bundling pattern was found 46 times in total, divided over the variants. It examines work items that are performed in succession and of the same work item attribute value. An example of bundling work items with the same case can be found in the quote "*Ten to half past twelve was a block where, with a colleague of mine, we're working on a book [...], we had a Teams call to update each other and discuss what we will do. Then, for the rest of that time block that we had reserved, we wrote in parallel.*" All items in the found pattern were related to the same case: writing a book, consisting of the work items 'updating each other' and 'writing'. Note that the involved actor is also the same for this particular sequence of work items. An example of bundling work items concerning the same task was mentioned in the following: "*I try, with varying success, to bundle the same type of activities as much as possible on one day. If you're already in 'meeting mode', then just keep it going all day*" where the participant bundled the task 'meeting.' The work items were of the same task type but were related to different cases.

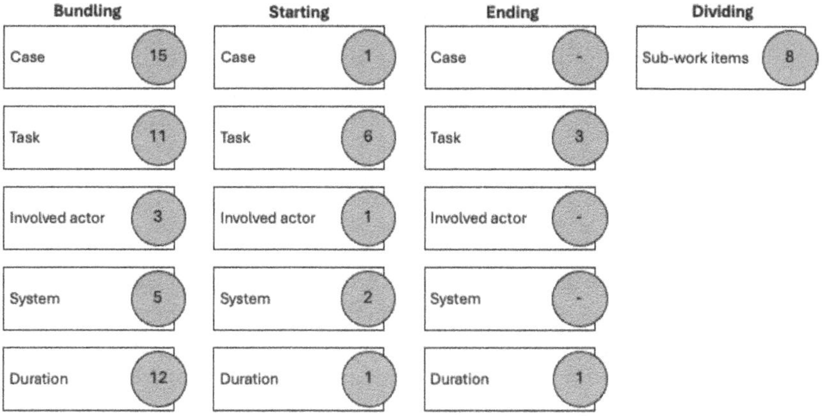

Fig. 2. Frequencies of Work Organization Patterns found in the Interviews.

Starting. The starting pattern was found 11 times, divided over the variants. It examines the first work item of each day, and whether repetition in attribute values of these starting work items can be found across days. An example of a starting pattern is *"I start every day by checking my email, going through all of it, and preparing what I have to do that day."* This relates to the system being used (Microsoft Outlook) and the task being performed (reading and sending emails). In calculating the frequencies, we only considered starting the day with a specific work item as a starting pattern when it happened at least twice in the work week discussed during the interview.

Ending. The ending pattern is very similar to the starting pattern, but now we examine what work item is performed at the end of the day. An example is one participant who mentioned they ended both Wednesday and Thursday with a meeting. Here, the work items had the same task type across days. In the interviews, this pattern did not occur often; only four instances were identified. For both the starting and ending patterns, the participants did not always explicitly mention this pattern; however, after analyzing the full work week, it was clear the participant started or ended different days in a similar way.

Dividing. For the dividing pattern, we examined work items divided into different sub-work items throughout the week, such as preparing or finishing a specific work item. The dividing pattern occurred eight times in the interviews. An example of dividing is *"After lunch [on Wednesday], I prepared for my appointment on Thursday."* This shows how one work item ('attending an appointment on Thursday') was divided by the participant into preparing for the appointment a day prior and attending the appointment itself. We consider the appointment itself and the preparation for the appointment sub-work items of the same work item because they are linked not only by case but also to a specific work item. The two sub-work items are performed in the same week but not in direct succession. Unlike the other patterns, the dividing pattern does not have different variants. Although specific attributes - such as involved actors or location - can be assigned to the work item in the dividing pattern, it is not necessary to identify the pattern. The dividing pattern is always related to a single case.

4 Detecting the Work Organization Patterns in AWT Data

In the second part of this study, we aimed to detect the work organization patterns found during the interviews in AWT data. We describe the approach to data collection and data analysis below.

4.1 Method

Data Collection. Three academic employees from the Faculty of Science of Utrecht University participated in collecting AWT data. Each participant received an explanation in person of the study's goal and how their data was being used and stored. The participants downloaded the Tockler application [12], which was used to collect the AWT data. The application starts automatically when the laptop is powered on. All data is stored in the application, so the researchers did not have direct access to the original data. The participants needed to share the data actively. Before we asked to share the data, extra information was provided, including regulations on how the data would be saved and used. All data was securely saved on the servers of Utrecht University, and all data shown in the results were pseudonymized. The participants shared as much data as they wished, meaning they could filter out any sensitive data before sharing. The participants could also request to filter out any data that was not needed to analyze work organization patterns. For example, one participant filtered out moments when they used the laptop for private use.

For participant 1, an assistant professor, data was collected from August 20, 2023, to August 19, 2024, i.e., an entire academic year. For the other two participants, participant 2, an associate professor, and participant 3, a full professor, data was collected from August 20, 2023, to January 28, 2024, i.e., one semester of the academic year. Participant 1's dataset consisted of 310 days, Participant 2's dataset 135 days, and Participant 3's dataset 136 days. The data was saved as CSV files and cleaned and analyzed using Python 3.11.5. All calculations and measurements were first formulated and tested on the data set of Participant 1, after which the procedures for detecting the work organization patterns were repeated on the following datasets.

An evaluation was then performed with Participant 1. First, we generated four random instances of the four work organization patterns detected automatically, representing each pattern variant. Then, we asked the participant to verify the accuracy of the detected patterns by examining the AWT data. We discuss the results of this evaluation in Sect. 4.2.

The Python script used to create the attributes and detect work organization patterns can be found on GitHub[2]. Due to the sensitive nature of the AWT data, we cannot share the full datasets used in this study. However, a curated dataset is publicly available [13], to which the script can be applied.

Preprocessing Procedure. The AWT data included five variables: *App*, *Title*, *Begin*, *End*, and *Type*. The data pre-processing procedure included confirming that all values for the *Start* and *End* variables were date values with an exact date and time (YY-MM-DD hh:mm:ss) and checking for any missing values. The column *Type* could be deleted,

[2] https://github.com/MariAJBraakman/AWTWorkOrganizationPatterns.

as it only showed that all rows were app track items, i.e., data collected by Tockler. There were some instances where Tockler did not collect the window title data. For Participant 1, 40 entrees were found where PowerPoint was active, but no Title was included, and for Participant 2, 12 entrees were found where Adobe Acrobat was active, but no Title was included; however, we included the entries in the analysis because it included information about which system was active at what time. No other missing values were found.

For most activities, we assumed that more than one system was used. We could see this in the data by the number of switches between systems during a period of continuous working. We divided the day into *timeslots* referring to periods of undisrupted computer work, assuming that while working in that timeslot, the participant was working on one work item. The resulting dataset formed the basis for the detection of work organization patterns. For each timeslot, it included the set of apps that were used, the set of window titles, the begin and end time of the timeslot, and the duration.

Table 2. Timeslot attributes

Timeslot Attribute	Description
Longest-duration App	The app that the participant spent most time in within the timeslot
Longest-duration Title	The title of the window that the participant spent the longest time in, within the timeslot
App corresponding to the longest duration title	The app corresponding to the longest-duration title, which could differ from the longest-duration app in that timeslot
Timeslot duration	The total duration of the timeslot, meaning the moment that the first window became active until there were no more windows active

Measuring Timeslot Attributes. To enable the detection of all patterns and variants, further enriching of the timeslot dataset was required. Four additional columns were added to the timeslot data, which we will refer to as *timeslot attributes*. A description is provided in Table 2, and we will elaborate further on our pattern detection based on these attributes below.

For detecting the *case* that corresponds with a timeslot, we examined the title within the timeslot that has the longest duration. Using this title, we distinguish whether the person was working on a case or a work item. When a window title only appears briefly, we consider it a sub-work item: a work item that has been divided and performed separately. If a title only appeared in a 14-day period, we considered it a *sub-work item*. As case, we considered the longest-duration title of a timeslot when it appeared over a period of 15 days or more. When it appeared only in a 14-day period or less, it was labeled a sub-work item. All unique cases and sub-work items were checked for generic titles, meaning titles that did not provide any information about a case or work item were excluded. An example is 'Document1 – Word', as all new documents in Word that do not have a name yet are named automatically 'Document1'. A list of generic cases was presented to Participant 1 to ensure no cases were falsely identified as too generic. This list of cases was not included when detecting the work organization patterns based on case.

The *task* that the person performed proved difficult to detect in AWT data. We did not possess a complete list of potential tasks for comparison and for the participants to provide a labelled dataset would be labor-intensive, as well as defeat the purpose of providing an automated approach. With the goal of providing interesting and accurate insights in an automated manner, we therefore omitted the detection of task variants.

For detecting the *involved actor* within timeslots, it is possible to extract the names from the longest-duration title. For example, the window titles collected for 'Microsoft Teams' provide information with whom someone is chatting or calling. However, from this title it is not possible to guarantee active communication is taking place. Moreover, systems have different formatting styles for the names in window titles. Due to various systems' different formatting, it was challenging to investigate the involved actor per timeslot. Like the task, we therefore chose to omit the detection of involved actors in this first exploration. More complex data preparation is necessary to analyze the involved actor attribute.

The timeslot attribute *system* was derived from the longest-duration app, or system, that the participant worked on during that timeslot. We categorized timeslots into three categories for the *duration*: short, medium, and long. For Participant 1, the average timeslot *duration* was 11 min with a standard deviation of 15 min, indicating many timeslots with a short *duration*. Therefore, we distinguish short durations (shorter than 5 min), medium durations (between 5 and 30 min), and long durations (longer than 30 min) for timeslots. These categories can be altered depending on the individual worker's work context. In this study, we kept the categories the same for all three participants to allow for comparison.

Measuring Work Organization Patterns. With the needed information about timeslots in place, the work organization patterns could be detected from the AWT data. To detect the *bundling pattern*, bundles were created for each variant. A bundle was detected when three or more consecutive timeslots carried the same attribute value, i.e., for the case, the system, or the duration. Three or more timeslots in succession were chosen because two timeslots in succession are more likely to happen as a coincidence, for example, when someone was interrupted and then continued with their work. To detect the *starting pattern*, the attribute values of the first timeslots of the day were analyzed. We identified a pattern when three or more days, not necessarily consecutively, began with a timeslot with the same attribute value. Similar to the starting pattern, we analyzed the attribute values of the day's final timeslots to detect an *ending pattern*: when three or more days ended with a timeslot that shared the same attribute, this was considered an instance of the ending pattern. To detect a *dividing pattern*, we investigated how often the same longest-duration title occurred within 14 days, assuming that these timeslots were continuations of the same work item. Therefore, all timeslots within that 14-day week with the same longest-duration title were considered sub-work items. A short description of the detection used for the work organization patterns can be found in Table 3.

Table 3. Description of Work Organization Patterns as Detected in AWT Data.

Work organization Pattern	Description
Bundling	When someone worked on the same case, in the same system, or for a similar duration, for three or more consecutive timeslots
Starting	When someone started three or more days (not necessarily consecutively) with work on the same case, in the same system, or for a similar duration
Ending	When someone ended three or more days (not necessarily consecutively) with work on the same case, in the same system, or for a similar duration
Dividing	When someone worked on the same short-lived (max 14 days) work item across two or more non-consecutive timeslots

4.2 Analysis

Below, we provide an overview of the results using the detection approach for the three datasets and discuss the results from the evaluation with Participant 1 of a random instance of a detected pattern.

Figure 3 shows the frequency of all work organization patterns detected in the AWT data of the three participants. For a consistent comparison of the three participants in the figure, all datasets were filtered to cover the period from August 20, 2023, to January 28, 2024. This figure shows how often an instance of a work organization pattern was detected. For example, we investigated short, medium, and long categories for the duration variant. For all three categories, we found three or more days that started or ended in one of these categories, resulting in the detection of these three patterns for each participant. Further investigation of the categories shows how often each duration category occurs, as seen in Table 4.

Example Bundling Pattern. The random instance of this pattern that was evaluated by Participant 1 entailed a series of five consecutive timeslots in which the participant supposedly worked on the same case on July 17th, 2024, between 13:27 and 14:50. Indeed, after consulting the raw AWT data, the participant confirmed having worked on one particular research paper for the duration of those five timeslots. However, the participant noted that the work on that project started a couple of minutes earlier and ended a little later, namely from 13:24 to 15:09. In the preceding and subsequent timeslots, the window title that was used to recognize the case was not the title with the longest duration, as the person mainly worked on another task during that period of time. Therefore, these timeslots were not included in the bundle, and rightly so, as including the complete timeslots would be inaccurate. This suggests that the detection approach accurately identifies correct bundles, but the inclusion of sections of other timeslots would be required for accurate time analyses.

Example Starting Pattern. The randomly generated instance of the starting pattern examined in the evaluation concerned three days in which Google Chrome was the dominant system used by the participant during the first timeslot of the day: the 6th of

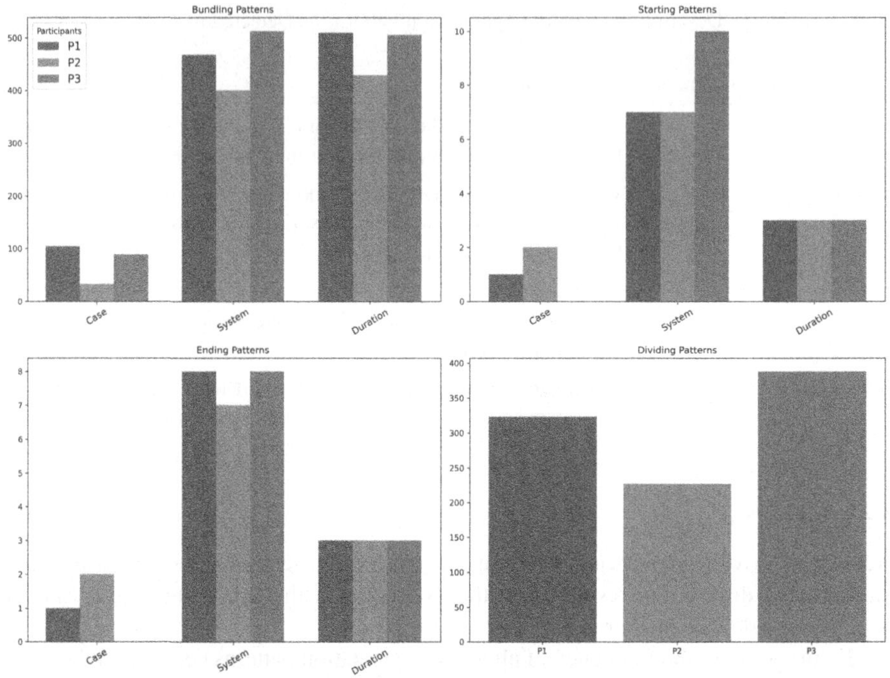

Fig. 3. Total Count of Detected Work Organization Patterns per Participant.

Table 4. Duration Pattern Variant

Participant	Bundling pattern	Starting Pattern	Ending pattern
P1 Short	339	92	69
P2	366	81	81
P3	427	109	88
P1 Medium	157	48	53
P2	61	38	21
P3	78	24	42
P1 Long	14	13	23
P2	2	16	31
P3	1	3	6

December, the 23rd of July, and the 18th of August. Indeed, the participant confirmed that Google Chrome was the main system used to perform the first task of the day. Notably, the participant continued working on the task and corresponding system where they ended the day before. Table 5 for example, contains a snippet of the data, which contains the final two window titles on the 17th of August and the first two window titles of the following day. The participant continued working on a Streamlit app at the start of the second day, mostly working in Google Chrome. Google Chrome was also the app that the participant started in on most days, followed by Microsoft Outlook and Adobe Acrobat.

Table 5. AWT data example from Participant 1

App	Title	Begin Time	End Time	Date
Google Chrome	Streamlit_app – Google Chrome	18:46	18:46	17-08-2024
LockApp.exe	Windows Default Lock Screen	18:46	18:48	17-08-2024
Google Chrome	Streamlit_app – Google Chrome	13:18	13:18	18-08-2024
Google Chrome	Simple Browser – determine-productive-type [Codespace: zany dollop] – Visual Code – Google Chrome	13:18	13:19	18-08-2024

Example Ending Pattern. The pattern evaluated by Participant 1 regarded a detected instance of the duration variant of the ending pattern. On average, the duration of the last timeslot of the day took 18.7 min. The participant was asked to verify whether the ending timeslots on the 30th of August, the 16th of October, and the 11th of January were indeed medium-length timeslots, i.e., between 5 and 30 min. The participant was able to confirm this from the data, as well as point out that all three of these timeslots concerned days in which the participant continued working after 5.30 pm – for the first two dates until between 10 and 11 at night - at which there is less disturbance and one can achieve longer continued periods of work.

Example Dividing Pattern. For Participant 1, 612 work items were found to be divided into sub-work items. The instance randomly selected to be verified concerned the work on a specific topic on the 6th of October, which continued twice on October 10. The participant described this as communicating about a research paper to an international collaborator via email. The first event represented the participant's first response to the email sent by the collaborator. After that, the participant spoke to a third person for advice, which the participant communicated on the 10th of October as a follow-up email. Indeed, this was considered an accurate instance of one work item (Participant 1 responding to the collaborator about the paper) being split into sub-work items.

5 Discussion

5.1 Theoretical Implications

This study has described work organization patterns and given an initial approach to automatically detect them. This detection approach can be used in different scientific research. Research in the fields of work, organizational psychology, and behavioral sciences has explored individual habits and organizational routines. Routines are defined as "*collective* recurrent activity patterns", whereas habits are *individual* recurrent activity patterns [14]. The patterns detected in the AWT data can be both. We mainly looked at the individual patterns that differed per participant. When the detection approach is applied on a larger scale including more people, general collective patterns for a team or department could be detected. In the data, long-duration timeslots are more likely to happen at the end of the day than at the start. This confirms earlier research where the duration of undisrupted work was longer in the evenings and weekends [4]. The focus is on individual recurrent activity patterns rather than the collective when looking at individual differences.

After detecting the work organization patterns, closer inspection of the pattern instances could give more insight into work organization differences. For example, in the datasets used in this study, we found that individuals in the highest position (full professor) exhibited the fewest patterns in the long-duration category, followed by associate professors and then assistant professors. While the sample size is too small to draw generalizable conclusions, this observation is noteworthy. It could indicate that the higher the position, the less undisrupted time they have. Alternatively, it could indicate the difference in tasks given to different positions or that the more experience the person has, the less time they spend on one work item. Further investigation of individual work organization patterns from various academic positions, using larger and more diverse datasets, could result in more insights into how different employees in positions organize their work.

5.2 Practical Implications

The work organization patterns described in this study show how AWT data can be used to explore and analyze individual work behavior and identify differences. Detecting work organization patterns provides opportunities for both the individual worker and the organization. First, individual employees can use the data to find patterns in how they organize their work and how that changes throughout the year. Additionally, this data can be combined with other data, such as performance rates or wearables, such as sleep or heart rate trackers, to see how they influence their physical and mental health [5]. This detection approach can be integrated into existing productivity tools to provide real-time feedback to employees, enabling them to self-assess and adjust their work organization practices. Second, organizations can use information from AWT data to find common work organization patterns, providing insight into how work is organized and performed within their organization. These insights can be used to integrate tailored tools to facilitate better optimization for individual employees, departments, or organizations. For an organization to use AWT data, it must recognize that this is sensitive information and take care to safeguard employee privacy accordingly.

5.3 Limitations and Future Work

As with any other research, there are limitations to consider. First, detecting the *task* and *involved actor* that corresponds with a timeslot was challenging and needed further work to be done accurately, possibly using Natural Language Processing techniques. Second, not all activities are performed on the computer, such as physical meetings or researching physical documents. These activities do not appear in the AWT data but are still part of the work organization. Observational studies could investigate how long someone is working away from the computer. Although observational methods would provide more information about the time away from the computer, AWT has the advantage that it is less intrusive for the participant and less labor intensive [15]. A short observational study and AWT combination could help indicate the average time away from the computer during office hours. Third, we defined the work organization patterns based on interviews with academic staff. The patterns were formulated in general terms to ensure they can be applied in other contexts, but other patterns may be found in other contexts. Additionally, focusing on one recent work week in the interviews could have limited the perspective on work organization patterns, as some patterns might only appear when investigating more weeks. Finally, we cannot guarantee that the timeslots equal the work items being performed. The evaluation showed evidence that for a random set of instances, this operationalization was reasonably accurate. However, the evaluation of the bundling pattern showed signs of false negatives, where parts of the timeslot surrounding the bundle also consisted of work being performed on that work item, but it did not show up as a separate timeslot. This occurs when an individual immediately follows a work item with another without disrupting computer work. Future work can investigate if there are more accurate approaches to detect when individuals switch between work items.

6 Conclusion

Overall, we have described work organization patterns and their variants, including the approach for detecting them within AWT data. This approach centers on how individual employees organize work and utilizes information from employees across various systems, providing a new method for gathering insights into the work system. Our findings offer a basis for further investigation and the possibility of providing additional information that could support future process optimization. We have shown an initial approach to automatically detecting work organization patterns and the challenges that can be investigated in future work, such as improving task detection accuracy, combining AWT data with other data sources, and exploring the relationship between work patterns and well-being indicators.

References

1. Trivedi, S., Patel, N.: Virtual employee monitoring: a review on tools, opportunities, challenges, and decision factors. Empirical Quests Manage. Essences, pp. 86–99 (2021)
2. Beerepoot, I., et al.: A window of opportunity: active window tracking for mining work practices. In: 5th International Conference on Process Mining (ICPM 2023), pp. 57–64 (2023)
3. Claessens, B.J.C., Van Eerde, W., Rutte, C.G., Roe, R.A.: A review of the time management literature. Pers. Rev. **36**, 255–276 (2007)
4. Beerepoot, I., Del Río Ortega, A., Resinas, M., Reijers, H.A.: Interruptibility during scientific research collaborations: the effect of pressure, proximity and quiet time. In: Companion Publication of the 2024 Conference on Computer-Supported Cooperative Work and Social Computing (CSCW 2024), pp. 179–184 (2024)
5. Vuori, V., Helander, N., Okkonen, J.: Digitalization in knowledge work: the dream of enhanced performance. Cogn. Technol. Work **21**(2), 237–252 (2019)

6. van der Aalst, W.M.P., ter Hofstede, A.H.M., Barros, A.P.: Workflow patterns. Distrib. Parallel Databases **14**, 5–51 (2003)
7. Russell, N., van der Aalst, W.M.P., ter Hofstede, A.H.M., Edmond, D.: Workflowresource patterns. In: Advanced Information Systems Engineering, pp. 216–232 (2004)
8. Rafnsdóttir, G.L., Heijstra, T.M.: Balancing work–family life in academia: the power of time. Gend. Work Organ. **20**(3), 283–296 (2013)
9. Peeters, M. C. W., Taris, T. W., de Jonge, J.: Introduction. In: An Introduction to Contemporary Work Psychology, pp. 1–30. Wiley (2013)
10. van der Aalst, W.M.P.: Process Mining: Discovery, Conformance and Enhancement of Business Processes. Springer, Heidelberg (2011)
11. NVivo 14. Lumivero (2023)
12. Erit, M.: Tockler (2021). https://github.com/MayGo/tockler
13. Beerepoot, I.: A month's worth of labelled active window tracking data. In: Proceedings of the Best BPM Dissertation Award, Doctoral Consortium, and Demonstration and Resources Forum Co-located 22nd International Conference on Business Process Management (BPM 2024) (2024)
14. Becker, M.C.: Organizational routines: a review of the literature. Ind. Corp. Change **13**, 643–678 (2004)
15. Šinik, T., Beerepoot, I., Reijers, H.A.: A peek into the working day: comparing techniques for recording employee behaviour. In: Research Challenges in Information Science (RCIS 2023), vol. 476, pp. 343–359 (2023)

Black Swan Theory for Navigating Trust in Mixed-Traffic Environments

Hind Bangui[1(✉)], Barbora Buhnova[1], and Mouzhi Ge[2]

[1] Faculty of Informatics, Masaryk University, Brno, Czech Republic
{hind.bangui,buhnova}@mail.muni.cz
[2] Deggendorf Institute of Technology, Deggendorf, Germany
mouzhi.ge@th-deg.de

Abstract. Connected and autonomous vehicles (CAVs) have revolutionized traffic systems, introducing mixed-driving environments where human-driving and driverless vehicles can coexist on the same roadways. This dynamic interaction has resulted in complex social driving behaviors, emphasizing the need for social trust management to balance interactions in mixed-driving environments effectively. This work presents our vision of a trust management framework designed for mixed driving environments. By gaining from the black swan theory, we address critical blind spots and vulnerabilities in current trust models. We offer insights into how trust relationships can be optimized in the face of uncertainty. Our approach aims to support reliable social networks and facilitate harmonious collaboration between human-driving and driverless vehicles, which promotes the safety and efficiency of mixed-traffic ecosystems.

Keywords: Trust management · Connected and autonomous vehicles · Mixed-driving environments

1 Introduction

Thanks to vehicle-to-everything (V2X) communications [8], connected and autonomous vehicles (CAVs) have been the predominant focus of intelligent transportation systems [8] to provide valuable services, such as data-driven mobility, congestion management, and road safety. Despite the promising applications of CAVs, they are still vulnerable to several cyberattacks [8] due to their dependability on V2X to obtain more information necessary for supporting their functionalities. To tackle security issues, trust management [3] has been adopted to build trustworthy vehicular environments and ensure the dissemination of reliable information among vehicles accurately.

Several trust models have been proposed in recent literature [3,8]. However, most solutions have been focused on evaluating the behavior of vehicles from a cybersecurity perspective [3,8] without considering the coexistence of CAVs and HDVs (Human-Driving Vehicles). Indeed, trust models must account for the

J. Grabis et al. (Eds.): RCIS 2025, LNBIP 547, pp. 87–102, 2025.
https://doi.org/10.1007/978-3-031-92474-3_6

presence and complexities of human behavior in mixed-driving environments while improving trust in both directions (automated and human driving vehicles). For instance, human drivers exhibit variability influenced by emotions that can affect driving style, such as aggressive and hesitant driving. Trust management can monitor observable human driving behaviors reflecting emotional states to instruct CAVs on understanding, predicting, and adapting to human emotions and behavior in real-time. However, current trust models are deprived of an understanding of human driver behavior due to the lack of human drivers' acceptability of CAVs [8,15,29], which opens questions about how we should deal with uncertainties in a mixed-driving environment and develop a sense of vulnerability in trust management.

To address the research gap, this work introduces our vision of a trust management framework for mixed-driving environments that aims to increase the understanding and awareness of trust management by adopting the Black Swan theory [30], which supports identifying and mitigating blind spots in the underlying contexts. Likewise, this work adapts the concept of black swans to design a gap between what we know or do not know in mixed-driving environments, e.g., the impact and consequences of sharing the same roadway by CAVs and HDVs. This can reshape trust management and the collection of trust evidence in mixed-driving environments.

The rest of the paper is organized as follows. Section 2 reviews the Black Swan theory and problem statement as a foundational theory background. Section 3 illustrates specific trust management scenarios that can be considered as Black Swan blindness. Then, Sect. 4 outlines how to approach knowing the unknown with the Black Swan theory. Section 5 further proposes the trust management framework based on Black Swan recommendations. Section 6 presents a primary validation of the proposed framework with application and is followed by discussions in Sect. 7. In the end, Sect. 8 concludes the paper and outlines future research directions.

2 Black Swans as Emerging Risks

The Black Swan is a rare and unpredictable event that surprisingly occurs and leads to catastrophic consequences despite all the plans and precautions. The Black Swan has been defined in [30] as: "First it is an outlier, as it lies outside the realm of regular expectations, because nothing in the past can convincingly point to its possibility. Second, it carries an extreme impact (unlike the bird). Third, in spite of its outlier status, human nature makes us concoct explanations after the fact, making it explainable and predictable". Thus, black swan theory centers around unforeseen events that come as a surprise. Generally, black swan type of events can be categorised as knowledge quadrant [1,10].

- Unknown unknown (meta-ignorance): Events that are completely unknown.
- Unknown known (tacit knowledge): Events not on the list of known events from the perspective of those who carried out a risk analysis, but the events are known by others.

- Known unknown (conscious ignorance): It is about knowing that types of risk exist, but there is no evidence available to conduct risk assessments. It is named Gray Swan and defined as "an unlikely but fully conceivable risk that lies well within the bounds of experience and has been openly discussed but becomes discounted and fails to generate mitigating actions." [17]
- Known known with negligible probability: Tangible events that have been identified and considered in the risk analysis, but they are judged to have a negligible probability of occurrence.

Black swans characterize rare, uncertain, and yet possible events that will always remain unpredictable and inescapable, even under the best circumstances. In other words, black swans are about unknown unknowns. However, it is possible to cope with black swans by turning them into gray swans, resulting in transforming "unknown unknowns" into "known unknowns". Accordingly, rare but possible events can be captured and distinguished from white swans. Thus, in [10,32], black swan events have been linked to the notion of emerging risks, as without exposure to a black swan event, there is no risk and then the vulnerability could not be discovered. From a risk perspective, the analysis of emerging risk definitions has been examined in [10] to describe the link between the black swan type of events and emerging risks. In this work, we will focus on the following sections to examine the black swan type of events within the context of trust management.

3 Hidden Blind Spots in Trust Models

3.1 Existing Trust Models

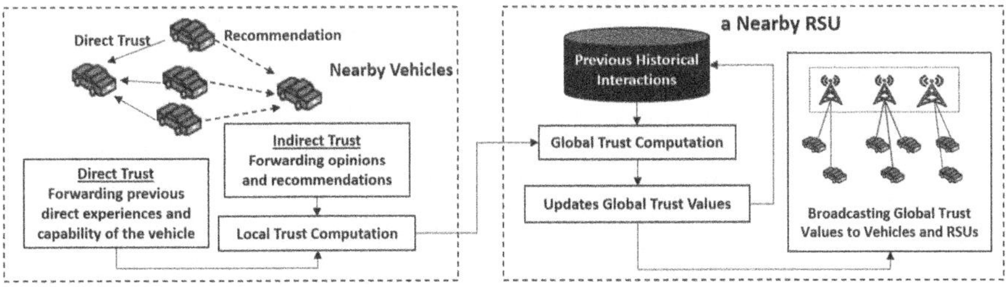

Fig. 1. RSU-aided distributed trust management in vehicular networks

Several trust management frameworks [3,11] in academic literature evaluate the trustworthiness of vehicles based on their behavior, history, and reputation. As depicted in Fig. 1, roadside units (or central authorities) mainly maintain and distribute trust values to mitigate malicious attacks (like on-off attacks), ensure data integrity, and improve overall security and safety in vehicular networks. Generally, trust management deals with intentional and unintentional attacks.

Intentional attacks target the degradation of network reliability or unfairly influence service outcomes. For instance, Sybil attacks create fake identities to manipulate trust, and bad-mouthing attacks propagate false information to reduce the

Table 1. Trust Black Swan Blindness

Intentional Attacks		
Trust Category	Attack Type	Description
Full Trust	Man-in-the-Middle Attacks	Intercepts communication between two parties to eavesdrop or manipulate messages
	Injection Attacks	Inserts malicious data into a trusted communication
	Modification Attacks	Alters the content of trusted communications or data
Selective Trust	Self-Promoting Attacks	Increases reputation falsely through self-generated feedback or actions
	Discriminatory (Selective behavior) Attacks	Exploits trust by selectively providing biased behavior towards certain entities
	Opportunistic Service Attacks	Provides service inconsistently to benefit from specific opportunities
	Traitor/On-Off Attacks	Alternates between good and malicious behavior to remain trusted
	Whitewashing Attacks	Erases or hides malicious behavior by resetting reputation
	Bad-Mouthing Attack	Deliberately reduces the trustworthiness of others by spreading false or biased information
	Ballot-Stuffing Attack	Inflates trustworthiness falsely through exaggerated or fake recommendations
	Sybil Attacks	Creates multiple fake identities to influence trust systems
	Impersonation Attacks	Pretends to be another entity to gain trust or access
	Spoofing/Masquerading Attacks	Misrepresents identity to deceive and gain trust
	Replay Attacks	Reuses valid data transmissions to gain unauthorized benefits
	Passcode and Key Attacks	Exploits or steals authentication credentials to gain access
Unintentional Attacks		
Selective Trust	Trust Bootstrapping	Exploits the reinitialization of trust for entities returning to a system
	Direct Trust Misuse	Exploits direct interactions for malicious purposes
	Indirect Trust Misuse	Misuses recommendations, reputation, or social beliefs to deceive others
	Hybrid Trust Misuse	Combines multiple trust misuse techniques to amplify effects
Distrust	Discrimination of Vendor Members	Unequal trust towards members of certain vendors
	Discrimination of Non-Friends	Bias against non-friends, limiting their access to trust-based systems
	Discrimination of New Members	Denial or restriction of trust for new entities

reputation of others. Unintentional attacks result from non-malicious actions compromising security, such as assigning default or adaptive trust values to new or rejoining entities.

3.2 Detecting Black Swan Blindness

Generally, a trust model addresses three different trust levels: Full trust, selective trust, and distrust (Table 1). Trust Black Swan blindness can be explored from the three main perspectives described below.

Full Trust. In the field of vehicular domain, trust has traditionally relied on the idea of certification [31]. This means that if a vehicle is certified, its trustworthiness is assumed without any further question, which is then referred to as *full trust.* In a broader sense, full trust comes from certain features of the vehicle (the "trustee") that matter to the person or system who trusts it (the "trustor"). These features can include certification from another authority, computing power, or even the reputation of the manufacturer. Although this approach simplifies the process of establishing trust, it comes with certain limitations. For example, it assumes that certified vehicles will always perform as expected in all scenarios. However, this may not account for unforeseen issues, such as evolving threats or unexpected behavior of the system. Also, this approach does not involve any actual trust model. Trustee features such as certifications or guarantees can help us understand what a system can do and how reliable it is. However, relying on these features alone without proper verifications may lead to two problems: 1) considered to be secure when there are actually risks, and 2) increasing the chance of security issues, such as an accidental or deliberate update installed by accident or on purpose. The issue with full trust is that this blind trust can miss important issues that are caused by biases or personal interests.

Distrust. On the other side, some researchers suggest using a zero-trust approach in vehicle ecosystems. This means not trusting anyone by default and only working with a pre-approved group, like members from the same vendor or a set of "trusted" partners. However, this approach has also drawbacks. It excludes others from using the system and benefiting from its services. This exclusion can lead to the case that someone may try to bypass the rules to access the system, which in turn malicious actions may be performed. The zero-trust approach can create a fragmented system by excluding certain groups and limiting collaboration. When people or organizations are excluded, this can reduce overall cooperation opportunities. Also, zero trust lacks to build trust gradually over time. Thus, it limits new partnerships and ideas. To address these issues, a more balanced approach is needed, for example, one that combines flexibility with security. By adjusting trust based on real-time information and context, systems can encourage collaboration while still maintaining security.

Selective Trust. The comprehensive perspective is based on a trust model that recognizes its own blind spots in how to make trust decisions. This can result in both false positives (trusting untrustworthy members) and false negatives (distrusting trustworthy members). These "trust Black Swan" scenarios may include situations like gaining a false positive reputation through authorized methods such as biased recommendations, forging identities as authorized entities, or selecting malicious members within a trusted group. Trust failures show the limitations of relying only on a static trust model. If a trust model does not consider risky issues, malicious actors may use its blind spots to gain trust, which will further weaken the system's trust integrity. To reduce these risks, trust models need to be flexible such as using real-time feedback and context to identify issues and prevent false trust. It is also important to regularly reassess trust to avoid trust manipulation and keep the system reliable.

Summary. In the three trust areas, black swans are often left unseen due to denying, constraining, or ignoring facts that reflect the risk existence, leading to the occurrence of an event and resulting consequences. For example, the current digital trust models focus on measuring qualitatively the level of trust in terms of performance, security, and reliability. However, they address relatively the complexity and diversity of mixed-driving environments as they lack human factors [7] that may cause unintentional attacks. For example, emotional states (such as anger) may lead to reckless interactions with autonomous vehicles, unintentionally creating risky situations. Similarly, human drivers may struggle to understand the actions of autonomous vehicles, leading to hesitation or distrust. Thus, there is a need for the black swan theory to understand emerging risks in mixed driving environments and support trust evidence collection to increase the maturity of digital trust management, guaranteeing safer and more trustworthy collaboration between autonomous and human drivers.

4 Black Swans Methodology

To support the vision of a trust management framework for mixed-driving environments that cover the identification and elimination of blind spots of trust management in case of unnoticed black swan events, we follow five steps that suggest turning "unknown unknowns" into "known unknowns" in [30]. These five steps are outlined below.

- GS-1 **Stop sweating the small predictions:** recommends making short-scale predictions instead of large-scale ones to minimize the consequences of risk related to black swan events and incur them.
- GS-2 **Maximize possibilities for positive black swans:** recommends maximizing exposure to positive black swan events to gain from their consequences by maximizing the possibilities of serendipity to capture black swans, such as being more socialized and proactive in forming friendships to detect black swans through discussions or recommendations.

GS-3 **Adopt the barbell strategy:** recommends developing a probabilistic analysis reflecting risk. It has been recommended in [30] to invest 85–90% into something extremely certain and safe with low risk and 10–15% into potential black swans with high risk.

GS-4 **Distinguish between positive and negative contingencies:** recommends taking more risks in positive black swan areas rather than negative ones to increase the chance of positive benefits.

GS-5 **Invest in preparedness not in prediction:** recommends building up resilience by framing risk scenarios and developing strategies that avoid and mitigate their potential negative impacts if they occur.

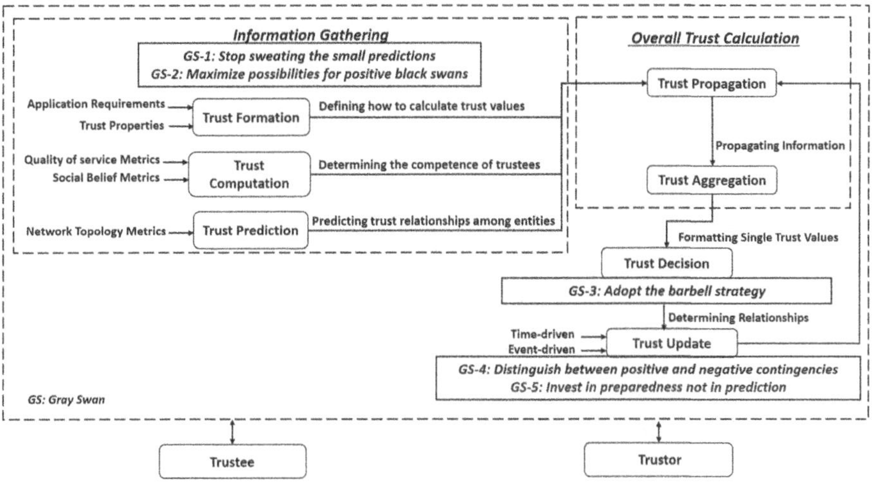

Fig. 2. Trust Management Framework based on Black Swan Recommendations

5 Proposed Trust Management Framework

This section presents the envisioned trust management framework, designed with the ultimate goal of enabling the trust model to notice and turn "unknown unknowns" (Black Swans) into "known unknowns" (Gray Swans) within the context of mixed-driving environments. It is hard to collect trust evidence from black swan events because their occurrence cannot be estimated from observed data, which causes blind spots in trust management. Thus, we gain from the five steps outlined in Sect. 4 to devise conceptual trust management that can address the impact of unforeseen events.

Figure 2 illustrates the envisioned trust management framework, compiled from the combination of trust-management components [3] with the projection of the five recommendations from the Black Swan theory to enable a trust model to detect "known unknowns" or gray swans. Below, we provide an explanation of how to employ the five recommendations to explore the blind spots and maintain the trustworthiness of trust management within the context of mixed-driving environments.

GS-1&2. Information Gathering

Description: Information gathering provides information about the competence of trustees in mixed-driving environments.

GS-1 Scenario: The trustworthiness and effectiveness of trust management models might be reduced due to the long-term use of trust evidence accumulated from direct and indirect past experiences. For example, a trust model may continuously assign the same trust value to a vehicle (or a trustee) based on its historical information without considering the possibility of changing its behavior. Thus, direct and indirect past experiences should be considered for short-term use as the trustworthiness of trustees is naturally suspicious and unconfident.

GS-2 Scenario: The use of recommendations of friends and friends-of-friends is crucial in trust management. However, until now, it is hard to determine and select the right friend as appropriate trust management policies still need to be designed to manage friendships among heterogeneous entities and distinguish between useful and useless relationships to guarantee trusted and reliable communication in mixed-driving environments. The selection of trusted nodes is, however, strongly context-dependent. There are multiple aspects to consider: (1) the nature of the communication (how sensitive the communicated information is and how serious consequences its misuse would have), (2) the situation in which the entity is, and (3) the privacy settings of an entity (the difference between the degree of trust in direct and transitive friends). Besides that, the method used to calculate trust has a significant impact on the final decision. Since the method can be defined for each entity in a different way, and thus even with identical input parameters, the entity can come to a different outcome. Thus, individual autonomous entities make the trust decision independently, making it impossible to hold others accountable in case of a wrong friend recommendation, especially when the entity's current context is unknown to the recommending entity. These facts highlight the need for trust management policies even more.

GS-3. Trust Decision

Description: Trust decision instructs the requesting trustor by providing the final decision on whether to trust or distrust the trustee in question.

Scenario: Minimizing the unintended and intended consequences caused by selecting inappropriate trustees could be handled in a trust decision by adopting the barbell strategy. For example, due to the lack of pre-defined static thresholds and the calculation lack of trustees' initial trust values, the trustworthiness of the trust decision could be suspicious e.g., considering a low threshold may classify malicious vehicles as honest and vice versa. Thus, merging risk assessment with a trust decision is crucial to enhance the trustworthiness of trust management. Moreover, the assessment of risks aids in distinguishing between trustees with low and high risk, resulting in enhancing trustworthiness evaluation by blocking expected malicious trustees from the social network and attempting to increase the probability of trustworthy and safer ones.

GS-4&5. Trust Update

Description: Trust update refers to the situation when trust values need to be updated.

Scenario: Trust update is the main guidance for ensuring the end-to-end trustworthiness of trust management since it reflects certain facts regarding the real behavior of trustees. Thus, trust update helps in preparing and preventing potential threats that may occur in social networks. However, trust update is still a fundamental challenge in trust management as the majority of trust models focus on ensuring security and reliability while ignoring their trustworthiness improvement. For example, when a vehicle is detected as malicious, the reward/punishment mechanism is used to decrease its reputation. After that, its trust value is updated and propagated over the network. However, the negative impacts caused by this malicious vehicle are not used to drive and learn how to improve the trustworthiness of trust models.

6 Trust Application

By considering the human presence in mixed-driving environments, this section aims to give an example of trust evidence collection and trustworthiness assessment that can support CAVs and HDVs to demonstrate their ability to coexist safely and foster mutual trust. Likewise, this example illustrates the expected blind spots in trust assessment.

6.1 Scenario

The reaction time [20,24,27,35] is one of evidence related to human drivers that is considered in trust management [4,5] as it is the time between the perception of a hazard or situation to the start of maneuvering or the start of control of the vehicle. Reaction times are affected by factors like emotional states (fatigue, stress, etc.) and how they are integrated into the risk assessment autonomous vehicles. In this work, we consider emotional states from the Russell Model (e.g., happy, sad, fear, anger) and their impact on the human driver's reaction time. Indeed, reaction time is linked to emotion recognition because emotional states directly influence cognitive functions. Using models like the Russell Circumplex Model (Fig. 4), we can assess how different emotions affect reaction time in a driving scenario. For instance, a tired driver (low arousal, negative valence) will react slower than an alert driver (high arousal, positive valence). This understanding can be used to improve driver safety systems in mixed-driving environments. We also consider other evidence related to vehicle-to-vehicle communication used in the literature [3,11], such as the speed of vehicles, the distance between them, and weather conditions.

6.2 Trust Evaluation

We give an example of trust evaluation using Fuzzy (Fig. 3), which is one of widely techniques used in trust management. [3,11,34]. We aggregate the following key variables (Trust Evidence): The speed of cars, the distance between them, weather conditions (Clear, Rainy, Foggy), and the emotional state of the human driver by considering emotional states from the Russell Model. Table 2 defines the fuzzy sets for each input variable.

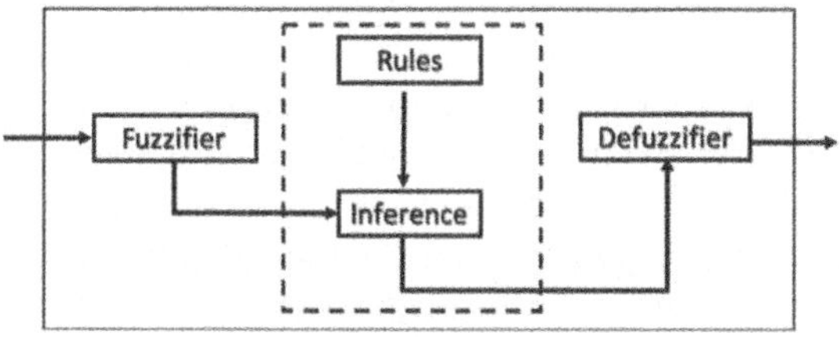

Fig. 3. Fuzzy Logic Controller Structure

Table 2. Parameter and their term sets.

Parameters	Term set
Speed of cars	
0–40 km/h	Slow
40–80 km/h	Moderate
80–120 km/h	Fast
Distance between cars	
0–20 m	Close
20–50 m	Medium
50–100 m	Far
Reaction Time	
1–2 s	Fast Reaction
2–3 s	Normal Reaction
3–5 s	Slow Reaction

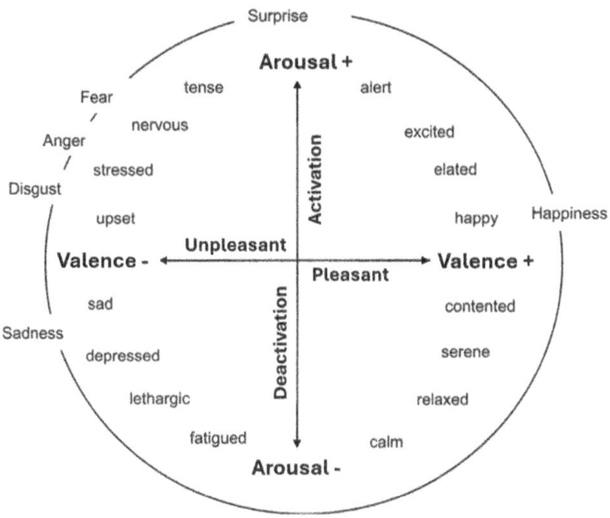

Fig. 4. Russell Model of Affect [12, 14]

According to the Russell Model [12, 14], emotional states are mapped to different reaction times. Below is the categorization of emotions:

- **High Arousal, Negative Valence**: Fear, Anger
  Impact: Increases reaction time.
- **Low Arousal, Negative Valence**: Sadness, Fatigue
  Impact: Significantly increases reaction time.
- **High Arousal, Positive Valence**: Happiness, Excitement
  Impact: Normal or faster reaction time.
- **Low Arousal, Positive Valence**: Calmness, Neutral
  Impact: Normal reaction time.

Table 3 maps emotions to their corresponding reaction times.

The following fuzzy rules are applied based on the emotion (reaction time), speed, distance, and weather conditions:

- **Rule 1**: *If the emotion is Fear or Anger and speed is Fast and distance is Close, then Risk is High.*
- **Rule 2**: *If the emotion is Happiness or Calmness and speed is Slow and distance is Far, then Risk is Low.*
- **Rule 3**: *If the emotion is Sadness or Fatigue and weather is Foggy and distance is Close, then Risk is High.*

The output of this trust assessment might help, e.g., an autonomous vehicle to act to increase safety by generating alerts to the human driver or taking over in dangerous situations.

Table 3. Emotion-based Reaction Time

Emotion	Reaction Time Impact
Fear	Slow (4–5 s)
Anger	Slow (4–5 s)
Sadness	Slow (3–4 s)
Fatigue	Slow (3–4 s)
Happiness	Fast (1.5–2 s)
Calmness	Fast (1.5–2 s)

6.3 Detecting Blind Spots

Current driving systems focus on supporting road safety by monitoring the external environment (like weather and road conditions) [3,11] and the state of human drivers to adjust their driving style (like drowsiness detection [23]). However, monitoring the state of human drivers is limited to detecting drowsiness [9], distraction [18], nervousness/anxiety [22], fear [6], happiness [6], and anger [16,28]. Actually, there are other more complex and nuanced emotions that are not fully considered in related work. For instance, Plutchik's Wheel [21] combines different emotions to create complex emotional responses. Within the context of this work, this type of complex emotional model can serve to understand and analyze emotions between human drivers/passengers and autonomous vehicles. For instance, angry human drivers are typically linked to riskier driving behaviors. Autonomous vehicles are programmed for safety and might respond by driving more cautiously, slowing down to reduce the ability to react swiftly to potential threats. However, according to Plutchik's Model, anger can be combined with disgust to form contempt or aggressiveness if it is combined with anticipation, which leads to a rethink about the meaning of trust evidence and the corrections that could be taken to adjust the trust assessment. Thus, full-scale emotion recognition is required to instruct trust management about emotional interaction between human drivers/passengers and autonomous vehicles in mixed-traffic environments.

7 Discussion

The envisioned trust management framework shows a promising direction in the identification and elimination of blind spots of trust management especially in case of unnoticed black swan events, helping the system become more resilient against similar future events.

The proposed trust management framework can be useful for mixed-driving environments, particularly it can be merged with Human-centric applications [33] that might provide a nuanced understanding of black swan events as they reflect human perception that can be used to enhance trust evidence. For example, in V2X communication that aims to connect human and physical entities [19], the proposed trust framework can use the perception or behavior of drivers/pedestrians and vehicle feedback as indirect trust evidence to maintain the trustworthiness of vehicular networks and punish misbehaved members. Similarly, it can form different types of relationships in V2X networks (like vehicle-pedestrian/driver interaction) [26] based on collecting behavioral data of pedestrians, drivers, and vehicles. Compared to the proposed trust framework, the majority of trust models [3,11] in vehicular networks focus mainly on building reliable interactions between vehicles, such as selecting the optimal routing path [3,11,25] to ensure trustworthy communication and maintain vehicle cooperation, while ignoring a significant part of trust evidence, which is an extensive data collected through human-centric approaches served to improve transport applications based on monitoring the behavior of drivers/pedestrians [33]. Thus, in our proposed trust framework we aim to extend trust evidence using human-centeric approaches that might be possible in vehicular networks as many studies have focused on exploring human-centric measurements to increase road safety in urban areas [13]. These explored human-centric measurements will improve the trust evidence collection, serve to turn black to gray swans in vehicular networks, and then calibrate the trustworthiness assessment correctly.

7.1 Implementation Possibility and Limitations

Implementing the proposed trust management framework in mixed driving environments in real world can be done using V2X communications and integrating trust management modules within roadside units. We consider that it is feasible to develop a modular software and deploy it within current intelligent transportation systems. This modular component could include data collection function to collect real-time data from sensors and human-centric monitoring systems. Using adaptive deep learning models and fuzzy logic techniques, the system would process input data such as weather conditions and driver emotions to continuously evaluate trust levels. For example, one pilot project can be implemented in a controlled urban area to validate the system performance, where connected vehicles and RSUs work together to generate dynamic trust scores. Meanwhile, deep learning models can be used to monitor driver emotion to predict and adapt to human behaviors. In this regard, developing a cloud-based backend for data aggregation and real-time processing would also help manage the high volume of data from multiple sources, while edge computing at RSUs can minimize latency for critical safety decisions.

However, some limitations need to be addressed before practical implementation. One of the main challenges is that it is difficult to maintain robustness in complex and mixed traffic scenarios. Thus, it is difficult to conduct field experiments as simulations for Black Swan events. Another limitation is the complexity of accurately detecting and interpreting human emotions, which can vary between individuals and between cultural contexts. Integrating emotion detections into real-time decision-making may also introduce computational latency. However, our recent experiments have shown that a high level of emotion detection accuracy can be achieved using CNN and transfer learning. Furthermore, we admit that collecting and processing human-centric data may raise privacy and ethical concerns [2]. Finally, the proposed framework may be vulnerable to adversarial attacks, where malicious attackers may possibly manipulate trust scores by injecting fake data. To mitigate these risks, robust security measures and anomaly detection algorithms can be used to ensure data integrity and system reliability.

8 Conclusion

This work introduces our vision of supporting the development of trust management that can deal with the heterogeneity of relationships in mixed-driving environments, such as vehicle-to-vehicle (V2V), vehicle-to-infrastructure (V2I), vehicle-to-driver (V2D), and vehicle-to-pedestrian (V2P), advancing them towards collaborative Social vehicular networks. By adopting the Black Swan Theory, we suggest increasing public acceptance of full autonomy in vehicles via supporting the ability of the underlying systems to identify and explore blind spots in trust management that can advance the socialization of objects in mixed-driving environments. Our research goal in this work is to spark a debate on how to gain from black swans to advance future development of trust management in mixed-driving environments.

We are now on a pathway toward a digital era where trust management is designed to deal with uncertainty and randomness [3]. Thus, it is critical to explore black swan theory to manage new requirements of interactive situations in ecosystems. For instance, preparedness might be elaborated more and improved through this theory in mixed-driving environments. Such a future research path might focus on developing methods for quantifying uncertainty, which could involve dynamic estimation of rare events' likelihood and using that knowledge to inform the individual system's decision-making processes. Moreover, sharing the computed information across the social network would enhance the collaboration between connected entities (vehicle-to-vehicle, vehicle-to-infrastructure) as well as between vehicles and humans (drivers or even pedestrians) [19]. Uncertainty quantification and its incorporation into decision-making processes could therefore not only help systems better prepare for an occurrence of a grey swan event but also become more predictable and transparent in their following actions, which would have an impact on increasing human trust in autonomous systems.

Acknowledgements. The work was supported by GAMU project "Forensic Support for Building Trust in Smart Software Ecosystems" (MUNI/G/1142/2022).

References

1. Bammer, G., Smithson, M.: Uncertainty and Risk: Multidisciplinary Perspectives. Routledge (2012)
2. Bangui, H., Buhnova, B., Ge, M.: Social internet of things: ethical AI principles in trust management. In: The 14th International Conference on Ambient Systems, Networks and Technologies (ANT 2023), Leuven, Belgium, 15–17 March 2023. Procedia Computer Science, vol. 220, pp. 553–560 (2023)
3. Bangui, H., Buhnova, B., Kusnirakova, D., Halasz, D.: Trust management in social internet of things across domains. Internet of Things (2023)
4. Bangui, H., Ge, M., Buhnova, B.: Deep-learning based reputation model for indirect trust management. In: The 14th International Conference on Ambient Systems, Networks and Technologies, 15–17 March 2023, Leuven, Belgium. Procedia Computer Science, vol. 220, pp. 405–412. Elsevier (2023)
5. Bangui, H., Ge, M., Buhnova, B.: When trustless meets trust: blockchain consensus review and reconsideration. **246**, 3351–3360 (2024)
6. Chen, J., Lin, X., Ma, W., Wang, Y., Tang, W.: EEG-based emotion recognition for road accidents in a simulated driving environment. Biomed. Signal Process. Control **87**, 105411 (2024)
7. Cheng, Z., Zhu, J., Feng, Z., Yang, M., Zhang, W., Chen, J.: Driving safety risk analysis and assessment in a mixed driving environment of connected and non-connected vehicles: a systematic survey. IEEE Trans. Intell. Transp. Syst. (2025)
8. Durlik, I., Miller, T., Kostecka, E., Zwierzewicz, Z., Łobodzińska, A.: Cybersecurity in autonomous vehicles—are we ready for the challenge? Electronics (2024)
9. El-Nabi, S.A., El-Shafai, W., El-Rabaie, E.S.M., Ramadan, K.F., Abd El-Samie, F.E., Mohsen, S.: Machine learning and deep learning techniques for driver fatigue and drowsiness detection: a review. Multimed. Tools Appl. **83**(3), 9441–9477 (2024)
10. Flage, R., Aven, T.: Emerging risk–conceptual definition and a relation to black swan type of events. Reliab. Eng. Syst. Saf. **144** (2015)
11. Hbaieb, A., Ayed, S., Chaari, L.: A survey of trust management in the internet of vehicles. Comput. Netw. **203**, 108558 (2022)
12. Joudeh, I.O., Cretu, A.M., Bouchard, S.: Predicting the arousal and valence values of emotional states using learned, predesigned, and deep visual features. Sensors **24**(13), 4398 (2024)
13. Kalantari, A.H., et al.: Who goes first? A distributed simulator study of vehicle–pedestrian interaction. Accid. Anal. Prev. (2023)
14. Karnati, M., Seal, A., Bhattacharjee, D., Yazidi, A., Krejcar, O.: Understanding deep learning techniques for recognition of human emotions using facial expressions: a comprehensive survey. IEEE Trans. Instrum. Meas. **72**, 1–31 (2023)
15. Li, X., You, Z., Ma, X., Pang, X., Min, X., Cui, H.: Effect of autonomous vehicles on car-following behavior of human drivers: analysis based on structural equation models. Phys. A **633**, 129360 (2024)
16. Love, S., Indrawan, Y., Larue, G.S.: Development of a nuanced approach to aggressive driving: an investigation into the self-regulatory and attitudinal profiles of aggressive drivers and their contextual role in negative road interactions. Transport. Res. F: Traffic Psychol. Behav. **102**, 142–154 (2024)
17. Mazarr, M.J., Mazarr, M.J.: The swans to worry about are gray. Rethinking Risk in National Security: Lessons of the Financial Crisis for Risk Management (2016)
18. Nicolls, M., Truelove, V., Stefanidis, K.: Examining the impact of interventions in reducing self-reported engagement in distracted driving: a systematic review. Accid. Anal. Prev. **202**, 107608 (2024)

19. Noor-A-Rahim, M., et al.: 6G for vehicle-to-everything (v2x) communications: enabling technologies, challenges, and opportunities. Proc. IEEE (2022)
20. Testaferrata de Noto, A.: Driver perception-reaction times in level 3 automated vehicles (2020)
21. Plutchik, R.: A general psychoevolutionary theory of emotion. Emotion Theory Res. Exp. **1** (1980)
22. Qiang, W., Li, F.: Work stressors and aggressive driving: the mediating roles of stress appraisals. Saf. Sci. **181**, 106656 (2025)
23. Ramzan, M., Khan, H.U., Awan, S.M., Ismail, A., Ilyas, M., Mahmood, A.: A survey on state-of-the-art drowsiness detection techniques. IEEE Access **7**, 61904–61919 (2019)
24. Roidl, E., Frehse, B., Höger, R.: Emotional states of drivers and the impact on speed, acceleration and traffic violations—a simulator study. Accid. Anal. Prev. **70**, 282–292 (2014)
25. Shokrollahi, S., Dehghan, M.: TGRV: a trust-based geographic routing protocol for VANETs. Ad Hoc Netw. **140**, 103062 (2023)
26. Stefanovic, C., Armada, A.G., Pratesi, M., Santucci, F.: Vehicular systems for 5G and beyond 5G. Driving 5G Mobile Communications with Artificial Intelligence Towards 6G, p. 183 (2023)
27. Steinhauser, K., et al.: Effects of emotions on driving behavior. Transport. Res. F: Traffic Psychol. Behav. **59**, 150–163 (2018)
28. Stephens, A.N., Collard, J., Koppel, S.: Don't sweat the small stuff; anger rumination and lack of forgiveness are related to aggressive driving behaviours. Curr. Psychol. **43**(7), 6338–6349 (2024)
29. Sun, L., Cheng, Z., Kong, D., Xu, Y., Wen, S., Zhang, K.: Modeling and analysis of human-machine mixed traffic flow considering the influence of the trust level toward autonomous vehicles. Simul. Model. Pract. Theory **125**, 102741 (2023)
30. Taleb, N.N.: The Black Swan: The Impact of the Highly Improbable, vol. 2. Random House (2007)
31. Ting, H., Kang, X., Li, T., Wang, H., Chu, C.K.: On the trust and trust modeling for the future fully-connected digital world: a comprehensive study. IEEE Access **9**, 106743–106783 (2021)
32. Zarghami, S.A., Dumrak, J.: Unearthing vulnerability of supply provision in logistics networks to the black swan events: applications of entropy theory and network analysis. Reliab. Eng. Syst. Saf. **215** (2021)
33. Zhao, Q., Li, G., Cai, J., Zhou, M., Feng, L.: A tutorial on internet of behaviors: concept, architecture, technology, applications, and challenges. IEEE Commun. Surv. Tutor. (2023)
34. Zhu, R., Boukerche, A., Long, L., Yang, Q.: Design guidelines on trust management for underwater wireless sensor networks. IEEE Commun. Surv. Tutor. (2024)
35. Zimasa, T., Jamson, S., Henson, B.: Are happy drivers safer drivers? evidence from hazard response times and eye tracking data. Transport. Res. F: Traffic Psychol. Behav. **46**, 14–23 (2017)

Friend, Foe, or Target? Domain Models as Risk Deterrents, Risk Sources, and Assets at Risk

Isadora Valle[1(✉)], Tiago Prince Sales[1], Eduardo Guerra[2], Ítalo Oliveira[1], Renata Guizzardi[1], Luiz Olavo Bonino da Silva Santos[1], Henderik Proper[3], and Giancarlo Guizzardi[1]

[1] University of Twente, Enschede, The Netherlands
{i.vallesousa,t.princesales,i.j.dasilvaoliveira,r.guizzardi, l.o.boninodasilvasantos,g.guizzardi}@utwente.nl
[2] Faculty of Engineering, Free University of Bozen-Bolzano, Bolzano, Italy
eduardo.guerra@unibz.it
[3] TU Wien, Vienna, Austria
henderik.proper@tuwien.ac.at

Abstract. Modelers and organizations often struggle to assess the benefits and drawbacks of modeling activities. This paper proposes addressing this challenge through a risk-oriented lens, leveraging the Common Ontology of ValuE and Risk (COVER) and the Reference Ontology for Security Engineering (ROSE). The proposal focuses on identifying assets at risk throughout the modeling process to clarify: when models mitigate risks and contribute to cost savings (models as risk deterrents), when models introduce risk to other assets (models as risk sources), or when they are vulnerable to risk events themselves (models as assets at risk), potentially generating additional costs. This perspective enables modelers and organizations to evaluate the benefits and costs of modeling practices, aligning investments with organizational goals, while helping researchers identify gaps for enhancing modeling languages, methods, and tools. The proposal is evaluated by analyzing case studies from the literature and interviews with nine professionals and researchers.

Keywords: Risk Assessment · Modeling · Domain Models · Conceptual Models · Return on Modeling Efforts

1 Introduction

Previous studies have highlighted and demonstrated that modelers and organizations face challenges in assessing the benefits and drawbacks of modeling activities. In [14], the authors introduced the concept of Return on Modeling Efforts (RoME), emphasizing the importance of systematically evaluating modeling costs and potential returns to guide enterprise modeling initiatives effectively. In [27], the same authors proposed further research to refine RoME, highlighting

J. Grabis et al. (Eds.): RCIS 2025, LNBIP 547, pp. 103–118, 2025.
https://doi.org/10.1007/978-3-031-92474-3_7

the need for theory-driven and practice-oriented approaches to address trade-offs between modeling efforts and their potential returns within specific contexts and purposes. The investigation reported in [33] revealed that modelers rely on subjective assessments of the returns from their modeling activities, largely due to challenges in quantifying the benefits and costs associated with the modeling process or its outcomes. Building on these findings, recent work has explored pain points in domain modeling and their relation with modeling costs [32]. In this paper, we contribute to the RoME research agenda by proposing risk assessment to support decision-making in domain modeling. More precisely, we advocate a critical examination of models and the modeling process through a risk-oriented lens, drawing upon the Common Ontology of ValuE and Risk (COVER) [28] and the Reference Ontology for Security Engineering (ROSE) [23].

Risk assessment identifies and evaluates potential risks, providing the foundation for cost-benefit analysis to consider expected and unforeseen outcomes. By framing modeling practices as an ecosystem of objects, events, agents, and their capabilities and vulnerabilities, we can evaluate real and hypothetical scenarios to identify the risks involved in modeling practices. This approach allows identifying assets at risk to determine (a) when models mitigate risks (models as risk deterrents), (b) when they might introduce risks to other organizational assets (models as risk sources), and (c) when they are vulnerable to risk events (models as assets at risk) [23,28,29]. When models act as risk deterrents, they can help reduce costs; however, when they act as risk sources or assets at risk, they may incur additional costs.

Our proposal facilitates effective trade-off evaluations uncovering options with high benefits and manageable risks. For modelers and organizations, it can aid in assessing the benefits and costs of models and modeling practices, ensuring that modeling investments are aligned with practical constraints and organizational goals. It can also help researchers and developers identify gaps that can lead to improvements in modeling languages, methods, and tools.

To illustrate the practical application of our proposal, consider the healthcare sector, which faces significant data interoperability challenges due to the rapid development of ICTs. This results in heterogeneous data across systems, ranging from patient records to billing information. Differences in data formats and terminologies create interoperability barriers, driving up costs, increasing error rates, and potentially compromising patient safety. To address these risks, the European Reference Network for Rare Neuromuscular Diseases (EURO-NMD) developed the FAIR-by-design EURO-NMD Registry to built-in interoperability founded on using accepted ontologies and classifications that promote data integration [1]. This registry integrates clinician-entered data, patient-reported outcomes, and quality-of-life measures, exemplifying how conceptual models can serve as security mechanisms to mitigate risks and control costs.

To explore the feasibility of our proposal, we analyze cases that evidence the practical use of domain models. These cases are drawn from the literature, as well as nine semi-structured interviews with professionals and researchers active in this field.[1] To the best of our knowledge, no existing literature addresses risks in conceptual modeling in the way proposed by this study. Instead, prior research largely focuses on the role of models as tools for managing risks in security [23], business [20], information technology [30], and other fields.

The rest of the paper unfolds as follows. Section 2 introduces our notions of domain models and risk. Section 3 presents an analysis of models as countermeasures, assets at risk, and threat objects. Section 4 makes some final considerations and discusses the direction of future work.

2 Background Knowledge

2.1 Structural Conceptual Models

The risk-oriented perspective proposed in this paper applies to various domain models. Nevertheless, this study focuses on its applicability in structural conceptual models, a category of domain models designed to represent a domain by capturing its structural regularities [15]. Such structural regularities involve types (classes), attributes, relationships, and constraints. As a result, they are versatile tools that serve a wide range of purposes, including problem-solving, decision-making, learning, and system development [33]. Examples of such models include those specified using Entity Relationship (ER) [7], UML Class Diagrams [24], OntoUML [15], and Object-Role Modeling (ORM) [16]. Henceforth, all references to "model(s)" or "modeling" in the remainder of this paper should be interpreted within the context of structural conceptual modeling.

2.2 Assumptions on the Nature of Risk

Despite considerable progress in clarifying the concept of risk, the term remains overloaded and conceptually ambiguous. Practitioners and researchers across various fields continue to define, perceive, and assess risks in different ways [2]. This study adopts the ontological assumptions about the nature of risk as defined by the Common Ontology of ValuE and Risk (COVER) [28]. The most relevant assumptions for this investigation are outlined below.

First, *risk is assessed in terms of its impact on goals.* For instance, if one is concerned with the risk of missing a train, this is because missing a train impacts one's goals, such as arriving on time at a meeting or saving money.

Second, *risk is experiential,* implying that risk is attributed to events rather than objects. In this view, a risk assessment targeting an object is driven by the overall risk from events that could affect it. For instance, consider the risks associated with a laptop. Assessing these risks involves examining: (i) the goals it enables (e.g., completing tasks, accessing documents), (ii) potential events that could disrupt these goals (e.g., hardware failure, data loss), and (iii) the circumstances that might lead to such events (e.g., accidental spills, power surges).

[1] These online interviews were conducted and anonymized by the primary author as part of a previous study detailed in [33].

Then the risk your phone is exposed to is the aggregation of the risk of hardware breakdowns, phishing attacks, and so on.

Third, *risk is contextual*, meaning that the risk an object is exposed to may vary even if its intrinsic properties remain constant. For example, while the characteristics of a car, such as an anti-lock braking system, can affect the risk of a car accident, external factors such as road conditions and weather conditions can increase or enable risks.

Finally, *risk is relative*, suggesting that an event may be perceived as risky by one agent but deemed non-risky by another. For instance, consider a hacker attack. From the target's perspective, the attack constitutes a risk to the security of their assets. Conversely, for the hacker, it represents a strategy to achieve their objectives rather than a risk.

2.3 Risk Terminology

Our analysis is based on a fragment of COVER [28] that addresses risk experience, which focuses on (unwanted) events and their underlying causes. We also incorporate the risk management perspective formalized by the Reference Ontology for Security Engineering (ROSE) [23]. From these ontologies, we adopt the following concepts:

- Intention: It is an intrinsic property characterized by a commitment to achieving a goal, where the goal represents the propositional content associated with this property. *Example:* Sarah's decision to attend the gym illustrates her intention to lose weight. *Note:* This definition closely aligns with the notion of an objective as defined in ISO 31073 [18].
- Agent: It is an entity with an Intention to achieve a specific goal within a given context. *Example:* A person committed to losing weight, or a company committed to increasing its profits.
- Threat Event: It is an event with the potential to cause loss or damage to an Agent's Intention. *Example:* An armed robbery threatens personal safety or property, while a factory fire threatens infrastructure and operations. *Note:* This definition closely aligns with the definition on NIST SP 800-30 Rev.1 [22] to the same term.
- Loss Event: It is an event that results in actual loss or damage to an Agent's Intention. *Example:* The theft of valuables during a robbery or the destruction of factory machinery in a fire.
- Risk Event: It is any event that can be categorized as a Threat Event or a Loss Event.
- Risk Experience: It is a complex event composed of Risk Events. *Note:* This definition closely aligns with the notion of a threat scenario as defined in NIST SP 800-30 Rev.1 [22].
- Risk Subject: It is an Agent whose Intention is at stake during an Loss Event. *Example:* A robbery victim at risk of losing safety or property, or a factory owner whose business is threatened by a fire.

- OBJECT AT RISK: It is an asset (person or thing) that may be harmed or damaged by a LOSS EVENT. *Example:* The personal belongings of a robbery victim or the machinery within a factory that could be destroyed by fire.
- THREAT OBJECT: It is an object (person or thing) bearing capabilities that, when manifested, can cause a THREAT EVENT. *Example:* The perpetrator responsible for committing a robbery or a short circuit in factory wiring that triggers a fire. *Note 1:* This concept is referred to as risk object in [5], hazard or risk source in ISO 31073 [18], and threat actor in NIST SP 800-221 [22].
- RISK ENABLER: It is an ancillary object (person or thing) with dispositions that enable or increase the likelihood of a RISK EVENT. *Example:* The weapon used by a robber to escalate the threat or the presence of flammable materials in a factory that increases the likelihood of a fire.
- THREATENING SITUATION: It is a circumstance or condition that favors the occurrence of a THREAT EVENT. *Example:* Walking alone on a dark street late at night heightens the risk of robbery, just as neglecting wiring maintenance increases the likelihood of a fire.
- SECURITY MECHANISM: It is an object (person or thing) intentionally designed to create value by systematically protecting an AGENT'S INTENTION from RISK EVENTS. *Example:* An anti-theft alarm protects an individual's intention to ensure safety by deterring theft attempts.
- CONTROL EVENT: It is an event that directly or indirectly mitigates the occurrence of RISK EVENTS. *Example:* A thief abandoning a robbery attempt upon hearing an alarm sound.
- CONTROL CHAIN EVENT: It is an event that can cause CONTROL EVENT. *Example:* An alarm emitting a loud sound as a warning of a theft attempt.
- VULNERABILITIES: They are dispositions inhering in a RISK ENABLER or an OBJECT AT RISK whose manifestations can triggers a RISK EVENT. *Example:* A house lacking an anti-theft alarm is more vulnerable and therefore more likely to be targeted by thieves. *Note:* This definition closely aligns with the definition on NIST SP 800-160v1r1 [22] and ISO 31073 [18] to the same term.

To represent these concepts clearly and facilitate the visualization of our analysis, we developed the diagrams presented in Sect. 3. Although we based these diagrams on the formalizations of COVER [29] and ROSE [23] and used symbols inspired by CORAS [21], they do not adhere to standard conceptual modeling conventions. This approach was driven by the lack of a formalized language capable of fully representing the COVER and ROSE concepts required for our analysis. As a result, we created a custom set of symbols that more effectively suit our goals.

3 Domain Modeling Through a Risk-Oriented Lens

In applying the assumptions and concepts discussed in Sect. 2 to domain modeling, we assert that models and elements of the modeling process are exposed to risk events that can affect the modeling goals proposed in [14] and validated in [33]. The impact of these events is contingent on the context and perspective from which they are analyzed. Therefore, in this section, we analyze several practical examples.

3.1 Risks Mitigated by Models

Examining the scenarios experienced by the interviewees and the application of models in practice, as evidenced in [3,8–11,17,34], models can be conceptualized as SECURITY MECHANISMS in certain circumstances. As such, models can prevent RISK EVENTS. For instance, an interviewee highlighted that an OntoUML model created by his development team to facilitate the implementation of three systems was subsequently used for training by the company's tax team. This use led to a more efficient training process, substantially reducing training costs. As shown in Fig. 1, in this scenario, the OntoUML model could be understood as a SECURITY MECHANISM to mitigate **the risk of lengthy and ineffective training of new employees**, which leads to higher training costs. In micro or small businesses, mitigating this risk alone may not justify the investment in developing a domain model. However, this interviewee works for a company that, by 2022, had a workforce of approximately 45,000 employees. Although the exact number of individuals who reaped the model's advantages remains undisclosed, given the size of this organization, one can imagine the financial impact that a reduction in training costs could have.

The prevalent use of models to support system development among the interviewees highlights the significant role of domain models as SECURITY MECHANISMS in software engineering. The results of the study conducted in [11] revealed that practitioners agreed regarding the effectiveness of conceptual models in facilitating various aspects of system development, including creating user-friendly systems, alignment with requirements, and adaptability to changing needs. In addition, they considered these models more effective than code in understanding system behavior and communicating system functionality to stakeholders. According to Liddle [19], model-driven methodologies potentially facilitate the early identification of defects (e.g. design flaws, omissions) and misunderstandings between stakeholders and developers. While adjustments can be made later in the development process to address oversights from the initial conceptualization phase, subsequent modifications become increasingly costly and disruptive as development progresses. Late-stage alterations incur greater expenses and yield less effective solutions. Consequently, in software engineering, model-centric approaches can leverage the model as a SECURITY MECHANISMS to mitigate **risks associated with technical challenges** such as domain debts [31], false agreement issues [13], and software entropy [6], **as well as economic concerns** like escalated rework costs, prolonged development timelines, and unnecessary expenditure of resources.

To illustrate the use of models as SECURITY MECHANISMS in software engineering, we will explore two examples retrieved from the interviews. In the first case, the interviewee was responsible for solving a system interoperability challenge. Specifically, the goal was to merge data from street cameras and a search

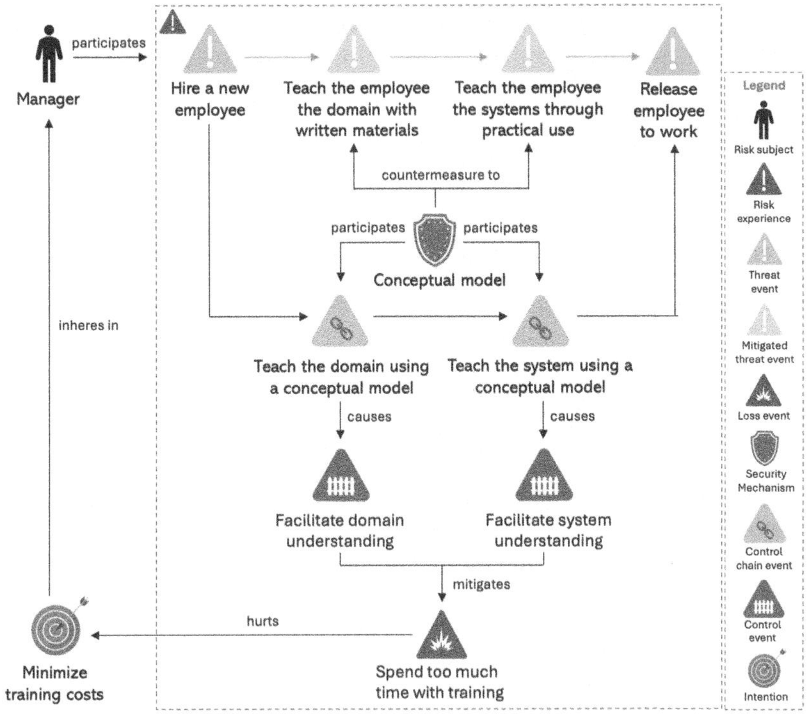

Fig. 1. Model as a SECURITY MECHANISM

engine to enable a police officer to run queries such as "How many people wearing black coats were in the downtown area of a given city on Christmas Eve morning?". However, his team encountered false agreement issues. An example of a false agreement in this context might occur when the police officer differentiates between coats and jackets when formulating search engine queries, while the camera system categorizes these terms identically in its recordings. To solve the problem, the interviewee developed conceptual models for both systems, which were crucial in mitigating **the risk of incorrect data integration**, thereby reducing rework, time, and costs in the process of systems interoperation.

In the second case, the interviewee was hired to document a system that was commercialized and maintained by a software company. The company used a UML model to facilitate system maintenance and to train novice programmers. On the one hand, the model potentially mitigated **the risks associated with direct changes to the code during system maintenance**. Such changes could lead to the risk of a trial-and-error cycle for developers, resulting in significant rework costs and a lengthy maintenance process. By using the model, developers could systematically analyze available options, perform testing, and select the optimal update path before implementing changes in the code. On the other hand, the model aided in mitigating risks that arose when a new programmer attempted to understand the system solely through practical use, such as **the risk of misunderstandings regarding data entry or relationships**. This method, while conventional, tends to be more time-consuming and error-prone than the alternative approach of studying the model and understanding how the system works analytically. Therefore, the model-driven approach was considered a safer and more cost-effective method in both situations.

3.2 Risks Created by Models

Upon reflection on the contexts in which models are employed, it becomes clear that certain risk events happen because of models. As the cases discussed in this session demonstrate, a model becomes a threat when it participates in risk events that could endanger other organizational assets.

Outdated models may be the source of several practical issues. For instance, the interviewee as cited in the preceding case, who developed a model for documenting an existing system, disclosed that the model became outdated upon his departure from the project. As depicted in Fig. 2, if programmers rely on this outdated model to learn about the domain covered by the system, they would be exposed to **the risk of an increased time to develop a new feature or fix a bug**, because of their inadequate understanding of the domain. As highlighted in [11], the fact that models become outdated and inconsistent with the underlying code represents a significant challenge in model-driven software development. For instance, using outdated models for maintenance exposes developers to **the risks of erroneous system updates, inadequate data integration, and incorrect domain interpretations**.

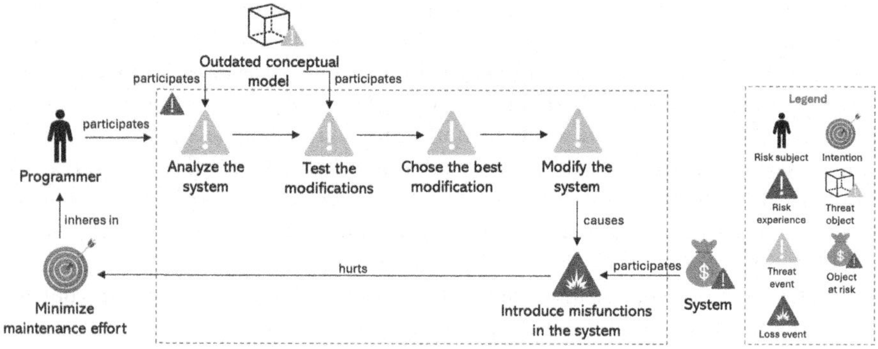

Fig. 2. Outdated Model as a THREAT OBJECT

Implementation dependent models can be understood as a threat because of the foundational design choices embedded in them. As noted in one interview, and supported by the study conducted in [26], models can pose risks when their development prioritizes implementation details over domain understanding. A participant in Petre's study [26] mentioned that a drawback of using

UML models emerges when their use diverts attention away from the primary focus on notational concerns. This participant said, "You start making decisions about how you are going to implement classes, inheritance, etc., when you should be mapping out what their business does". Adopting an implementation-dependent model may expose programmers to **the risk of limited system maintenance and reuse** because this model lacks critical information needed for system updates, data integration, and systems interoperability. They can also result in accrued domain debt [31] during the development process or in systems that are either unusable by end-users or only partially satisfy stakeholder needs.

Inadequately constructed or maintained models can present risks that extend beyond the realm of systems development and maintenance; they can also affect scenarios involving, for example, data integration and knowledge acquisition. Consequently, when examining a model that lacks comprehensiveness, a data analyst may be exposed to the **risk of erroneously integrating data** because of misinterpretations of data relationships. Similarly, a manager may be exposed to **the risk of acquiring inaccurate domain knowledge** by relying on a poorly structured model to understand a domain.

While poorly constructed and maintained models are typically the primary source of risks, *well-developed and maintained models* can also contribute to risks under certain circumstances. For instance, they may expose organizations to the **risk of disclosing sensitive information**. Consider once again the situation in which the interviewee was hired to document an existing system that was commercialized and maintained by a software company. He revealed a frustration stemming from the company's reluctance to use the model as a communication medium with external stakeholders. However, he acknowledged the company's concern that the potential value of the model in this regard was outweighed by the associated risk of potentially revealing sensitive information, such as how the system implemented the domain constraints. This scenario is illustrated in Fig. 3, which demonstrates how the model can assume the role of a THREAT OBJECT in contexts where the information it contains is an asset to the organization.

Finally, *the modeling process* itself can be considered a threat in certain circumstances. For instance, organizations can be exposed to **the risk of an extensive investment of time and resources in creating or maintaining a model** that could jeopardize a project's feasibility. This argument is often used by those who oppose the practical use of models, particularly in software development [11,19].

3.3 Models at Risk

Our analysis of the contexts in which models are used reveals situations in which the model itself is at risk. This stems from the fundamental premise that the model's value depends on its accuracy, completeness, usability, availability, and up-to-dateness while emerging from its use. Consequently, any circumstance that poses a risk to those qualities can be understood as a threat. When such risks

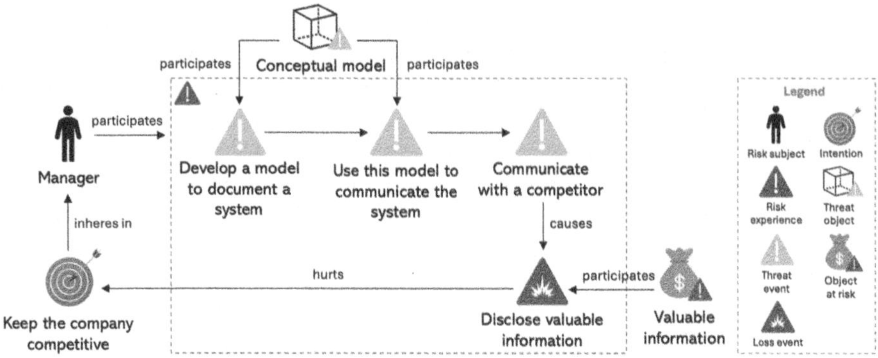

Fig. 3. Valuable Model as a THREAT OBJECT

materialize, the model will fail to meet the goals set by the user or the modeler. In the rest of this section, we discuss some risks models may be exposed to.

Neglecting model maintenance poses a significant threat that can lead to **the risk of losing control over model updates**. According to Valle [33], such occurrences are prevalent in projects undertaken by organizations lacking a modeling culture. Given that modeling expertise typically resides with the modeler, the model tends to be neglected once their involvement concludes. As a result, the model may become outdated, reducing its effectiveness and usefulness in various activities. As discussed in Sect. 3.2, outdated models have the potential to precipitate substantial practical issues, contributing to erroneous decision-making or flawed knowledge acquisition. Therefore, an outdated model has minimal utility.

The unavailability of domain experts is another potential threat that can affect model use and reuse. Figure 4 illustrates a scenario in which this situation was a threat to the project managed by one interviewee. He worked in the healthcare domain and had limited access to domain experts to validate the model. Because of this limitation, his team had to validate the model based on their own knowledge and information obtained from technical reports and the company's website. This led to uncertainties regarding the model's quality and usability, thereby increasing **the risk of the model being unsuitable for use or not being used at all**. The interviewee said: "I had to admit that if we would want to really know if the model would apply to the health care institutes, we should do some extra research." Because of this issue, his team started a new project centered on improving model communication and comprehension, in which they developed a thesaurus to define the domain's key concepts. While necessary and beneficial to facilitate later validation of the model with domain experts, this additional effort required extra investments.

Inexperienced modelers are a potential threat to models and modeling effectiveness, as their lack of modeling skills can increase **the risk of models becoming incorrect or difficult to understand and use**. For instance, two interviewees highlighted the importance of a good diagram layout to increase

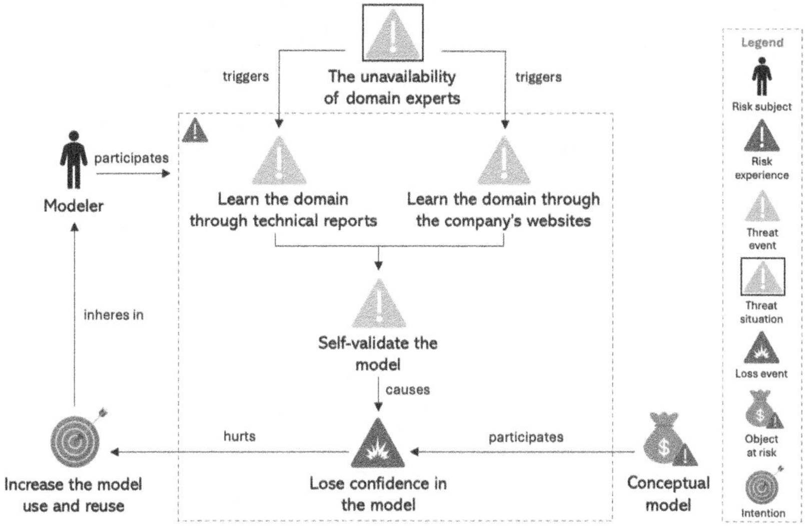

Fig. 4. Model as an OBJECT AT RISK with a THREATENING SITUATION example

the model's comprehensibility. One of them said: "For people that are not used to reading models, I am sure that the position, the presentation, and the colors are really important. Even so important that they will say that they like it very much or don't like it at all." Therefore, the modeler's lack of familiarity with diagram layout standards [4] can lead to the creation of visually inappropriate diagrams, resulting in models that are difficult to interpret and understand. For example, diagrams with crossed lines, curved lines, or overlapping classes.

Users with limited modeling experience can also pose a threat, as they may lack the necessary skills to understand and use the model appropriately. As a result, there is **the risk of models being used incorrectly or for purposes inappropriate to their design**. In such circumstances, users may perceive the model as inadequate to meet their needs, thereby diminishing its value. For instance, individuals attempting to use an ER model for process understanding may experience frustration and fail to achieve their goals, possibly leading to the conclusion that the model is ineffective. However, while an ER model excels at conceptualizing data-oriented domains, its applicability may be limited in more process-oriented domains, where a process model would be a more appropriate alternative. To mitigate the risks emerging from users lacking modeling skills, one interviewee said that he devised training courses to instruct users on understanding and utilizing the OntoUML model developed by his team. Moreover, they progressively engaged these users in the process of model creation.

Model complexity poses a threat when models reach a level of complexity where the effort required for their development or maintenance outweighs their potential benefits. As outlined in [12], the maintainability of UML class diagrams correlates directly with the complexity of the models. This complexity can be measured by factors including the total number of classes, associations, aggregations, attributes, and similar elements, as well as the organization and comprehensibility of these elements. One interviewee, who had developed a model to guide the development of a system, mentioned during the interview **the risk of the model becoming obsolete** if it became too complex because of the high cost and effort involved in using and maintaining it. He said: "The model generates a cost over time, right? In other words, when the model starts to get too complex, people do not want to work on the model. They think it will be more work to tinker with the model than to do it without a model."

Limited modeling tools also may pose a threat to models. For instance, if a modeler manually verifies the model, there is a possibility that certain errors will escape detection. Consequently, the resulting model derived from such verification would be exposed to **the risk of being semantically unreliable**, rendering any subsequent use prone to errors. As shown in Fig. 5, if manual verification becomes necessary due to the limitations of the modeling tool to support automated verification, it can be argued that the modeling tool contributed to the occurrence of verification errors by assuming the role of THREAT ENABLER. One interviewee mentioned another instance where modeling tools could also be considered a THREAT ENABLER facilitating the exposure of models to **the risk of having a poor layout**. According to him, "Tools are very bad at getting the model layout right. Getting the lines the way you want them. Doing all that seems like such a silly thing, something that the tools should have fixed by now, but they haven't." Beyond the tool itself, inappropriate modeling methods or languages can also act as potential THREAT ENABLERS facilitating the occurrence of certain Risk Events.

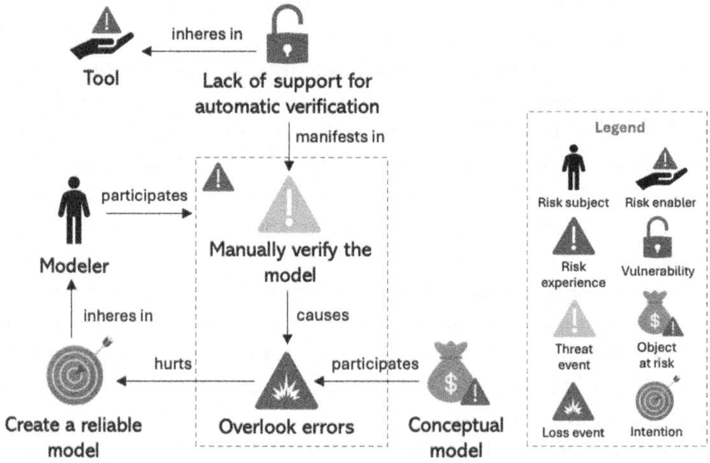

Fig. 5. Model as an OBJECT AT RISK with a RISK ENABLER example

The risks inherent in insufficient model assessment are a concern highlighted in [25]. Model and its constraints require validation[2] and verification[3] before implementation, as this practice can prevent numerous design flaws and implementation errors. Particularly for mission-critical systems, error detection during modeling can prevent failures that could lead to significant damage. While various approaches to model validation and verification attempt to mitigate the associated risks, these tasks remain inherently risky and dependent on external factors such as the model's quality, the modeler's expertise, and the modeling tools' capabilities.

4 Final Considerations

Managing models from a risk-centric perspective is essential for analyzing model benefits and costs, planning modeling investments, and integrating modeling practices across projects and organizations. Treating models as assets facilitates evaluating the cost-benefit ratio and return on modeling efforts for model-based solutions, guiding the allocation of resources to increase model value while reducing risk and cost. Accordingly, investments should focus on enabling models to serve as countermeasures against risk events and mitigating risks that could undermine the value of models or those introduced by the models themselves.

Building on the foundations of COVER [28] and ROSE [23], we identified that models can assume three distinct roles: as a SECURITY MECHANISM, where its capabilities, when manifested, prevent risk events from occurring (e.g., using data models to avoid system interoperability issues); as a THREAT OBJECT, where its capabilities, when manifested, lead to risk events that could jeopardize goals (e.g., using an outdated model to guide system maintenance, resulting in incorrect decision-making); and as an OBJECT AT RISK, where the model's vulnerabilities can be manifested in threatening situations composing the modeling process (e.g., using a modeling tool that lacks layout functionalities, which may result in poorly designed diagrams and reduce the model's comprehensibility). Figure 6 presents a non-exhaustive summary of examples from this paper, illustrating circumstances under which models assume these roles.

Thinking of a model as a SECURITY MECHANISM helps one to identify the various situations in which it is useful to mitigate risk and avoid additional costs. It encourages modelers and organizations to think about the model beyond its original development objectives. This broader perspective enhances the model's value and helps amortize its development costs. For instance, if developing a model M to achieve objective O1 costs C, and the same model M can later be used to achieve objectives O2 and O3 without further investment, the model's overall value increases while its development cost is diluted. Therefore, this strategy is also a valuable tool for convincing skeptical sponsors and coworkers of the merits of investing in modeling efforts.

[2] Have we built the correct model, according to the pre-specified requirements?.
[3] Have we built the model correctly, without syntactic and semantic flaws?.

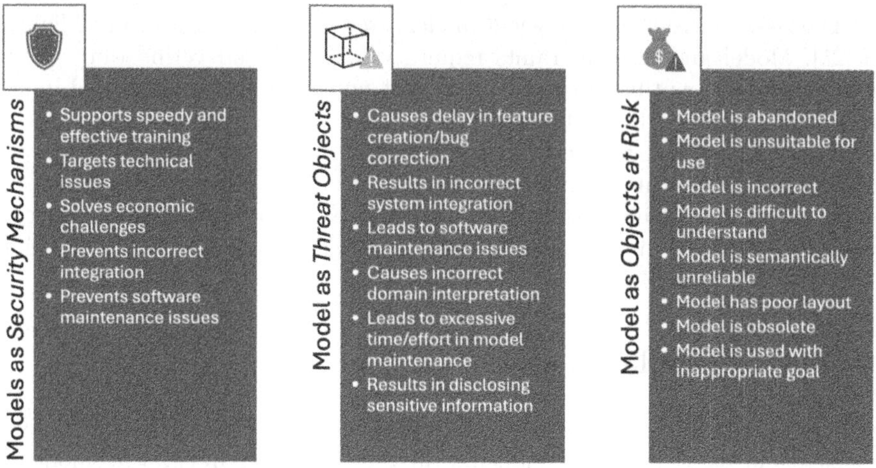

Fig. 6. Non-exhaustive summary of examples illustrating circumstances under which models play the role of Security Mechanisms, Threat Objects, or Objects at Risk

Posing models as THREAT OBJECTS allows one to identify circumstances in which they should be avoided because of their potential to create more problems than solutions. Models become threats when they participate in risk events that could endanger other organizational assets, such as systems or sensitive information. Although threats often stem from poorly constructed and maintained models, well-developed models can also pose risks under certain circumstances. Therefore, it is imperative to ensure that the appropriate model is used for the intended purpose.

Viewing a model as an OBJECT AT RISK facilitates the identification of threats that may undermine its value and increase its costs. These costs may arise either from complications inherent in the modeling process or from challenges encountered in using the model. The risk perspective provides valuable insights into the vulnerabilities of the model, the modeling process, the modeler, and the tools and languages used. By addressing these vulnerabilities, modelers and organizations can improve model quality and better plan modeling initiatives, by taking proactive measures to mitigate the risks of failure or adverse outcomes. Additionally, their identification could assist researchers in refining modeling languages, methods, tools, and training programs.

Through this study, we learned about several gaps that could be exploited to enhance the efficacy of modeling initiatives. Our risk-centered analysis underscored opportunities for improving modeling tools, languages, and methods that developers and researchers can explore. Additionally, it revealed alternative approaches and perspectives that organizations and modelers can consider when planning and executing their modeling efforts. We also recognize that risk analysis centered on models and the modeling process is not trivial, underscoring the potential value of a dedicated language to guide such analysis.

Future research could investigate how the risks identified in this study can be translated into quantifiable costs, and how the costs of domain models change over the life cycle of the model. In addition, an analysis of the risks and costs associated with not using these models in practical projects could be conducted. Finally, the formalization of a structured language tailored to modeling risk and security situations would be beneficial.

References

1. Atalaia, A., et al.: EURO-NMD registry: federated FAIR infrastructure, innovative technologies and concepts of a patient-centred registry for rare neuromuscular disorders. Orphanet J. Rare Dis. **19**(1), 66 (2024)
2. Aven, T., et al.: On the ontological status of the concept of risk. Saf. Sci. **49**(8–9), 1074–1079 (2011)
3. Batra, D., Marakas, G.M.: Conceptual data modelling in theory and practice. Eur. J. Inf. Syst. **4**(3), 185–193 (1995)
4. Bergström, G., et al.: Evaluating the layout quality of UML class diagrams using machine learning. J. Syst. Softw. **192**, 111413 (2022)
5. Boholm, Å., Corvellec, H.: A relational theory of risk. J. Risk Res. **14**(2), 175–190 (2011)
6. Canfora, G., et al.: How changes affect software entropy: an empirical study. Empir. Softw. Eng. **19**, 1–38 (2014)
7. Chen, P.: The entity-relationship model: toward a unified view of data. ACM Trans. Database Syst. **1**(1), 9–36 (1976)
8. Davies, I., et al.: How do practitioners use conceptual modeling in practice? Data Knowl. Eng. **58**(3), 358–380 (2006)
9. Dobing, B., Parsons, J.: How UML is used. Commun. ACM **49**(5), 109–113 (2006)
10. Fettke, P.: How conceptual modeling is used. Commun. Assoc. Inf. Syst. **25**(1), 43 (2009)
11. Forward, A., Lethbridge, T.C.: Problems and opportunities for model-centric versus code-centric software development: a survey of software professionals. In: MISE, pp. 27–32 (2008)
12. Genero, M., et al.: Assessing object-oriented conceptual models maintainability. In: Advanced Conceptual Modeling Techniques: ER 2002 Workshops, pp. 288–299. Springer (2003)
13. Guarino, N.: Formal ontology in information systems: Proceedings of the first international conference (FOIS'98), vol. 46. IOS press (1998)
14. Guizzardi, G., Proper, H.A.: On understanding the value of domain modeling. In: Proceedings of 15th VMBO (2021)
15. Guizzardi, G., et al.: Towards ontological foundations for conceptual modeling: the unified foundational ontology (UFO) story. Appl. Ontol. **10**(3–4), 259–271 (2015)
16. Halpin, T.A., Morgan, T.: Information Modeling and Relational Databases. Morgan Kaufmann, Burlington (2010)
17. Ho-Quang, T., et al.: Practices and perceptions of UML use in open source projects. In: International Conference on Software Engineering, pp. 203–212 (2017)
18. ISO: ISO 31073:2022 - Risk management – Vocabulary (2022)
19. Liddle, S.W.: Model-driven software development. In: Handbook of Conceptual Modeling: Theory, Practice, and Research Challenges, pp. 17–54. Springer (2011)

20. Lu, Y., Marais, K.B., Zhang, S.G.: Conceptual modeling of training and organizational risk dynamics. Procedia Eng. **80**, 313–328 (2014)
21. Lund, M.S., et al.: A guided tour of the CORAS method. Model-driven risk analysis: The CORAS approach, pp. 23–43 (2011)
22. National Institute of Standards and Technology (NIST): NIST Computer Security Resource Center (CSRC) Glossary (2025)
23. Oliveira, Í., et al.: An ontology of security from a risk treatment perspective. In: Conceptual Modeling. ER 2022, vol. 13607, pp. 365–379. Springer, Cham (2022)
24. OMG: UML 2.0 superstructure specification – Final adopted specification. Technical report ptc/03-08-02, Object Management Group (August 2003)
25. Pakalnickiene, E., Nemuraite, L.: Checking of conceptual models with integrity constraints. Inf. Technol. Control **36**(3) (2007)
26. Petre, M.: UML in practice. In: 2013 35th International Conference on Software Engineering (ICSE), pp. 722–731. IEEE (2013)
27. Proper, H.A., Guizzardi, G.: Modeling for enterprises; let's go to RoME ViA RiME. In: The Practice of Enterprise Modeling 2022 Forum, vol. 3327 (2022)
28. Sales, T.P., et al.: The common ontology of value and risk. In: Conceptual Modeling: 37th International Conference, ER, pp. 121–135. Springer (2018)
29. Sales, T.P., et al.: Ontological analysis and redesign of risk modeling in archimate. In: 22nd EDOC, pp. 154–163. IEEE (2018)
30. Strecker, S., Heise, D., Frank, U.: RiskM: a multi-perspective modeling method for IT risk assessment. Inf. Syst. Front. **13**, 595–611 (2011)
31. Störrle, H., Ciolkowski, M.: Stepping away from the lamppost: domain-level technical debt. In: 45th SEAA, pp. 325–332 (2019)
32. Valle, I., et al.: Unraveling the pain points of domain modeling, Preprint on SSNR at https://ssrn.com/abstract=5045287
33. Valle, I., et al.: What do I get from modeling? An empirical study on using structural conceptual models. In: Enterprise Design, Operations, and Computing, pp. 21–38. Springer, Cham (2023)
34. Verbruggen, C., Snoeck, M.: Practitioners' experiences with model-driven engineering: a meta-review. Softw. Syst. Model. **22**(1), 111–129 (2023)

Conceptual Modelling and Ontologies

Towards an Ontology of Type-Level Phenomena for System Modeling

Rodrigo Fernandes Calhau[1,2,3(✉)], João Paulo A. Almeida[2], Giancarlo Guizzardi[3], Raquel Hoffmann[4], Luís Ferreira Pires[3], James Logan[5], and João Rebelo Moreira[3]

[1] LEDS Research Group, Federal Institute of Espírito Santo, Vitória, Brazil
calhau@ifes.edu.br

[2] Ontology and Conceptual Modeling Research Group, Federal University of Espírito Santo, Vitória, Brazil
jpalmeida@ieee.org

[3] Semantics, Cybersecurity and Services (SCS) Group, University of Twente, Enschede, The Netherlands
{g.guizzardi,l.ferreirapires,j.luizrebelomoreira}@utwente.nl

[4] System Design and Management Graduate School, Keio University, Minato, Japan
raquel.hoffmann@keio.jp

[5] Ontogenesis Solutions, LLC, Herndon, USA
jlogan@ontogenesis-solutions.com

Abstract. In this paper, we propose a well-founded domain-independent system ontology based on the Unified Foundational Ontology (UFO) and the associated Multi-Level Theory (MLT). In our ontology, (composite) system types are designed by defining component types, connection types, and other type-level phenomena. We discuss how the proposed ontology can be used to guide the modeling of systems in specific domains. We position our work with respect to a number of ontologies in the literature that address system-related notions, as well as with respect to system modeling approaches.

Keywords: systems · ontologies · system modeling · conceptual modeling

1 Introduction

Modern system engineering faces increasing challenges due to the inherent complexity of emerging technologies (e.g., AI, IoT, Advanced Robotics, Software Agents) and their combination. Among these challenges, there is the growing demand for integrating fundamentally different types of components forming a system. For example, combining hardware and software components forming cyber-physical systems, or combining computational, organizational and human parts forming cyber-social systems.

Modern system engineering requires better theories, tools, techniques, and methods for their development and maintenance. In particular, it requires proper

J. Grabis et al. (Eds.): RCIS 2025, LNBIP 547, pp. 121–139, 2025.
https://doi.org/10.1007/978-3-031-92474-3_8

support for the *modeling* of: complex systems, their composition (their parts), their structure (the interrelations between the parts), as well as capabilities and vulnerabilities that emerge from the interaction between these parts.

Modeling in this context can be challenging for many reasons. For example, understanding system decomposition (i.e., how a system is decomposed into parts) is linked to complex phenomena such as (property and behavioral) emergence (e.g., a system with safe parts does not necessarily equal a safe system, since this depends on the kind of relationships and interactions among these parts).

Systems science researchers have studied these phenomena since General Systems Theory (GST) [1,8,12], identifying common principles applicable to different types of systems. However, these principles do not always result in practical guidelines that explicitly support complex system modeling.

Well-founded reference ontologies have been successfully employed to improve the expressivity, *domain appropriateness* and *comprehensibility appropriateness* of models and modeling languages in different domains (e.g. [4,26,32]). There is evidence that they can also play a fundamental role in providing explicit foundations for system modeling [39,42]. In fact, many ontologies have been proposed over the years to model different types of systems [3,5,33,44]. However, these tend to focus on issues that are specific to individual fields, and fail to address a broader system perspective.

In this paper, we propose a *well-founded multi-level system ontology*, and explore some of its *implications for ontology-based systems modeling*. Our ontology incorporates concepts from formal system ontology [12], GST [1,8] and Systems Engineering [21]. We build on a *system core ontology* based on GST to explain the phenomenon of capability emergence [13,14]. We extend that work to approach *type-level* phenomena in this domain. For example, (composite) system *types* are designed by defining component *types*, connection *types*, and other relevant concepts. These types are specified at varying levels of abstraction in the system engineering lifecycle, ranging from abstract functional specifications to detailed (physical) design.

We base our ontology on the Unified Foundational Ontology (UFO) [27]. UFO offers a set of domain-independent formal micro-theories that have been successfully applied in a variety of domains and tasks [25], in particular, in the revision of conceptual modeling constructs and languages. One of these micro-theories that has been incorporated into UFO is MLT (Multi-Level Theory) [16,17], which is used here to support the proper modeling of important reified types (e.g., system, component, and connection types). To represent the proposed systems ontology, we employ the UFO-based conceptual modeling language OntoUML [26].

We have two ultimate application goals in mind for our ontology, namely, to use it to: (i) guide the definition of domain-specific system models by combining, specializing, instantiating, and refining the axioms of the ontology; (ii) establish a semantic basis for system modeling language revision, integration and reengineering.

The remainder of the paper is further structured as follows: Sect. 2 provides germane background on system theory; Sect. 3 introduces UFO and a preliminary system core ontology [14], both of which are employed in Sect. 4 to develop the multi-level system ontology; Sect. 5 elaborates on how this ontology can be used to support system modeling; Sect. 6 discusses some related work; Finally, Sect. 7 presents final considerations.

2 Systems and Their Components

Despite the significant variation in definitions [21], the majority of authors writing on the topic use the notion of system in a sense that reflects the original formulation of General-System Theory (GST) [1,8,12], namely, taking the term to refer to a "complex" (or "structured") whole made up of "linked" (or interacting) parts. Accordingly, a system is viewed as a group of items that, via their relationships (or interactions), produce something novel, like behavior or emergent properties [8,21]. To support this understanding, Dori and Sillitto [21] provide the generic definition of a system as follows (based on a thorough literature review that examined over a hundred definitions): *"A system is a set (...) of parts (...) combined (...) in a way that creates (...) properties (...) not possessed (...) by the separated (...) parts (...)."* (we use ellipses ... here to simplify the presentation of the definition and omit the many alternative terms they have captured in their proposal.)

The generic definition provided by Dori and Sillitto [21] capture the main facets of the "system" concept. In addition to referring to the existence of properties that characterize the system *as a whole*, i.e., properties that go beyond the properties of parts in isolation, the definition also alludes to the inevitable *plurality* of things that in *combination* form a system ("a set of parts combined...").

Bunge [12] was one of the philosophers who dediated several writings to this notion. He used the term "global properties" to refer to system properties and divided them into: *emergent* and *resultant*. Properties that can be broken down, clarified, or reduced to the characteristics of a system's constituent parts are termed *resultant*. For instance, the simple sum of the mass of each constituent part of a system determines the system's total mass. Conversely, emergent properties are those that are not present in isolation in the separated parts, even though they are somehow grounded on the properties of parts. An example of emergent property is a ship's buoyancy. The buoyancy of the whole ship cannot be explained away by the buoyancy of the ship's individual components. In fact, most sections of a ship's steel hull are not buoyant by themselves. Buoyancy emerges from the combination of the various parts (including many non-buoyant ones).

Systems are created or developed to facilitate specific *functions* in their context. The notion of *function* is typically present in a teleological view of systems [37]. Accordingly, *functions* are demonstrated by a targeted outcome or behavior (action, process) [1,18]. The description of the *functions* of a system frequently begins by identifying *macro-functions* of the whole system, which are broken down into *sub-functions* (which are typically assigned to the components) [18,37], and so on recursively.

System structure, according to Bunge [12], is the collection of all connections among a system's components. He considers that the relationships between system elements can be either *"bonding"* or *"non-bonding"* [12]. Bonding relationships, also known as connections, are those in which one element influences and changes the other in some way [12]. The movement of information, material, or energy between elements can also be used to describe these relationships [12]. The *bonding structure* (or "configuration") [12] of a system is shaped by the set of *bonding* relationships that link all of its elements at a given time. The form, organization, or arrangement of the system is represented by the *bonding structure* in this context [21]. Bunge [12] argues that for an object to be regarded as a system, it must have a *bonding structure*. Similarly, Ackoff [1] considers that every component of every system must be directly or indirectly related to every other component. Therefore, there cannot be an isolated subset of elements in a system. Accordingly, elements that are wholly independent or unrelated should not be regarded as a system [37]. To put it another way, systems are *integral wholes* in the sense of Simons [36], so they need a unifying condition that identifies and binds their components to each another.

3 Foundational Baseline

Our reference ontology of systems is designed as a specialization of domain-independent notions defined by the Unified Foundational Ontology (UFO). Here we focus on those notions that are germane to the purpuses of this paper. For a full presentation of UFO, including axiomatization, we refer to [10,26,27]. UFO has two main taxonomies: one concerning types of *individuals* and another concerning types of *types*.

UFO categorizes individuals into *endurants*, *events*, and *situations*. *Endurants* are individuals that endure in time and may change qualitatively while keeping their identity. *Events* are individuals that 'occur' or 'happen' in time (e.g. 1970 Football World Cup Final in Mexico). *Situations* are particular configurations of a certain part of reality. There are two categories for *Endurants* in UFO: *objects* and *moments*. *Objects* are *endurants* that exist independently of another *individual* (e.g., Elvis Presley, the Earth, the Rolling Stones) . In contrast, *moments* (objectified or particularized properties) rely on another *individual* to exist (e.g., the yellowish color of the Sun, Martha's headache, Gerard's height). A particular kind of *moment* that can link two or more entities is a *relator* (e.g., Bonnie and Clyde's marriage, Paul McCartney and John Lennon's partnership). It reifies the moments that objects (the *relata*) have in the context of a relationship. An *intrinsic moment*, in contrast, depends existentially on a single individual. This includes both *modes* and *qualities*, the latter of which can be conceived of in a value space, such as color, height, weight, etc. *Modes* , in their turn, are intrinsic moments that are not directly projected into a value space. They encompass *dispositions*, which are moments that can manifest themselves through *events* in specific *situations*: examples include Pelé's ability to play football, which manifested in the 1970 World Cup; and Apollo 11's capability to travel in space manifested itself in the journey to the Moon in 1969.

UFO does not include the category *system*. It does, however, include the mereologically comparable notion of *functional complex*. Functional complexes are *objects* whose components play *functional roles* in its structure. In this case, whole and parts are bound by the *"component of"* relation [26]. A relation that is irreflexive, anti-symmetric and non-transitive [26], where non-transitivity means that transitivity may hold for certain cases but not for others. For example, the heart of the human being who plays the role of lead singer in a band, while a component of the person, is not a component of the band. In contrast, the piston of an engine that is a component of a car is also considered to be a component of that car. Moreover, like all other proper parthood relations, "component of" includes the *weak supplementation principle*, i.e., if x is a part of y then there must be a z mereologically disjoint of x, which is also part of y.

Differently from *individuals*, *types* are predicative entities which establish common properties for their instances. According to the ontological nature of their instances, the UFO taxonomy of types includes *endurant types* (e.g., "Person", "Marriage"), *event types* (e.g., "Wedding", "Football Match"), *object types* (e.g., "Organization", "Car"), *moment types* (e.g., "Age", "Enrollment"), etc. Additionally, UFO classifies types based on their modal metaproperties, dividing them into: *rigid*, *semi-rigid*, and *anti-rigid*. *Rigid* types, such as *kinds* and *subkinds* (e.g., "Person", "Car", "Pineapple"), are those that apply to their instances invariably over the course of their entire existence. An endurant must instantiate exactly one kind since kinds give their instances a principle of identity. A kind, e.g., "Dog", passes on an identity principle to its subkinds, e.g., "Poodle", "Collie". In turn, abstract rigid types called *Categories* capture common attributes of endurants of different kinds (e.g., "Physical Object").

Types that classify their instances contingently (or "dynamically") are known as "anti-rigid" types. These anti-rigid types, whose contingent classification conditions are relational (e.g., "Supplier", "Husband", "Employer", and "Customer") [26], are of particular interest to us here, since components of a system play distinct roles on it based on their relationships. To generalize roles and role mixins, we use 'role (mixin)'. An example of a role mixin is "Customer" that subsumes "Personal Customer" and "Organizational Customer". Role mixins are relational, anti-rigid types that represent common relational properties of objects that instantiate different kinds.

Types encompass *relations* among two or more *entities*. In UFO, a special type of relationship termed *material relation* requires the explicit representation of a link binding it to a corresponding relator type from that relation is derived. For example, the "married with" is derived from the relator type "Marriage"; the "enrolled in" relation from *enrollment*, and so on. At the instance level, a relationship exists between two or more *relata* (e.g. John is "married with" Mary) if there is a particular relator of a suitable type that binds them (in this case, a particular marriage). In material relations, the types of the associated relata are typically *roles* (in this case, the role "spouse").

In [14], some of us proposed a preliminary system core ontology based on UFO that discusses the primary distinctions among systems and their constituent parts from an individual standpoint. GST and Bunge's system ontology (in particular, his notions of composition, environment, structure, and mechanism model (CESM)) were used to identify the requirements for that ontology [12]. In order to distinguish *systems* from "simple" *objects*, that ontology takes into account the following factors: (i) the degree of complexity (number of components); (ii) the integrated structure; (iii) the complementarity and heterogeneity of the functional roles (and capabilities) of components; and (iv) the emergence of new properties and behaviors.

Figure 1 presents a fragment of the system ontology in OntoUML [26]. "System" is considered a *«category»* of *objects*, characterized by *system moments*, connected with external entities and with at least two proper parts (*components*). Note that the entities in the systems ontology specialize notions of UFO in its taxonomy of individuals, namely, the notions of "Object", "Moment" and "Situation".

Components are those objects that are part of a system. This includes entities usually referred to as *units*, *modules*, *interfaces*, etc. When a system is part of another system, it is termed a *subsystem* as depicted in Fig. 1. This enables the recursive (hierarchical) decomposition of systems. The relation "has proper part" is of type *«componentOf»* (and hence, non-transitive, irreflexive, asymmetric, and obeying weak supplementation). It is also a case of non-exclusive parthood (hence the UML aggregation symbol), i.e., components can be part of more than one system.

Structure is captured via *bonding* and *non-bonding relations*, following Bunge [12]. Bonding relations are *«material»* relations. They can be used to: connect a system or a system component with *external entities*, i.e., entities in the environment ("is externally connected with"); connect components within the same system ("is internally connected with"). Figure 1 only shows the reification of internal connection relators, but the same applies to external connections. Non-bonding relations, in turn, concerns system constraints and are represented using comparative relations from UFO (not illustrated in the figure).

The ontology addresses global properties [11] through *system moments*, as shown in Fig. 1. It adopts the distinction between *emergent* and *resultant properties* discussed in [11]. In this view, emergence is not a simple phenomenon that appears from one and only factor. For example, (emergent) *system moments* are said to stand in specific *system situations* ("stands in" relation). This is meant to represent the complexity of the settings of emergence, mainly: system constraints, connections, component aspects (i.e., capabilities). Additionally, *system moments* "emerge from" (component) *moments* (including emerging *dispositions*, *qualities*, *relators*). The notion of *function* is reflected in the ontology via the notion of *capability*. We see *functions* as specialized *capabilities* (*dispositions*) that bring benefits to other entities in the system context [21,30]. Both system

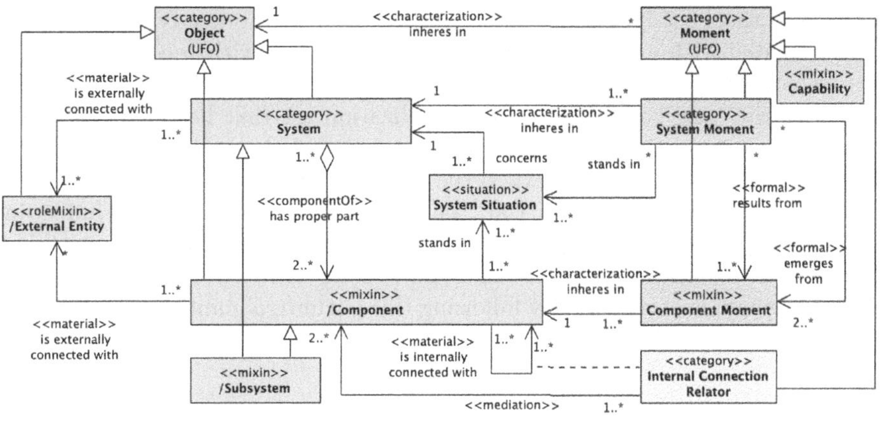

Fig. 1. Well-founded System Ontology proposed in [14]

and component moments may be capabilities too, i.e., one may talk about system capabilities and component capabilities.

4 Types of Systems, Components, and Connections

The ontology presented in the previous section captures invariant features of systems, their parts (components), and their properties (system and component moments). However, it is an ontology of individuals and, as such, it does not capture an important type-level dimension of this domain. If we understand the system design task as one of producing a system type, along with component types, connector types, etc., it becomes clear that the universe of inquiry of system design encompasses not only individuals but also types as first-class citizens.

System *types* are designed by establishing conditions for conforming systems, including their characterizing properties, the *types* of components they require, the *roles* components play in a conforming system, the *types* of connections they establish in the scope of the system, etc. For example, the "Porsche 356 Carrera GT Speedster" (a system type) features a "1.6-Liter Four-Cam Engine", which plays the role of "Main Engine" in cars of that type. Instances of this type can be connected with external entities (e.g., with a "Driver" or with a "Gas Pump Nozzle") through certain types of connections.

Figure 2 details the taxonomy of the system type ontology, focusing on the subtypes of *object type* we are concerned with: *system type*, *component type*, and their specializations. An important feature of the ontology is the use of high-order types, i.e., types whose instances are themselves types. For example, instances of "System Type" are types that specialize the base type "System" (e.g., "Car"). This modeling strategy is based on Odell's powertype pattern as incorporated in the OntoUML profile, following its axiomatization in the MLT multi-level theory [15]. The pattern is formed by a high-order type stereotyped *«type»* related with a base type through an association with specialized semantics stereotyped *«instantiation»*. In the model, instances of the high-order type specialize the base type through a generalization set that has that high-order type as powertype (which in UML is represented with a colon ':' followed by that high-order type's name next to the generalization set). So, when designing a particular system type, e.g., "Car", one would specialize "System", indicating "System Type" as a powertype of the specialization's generalization set. This makes "Car" an instance of "System Type" (along with any other subclasses in the same generalization set). By following this pattern, a number of rules can be enforced on the model [15].

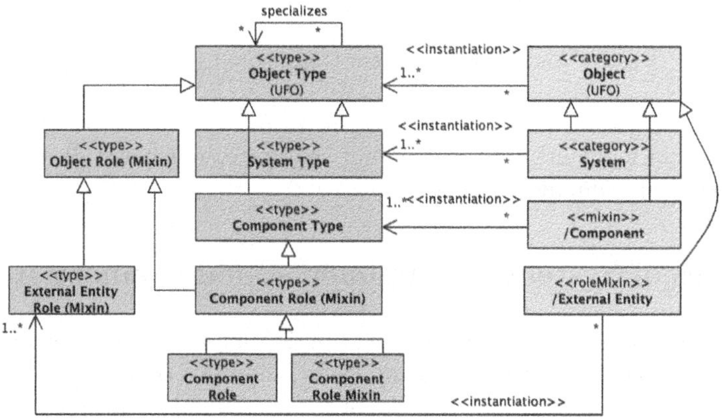

Fig. 2. System Type Ontology - Taxonomy

When defining a component type, such as a component role, one may establish the supertypes of that type in order to determine the conditions for componenthood. For example, a "Main Engine" (a "Component Role" in the scope of a "Car") is necessarily an "Engine" and hence, "Main Engine" specializes "Engine". More complex patterns can also be used to establish requirements for component role playing. For example, the role of "Blood Pump Component" in the human body's circulatory system may be played by a "Heart" or by an "ECMO Machine". In this case, we can establish a super type of "Heart" and "ECMO Machine" focusing specifically on the required function of that component (i.e., "Blood Pump"). The "Component Type" is the locus of any other requirements that are imposed to objects as components of a system.

Figure 3 shows a fragment of the system type ontology, focusing on system composition at the type level. A "System Type" "is formed by" at least one "Component Type" and "is externally bonded to" "External Entity Roles (Mixins)". A "Component Type" "is internally bonded to" at least one "Component Type", which may be "externally bonded to" zero or more "External Entity Roles (Mixins)".

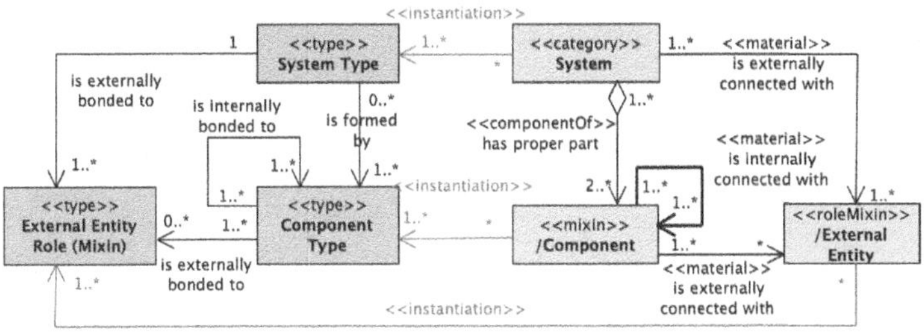

Fig. 3. System Type Ontology - System Composition and Structure

The association "is formed by" is an example of association with *deep characterization*: it affects not only the classification level below, in that instances of "System Type" must be related to at least an instance of "Component Type", but it also affects the instances of subtypes of "System" and of "Component". This is because "is formed by" is a *regularity property* [24]: it regulates "has proper part" by constraining the types that may be related through this association. Note that UML class diagrams do not have the expressive means to declare this kind of regulation in general. Hence, an additional constraint would be needed here. A syntactic constraint for models specializing the system type ontology is recommended: only types that are instances of "System Type" and "Component Type" (and hence specializations of "System" and "Component") may have associations that specialize "has proper part" in this ontology. The association "has proper part" is considered *abstract* in an analogous way to abstract classes; it can only be used to link instances of "System" and "Component" through associations that specialize it (e.g., "has main engine", "has chassis"). Specializations of "has proper part" determine the compatible system and component types (e.g., "Car" and "Engine", "Car" and "Chassis").

The same rules apply to "is internally bonded to", as it regulates "is internally connected with" (and analogoulsy "is externally bonded to"). The semantics of "is internally bonded to" and of "is internally connected with" require that only component roles that form the same system type may be internally bound to each other; likewise, a component can only be internally connected with other components when their roles are internally bonded to each other.

Figure 4 shows a fragment of the system type ontology focusing on *Connection Relators*, from which connections are derived. As illustrated, "Relator" (UFO) instantiates "Relator Type" (UFO). At the type level, "Relator Type" has mediation with "Object Role (Mixin)", which includes "Component Role (Mixin)" and "External Entity Role (Mixin)", as previously presented. At the individual level, Relator (UFO) includes both "External Connection Relator", which mediates the relationship between components and external entities, and "Internal Connection Relator", which mediates the relationship between components.

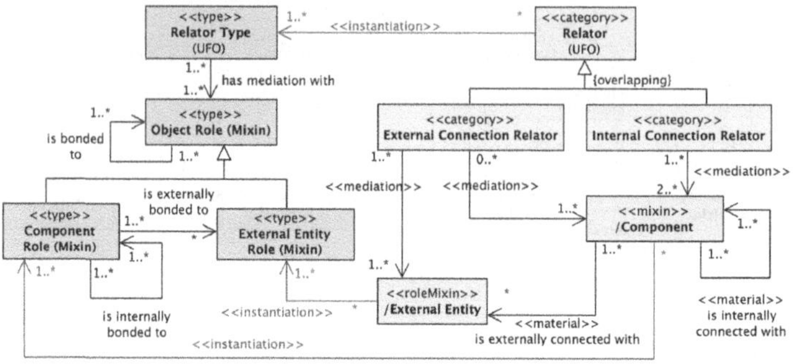

Fig. 4. System Type Ontology - Connection Type

5 Modeling Systems and Their Components

In order to show some consequences of the ontological analysis and the ontology proposed in the previous section, we present in this section an application related to system modeling. Such a case is specifically related to system components and the detailing of connections. We propose some (integrated) viewpoints for ontology-based conceptual modeling. Then, we present an example of modeling using these views in OntoUML. We propose a number of viewpoints in alignment with Bunge's CESM (Component, Environment, Structure, and Mechanism) model [11]. According to this model, a system has a *composition* (formed by the set of components of the system), an *environment* (formed by the external entities that interact with the system), a *structure* (formed by the relationships/connections between its parts), and a *mechanism* (formed by the events/processes which happened in the system). Besides the viewpoints from CESM, we also included the *property* and *emergence* viewpoints of the system.

A ***system environment viewpoint*** reveals the types of external connections between the system and *external entities*. It identifies *system roles* and *external entity roles* that are played by the system and external entities, respectively, when they are connected. Supertypes of system role and external entity roles reveal the types of objects involved. This viewpoint may involve the reification of external connection relators. As mentioned, the ***system composition viewpoint*** focuses on the internal perspective at the component level, detailing the types of system components that form a system type. The viewpoint identifies component types (often roles) in the context of the system and specifies the types of objects that may be components of instances of the system type. A ***system structure viewpoint*** focuses on the connections between the components of a system. The main focus is on the internal perspective, however, the external perspective can also be considered. All component types must be connected directly or indirectly. Internal connections can only occur between component types of the same system type. External connections only occur with entities external to the system. A ***system property viewpoint*** reveals system and component moments such as qualities and dispositions (e.g., capabilities and vulnerabilities). A ***system mechanism viewpoint*** reveals system and component events as a result of the manifestation of their properties (dispositions). Finally, a ***system emergence viewpoint*** reveals how system moment types and/or system connection relator types arise from or are derived from component moment types and/or component connection relator types.

To illustrate the application of the system ontology and proposed modeling viewpoints, we consider here a motorcycle as an example of a cyber-physical system, i.e., a physical system that can include cybernetic components. This is a recurrent example in the model-based systems engineering literature [9,23,43,45]. To illustrate the proposal, we will consider the motorcycle-biker system as part of an organization that needs to track and control a fleet of motorcycles. To do this, the organization needs a *fleet tracker* as its component. So, as a result, the motorcycle must have capabilities that allow it to be tracked by the fleet tracker.

This scenario can be represented by the *system environment viewpoint* as depicted in Fig. 5. In this case, the environment viewpoint can be understood from two different perspectives: (a) considering the motorcycle and the fleet tracker as parts of an organization system;(b) from the perspective of the motorcycle as a system, without considering the organization. The distinctions from system type ontology help clarifying these different perspective. In the first one, illustrated in Fig. 5 (a), 'Organization' specializes 'System' and both 'Motorcycle' and 'Fleet Tracker' specialize 'Component' as parts of that organization. In this case, the connection between an instance of 'Motorcycle' and the 'Fleet Tracker' is internal, with 'Tracking Connection Relator' specializing 'Internal Connection Relator' (not depicted). In the second perspective, in Fig. 5 (b), 'Motorcycle' specializes 'System' and 'Fleet Tracker' specializes 'External Entity'. In this case, the connection between a motorcycle and the fleet tracker is external, with 'Tracking Connection Relator' specializing 'External Connection Relator'. In short, a motorcycle is a subsystem because it is both a system and a component. 'Fleet Tracker' is both an 'External Entity' and a 'Component' depending on the perspective.

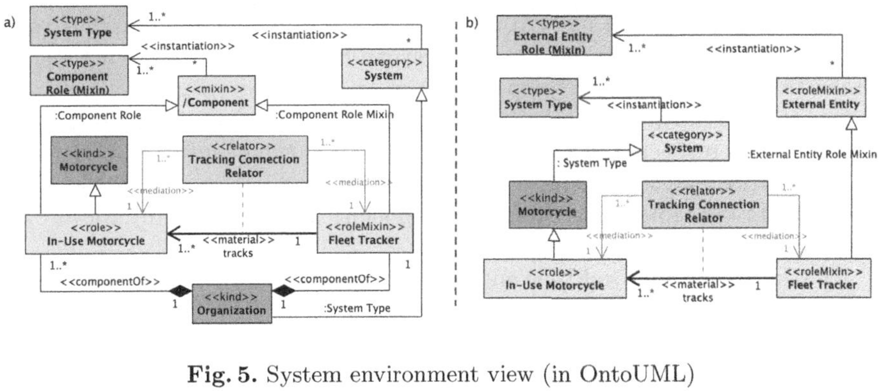

Fig. 5. System environment view (in OntoUML)

As shown, in both cases, the connections are mediated by a relator. In particular, the connection between the 'Fleet Tracker' and the 'In-Use Motorcycle' is mediated by the 'Tracking Connection Relator', from which the 'tracks' relation is derived. Reification allows us to treat the (external/internal) connection relator as an endurant, ascribing to it temporal properties (it is created/destroyed at certain points in time) and qualities (representing its characteristics). Moreover, these relators can serve as the basis for further decomposition (when parts of the connection relator are revealed) and can enable the detailing of emergent phenomena (when component connection relators are considered).

To allow the motorcycle to be tracked by the Fleet Tracker, the motorcycle must have appropriate components that enable this function. Figure 6 identifies the motorcycle component roles, forming the 'Motorcycle' system type (a kind). The composition of the motorcycle is divided into two levels: motorcycle subsystems (braking, tracking, acceleration, steering, etc.) and components, which are parts of these subsystems. The tracking subsystem is formed by 'Microcontroller Component' and 'GPS Tracker Component'. As depicted, there is an explicit separation between component types (in this case component roles) and their allowed types (Handlebar, Wheel, Frame, etc.). This allows us to identify component roles that may be played by objects of the same type (the case of 'Wheel' and 'Motorcycle Rear Wheel Component' and 'Motorcycle Front Wheel Component'.) A particular 'Wheel' is only contingently a motorcycle component. This separation of object type and component type allows us to stipulate requirements for objects as components in the scope of a certain system and in a certain role. For example, different tire pressures may be required for wheels functioning as 'Motorcycle Rear Wheel Component' or 'Motorcycle Front Wheel Component'. This viewpoint is also the locus for the specification of certain constraints, e.g., it is not the case that the same wheel may play both roles simultaneously in a motorcycle. Nevertheless, since roles classify objects dynamically, the possibility of replacement of components exists, unless constraints are specified. For example, one could rule out the replacement of the 'Motorcycle Frame Component' by deeming it an 'essential part' [26]. Note that we have omitted the powertypes (that were revealed in Fig. 5) for the sake of readability.

When certain component roles can be played by objects of different kinds, the OntoUML role mixin pattern can be employed [26]. This can be illustrated in the motorcycle example. The motorcycle can have up to three distinct storage compartments: side storage compartment, rear storage compartment, and seat storage compartment. These subtypes of the storage compartment role mixin can be played, respectively, by instances of the rigid types: side bag, case (box), and seat. Note that a seat can play two roles: that of a Motorcycle Seat Component and of a Motorcycle Storage Component (more specifically Motorcycle Seat Storage Component).

The separation of object type and component type allows us to identify object types whose specializations form distinct system types. For example, an engine may play the role of 'Motorcycle Engine Component' or 'Tricycle Engine Component'. A constraint in this model is required to state that the same engine cannot

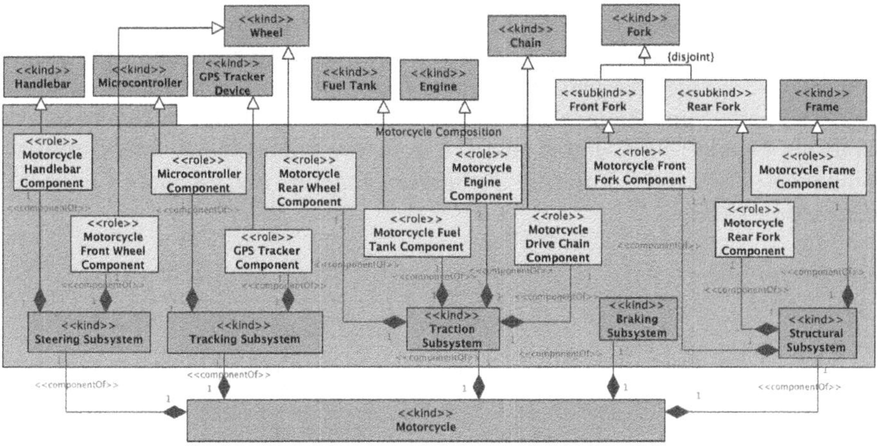

Fig. 6. System component view

play both roles simultaneously. The tricycle shares the same *type of engine* as the motorcycle. Given the dynamic aspect of roles, it is even possible for an engine to be swapped from a motorcycle to a tricycle (and vice versa).

Still considering the tracking issue, to allow the motorcycle to be tracked, it must have appropriate internal connections that enable such capability. Figure 7 shows a *system structure view*. Each of the motorcycle components is connected with at least one different component. Furthermore, the connection with the Fleet Tracker is now refined, detailing how they happen at the component level. Two connections are identified: the Fleet Tracker "controls remotely" the motorcycle tracking subsystem (to block the motorcycle's engine, if necessary), and the 'GPS Tracker Component' "informs the location" of the motorcycle to the Fleet Tracker controls it. Although the example focuses on connections between components, the subsystems presented above are also connected (through the connections between their components).

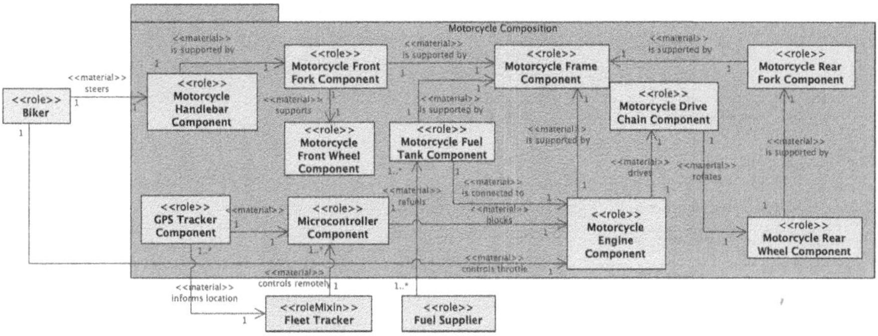

Fig. 7. System structure view

To understand the emergence of the tracking capability, we leverage the reification of connections in the emergence view, in order to show how system-level connections emerge. We first represent *connection relators*, as shown in Fig. 8, for each of the connections with the Fleet Tracker. As shown in the figure, the 'Tracking Connection Relator' emerges from two connection relators from the component level: *'GPS Tracking Connection Relator'* and *'Remote Controlling Connection Relator'*.

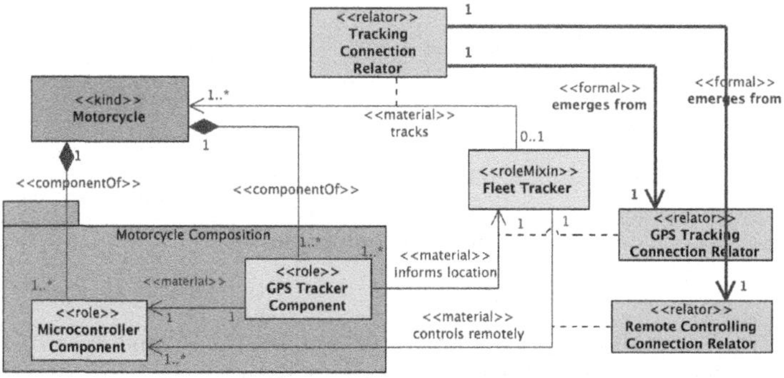

Fig. 8. System emergence view (connections)

Along with the emergence of connections presented above, there is also the emergence of capabilities. This is illustrated in Fig. 9. As depicted, the motorcycle has the susceptibility to be tracked (a capability). This capability emerges from the capabilities of its parts, such as 'Motorcycle Controlling Capability' and 'Location Sending Capability'. On the other hand, the Fleet Tracker has the following capabilities: 'Remote Motorcycle Locating Capability' and 'Remote Motorcycle Controlling Capability'. As a result of the motorcycle's capability to be tracked and the capabilities of the Fleet Tracker, the 'Tracking Capability' of the whole system (the organization) emerges.

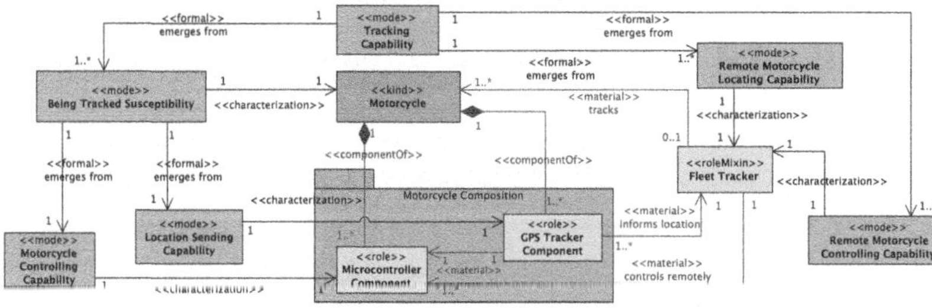

Fig. 9. System emergence view (capabilities)

6 Related Work

A number of ontologies have addressed systems and their components and connections. We review the most important of the proposals in the sequel, positioning them with respect to their account of the various notions we have discussed in this paper (composition, functional decomposition, structure), especially in the type perspective. We focus on system ontologies that approach systems in a general sense, based on system science.

Based on Bunge's work, Wand et al. [40] proposed the Bunge-Wand-Weber (BWW) ontology. BWW has been an influential conceptual framework in the IS field, serving as a basis for a number of conceptual modeling efforts in the area [29]. BWW considers important distinctions (system, subsystem, etc) and also composition and structure. However, it does not address some aspects discussed here. Moreover, there is no way of capturing the *system type* perspective (including system types, component types, and component roles) in that framework. Doing so requires treating types and type relations as first-class citizens (higher-order types and higher-order relations), something which is not available to the BWW framework.

Lukyanenko et al. [29] present a new ontology called Bunge's Systemist Ontology (BSO). BSO is one of the most related works, since is also based on Bunge's work. BSO focuses on the concept of a *system*, which is the elementary unit of the ontology [29]. Compared with BWW, BSO further details the emergence phenomenon. BSO extends Bunge's CESM model to support conceptual modeling [29], adding emergent properties. This is similar to this work, which also added a viewpoint regarding emergence. One of the differences between this work and BSO is that here we consider multi-level distinctions such as system type and component type.

OntoCAPE is an ontology developed in the context of an open initiative that technically supports, through the use of standards, the problem of interoperability in engineering. OntoCAPE defines in its ontology some established concepts which are strongly based on GST concepts, such as: *system*, *subsystem*, *super-system*, *element*, *property*, *connection*, *requirement*, *function*, and *behavior* [31]. However, it lacks support for representing a *system-type* perspective, the phenomenon of emergence, component roles and system roles.

In [21] and [20], the authors propose what they call an "ontological framework", in which systems are modeled using the Object-Process Methodology (OPM) [19]. The OPM notation is based on an ontology that allows the representation of several important aspects related to systems such as whole, part, system, subsystem, and etc. Their approach lacks the support for modeling: the phenomenon of emergence, the system-type perspective, as well as a notion of component role.

Another relevant ontology in this context is the Industrial Ontology Foundry (IOF) [38]. IOF, like our ontology, is based on a foundational ontology, namely, BFO [34]. By leveraging BFO's support, it provides well-founded definition for concepts such as system, engineered system, mechanical system, function, capability, component, and component role. IOF is one of a few works considering capabilities and roles. However, it lacks support for the representation of emergence, as well as for explicitly representing connections as relators, as there is no equivalent notion to relator in BFO. The type perspective is not considered as well.

From the standpoint of system structure, the majority of the works take into account a type of structural relationship between systems or components through the following definitions: *relation* [22], *structural relation* [33], *connector* [19], *connection* [28,31], *binding mutual* [40], *link* [5], or *interaction* [2,7]. Concepts about mediators of some sort are also taken into account by other ontologies as interfaces, ports, or connection points [6,28,35,41]. Most ontologies do not explicitly consider external connections (i.e., between a component and an external entity). Generally, this kind of relationship is represented only between systems and not between components of a system.

7 Final Remarks

In this paper, we have present a well-founded multi-level systems ontology covering various aspects of systems, such as their environment, composition, functions, component roles, among other elements. The ontology is grounded on UFO [26] and was proposed based on GST principles, allowing the broad representation of systems including their composition, function, structure, and properties.

A key aspect in UFO (including MLT) is the modeling of types, doing justice to the distinction among different types of types (e.g., kinds, roles, role mixins), as well as supporting higher-order types. With the support of these theories, we could account not just for systems and components, but also for system types and component types, as well as their relations.

Important domain information often remains tacit in process standards, data item descriptions, and representation approaches used in system engineering. Transitioning from tacit to explicit information in this area is challenging, but key to advancing the theory and practice in the field. The ontological analysis and the resulting artifacts proposed here contribute to addressing this problem by making system concepts and constraints explicit. As a modeling contribution, we propose *modeling views and patterns* extracted from this ontology, and which can be reused and extended for support modeling in application domains.

In the future, we intend to extend this work to also consider behavioral aspects addressing the *Mechanism Viewpoint*. We also intend to provide clear guidelines for understanding and modeling system (de)composition in existing modeling notations (e.g., UML, SysML), with special attention to the emergence of desired capabilities.

A complete formalization of the proposed ontology in accordance with the axiomatization of UFO ([27]) is planned as future work. This formalization will also enable us to support automated verification, validation and reasoning over system models extending this ontology. In system modeling, a number of constraints must be attended in order to guarantee the consistency of complex system specifications (e.g., if a system is connected to an external entity, at least one of its components must also be connected to it; certain component connections demand components of certain types). This is especially important for creating constraints for some classifications of systems, due to their nature. For example, cyber-physical systems must be formed only by cybernetic and physical components, and must not be formed by social components; socio-technical systems must necessarily have human components (people); physical systems cannot emerge from social components; or, components of physical systems must not have social connections but only physical connections. There are other more general formalizations that can be useful in modeling all types of systems, generating important constraints. For example, a system cannot emerge from itself or from the system of which it is a part; the emergence relationship cannot be cyclic; and so on.

Acknowledgment. This study was supported in part by CNPq (443130/2023-0, 313412/2023-5), FAPES (1022/2022) and the DSYNE INTPART network (Research Council of Norway project number 309404).

References

1. Ackoff, R.L.: Towards a system of systems concepts. Manag. Sci. **17**(11), 661–671 (1971)
2. Akkiyat, I., Souissi, N.: Process meta model extended for the improvement cycle in a SoS context. In: 2019 4th World Conference on Complex Systems (WCCS). IEEE (April 2019)
3. Aquino, N.M.R.: A smart assessment of business processes for enterprises decision support. Université de Lorraine (2021). ph.D. thesis
4. Azevedo, C., Almeida, J.P.A., van Sinderen, M., Ferreira Pires, L.: Towards capturing strategic planning in EA. In: 2015 IEEE 19th International Enterprise Distributed Object Computing Conference, pp. 159–133. IEEE (2015). https://doi.org/10.1109/EDOC.2015.31
5. Baek, Y.M., et al.: A meta-model for representing system-of-systems ontologies. In: 6th IEEE/ACM International Workshop on Software Engineering for Systems-of-Systems, SESoS@ICSE 2018, Gothenburg, Sweden, 29 May 2018, pp. 1–7. ACM (2018)
6. Bakirtzis, G., et al.: An ontological metamodel for cyber-physical system safety, security, and resilience coengineering. Softw. Syst. Model. **21**(1), 113–137 (2021)
7. Banzhaf, W., et al.: Defining and simulating open-ended novelty: requirements, guidelines, and challenges. Theory Biosci. **135**(3), 131–161 (2016)
8. von Bertalanffy, L.: General System Theory. George Braziller, New York (1968)
9. Biggs, G., et al.: Integrating safety and reliability analysis into mbse: overview of the new proposed omg standard **28**(1), 1322–1336 (2018)
10. Botti Benevides, A., Bourguet, J.R., Guizzardi, G., Peñaloza, R., Almeida, J.P.A.: Representing a reference foundational ontology of events in sroiq. Appl. Ontol. **14**(3), 293–334 (2019)
11. Bunge, M.: Treatise on Basic Philosophy: The Furniture of the World. Ontology I. Reidel Pub. (1977)
12. Bunge, M.: Treatise on Basic Philosophy. Ontology II: A World of Systems. Springer Netherlands, Dordrecht, Netherlands (1979)

13. Calhau, R.F., et al.: Exploring system behavior in a system ontology. In: 2023 ER Forum, Demo and Posters, ER Forum 2023, co-located with 42th International Conference on Conceptual Modeling, ER 2023 (2023)
14. Calhau, R.F., et al.: A system core ontology for capability emergence modeling. In: Enterprise Design, Operations, and Computing, pp. 3–20. Springer Nature, Switzerland (October 2023)
15. Carvalho, V., et al.: Using a Well-Founded Multi-Level Theory to Support the Analysis and Representation of the Powertype Pattern in Conceptual Modeling. In: Proceedings of the CAiSE 2016. LNCS, vol. 9694, pp. 309–324. Springer (2016)
16. Carvalho, V., et al.: Multi-level ontology-based conceptual modeling. Data Knowl. Eng. **109**, 3–24 (2017)
17. Carvalho, V., et al.: Toward a well-founded theory for multi-level conceptual modeling. Softw. Syst. Model. **17**, 205–231 (2018)
18. Compagno, F., et al.: Towards a formal ontology of engineering functions, behaviours, and capabilities. Semantic Web Journal (forthcoming)
19. Dori, D.: Object-process methodology. In: Encyclopedia of Knowledge Management, Second Edition, pp. 1208–1220. IGI Global (2011)
20. Dori, D.: Object-process methodology for structure-behavior co-design. In: Handbook of Conceptual Modeling, pp. 209–258. Springer Berlin Heidelberg (2011)
21. Dori, D., et al.: What is a system? An ontological framework. Syst. Eng. **20**(3), 207–219 (2017)
22. Duan, H., et al.: The development of a digital twin concept system. Digital Twin **2**, 10 (2022)
23. Eckl, C., et al.: Using the "model-based systems engineering" technique for multidisciplinary system development. In: ISPE CE, pp. 100–109 (2015)
24. Guizzardi, G., et al.: Towards an ontological analysis of powertypes. In: Proceedings of the Joint Ontology Workshops 2015 Episode 1: The Argentine Winter of Ontology. CEUR Workshop Proceedings, vol. 1517. CEUR-WS.org (2015)
25. Guizzardi, G., et al.: Towards ontological foundations for conceptual modeling: the unified foundational ontology (UFO) story. Appl. Ontol. (Online) **10**, 259–271 (2015)
26. Guizzardi, G.: Ontological Foundations for Structural Conceptual Models. No. 15 in Telematica Instituut Fundamental Research Series, Telematica Instituut, The Netherlands (2005)
27. Guizzardi, G., et al.: UFO: unified foundational ontology. Appl. Ontol. **17**(1), 167–210 (2022)
28. Lefrançois, M.: Planned ETSI SAREF Extensions based on the W3C&OGC SOSA/SSN-compatible SEAS ontology paaerns. In: Workshop on Semantic Interoperability and Standardization in the IoT, SIS-IoT, p. 11p. Amsterdam, Netherlands (September 2017)
29. Lukyanenko, R., Storey, V.C., Pastor, O.: System: a core conceptual modeling construct for capturing complexity. Data Knowl. Eng. **141**, 102062 (2022)
30. Merrell, E., et al.: Capabilities (2022)
31. Morbach, J., et al.: Onto CAPE 2.0 —a (re-)usable ontology for computer-aided process engineering. In: Computer Aided Chemical Engineering, pp. 991–996. Elsevier (2008)
32. Nardi, J.C., et al.: A commitment-based reference ontology for services. Inf. Syst. **54**, 263–288 (2015). https://doi.org/10.1016/j.is.2015.01.012
33. Naudet, Y., et al.: Towards a systemic formalisation of interoperability. Comput. Ind. **61**(2), 176–185 (2010)

34. Otte, J.N., et al.: BFO: basic formal ontology1. Appl. Ontol. **17**(1), 17–43 (2022)
35. Pfister, F., et al.: A design pattern meta model for systems engineering. IFAC Proc. Vol. **44**(1), 11967–11972 (2011)
36. Simons, P.: Parts: A Study in Ontology. Oxford University Press, Oxford (August 2000)
37. Skyttner, L.: General systems theory: ideas & applications. World Scientific (2001)
38. Smith, B., et al.: A first-order logic formalization of the industrial ontologies foundry signature using basic formal ontology. In: Proceedings of the Joint Ontology Workshops (JOWO) 2019 Episode V:. CEUR Workshop Proceedings, vol. 2518. CEUR-WS.org (2019)
39. Van Ruijven, L.: Ontology and model-based systems engineering. Procedia Comput. Sci. **8**, 194–200 (2012)
40. Wand, Y., et al.: On the deep structure of information systems. Inf. Syst. J. **5**(3), 203–223 (1995)
41. Yang, D., et al.: Development of a product configuration system with an ontology-based approach. Comput. Aided Des. **40**(8), 863–878 (2008)
42. Yang, L., Cormican, K., Yu, M.: Ontology-based systems engineering: a state-of-the-art review. Comput. Ind. **111**, 148–171 (2019)
43. Yildirim, U., Campean, F.: Functional modelling of complex multi-disciplinary systems using the enhanced sequence diagram. Res. Eng. Des. **31**(4), 429–448 (2020)
44. Yilma, B.A., et al.: Systemic formalisation of cyber-physical-social system (CPSS): a systematic literature review. Comput. Ind. **129**, 103458 (2021)
45. Zheng, P.: One-of-a-Kind Product Development for Mass Personalisation in a Cloud-Based Environment. Ph.D. thesis, ResearchSpace@ Auckland (2017)

Low-Code Browser Front-End Automation Using RDF Graphs and a Domain-Specific Language for UX Representation

Ştefan Uifălean and Robert Andrei Buchmann(✉)

OMILAB@Faculty of Economics and Business Administration, Babeş-Bolyai University, Cluj-Napoca, Romania
{stefan.uifalean,robert.buchmann}@econ.ubbcluj.ro

Abstract. Software development companies are tackling front-end automation primarily from a software testing perspective or a robotic process automation perspective. Industry solutions still lack a knowledge-focused approach to managing automatable front-end workflows. This paper reports on a Design Science Research treatment employing semantic graphs derived from UX (user experience) flow models repurposed for low-code browser automation.

The reported work aims to bridge the interoperability gap between UX design, front-end automation and Knowledge Management, using semantic graphs as means of externalization/combination. We developed a visual designer that captures the human-computer interaction flow in procedural UX descriptions with the help of a domain-specific modeling language. The tool further renders visual models executable through a platform-specific automation script, while at the same time maintaining them as RDF graphs accumulating in a knowledge repository. Execution creates a browser instance that replicates user actions within a limited context. The engine relies on Selenium WebDriver, a popular Test Automation framework, to generate automation out of the procedural UX knowledge captured in visual models. As a technical evaluation we demonstrate the feasibility of a low-code automation scenario using the popular browser-based mail client Microsoft Outlook.

Keywords: Front-end Automation · Low-Code Development · Knowledge Graphs · Model-Driven Engineering · Domain-specific Modeling

1 Introduction

Software users interact with apps through the Graphical User Interface (GUI), by executing action flows that build up into the so-called user experience (UX), together with other aspects such as look-and-feel, layout etc. UX scenarios are orchestrations of front-end elements access, are typically planned as mockup workflows in the design phase and tested to satisfy expected results during the Software Development Life Cycle (SDLC). Test Automation employs end-to-end usage flow scenarios for regression testing, by mimicking UX steps programmatically to ensure time-efficient coverage. Furthermore,

J. Grabis et al. (Eds.): RCIS 2025, LNBIP 547, pp. 140–155, 2025.
https://doi.org/10.1007/978-3-031-92474-3_9

GUI Automation inspired software platforms that autonomously interact with GUIs of existing software based on scripted behavior. Robotic Process Automation (RPA) [1] enables modern organizations to automate manual front-end processes across multiple software systems – e.g. a success story is that of Uber constructing UIPath [2] bots to take the task of invoice processing off the shoulders of employees [3].

GUI Automation works with the core UI interface-building ingredients that facilitate both front-end Test Automation and RPA, but a holistic knowledge-based view is missing from available tooling. They are more interested in building the automation artifacts/scripts rather than facilitating knowledge capture, traceability and semantic contextualization capabilities for the automated workflows – e.g. it is challenging to trace GUI/process automation deployments back to the original UX flow.

This paper reports on a Design Science Research (DSR) project initiated in a Software Outsourcing enterprise seeking an experimental, knowledge-based process automation solution for repetitive tasks over their internal systems. The solution must enable a knowledge management capability for maintaining design decisions and a semantic context for the automated workflows and for this reason we adopt a domain-specific modeling language (DSML) as the entry point for the captured knowledge. We then adopt an Open-Source software testing framework (Selenium[1]) as the execution engine for UX flows described with this DSML and facilitate the model interchange with the help of RDF graphs.

An early draft of the DSML – a BPMN adaptation using UX-specific task taxonomies - was first introduced in a workshop scope in [4] and the current paper reports on the model-driven engineering efforts developed based on its constructs. The metamodel bridges a gap between RPA process designs and UX designs, by conceptualizing and semantically linking UI wireframes to the associated usage flow and data resources to be accessed by an automation bot. RDF graphs exported from the diagrammatic designs allow for navigating alternate paths of the usage workflow as they enable users to automate interactions with a Web browser. By resorting to the ADOxx metamodeling platform, we deploy a tool for modeling such UX flows in ways that inform the browser automation engine, which translates graphs into platform-specific (browser-specific) scripts executed through the Selenium framework for Web Automation. The modeling tool and the engine enable modelers to create and run simple automation tasks supporting clicks, keystrokes and parameters referencing external data dependencies. This approach repurposes a workhorse SDLC method popular in Test Automation to interoperate with UX flow diagrams using semantic graphs as a mediator.

The remainder of the article is organized as follows: Sect. 2 places the work in the context of related research. Section 3 formulates the problem in the DSR frame. Section 4 describes design and development decisions for the modeling tool and Sect. 6 does the same for the front-end browser automation engine. Section 5 discusses a demonstration scenario. Section 7 is a discussion reflecting on the proposed treatment. Conclusions end this paper.

[1] https://www.selenium.dev/.

2 Related Work

The use of semantic graphs and knowledge representation for automation predates the advent of knowledge graphs [5] however the conceptualization of UI building blocks with the purpose of exposing them as knowledge structures has been a rather recent concern, typically motivated by the RPA uptake. From a software engineering perspective, the recent proposal in [6] provides a mapping between UI building blocks and back-end data access procedures/requests, however the conceptual model of the mapping is mostly implicit or hardcoded. To isolate it, design-oriented works proposed a conceptual framework for RPA bots in [7], although such efforts do not have in scope a model-driven engineering approach but mostly a process discovery interest. From the same research collective, the challenge of ontology-based abstraction of RPA bots provided results in a conceptual modeling frame [7], focusing on the ontological level.

Closer to our approach, in terms of having a focus on language engineering and metamodeling, are studies that promote metamodels for the automation of apps and services – [8] is one of the many reports promoting the ITLingo application-specific text-based language, whereas in [9] the authors follow the same approach as us by using the ADOxx metamodeling platform to extend a BPMN implementation – their approach aims to automate diverse data access protocols (CoAP, HTTP).

Such recent works show a trend of convergence between traditional model-driven engineering and metamodel-based low-code automation platforms, a trend that was explicitly recognized in [10]. To this agenda, we add the proposal of employing semantic graphs as a knowledge mediator and interchange format between the design decisions and the run-time automation. The traditional practices of abstraction, knowledge engineering and modeling method engineering are thus called to contribute not only to conceptualizations, but also to concern separation and abstracting building blocks in visual interaction environments. In this larger context, we also frame our work against what we consider to be a false dichotomy between "general-purpose" and "domain-specific" [11] – instead, as active members in the OMILAB community of practice [12], we invoke a "multi-specificity" view on visual languages. This has been argued for in the Purpose-Specificity Framework [13] that accounts for the diversity of purposes and degrees of specificity that manifest in model-driven engineering – based on this agenda, we take the approach of "contaminating" established conceptualizations (here, BPMN) with the needed constructs to bring the visual designs and low-code configurations as close as needed to the execution environment involved in automation use cases. In our case, BPMN is the foundation for a DSML that brings its own taxonomy of tasks and workflow assets. By employing semantic RDF graphs as a mediator, the knowledge graphs' structural flexibility is transferred to diagrammatic designs, while using the DSML metamodel as a constraining schema for the resulting semantic graphs.

Extending or adapting BPMN to diverse application areas is a common practice surveyed in [14], however most such initiatives remain in the scope of business processes – tailored for decentralization, for service ecosystems, for cross-organizational governance, for cyber-physical environments, or for particular sectors (healthcare, manufacturing). We only inherit its basic workflow patterns, to enable a procedural description of UX-specific actions – recognizing a UI element, acting upon it, providing content to it etc. – and mappings to the involved assets, i.e. the UI elements and content objects that are involved in the user action.

The conversion of BPMN (and BPMN-inspired) process descriptions to RDF graphs is also a recent preoccupation, traditionally proposed as a means to populate enterprise knowledge graphs with procedural knowledge already available in the form of BPMN [15] or in adaptations for domain-specific business process management systems [16]. This idea is hereby repurposed towards a knowledge management capability in the technologically focused Selenium-based automation environment, building on our earlier efforts of capturing semantic aspects of RPA in machine-readable form [4].

State-of-the-Art GUI automation methods range from traditional scripting to machine learning. Our work fits in the first category taking a deterministic design-first approach, by applying ingredients from model-driven engineering and symbolic knowledge representation to generate scripting snippets from a visual environment, while also cumulating a repository of semantic workflow graphs.

An effort of exploring GUI automation within the boundaries of software testing frameworks has been reported in [17], where the aim is to extend the coverage of automated tests. They demonstrate using Selenium the in-browser testing of scenarios such as generating a parcel order as part of a logistics app user flow.

The low-code paradigm has been a staple in commercial tools for front-end automation using visual scripting. The authors of [18] compare some of them considering platform openness and performance, assessing a high score for UIPath - a commercial triumph based on community-sourced reusable components contributing to a centrally managed ecosystem. The work of [19] presents demonstration scenarios of document-centered digital processes that involve the Microsoft 365 ecosystem, where Microsoft Power Automate [20] evolves as a challenger platform. It shifts focus on the visual building blocks and "Connector" functionalities with pre-built logic for Microsoft's tools (Teams, Office). The authors of [21] are closer to our model-first approach, as they employ UIPath for a report generation scenario that derives knowledge from automation design decisions.

A solution proposed as workaround for having to deal with Web page DOM inspection while scripting UI interaction is Programming by Demonstration (PbD). The authors of [22] propose Rousillon, a tool that supports distributed scrapping of data in loops. Ringer is a solution proposed in [23] by a team at Microsoft that addresses non-determinism caused by the changes in the underlying DOM representation of a Web page through the recording of DOM events. It features a record and replay functionality that mimics user interaction. To simplify the effort of finding tasks that are suitable for GUI automation, the work of [24] proposes mining front-end interaction logs to identify automatable routines and provide the user with an option to generate scripts for them. Unlike PbD tools that require a human example, the processes are discovered automatically over a whole day of interaction. The compiler support is limited to generation of bots for the UIPath platform, and they demonstrate automation for an Excel to browser-based form data submission.

3 Problem Statement

Traditional script-based automation derived from front-end Test Automation requires a programming language and skills to manipulate the front-end's DOM representations that are not consistent. The Programming-by-Design method follows user example at the expense of programming customization. A common feature in commercial automation solutions is flow visualization, but resulting flows do not follow an interoperability standard, thus limiting the user to specific tooling. Moreover, they are typically not interested in exposing the UX flow mappings to a knowledge management capability interested in their traceability or reasoning via semantic queries.

We fill this gap by enabling flow design through means that resemble BPMN, while aligning with the human cognition process during a think-aloud performance – i.e. explicitly recognizing the elements and acknowledging intentions of accessing them. As advocated in UI/UX engineering, the most comfortable user flow is driven by perceived graphical elements and granular interactions with them. At the same time, externalizing such flow designs to semantic RDF graphs generates an asset that is equally an enabler for executability (at run-time) and for semantic traceability (at design-time). The stakeholders for which we introduce the DSR treatment include end-users that want to automate mundane browser-based tasks, UI/UX designers looking to document their task flows or testers that validate automated executions.

By taking the extra step of serializing UX flows into semantic graphs, we ensure interoperability via the standard graph-based RDF format while also enabling the creation of a knowledge repository for subsequent analytics and traceability.

We formulate the project's problem statement in the tradition of the DSR template:

> *Improve* ***front-end browser automation projects management*** *(problem context)*
> *...by treating it with* ***a Selenium-based automation engine informed by semantic graphs that describe UX action flows and their context*** *(artifact)*
> *...to satisfy a requirement for* ***low-code development of knowledge repositories for automation design decisions and artifact descriptions*** *(requirements)*
> *...in order to* ***enable a capability of managing UX design knowledge for both design-time semantic traceability analysis and run-time executability*** *(goals)*

The visual UX flows are required to be created with a minimal domain-specific modeling language (DSML) built around a taxonomy of user actions. The interoperability with the model-driven engine relies on a custom Object-To-Graph Mapper that navigates the RDF graph of a diagram and one additional API for on-the-fly front-end browser automation. Selenium was a technological requirement for the execution environment. The automation component enables multiple ways of identifying Web Elements on the screen, including XPath, through design-time configurations, in case an obfuscated implementation leads to changing underlying implementations.

Ultimately, the automation project manager also wants to harness the UX flows as a knowledge base amenable to semantic queries and reasoning. Graph representations of front-end browser automation processes should also account for changes to the user experience while completing the UI-based activity. The requirement of including information on actions (tasks) related to web interaction in Knowledge Graph form contributes to the first attempt towards the main goal of [25] - where graphs are also used to inform model-driven RPA.

The diagram-to-RDF conversion relies on the ADOxx metamodeling platform where the UX flow design DSML was implemented – a plug-in for converting the internal ADOxx constructs to RDF patterns was presented in [26] and successfully applied in past research projects as a mediator for model-driven orchestration of runtime resources, originally in the Comvantage EU project [27].

4 Design and Development

4.1 Methodology and Preliminary Work

We have aligned our work process to the DSR methodology [28]. Peffers' DSR Process in [29] is been used as a reference research workflow as reflected in this paper's structure. For the *Design and development* phase of the overall DSR process, we nested the Agile Modeling Method Engineering [30] methodology (AMME) which prescribes modeling tool development within the ADOxx ecosystem. Relative to Hevner's three-cycle view on DSR [31], Fig. 1 shows also this nesting in the Design Cycle. AMME was specifically designed to be responsive (in the traditional "agile" sense) to evolving requirements.

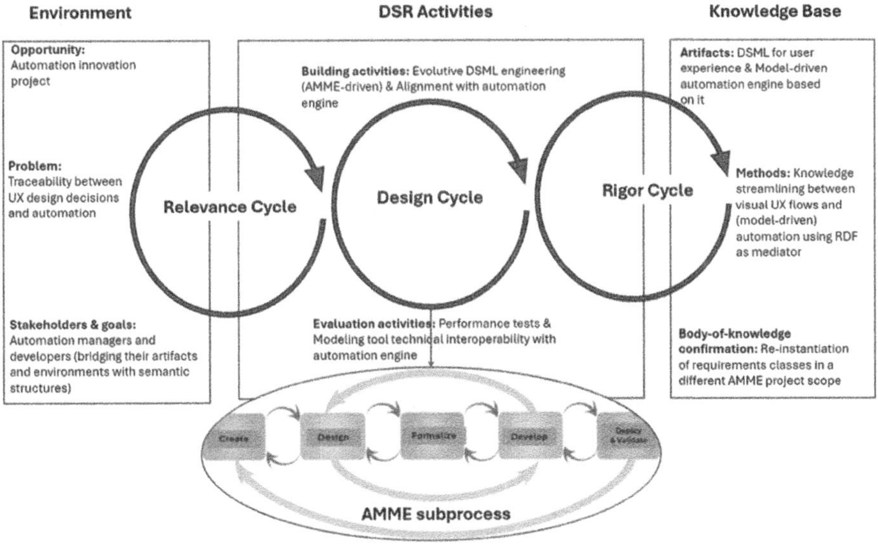

Fig. 1. Hybridizing Hevner's DSR view and AMME

As exemplified in past project experience reports, see [27], requirements in AMME projects may emerge from language requirements (e.g. preferred notations or a need for some specific first-class modeling construct), from modeling procedure requirements (e.g. on how to facilitate model creation and subsequent use), from tool requirements (e.g. tool usability), or from model-driven systems (that expect to retrieve certain properties via interoperability channels). These categories are also related to the quality criteria of traditional model quality frameworks such as SEQUAL [32], although they are pragmatized as explicit backlog item categories during modeling tool development along a flavor of Agile management.

In this project, requirements evolved from a need for workflow-inspired UX documentation, towards active documentation that is also adequate semantic queries and reasoning, finally to model-driven engineering of automation enablers.

An early DSR iteration focused on the DSML's metamodel design was presented in a workshop scope [4]. With this DSML, the UX flow depicted the interaction paths a user takes when using a software product, including alternate paths – i.e. a user-centric visual representation of the steps to achieve a specific goal within a website or app, enriched with meta information on elements they interact with. We tailored the proposed metamodel to include data flows in order to inform RPA projects with data requirements documentation, which eliminated the need for separate DFD designs (Data Flow Diagrams). The modeling tool was implemented in ADOxx and applicability demonstrations were based on scenarios provided by the beneficiary enterprise, for instance to model the mundane flow of creating new roles in an internal project management tool for the on-boarding phase.

The current iteration focused on the model-driven automation demonstrator by refining the DSML developed in the prior iteration to also capture run-time relevant details, and aligning the model interchange with the automation engine based on Selenium. It also improved focus on the kinds of SPARQL queries expected to showcase the competency of the resulting repository for potential report generation.

The rigor cycle of the project contributes as suggested on the right side of Fig. 1, with (a) artifacts (a modeling tool that repurposes and tailors BPMN for UX flow designs), (b) a method/treatment (of knowledge streamlining between diagrammatic UX flows and semantic graphs) and (c) a confirmation (re-instantiation) of requirements classes previously detected by past AMME projects [27], for which dedicated requirements engineering practices must be devised, considering the dependencies and propagations manifesting in a modeling method, tool and language (these are gradually refined through our own experience, but not in the scope of this paper). Less relevant to us was the goal of theory building, due to the pragmatic nature of the project not leaving room for this type of activity, although in future work we plan to reflect back on the experience and how it informs established notation design theories such as Moody's principles [33].

AMME was therefore the driving methodology to engineer the DSML and the knowledge streamlining it enables down to the software components requiring it downstream. AMME starts with the "Create" phase by acquiring domain-specific constructs and eliciting the modeling method requirements. Brainstorming sessions organized according to the Design Thinking framework drove this elicitation effort towards expanding metamodel designs in a way that bridges the way tasks are carried out by bot instances and how they are cognized by a human performer. Human performers were asked to think aloud while executing simple front-end tasks and, in AMME's "Design" phase, the conceptualization of the metamodel refined and classified the granular steps recognized during think-aloud protocols as "task types" (to replace the BPMN task taxonomy) and their involved assets – UI elements and data sources (to replace the BPMN data objects with retrievable content). This resulted in integrating a UI wireframe approach with a BPMN-inspired UX flow language.

In AMME's "Develop" phase, the metamodel was made operational in an ADOxx-based tool, and in parallel the model-driven components developed through the Spring Java framework were developed and aligned to the diagrammatic content and annotations. Along several AMME micro-iterations (AMME is itself a cycle further incorporating granular development sprints), we also refined the visual notation to increase legibility and semantic transparency, adding separate attributes for instance labels to overcome the ADOxx limitation of unique naming. Some concepts were modified to accustom for the automated traversal requirement of the automation engine - *Start* and *End* events not marking Activity boundaries have further been specialized in *Process-level States* to identify easier where the process starts and to detect the end of the automation. The final phase in AMME iterations is "Deploy/Validate". Feasibility will be shown in Sect. 5, the pragmatic focus of validation being placed on this, together with technical interoperability between the modeling environment and model-driven component and the competency enabled for SPARQL queries navigating the flows. It is left for post-project work to reflect on the DSML quality in light of the SEQUAL framework [32].

4.2 Metamodel

Diagrammatic models capture a non-linear process using multiple paths, alternatives and loops, patterns common in both flowchart diagrams and RPA. User actions are classified by a domain-specific "task" taxonomy, starting from a distinction between active but also passive (i.e., discerning a visual element before acting upon it or making a decision) or invisible (happening outside the targeted UI). Tasks are linked to a visual/mockup description of the UI elements associated as a workflow resource – an approach to integrate wireframe designs with UX flows. Required input content is attached to each step in a retrievable manner (from files or HTTP resources) without bothering the modeler with jargon like variables or mappings.

As shown in Fig. 2 we mapped concepts describing task flows under the *Workflow-Element* class hierarchy. They refer to steps that may follow one another linearly, through *followedBy* relations. They have been specialized in 3 abstract classes (orange colored) and one concrete (in green). An *Action* is a user flow Task describing an atomic user-centric interaction with the front-end. *Start* and *End States* mark the consistent entry and end point of the scenario, and we designed them to future-proof event-based functionality (similar to BPMN events). *Activity* is a scope/encapsulation of tasks around a logical subflow.

A typical flow made with the intention of front-end browser automation, begins from a *Process-level StartState*. The element that should follow is an *Invisible Action* of *access*ing the *webpage* whose URL in the form of a linked property to be manipulated by the execution engine (read from where it is stored and passed to a browser address entering method in the Selenium framework). The rest of the flow is designable to fit the UX flow to the particularities of the UI interaction for any app. Designers recommend structuring the linear flow into modules that address the limited time span of the human mind. For this we implemented a containment concept - we envisaged that a user centric

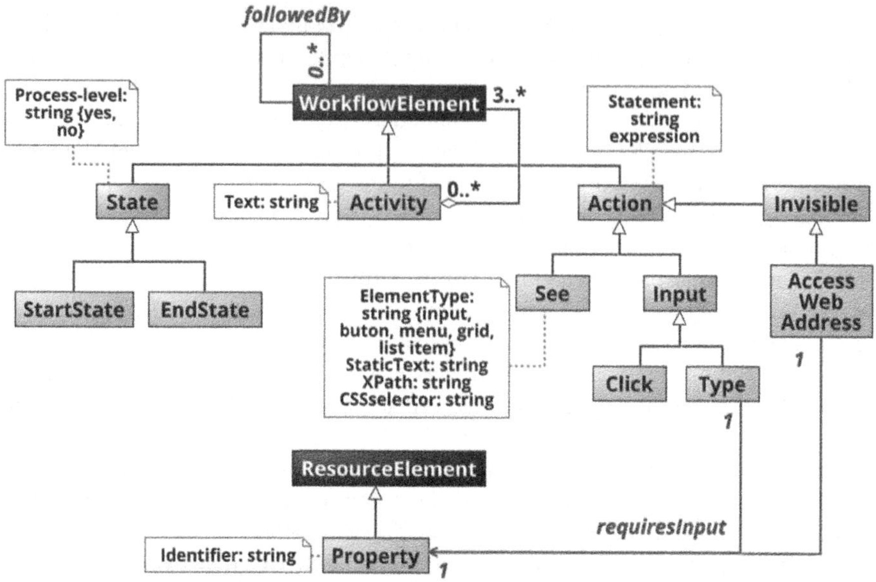

Fig. 2. Metamodel of the Task Flow conceptual models

Activity first identifies an element on the screen through a *See Action* that instructs the bot to retrieve the UI Element programmatically and to focus on it. Thus, up until a new *See Action* is reached, any succedent *Input Action* implicitly targets that element (denoted as context in our model-driven engine). User actions inherited from Task Flow diagrams are *Click* and *Type*. We have decluttered resulting models by the standardized agreement that both refer to the context element. It eliminated the need to specify the *Click* target through a *Property* - a singular text dependency given an identifier. *Type* actions use a *Property* to keystroke referenced text.

5 Demonstration

To validate feasibility and identify obstacles in the context of beneficiary's scenarios, an exemplary candidate for front-end browser automation is the sending of a dummy mail using the front-end of the *Microsoft Outlook* online service. We executed the task manually several times and used the deployed ADOxx modeling tool to refine a granular UX flow. An excerpt of the design is rendered in Fig. 3, highlighting the subflow of writing the body of the email through an *Activity* concept. It can be followed by an *End State* which is at *Process-level* (dynamical notation indicated by the graph symbol inside the circle) – informing both the human and machine of the flow's end point. All flow elements follow sequentially, connected by the *followedBy* edge.

The Activity has the label "*Fill Body field*" to increase comprehensibility by the UI designer/low-code developer. The user first *See*s an input with the static text "*Body*". They type the text stored in the *body Property* (indicated by the *requiresInput* relation between the *Action* and the *Property* nodes). Afterwards, the *Action* of *See*ing the input with the text "*Subject*" and *Click*ing it changes focus on that field and was appended to the model as an overhaul after several failed automated executions. The "*Send*" button situated close to it was masked by a pop-up which we suspect to be left intentionally by Microsoft to discourage automation and scrappers. By adding the intermediary Action, which is not part of the by-design human flow but eliminates the problem in the context of automation, we validate two aspects that our treatment addresses. UX Flow models used as documentation by UI/UX designers can be updated to accustom for changes in the requirements and they can be continually repurposed for orchestrating GUI automation. The hollow *End State* denotes the end of the *Activity, followedBy* a *See Action* that moves the user's attention (or bot's context) to the "*Send*" *button*. Then, the bot *Click*s on the element in the context, and the scenario ends.

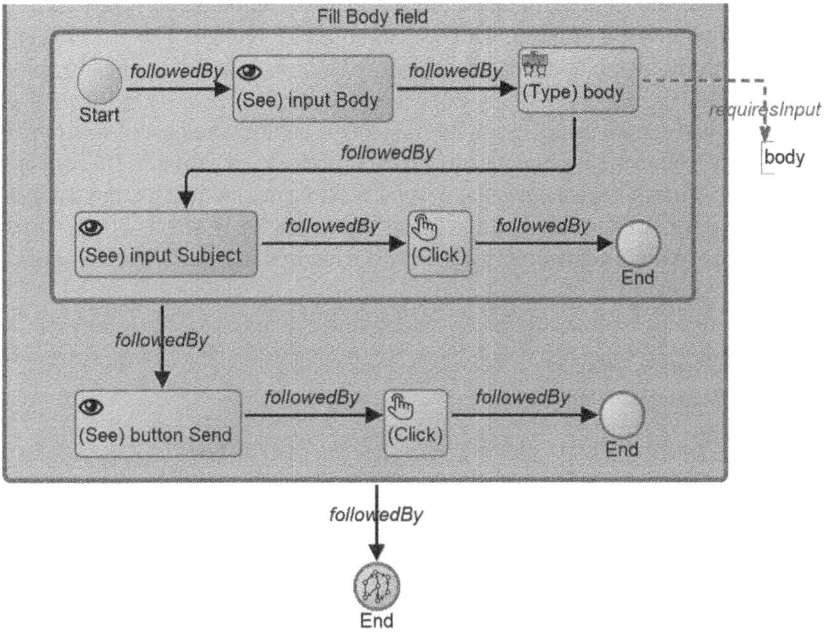

Fig. 3. Excerpt from the Outlook mail sending Task Flow model

After several runs of the automation artifact on the Knowledge Graph exported from this model the results were non-deterministic. In most of the cases Selenium failed to identify UI Elements in the DOM. We traced the problem to a very solid anti-automation protocol in Outlook's front-end. The HTML elements have dynamically allocated IDs and varying DOM representation with each request, thus leaving XPath and id-based identification possible in just a few cases. We overcome this upon more specific HTML inspection of the web page and identifying the attributes that are constant between requests (which are not always IDs, but also custom ones). For the "*(See) button Send*" Action before the flow ends, we applied a common practice in test automation and used the *CSS Selector* field.

6 The Automation Engine

We made the RDF graph for the reported exemplary scenario and the code for the automation engine web app Open-Source (MIT License) available in the public repository [34]. To replicate the setup, the RDF serialization of the ADOxx UX Flow model is imported into a triplestore repository - we use Ontotext GraphDB [35]) It can be consulted in the project structure under *src/main/resources/model/TaskFlow_OutlookSendMail.trig.*[2]

In Fig. 4 we depict how individual components of our artifact are coupled to provide the low-code automation functionality. The order of operational workflow is labeled in red. The citizen developer is first a modeler who designs the UX Flow using the domain-specific modeling tool (a). The serialized open model is converted into triples in.TRIG format using a dedicated tool described in the work of [26] and uploaded in a remote GraphDB repository (b). Next, the automation engine is run (c) by calling the endpoint address inside a browser, which sends an HTTP request. The process can also start directly from this point as the user runs several iterations on an already uploaded UX flow.

The main flow carried in the Spring Boot app of the engine commences with retrieving the StartState (I) semantically annotated with the Process-level attribute. *The Automation Controller* forwards this to a *Knowledge Service* residing in a module for domain-specific model-driven development based on the Knowledge Graph. A *SPARQL* [36] *query* from a repository is sent using a connection to the SPARQL endpoint of the GraphDB repository:

```
SELECT ?iri WHERE {

    ?iri rdf:type ns0:StartState.
    ?iri ns0:Process-level "yes"

}
```

The returned record is parsed into an IRI which is sent to a Object-to-Graph-Mapper (OGM) framework in the *OGM component* that we have custom-made to *map* returned resources from GraphDB into Plain Old Java Objects (POJOs). These correspond to the same classes from the metamodel (Fig. 2). This enables working with UX Flow semantic graphs while remaining in the Object-Oriented paradigm with nodes provided similar to Labeled Property Graphs (*LPG*) alternate graph method (hence, the eponymous name of the *service*). The *Node* object (a Start State at the beginning of the automation) is stored in a *step* variable which contextualizes the current task in the flow traversal. Coded business logic checks and guarantees the step to be either a See, Click, Type or Access Web Address kind of Action. This represents a loop starting point from where the engine sends the *step* to the Execution Service and after to the Traversal Service which also

[2] The resulting remote repository link has to be updated for the field "*graphdb.repository.url*" in the Spring Boot app configuration file "*src/main/resources/application.properties*". Update the *_properties_* file in the *resources* with valid credentials. Next, build the Java project which opens an Embedded Tomcat server to listen for requests. To use the automation engine, make a request to the endpoint "http://localhost:8080/automate" which executes one iteration of browser-based GUI Automation on the referenced graph.

returns an updated valid step. The cycle iterates until the *Process-Level End State* is placed inside the *step*.

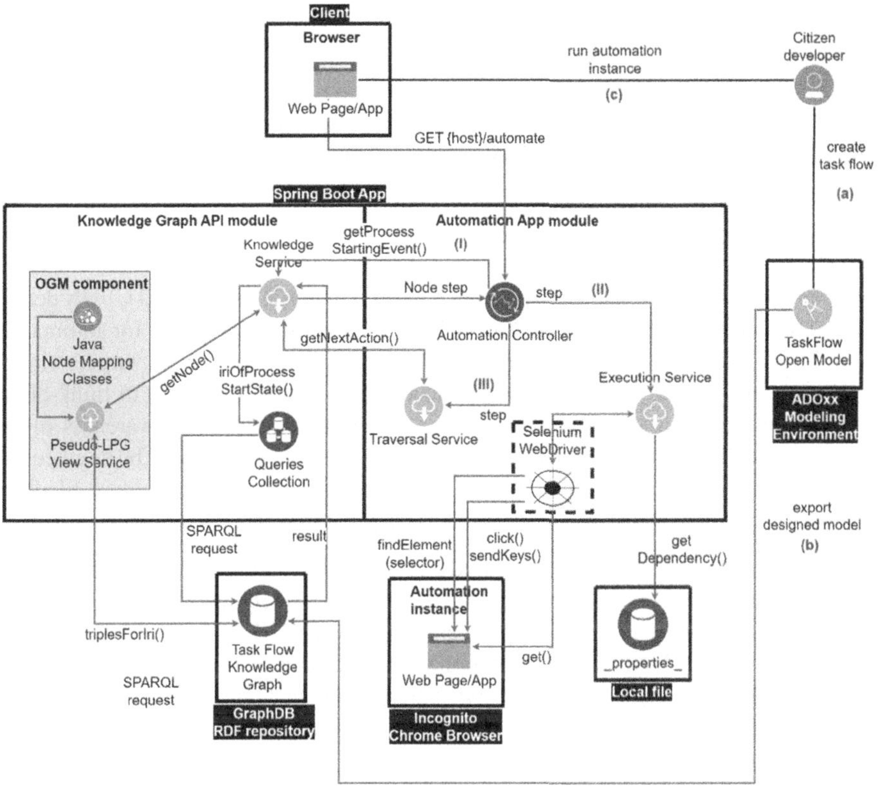

Fig. 4. Architectural integration diagram

The execution service decides, based on an if-else logic considering the Action type, what method to call on the browser DOM using the dedicated WebDriver for the browser type (ChromeDriver in our case). For the *Invisible Action* of sending a browser request for the page, the *get()* method is called out of the Web Driver instance using the *dependency* value of the URL extracted from a *local file*. For a *See Action*, whichever field from *XPath* or *CSSSelector* is non-empty is sent as argument to the *findElement()* Selenium API method, after waiting for the DOM resource to be interactable with. The context variable (*target*) is populated with the return and used to call either *click()* or *sendKeys()* from Selenium.

To provide a context for evaluating the artifact from a time-efficiency perspective, we have conducted some testing on the exemplary scenario automated execution time. We have benchmarked the total execution time of the automation of the exemplary Send Mail scenario. With caching disabled, Chrome measured a mean time of 21.98 s to complete one automation instance of the task flow, which was deemed satisfactory as we don't aim to compete with industry grade commercial platforms, but mostly to demonstrate the dual approach of treating low-code automation designs as relevant to both run-time needs and knowledge management needs – the latter was in the focus on the previously mentioned workshop paper [4].

7 Discussion

We critically assessed our treatment through the lens of the reviewed literature and a SWOT analysis derived from the industry partner of the project and its end-users involved in both requirements elicitation and tests.

Strengths. We repurposed a workhorse method in Software Testing to serve the purpose of GUI Automation, inspired by the work of [37] reporting that there is a subset of commercial RPA tools which evolved from UI testing tools. We integrated UI/UX design with GUI Automation through enabling task flows to be used by engines for automating scenarios in the browser. The engine, following a low-code paradigm, targets citizen developers by providing a visual DSML to describe flows according to think-aloud protocols, detecting both active (e.g. clicks) and passive steps (e.g. awareness of the presence of a UI element as cognition context for a set of interactions). The engineering effort respects Separation of Concerns, as required data is separated into a text file from the flow description, where it is referenced using diagrammatic placeholders.

Weaknesses. The automation app is based on scripts for the front-end, and as a result it is prone to errors analyzed in the work of [38]. Specifically, selecting UI Elements through queries is recognized as a common problem when doing DOM-based identification of elements using the HTML implementation. A workaround is based on CSS Selectors that are more customizable to front-end patterns, which helped us overcome heavy obfuscation of the Outlook web app. The model has to be modified with each change in the UI, but not the engine.

Opportunities. Additional assets and flows in the UI/UX designer's toolkit can be integrated progressively in order to extend the modeling language beyond UX flows, towards a contextualization with enterprise models where those tasks must be executed. A management platform can be built to support RPA project managers with prioritizing and selecting the next process to be automated, based on how these flows impact the layers of an Enterprise Architecture.

Threats. The authors of [39] trained neural networks based on Programming-By-Demonstration to build GUI automation agents. However, we believe that the stochastic AI approach must be complemented by deterministic designs and rules, possibly leading to future neuro-symbolic hybridization approaches to GUI automation.

8 Conclusions and Future Work

This paper proposes a low-code automation solution based on serialized RDF graphs derived from UX models described as workflows that repurpose BPMN for the specificity of task-oriented human-computer interactions. The automation engine was deployed as a Spring Boot app that executes the flow automatically using the open-source Selenium framework for web automation. We demonstrated feasibility on a scenario of sending a mail using the Outlook browser client. This advances the idea of employing RDF graphs for automation - to inform model-driven automation engines with design-time semantically rich descriptions. Common patterns in UI/UX design may eventually be also used as input for graph-based machine learning.

In the future, we will focus on providing specific implementations for UI/UX deliverables that focuses on the bigger picture, where the RDF graphs would further provide for integration of UX flows with Enterprise Architecture elements captured in Archimate models and semantic extensions to achieve the bridging with the enterprise context on multiple levels of granularity.

Acknowledgments. The presented work has received financial support through the project: Integrated system for automating business processes using artificial intelligence, POC/163/1/3/121075 - a Project Cofinanced by the European Regional Development Fund (ERDF) through the Competitiveness Operational Programme 2014–2020.

References

1. van der Aalst, W.M.P., Bichler, M., Heinzl, A.: Robotic process automation. Bus. Inf. Syst. Eng. **60**, 269–272 (2018). https://doi.org/10.1007/s12599-018-0542-4
2. UiPath: The Business Automation Platform. https://www.uipath.com/. Accessed 01 Mar 2025
3. Korzynski, P., Kozminski, A.K., Baczynska, A.: Navigating leadership challenges with technology: uncovering the potential of ChatGPT, virtual reality, human capital management systems, robotic process automation, and social media. Int. Entrep. Rev. **9**, 7–18 (2023). https://doi.org/10.15678/IER.2023.0902.01
4. Uifălean, Ş.: Employing knowledge graphs for capturing semantic aspects of robotic process automation. In: Ruiz, M., Soffer, P. (eds.) CAiSE 2023. LNBIP, vol. 482, pp. 152–162, Springer, Cham (2023). https://doi.org/10.1007/978-3-031-34985-0_16
5. Ghiran, A.M., Silaghi, G.C., Tomai, N.: Ontology-based tools for automating integration and validation of firewall rules. In: Abramowicz, W. (ed.) BIS 2009. LNBIP, vol. 21, pp. 37–48. Springer, Heidelberg (2009). https://doi.org/10.1007/978-3-642-01190-0_4
6. Sterca, A., Niculescu, V., Chisăliţă-Creţu, M.C., Osman, C.C.: Primary building blocks for web automation. In: Zhang, F., Wang, H., Barhamgi, M., Chen, L., Zhou, R. (eds.) WISE 2023. LNCS, vol 14306, pp. 376–386. Springer, Singapore (2023). https://doi.org/10.1007/978-981-99-7254-8_29
7. Hohenadl, T., Völker, M., Stummeyer, C., Weske, M.: Democratizing robotic process mining: a conceptual framework for user actions, tasks, and RPA bots. In: Di Ciccio, C., et al. (eds.) BPM 2024. LNBIP, vol. 527, pp. 185–199. Springer, Cham (2024). https://doi.org/10.1007/978-3-031-70445-1_12
8. Vilão, P.B., Rodrigues da Silva, A.: Rapid production of enterprise applications in a low-code environment: comparing the ITLingo-ASL and PowerApps metamodels. In: Proceedings of the 29th Information Systems Development, Organizational Aspects and Societal Trends, ISD2023. AIS (2023). https://doi.org/10.62036/ISD.2023.49
9. Chis, A., Ghiran, A.M.: BPMN extension for multi-protocol DataOrchestration. In: Karagiannis, D., Lee, M., Hinkelmann, K., Utz, W. (eds.) Domain-Specific Conceptual Modeling, pp. 639–656. Springer, Cham (2022). https://doi.org/10.1007/978-3-030-93547-4_28

10. Cabot, J.: Positioning of the low-code movement within the field of model-driven engineering. In: The 23rd ACM/IEEE International Conference on Model Driven Engineering Languages and Systems: Companion Proceedings, pp. 1–3. ACM (2020). https://doi.org/10.1145/3417990.3420210
11. Kosar, T., et al.: Comparing general-purpose and domain-specific languages: an empirical study. ComSIS **7**(2), 247–264 (2010). https://doi.org/10.2298/CSIS1002247K
12. Vaidian, I., Jurczuk, A., Misiak, Z., Neidow, M., Petry, M., Nemetz, M.: Challenging digital innovation through the OMILAB community of practice. In: In: Karagiannis, D., Lee, M., Hinkelmann, K., Utz, W. (eds.) Domain-Specific Conceptual Modeling, pp. 41–64. Springer, Cham (2022). https://doi.org/10.1007/978-3-030-93547-4_3
13. Buchmann, R.A.: The purpose-specificity framework for domain-specific conceptual modeling. In: Karagiannis, D., Lee, M., Hinkelmann, K., Utz, W. (eds.) Domain-Specific Conceptual Modeling, pp. 67–92. Springer, Cham (2022). https://doi.org/10.1007/978-3-030-93547-4_4
14. Zarour, K., Benmerzoug, D., Guermouche, N., Drira, K.: A systematic literature review on BPMN extensions. Bus. Process. Manag. J. **26**(6), 1473–1503 (2020). https://doi.org/10.1108/BPMJ-01-2019-0040
15. Bachhofner, S., Kiesling, E., Revoredo, K., Waibel, P., Polleres, A.: Automated process knowledge graph construction from BPMN models. In: Strauss, C., Cuzzocrea, A., Kotsis, G., Tjoa, A.M., Khalil, I. (eds.) DEXA 2022. LNCS, vol. 13426, pp. 32–47. Springer, Cham (2022). https://doi.org/10.1007/978-3-031-12423-5_3
16. Cinpoeru, M., Ghiran, A.M., Harkai, A., Buchmann, R.A.: Model-driven context configuration in business process management systems: an approach based on knowledge graphs. In: Pańkowska, M., Sandkuhl, K. (eds.) BIR 2019. LNBIP, vol. 365, pp. 189–203. Springer (2019). https://doi.org/10.1007/978-3-030-31143-8_14
17. Khankhoje, R.: Bridging the gap: Selenium and RPA for unparalleled automation. Int. J. Cybern. Inform. **13**(13), 23 (2024). https://doi.org/10.2139/ssrn.4701292
18. Issac, R., Muni, R., Desai, K.: Delineated analysis of robotic process automation tools. In: Proceedings of the Second International Conference on Advances in Electronics, Computers and Communications ICAECC 2018, pp. 1–5 (2018). https://doi.org/10.1109/ICAECC.2018.8479511
19. Davlatov, N.: Robotic process automation : optimization of business processes with a cost-effective automation tool. Bachelor's thesis, Metropolia University of Applied Sciences, (2023) http://www.theseus.fi/handle/10024/818914. Accessed 01 Mar 2025
20. Microsoft Power Automate – Process Automation Platform. https://www.microsoft.com/en-us/power-platform/products/power-automate. Accessed 01 Mar 2025
21. Hindel, J., Cabrera, L.M., Stierle, M.: Robotic process automation: hype or hope?. In: Wirtschaftsinformatik (Zentrale Tracks), pp. 1750–1762 (2021). https://doi.org/10.30844/wi_2020_r6-hindel
22. Chasins, S.E., Mueller, M., Bodik, R.: Rousillon: scraping distributed hierarchical web data. In: Proceedings of the 31st Annual ACM Symposium on User Interface Software and Technology, pp. 963–975. ACM (2018). https://doi.org/10.1145/3242587.3242661
23. Barman, S., Chasins, S., Bodik, R., Gulwani, S.: Ringer: web automation by demonstration. SIGPLAN Not. **51**, 748–764 (2016). https://doi.org/10.1145/3022671.2984020
24. Leno, V., Deviatykh, S., Polyvyanyy, A., La Rosa, M., Dumas, M., Maggi, F.: Robidium: automated synthesis of robotic process automation scripts from UI Logs. In: Proceedings of the ICPM Doctoral Consortium and Tool Demonstration Track 2020, CEUR-WS 2673 (2020). https://ceur-ws.org/Vol-2673/paperDR08.pdf
25. Riva, O., Kace, J.: Etna: harvesting action graphs from websites. In: Proceedings of the 34th Annual ACM Symposium on User Interface Software and Technology, Virtual Event USA, pp. 312–331. ACM (2021). https://doi.org/10.1145/3472749.3474752

26. Karagiannis, D., Buchmann, R.A.: A proposal for deploying hybrid knowledge bases: the ADOxx-to-GraphDB interoperability case. In: Proceedings of the 51st Hawaii International Conference on System Sciences, pp. 4055–4064. AIS (2018). https://doi.org/10.24251/HICSS.2018.510
27. Buchmann, R.A., Karagiannis, D.: Agile modelling method engineering: lessons learned in the ComVantage research project. In: Ralyté, J., España, S., Pastor, Ó. (eds.) PoEM 2015. LNBIP, vol. 235, pp. 356–373. Springer, Cham (2015). https://doi.org/10.1007/978-3-319-25897-3_23
28. Wieringa, R.J.: Design Science Methodology for Information Systems and Software Engineering. Springer, Cham (2014). https://doi.org/10.1007/978-3-662-43839-8
29. Peffers, K., et al.: The design science research process: a model for producing and presenting information systems research (2006). https://jyx.jyu.fi/handle/123456789/63435. Accessed 01 Mar 2025
30. Karagiannis, D.: Agile modeling method engineering. In: Proceedings of the 19th Panhellenic Conference on Informatics, pp. 5–10. ACM (2015). https://doi.org/10.1145/2801948.2802040
31. Hevner, A.R.: A three-cycle view of design science research. Scand. J. Inf. Syst. **19**(2), 4 (2007). https://aisel.aisnet.org/sjis/vol19/iss2/4
32. Krogstie, J., Sindre, G., Jørgensen, H.: Process models representing knowledge for action: a revised quality framework. Eur. J. Inf. Syst. **15**, 91–102 (2006)
33. Moody, D.L.: The "Physics" of notations: towards a scientific basis for constructing visual notations in software engineering. IEEE Trans. Softw. Eng. **35**(5), 756–779 (2009). https://doi.org/10.1109/TSE.2009.67
34. Uifălean, Ş.: Open-source spring boot app implementation of a low-code front-end browser automation engine driven by Task Flow models. https://github.com/stefanuifalean/taskflow-automator. Accessed 01 Mar 2025
35. GraphDB by Ontotext (2025). https://graphdb.ontotext.com/. Accessed 01 Mar 2025
36. SPARQL 1.1 Query Language (2013). https://www.w3.org/TR/2013/REC-sparql11-query-20130321/. Accessed 01 Mar 2025
37. Cernat, M., Staicu, A.-N., Stefanescu, A.: Improving UI test automation using robotic process automation. In: Proceedings of the 15th International Conference on Software Technologies ICSOFT, pp. 260–267. SciTePress (2020). https://doi.org/10.5220/0009911202600267
38. Krosnick, R., Oney, S.: Understanding the challenges and needs of programmers writing web automation scripts. In: Proceedings of the IEEE Symposium on Visual Languages and Human-Centric Computing (VL/HCC), pp. 1–9. IEEE (2021) https://doi.org/10.1109/VL/HCC51201.2021.9576476
39. Li, W., Hsu, F.-L., Bishop, W., Campbell-Ajala, F., Riva, O., Lin, M.: UINav: a maker of UI automation agents. (2023). http://arxiv.org/abs/2312.10170. Accessed 01 Mar 2025

Risk Response
An Adversarial Discourse Game and Group Modelling Tool

Max Willis[1(✉)], Greta Adamo[2,1(✉)], and Anna Sperotto[3,4]

[1] Universitat Politècnica de València, València, Spain
max@maxwillis.net
[2] Free University of Bozen-Bolzano, Bolzano, Italy
Greta.Adamo@unibz.it
[3] Cà Foscari University of Venice, Venice, Italy
[4] Basque Center for Climate Change, Leioa, Spain
anna.sperotto@bc3research.org

Abstract. Communication of risk, e.g. posed by floods, heatwaves, or infrastructure failure, intends to reduce impacts, limit damage and ultimately save lives, as well as foster people's adaptive and coping capacities to future events. Yet risk is a complex, semantically overloaded and at times ambiguous notion, which affects its understanding and management. In addition, risk communication is influenced by factors such as expert/non-expert knowledge exchange, imbalances of top-down approaches, trust, and empowerment. In response to this, participatory and community-oriented strategies, including game-based interactions, have been implemented to support risk management through stakeholder engagement. Joining this conversation about participation in risk communication, this paper presents *Risk Response*, an adversarial discourse game that facilitates discussion of future environmental, social and technological risks, and appropriate responses to them. The game offers a card-based interaction which mediates (i) conversational play (LEVEL I), in which players speculate on risks and imagine responses, and (ii) group modelling play (LEVEL II), wherein teams engage in semantically enriched conceptual modelling using ontology-based formalisations. This article describes the background, design and play of the game, with a preliminary evaluation of in-game group modelling enacted during a workshop that demonstrates semantic alignment using ontologies in participatory risk communication.

Keywords: Risk · Game and play · Participatory practices · Group modelling · Ontologies

1 Introduction and Motivations

Risks can be of many kinds, e.g. floods, earthquakes, heatwaves, fires, data breaches, identity theft, and pandemics. Hence risk analysis, assessment, and management are constitutive of several fields, e.g. environmental and climate

J. Grabis et al. (Eds.): RCIS 2025, LNBIP 547, pp. 156–172, 2025.
https://doi.org/10.1007/978-3-031-92474-3_10

change sciences, finance, engineering, cybersecurity, Artificial Intelligence and public health, and present a long and articulated history [7]. Despite its established literature, understandings of risk and risk management are hindered by overloaded and unclear semantics [1,38], which complicates their communication to ensure effective planning and responses [8,27,43,50]. As the primary goal of risk communication is to mitigate and reduce impacts and losses, understanding and crafting appropriate information dissemination strategies is a priority to reach widely diversified audiences [43,50]. For instance, to reduce and mitigate the disaster risks of hydrological or seismic hazards, mastering safety drills can prepare civilians to respond, during and after such events, and awareness of evacuation routes and sites of refuge can ensure that at-risk populations are able to exploit these resources. However, risk and response communication can at times be ineffective, owing to mismatches in communication styles, information overload, perceived power imbalances associated with top-down, expert-driven directives, overly technical communication, mistrust of authorities, and uneven sharing of responsibilities [27,41,43,50]. Consequently, the necessity to identify, develop, and evaluate participatory and community-oriented practices for effective risk information delivery and communication becomes an imperative. This requires a shift in communication in which publics are addressed not merely as deficient receivers of information, rather as active participants with agency to shape the interaction [41,43,50]. This objective has been widely addressed using game and play techniques (e.g. [28,41]), and in this paper, we present *Risk Response*, an adversarial discourse game that contributes to risk communication and participation discussions by (*a*) proposing a participant-initiated, risk-agnostic scenario generator to ensure flexibility and ownership of the discourse, (*b*) balancing stakeholder communication and information dissemination through adversarial discourse, and (*c*) facilitating in-game, *ontology-based group modelling* to formalise the semantics of pluralistic risk and response discussions. Risk Response examines environmental, social and technological risks, and accommodates a wide range of stakeholders as participants and gameplay contexts, offering two levels of play in a card-based format. Level I engages participants in *conversational play*, derived from common-sense, collective intelligence and speculation about future risks and responses. Level II leads participants in *group modelling play*, introducing ontology-based conceptual models and the collaborative design of model artefacts to align the semantics of risk and response discourse. This work describes the premise, design and play of the game, with a preliminary evaluation of group models and gameplay from a workshop held at the *18th Research Challenges in Information Science* (RCIS 2024)[1] in Guimarães (PT). The game and modelling activities can be useful for organisations for risk communication, and for development of information systems to identify semantic and thematic congruences in map- and graph-based climate change and risk analyses.

The paper is structured as follows: Sect. 2 introduces background knowledge on risk and its communication. Section 3 details the Risk Response game, Sect. 4

[1] https://www.rcis-conf.com/rcis2024/.

offers an initial evaluation, followed by related works, discussion and limitations, in Sect. 6 and 5, conclusions and future works in Sect. 7.

2 Background Knowledge

Risk and Related Notions. The significance of the notion of risk and its definitions is an established discourse, e.g. [7,9,38]. Many fields have dealt with risk, its assessment, and management [7], leading to the development of various types of risks and associated concepts, each tailored differently depending on the context of application. It is beyond the scope of this article to dissect the concept of risk and its understanding, hence we begin with a general, domain-independent definition of risk as *"a situation or event where something of human value (including humans themselves) has been put at stake and where the outcome is uncertain"* [37], after Sales et al. [38]. The latter outlines the three core tenets of this definition as (i) the dependencies between risk and the possible effects on an agent's *interest* (i.e. what agent x values, for example their health and well-being), (ii) the connection between risk and *uncertainty*, particularly for its assessment, and (iii) that risk concerns *possibilities* and futures.

The scientific community has not yet agreed on the semantics of many of risk-related terms, a problem highlighted in [39], and there is a need to establish common grounds, especially in multi-stakeholder settings where mutual agreement and shared understandings play a crucial role. Despite the different definitions used, the international community, via institutions such as the United Nations Office for Disaster Risk Reduction (UNDRR)[2] and the Intergovernmental Panel on Climate Change (IPCC),[3] aim to agree on a set of "key risk components," as follows. (i) **Hazard** is the occurrence of an event or process that might have negative consequences; hazards can be of different types, such as meteorological, hydrological, biological, technological, and societal, and can be either natural or human-made [17]. A typical hazard could be the occurrence of extreme rainfall in a short time-period, such as that which facilitated the disruptive flood that occurred in Valencia in October 2024. (ii) **Exposure** describes the presence of elements (e.g. people, infrastructure, ecosystems) that might be affected by a hazard. In the Valencia example, exposure includes the people, households and businesses located in the municipalities affected by the flood, such as Paiporta. (iii) **Vulnerability** relates to the susceptibility of the exposed elements to suffer damage or loss from a hazard, influenced by factors such as their capacity to respond or to adapt. In the Valencia flood, vulnerabilities included factors such proximity to water bodies, the age and mobility of affected populations, and a mismatch between weather forecasting and activation of warning systems. Accordingly, risks are future, possible, multi-scale dynamics resulting from the complex interaction among the key risk components.

[2] https://www.undrr.org/drr-glossary/terminology.

[3] https://apps.ipcc.ch/glossary/.

The occurrence of an adverse event can result in a system experiencing actual consequences or negative outcomes, which can induce critical socio-environmental disruptions, such as loss of life, infrastructure damage, or failure of system function. Such consequences are often referred to more generally as **impacts** (IPCC) [4] or **disasters** (UNDRR).[4] Once risk and its consequences have been assessed, risk management comes into play. This can be defined as coordinated efforts to monitor, minimize and control the probability or adverse impact of events through the implementation of appropriate response measures [24]. Response measures can be loosely differentiated into **risk reduction measures**, which aim on reducing the likelihood or severity of risks, and **adaptation measures**, which focus on adjusting systems, infrastructure, and practices to cope with the impacts. For example, while appropriate river bank management can mitigate the risk of flood events by reducing the likelihood that extreme rainfall becomes a hazard, early warning systems and effective risk communication plans can help adapt to the impacts of potential floods by preventing or alleviating the associated losses, i.e. reducing exposure and vulnerability.

Risk Communication and Participatory Practices. Risk communication is an evolving and dynamic field rooted in risk management [11,42], with branches such as health, environmental and emergency management [42]. To be effective, it must overcome some of the core challenges of science communication, which suffers from a range of issues related to the dominance, and rejection of expert-led narratives in public discourse (see [15] for an overview). Three paradigms of risk communication have been identified, which can be seen as engagement scales [11,27,43]. These paradigms are not exclusive XOR, neither are they an historical evolution, rather are overlapping styles of communication based on particular (often site- or community-specific) exigencies, requirements, and contexts [27]. In the *deficit* model, experts inform citizens of the risk, or potential danger of x, e.g. the public needs to be informed about the nature of COVID-19 in order that they prevent transmission, and health experts as the gatekeepers of (technical) knowledge must translate and transmit that knowledge to the public [15,27,43]. This mode of engagement can be experienced by non-scientific stakeholders as overly technocratic and can lead to disconnect, even willful disbelief due to mistrust of scientific establishments. In the *dialogue* model, attempts are made to improve information delivery by engaging stakeholders in discussion, for example with interviews, surveys and focus groups, through which citizens are "given opportunity" to share their perspectives [43]. This mode of communication, ostensibly inviting stakeholders to engage in risk discourse, can be perceived as simply another way for experts to exert the dominance of their information, and persuade stakeholders to correct courses of thought and action [27].

A third, increasingly adopted, paradigm of risk communication is the *participation* model, which emphasises community-engagement and focusses more on developing relationships between the experts, policy-makers, civil society and publics to ensure mutual-learning and understanding, and empowerment

[4] https://www.undrr.org/terminology/disaster.

of people experiencing risk on the ground [43]. Such community-oriented and stakeholder-focussed communication initiatives adopt methods from social science, *à la* Action Research and co-production, to gather together the plurality of perspectives, lived experiences, and skills for participatory sense-making on the subject [48]. Local communities can be engaged inside and outside traditional spaces, such as in cafes, bars and community centers, using more humanistic means of communication including poetry, story-telling, drawing, music, theater and games [43]. As risk discourses are decidedly futures-oriented, the participatory mode of communication incorporates many aspects of *futuring* [20], such as speculative enactment, counterfactual histories, and design fiction. Serious games and gamification have been extensively applied in this context to facilitate mutual learning about risk, and disaster risk management in particular [41]. Role-playing and simulation offer proven potentials to mediate multi-stakeholder, multi-perspective engagement [28], and as games themselves are less formal communication styles, they allow for a more horizontal worldview sharing between policy-makers, volunteers, scientists and publics [41]. However games applied to risk contexts also present some limitations [41], hindering their applicability and scientific relevance. For example (*a*) many game designs are overly specific, targeting a singular hazard or a particular community, and as such are less applicable to other situations. This specificity can lead to (*b*) shortfalls in addressing broader ecological, social and technological understandings of risk and risk management. In addition, (*c*) many risk-orientated games lack proper testing and evaluation methodologies [41].

In redress of the aforementioned limitations of games for risk communication, we present the adversarial discourse game called Risk Response, and in the following discuss its origins, interaction design, material components, rules and gameplay, with a preliminary empirical evaluation.

3 Risk Response Game

Risk Response is an adversarial discourse game that examines future environmental, social and technological risks, and human mitigation and adaptation responses to them. Gameplay in Risk Response is designed not to inform or educate on risk, rather to **enact the collectivism** which underpins the intergovernmental bodies' science-based risk investigation, to **spark intuition** on the interrelation of adaptation and mitigation paradigms, and ultimately to **prompt semantic alignment** in risk communication through ontology-based group modelling. The game aligns with initiatives to foster *Futures Literacy* and anticipation, acknowledged by UNESCO as fundamental skills for individuals, communities and societies to navigate the challenges of the emerging future [46]. Risk Response was not developed following an established game design framework, yet it draws practical inspiration from renowned game concepts, such as Nealen's minimalist game design analysis [33], maintaining simplistic game rules and mechanics that facilitate a broad range of play possibilities. Cook's loops and arcs concept is also applied [16], to scaffold learning and player engagement, as are Bogost's theoretical groundings on procedural rhetorics [13] reinforce perspective shifts through implicit communication within the game artefact, its rules, and gameplay contexts. Risk Response also draws from *speculative design* [5] to drive critical interactions that embody and enact futuring concepts as in [6,18,20,46] through game and play.

The game's design follow dual objectives, the first focused on the *game artefact*, e.g. the game must engage a wide variety of participants and perspectives, it must be straightforward to learn, and fun to play. The second track of constraints addresses the *game objectives* of (i) making complex notions of future environmental, social and technological risks understandable and accessible, (ii) driving discourse, using explicit futures concepts, on the interrelatedness of various risk mitigation and adaptation paradigms and (iii) aligning participant perspectives with shared semantics via ontology-based group modelling, which also provides for (iv) an evaluation mechanism to analyze engagement and sense-making potentials of the game (see e.g. [48]).

3.1 Game Cards, Rules, and Modes of Play

The Risk Response game design was inspired from the second and third authors' collaborative work on an *ontological unpacking of climate change risk* [1], which aimed at clarifying the semantics of climate change risk and related definitions, e.g. impact and uncertainty, proposed by the IPCC in their *6th Assessment Report* [29]. To aid in sense-making of the complex ontological analysis, the game was prototyped, and following repeated playtests, expanded with the futures perspectives of the IPCC's *5th Assessment Report* [26], the UNDRR 2021 *Hazard Information Profiles* (HIPs) [45], and various futures literature (e.g. [46,47]). The game cards are divided into three groups, the first includes 13 *Risk or Impact* cards, each labeled at the top with a general rubric, or framing of a potential risk, such as *Fire*, *Technology*, or *Social* risks. On each risk card is a set of examples; *Fire* risk or impact examples include *Atmospheric Temperature*, *Stronger UV Rays*, *Forest Fires*, *Urban Heat*, and suggestions for *Social* risks or impact include *Economic Instability*, *Resource Conflict*, *Intangible Cultural Heritage*, and *Migration*, among others. Each card's list of examples ends with *etc.*

The second set of 12 cards are response cards, labeled *Mitigate or Adapt*, categorised into rubrics such as *Citizens*, *Governance* and *Technology*, with suggestions. For example, *Citizens* responses include *Reduce-Reuse-Recycle*, *Green Mobility*, *Activism* and *Volunteering*, and *Population Control*, while the *Technology* card suggests *ICT for development*, *Earth Observation*, *Geosciences & GIS*, and *AI Solutions*, among others. Each list again ends with *etc.* The third set of 15 cards are the game's *Discourse cards*, with progress discussion contributions, e.g *Sustainable Action* and *Resilience*, inhibiting directions, such as *Immutable Status Quo* and *Hidden Danger*, and several terminal discourse cards, such as *Futures Judgment*, *Catastrophe*, and *Move to Mars*. Each card contains instructions on how the concept can contribute to the game discussion, e.g. *Stroke of Luck* could be used against overly realistic play, and *Catastrophe* can be played to counter overly optimistic discussion. Table 1 lists the complete set of cards.

Table 1. *Risk Response* cards.

Risk or Impact	*Mitigate or Adapt*	*Discourse card*
Air	*Citizens*	*Resilience*
Fire	*Governance*	*Equilibrium*
Water	*Organisations*	*Blue-Green Future*
Earth	*Education*	*Sustainable Action*
Ocean	*Health*	*Anticipation*
Sound and Light	*Society*	*Knowledge Commons*
Vegetation	*Economy*	*Immutable Status Quo*
Animal	*Food Systems*	*Hidden Danger*
Food	*Services*	*Uncertainty*
Social	*Environment*	*Response Outcome Risk*
Technology	*Resource*	*Future Judgment*
Built Environment	*Technology*	*Catastrophe*
Extraterrestrial		*Stroke of Luck*
		Time Machine
		Move to Mars

Additional blank risk cards are also available, for gameplay in participatory project development settings where particular themes or site-specific risks need to be analyzed.

Finally, the *Futures Cone* completes the game's material aspect (Fig. 1 Left). This speculative design tool describes potential future trajectories moving from the present, on the left, to the future on the right. Concentric circles, (cones) demarcate the temporal areas of Possible, Preferable, Plausible and Probable futures. At the outermost fringes of the Futures Cone are located Wildcards, that are low-probability events, and Preposterous futures, which although they are highly unlikely to happen, remain within human imagination. The Futures Cone has undergone many developments and reinterpretations since its introduction to defense studies in the 1980's [21]; the version used in Risk Response has been adapted by the first author from [47].

Game Rules. The game's basic format consists of: players forming two teams, *Team Risk* and *Team Humanity.* Team Risk proposes a risk and outlines its impact, after which Team Humanity proposes a response, based on cards randomly drawn from the decks (see a sample of Risk Response cards in Fig. 1 Right). Team Risk is implored to play impartially, imagining and portraying their proposed risk as data-driven and inevitable. Team Humanity is encouraged to an impassioned play, focusing on visionary, innovative potentials that humanity has can use to manage risk. The two teams debate the plausibility of Team Humanity's response, and spectators are invited to join in the discussion,

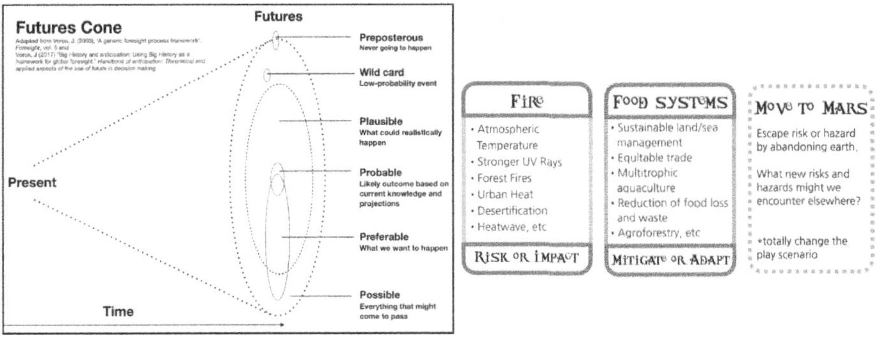

Fig. 1. Futures Cone (Left); Sample Risk Response cards (Right).

referring to the Futures Cone. A subsequent vote on the possible future, and whether the response outcome proposed by Team Humanity is Plausible, Probable or Preposterous, decides the winner of the round, if that team is convincing in their argument, or supported by terminal discourse cards. Teams are then reconfigured, typically by revolving one player at a time to the right, the cards are shuffled and redistributed for the following rounds, which continue until each participant has had the opportunity to play on both Risk and Humanity teams.

The gameplay rules are: Team Risk randomly draws ONE (1) *Risk or Impact* card, and must devise a risk proposal within that rubric, although not necessarily choosing from the suggestions (hence the *etc* at the end of each list). Team Humanity randomly draws TWO (2) *Mitigate or Adapt* cards, and their response to the proposed risk must be framed in relation between those two rubrics. Each player, and any spectator who wishes to participate, randomly draws ONE (1) *Discourse card*. These cards can be played by any player, or spectator, at any time to influence the discussion. Several of the discourse wildcards offer final decisions which allow to break an impasse, or when the futures vote is tied.

Conversational Vs Group Modelling Play (LEVELS I AND II). In conversational play (LEVEL I), the game is played based on common-sense, drawing from players's collective intelligence to imagine and describe risks and responses. This level represents the first loop in which players are introduced to the rules and play possibilities provided by the different cards and mechanics. In these first gameplay encounters, questions and challenges to the minimal rules are sounded out, and the teams decide for example on the style of voting, e.g. a show of hands, unanimous or simple majority. After playing at least one round in conversational mode, which lasts approximately 20 minutes, and depending on the context of play (e.g. if the game is instantiated in a formal workshop, at a conference social event, or an informal game night among friends and colleagues), players can advance to the next level.

The second level of play is ontology-based group modelling (LEVEL II), which follows the same team setting, risk and response proposals, Futures Cone discussion, and plausibility vote, yet with the addition of a printed *reference seman-*

tic model/framework, such as ontology-based conceptual models or Knowledge Graphs. Example of those could be the *social-ecologial systems (SESs) integrated conceptual model* [2,3] and the *Common Ontology of Value and Risk* (COVER) [38]. For this mode, one new game rule is implemented: when Team Risk proposes their risk they must create a conceptual model following the reference model, describing some aspects of their risk, e.g. what is the hazard and who or what is potentially affected (exposure). Likewise, Team Humanity must model their response, using the same reference model. Teams can then use their models to elaborate, articulate and argue for their risk and responses in the discourse. At the conclusion of each round, which typically lasts about an hour, the risk and response models are labeled and collected for later analysis. Figure 2 captures group modelling in action.

Fig. 2. Risk Response group modelling play (LEVEL II).

4 Preliminary Evaluation

This section highlights preliminary evaluation results obtained from the participatory workshop *Exploring ontology-based design interactions for sustainability research*,[5] organised and facilitated by the first two authors at RCIS 2024. The **goal** of this evaluation is to examine the use, by participants, of ontological notions to effect semantic alignment in risk communication.

4.1 Context and Participants

The workshop introduced and investigated ontology-based participatory sensemaking approaches for social-ecological systems research (see [48]), with activities centered around an ontology-based SESs integrated conceptual model, from [2,3]. These included an icebreaker to discuss relations in the model, graphical scenario analysis using the conceptual model, and Risk Response gameplay. The

[5] https://humanfactorsinsemantics.net/RCIS2024.html.

ontology-based SESs integrated model captures main elements of two social-ecological systems paradigms, the *social-ecological system framework* [30] and the *ecosystem services cascade* [36], via ontological analysis, using for example the *Unified Foundational Ontology* [25], for their ontological disambiguation and integration. The model includes concepts such as *ecosystem service*, *structure*, and *function*, *actor*, *social actor*, *organisation*, *governance*, as well as *resource system*, *natural resource human-made resource*, among others. It has been refashioned in the *Web Ontology Language* (OWL2) and can be found in GitHub.

Forty-five (45) workshop participants engaged in Risk Response gameplay, including a large, multicultural contingent of masters students from information systems studies. Four (4) groups of 8 to 12 participants formed teams of 5 or 6 players each for Team Risk and Team Humanity. After playing a round in LEVEL I conversational play, the four groups switched roles for a round of (LEVEL II) group modelling play, each completing the **tasks** of risk and response modelling following the SESs integrated conceptual model as a semantic reference.

4.2 Results

Our frames of analysis elucidate *quality dimensions* in conceptual models, in line with longstanding discussions in Software Engineering and Information Systems literature (e.g. [12,22,34]). We focus on *modelling styles and techniques*, correct use of *model concepts and relations*, *cross-model semantic alignment*, and ethnographic observations of *models in communication*. Primary data are the group models created during gameplay, photographic evidences and observations of the first two authors as workshop facilitators. In the following report, reference model components are named in lower-case *italic*, and components introduced by participants are written, first letter capitalised, in `Courier`.

Modelling Styles and Techniques. Several distinct modelling practices can be observed in the group models. In Groups 1 and 2, Team Risk first created a model of their proposed risk and Team Humanity created a corresponding model of the proposed response ("dual modelling"). Team Risk in Groups 3 and 4 first created a model of the proposed risk, and Team Humanity *extended* that model with their response ("singular modelling"). The use of relations arrows also differed between groups: for example in Group 4's model, we find mostly unlabelled relations following the reference model standard of defining subsumption using closed-head arrows, as in UML, while in Group 1's risk model, variant arrows are used, yet these are labelled with the proper relations from the reference model. The other groups and teams employed a variety of open arrows and connecting lines. In all of the group models, response modelling generally follows the style for concepts and relations set by the risk teams, and in the groups practicing singular modelling, players differentiated between risk and response using different colors of ink. As participants were not given detailled modelling instructions, the emergence of similar practices among different groups is noteworthy.

Correct Concepts and Relations. We considered as evaluation parameter the uptake and correct use of concepts from the SESs reference framework, first examining the group models for *explicit semantics*, i.e. terminology adopted directly from the reference model, either as labels for named concepts, or as primary class or sub-class entities depicted in their models. Secondly, the group model concepts were examined for correct *implicit semantics*, identifiable by their named relations and position in the model. Considering the risk model of Group 1 (dual modelling), we find only one reference ontology concept, in the depiction of `Humanity` *has* the *goal* of `Survival`. This is counted as one marker of explicit correctness. In that model, thirteen (13) additional concepts describe impacts of Vegetation risks, such as `Deforestation` and related `Habitat Loss`, none of which are found in the reference model. However, the proper use of five (5) reference model relations, several applied multiple times, reveal eight (8) more correct uses of implicit semantics. For example, `Pesticide Use` is identifiable as an *activity*, because it is *performed_ by* `Humanity`, itself properly placed as a *social actor*, due to the *creates* and *recognized_ by* relations drawn to *human-made resources* and *natural resources.* Figure 3 depicts the most common, correctly used reference model concepts, of which *activity* and *goal* are predominant, followed by *actor*, *social actor*, *natural-resource* and *human-made resource.*

Cross-model Semantic Alignment. In group modelling, cross-model analysis makes possible to identify emergent shared understandings, wherein model concepts and relations are re-fashioned or corrected in subsequent collaborative modelling operations. For instance, the risk model of Group 2 (dual modelling approach) describes Air pollution using four (4) concepts from the reference model: `Humans` as *social actors*, and `Machines` as *actors* are connected to *activities* which `Produce Waste` with the *goal* of `Air Pollution`. This model incorporated no named relations. In reply, Group 2 Team Humainty reformulated

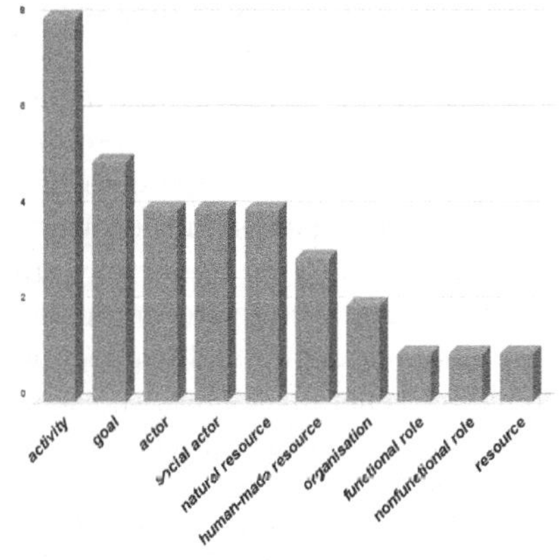

Fig. 3. Correct categorisation.

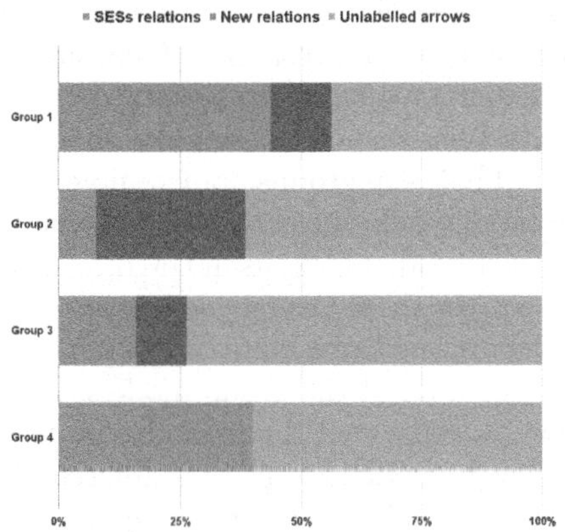

Fig. 4. Use of relations.

the same four reference concepts, to show `Human` and `Electric Machines` as *social actors* related to *activities* with the *goal* of `Clean Air`. Players modelled *resources* and `government` (echoing *governance*) in a response that demonstrates multiple correct implicit semantics, for example of `Policy Making`, recalling the reference model's *policy*. Though not all model concepts and relations are correct, the response model offers some semantic clarification of the risk model, in particular concerning governance, policy and regulation, that directly reflects participants' engagement with the ontology of the reference model.

Group 3 used singular modelling, and iteratively built on each other's ideas concerning Vegetation risk. Team Risk modelled the risk posed by Genetically Modified Organisms (GMO's) relating to `Government` and *policies*, with a new relation `Enforced_by`, and several *organisations*, using 10 unlabelled relations arrows to draw connections. Group 3 Team Humanity added two `Activism` elements, and described future impacts of *policies*, *organisations* and `Consumers`. Their relations clarify the original risk model: `Activism` is an *activity* that is *performed_by* the *social actors* of `Consumers` and `Food Regulation Organisation` and furthermore that `Disease Prevention and Control` is a *goal associated to* the *activity* of `Governnment`, which though is misspelled, is in complete semantic alignment with *governance* from the reference model. In these cases the correct use of relations facilitates semantic alignment, and serves to correctly categorise concepts from the first model, according to the reference model. Figure 4 shows the different group models' correct use of relations, newly introduced relations, and unnamed modelling arrows. Groups 1, 3 & 4 all used more relations from the reference model, with only Group 2 requiring more imagined relations to communicate their idea. Group 4's model is noteworthy as its only relations come from the reference model, and while there are many variant arrows and connecting lines, it was this group that used explicit *is_a* closed arrowhead relations connectors, strictly aligning with the reference model's composition.

Models in Communicative Action. One of the goals of conceptual models is to foster communication and understanding [32], topics widely discussed in literature. We isolate from participant observations, a particular kind of *embodied* communication involving the models themselves. During the preparation of the group models, and gameplay discussions, players frequently gestured and traced relations across the group and reference models. This physical engagement with cognitive and material artefacts to effect communication, also known as embodied sense-making (see [48]), is less present in LEVEL 1, conversational play, and emerges with the introduction and creation of model artefacts in LEVEL 2 group modelling play. Such communication is further evident in the distinct practices adopted by players, of using the models as communication tools. During discussion and voting, players most often placed their models on the table, together with the Futures Cone, pointing out concepts and relations as they debated possible and plausible futures. In one group, a player held their team's model up for the other players to see, explaining each model concept and relation; in another group, a player held their team's model facing themselves, explaining their ideas using the model as a guide, rather than explaining the model. The emergence among four groups of players, of multiple, distinct embodied communication styles involving the model artefacts suggests that the game-based group modelling activity is broadly inclusive of participants' differing expressive exigencies discursive styles, and demonstrates the communicative affordances of conceptual models as discursive artefacts.

5 Discussion and Limitations

In discussion we offer some critical reflections on Risk Response, its relation to risk communication and some limitations of the group modelling activity and evaluation. The game's execution activates collective intelligence that is central to the intergovernmental bodies which have produced risk-related documents on which the game is based. Participants generate the risks for discussion, brainstorm mitigation and adaptation responses, then speculate and vote on the possible or plausible futures, thereby up-ending and reinterpreting the conventional science communication and policy-decision paradigms. This aspect figures into the previously discussed transformation from information deficit, to dialogue and participation in risk communication.

In LEVEL 2 group modelling play, the game's minimal rules and instructions allow players to decide if one, or multiple players will draw the model, and how to interpret reference ontology components. While this offers players flexibility to interpret modelling as a communicative act, it poses some limitations to the execution and evaluation of models produced in-game. These models do not capture the discourse of gameplay, argumentation of players, evolution of ideas, nor the engagement and intrigue that can be triggered by the Futures Cone and plausibility vote. All of these have been observed by the authors as workshop facilitators, especially when teams try to convince eachother of their imagined future. Neither do the models record the use of, and envelopment of spectators provided by the discourse cards. Photographs taken during the workshop offer additional insights into group modelling and discursive processes during gameplay, however the authors deliberately refrain from video or audio surveillance. This omission, while retaining a familiar, relaxed environment for participants, requires that some analyses remain interpretation, based on tacit knowledge derived from the authors' many previous ethnographic studies via participatory practices and workshop facilitation, enacting conceptual- and cognitive modelling towards interaction design research (e.g. [49]).The researchers' implicit bias can endanger evaluation validity, and as players are given the directive of modelling a response to a proposed risk using a reference model, it is almost assured that their models will reflect the ideas of both the reference model and those of other participants. However, the completed models (and model pairs) of all four groups depict active engagement between teams, and elaboration of each other's ideas using the structures of the reference models as they further define, semantically align, and communicate concepts and their relations.

6 Related Works

The tool and approach proposed in this paper combines gameplay, group modelling, ontology-based semantic alignment, and in-game evaluation strategies, characteristics that cannot be found in other related games and research literature. Yet Risk Response does not emerge from a vacuum, and numerous educational and/or serious games addressing risk exist (see [41] for a detailled review). These are largely simulation boardgames concerned with Disaster Risk Management, that leverage role-play for multi-party discourse and information sharing. Typical examples of the genre are *Hazagora* [31], a serious game for students and administrators to increase their awareness of geohazards and disaster risk reduction, and *ANYCaRE* [44], a role-play game focused on crisis management, forecasting, and early warning of flash floods, or strong winds. A more academic entrant is the *RAMSETE* series [19] of information elicitation games on disaster risk and climate change adaptation, set within European policy space. Evaluations of these games, when are performed, rely on post-experience questionnaires and notes taken by facilitators. In contrast Risk Response allows evaluation of models produced *during* gameplay. In addition, while most risk games revolve around group discussions and collaborative action [41], none that we have found explicitly aim to align, or analyze the semantics of such discussions.

The intermittent academic investigation of games for participatory modelling stems can be traced to Bousquet et al. [14] seminal work, wherein stakeholders role-play and roughly craft models of resource use and conflicts, regarding for example irrigation systems and agroforestry, which are then formalized as computer simulations. Perez et. al. subsequently offered *ComMod* [35] in which role-playing, model creation, and simulations are blended in a knowledge elicitation practice. More recently Bakhanova et.al [10] acknowledge many potential benefits of games for participatory modelling, however the in-game modelling is far from established practice. Risk Response, by combining game, discourse and group modelling, and by integrating ontologies for semantic alignment in player discussions, opens a new direction in this field.

7 Conclusions and Future Works

This article contributes to discussions on participatory risk communication by introducing the adversarial discourse game Risk Response, a stakeholder engagement and group modelling tool for examining future risks and associated responses. The game and play are presented, with a preliminary evaluation that elaborates the affordances of game-based group modelling for semantic alignment. An important next step will be to enact more thorough evaluations, in real-world risk analysis settings, playing with domain experts and local stakeholders. Dedicated cards formulated around site-specific risks and localized scientific research can be designed to reflect participants' lived experiences and further operationalize ownership of risk discourse.

The development, playtesting, and evaluation of this game and group modelling outcomes, revealed some potentially useful game revisions, such as extending the socio-political and technological risk suggestions in the cards with themes that repeatedly emerge in gameplay sessions, e.g. loss of democracy or human rights, misinformation, and risks posed by robots and automation. Additional *Discourse cards* will add compound risk [40] and multi-hazard [23] concepts to the game, guiding players to imagine aggregate, and cascading risks. In addition, the communicative potentials of group modelling in the game will be sharpened to include map and graph analysis, collating players' identified interrelations between risks, impacts and responses and plotting them in real time.

Acknowledgments. Authors acknowledge the financial support the project D2G2 (DFG grant n. 500249124). We would also like to thank the reviewers for their insightful feedback, our RCIS 2024 workshop attendees for their participation, and the conference organisers for graciously enabling this ongoing research.

References

1. Adamo, G., Sperotto, A., Fumagalli, M., Mosca, A., Sales, T.P., Guizzardi, G.: Unpacking the semantics of risk in climate change discourses (2024)
2. Adamo, G., Willis, M.: Conceptual integration for social-ecological systems - an ontological approach. In: Research Challenges in Information Science - 16th International Conference, RCIS 2022, Barcelona, Spain, 17–20 May 2022, Proceedings. Lect. Notes Bus. Inf. Process., vol. 446, pp. 321–337. Springer (2022)
3. Adamo, G., Willis, M.: The omnipresent role of technology in social-ecological systems - ontological discussion and updated integrated framework. In: Research Challenges in Information Science: Information Science and the Connected World - 17th International Conference, RCIS 2023, Corfu, Greece, 23–26 May 2023, Proceedings. Lect. Notes Bus. Inf. Process., vol. 476, pp. 87–102. Springer (2023)
4. Andy, R., et al.: The concept of risk in the ipcc sixth assessment report: a summary of cross- working group discussions. guidance for ipcc authors. Technical report (2020)
5. Auger, J.: Speculative design: crafting the speculation. Digit. Creat. **24**(1), 11–35 (2013)
6. Auger, J., Hanna, J., Encinas, E.: Reconstrained Design: Confronting Oblique Design Constraints. Design+ Power. Nordes, Oslo (2017)
7. Aven, T.: Risk assessment and risk management: review of recent advances on their foundation. Eur. J. Oper. Res. **253**(1), 1–13 (2016)
8. Aven, T.: Perspectives on the nexus between good risk communication and high scientific risk analysis quality. Reliab. Eng. Syst. Saf. **178**, 290–296 (2018)
9. Aven, T.: Is the definition of risk still contested? (2023)
10. Bakhanova, E., Garcia, J.A., Raffe, W.L., Voinov, A.: Targeting social learning and engagement: what serious games and gamification can offer to participatory modeling. Environ. Model. Softw. **134**, 104846 (2020)
11. Balog-Way, D., McComas, K., Besley, J.: The evolving field of risk communication. Risk Anal. **40**(S1), 2240–2262 (2020)
12. Becker, J., Niehaves, B., Pfeiffer, D.: Ontological evaluation of conceptual models. Scand. J. Inf. Syst. **20**(2), 4 (2008)

13. Bogost, I.: Persuasive Games: The Expressive Power of Videogames. The MIT Press, Cambridge (June 2007)
14. Bousquet, F., Barreteau, O., d'Aquino, P., Etienne, M., Boissau, S., Aubert, S.: Multi-agent systems and role games: collective learning processes for ecosystem management. In: Complexity and Ecosystem Management, pp. 248–285. Edward Elgar Publishing (2002)
15. Bucchi, M.: Of deficits, deviations and dialogues: theories of public communication of science. In: Handbook of Public Communication of Science and Technology, pp. 71–90. Routledge (2008)
16. Cook, D.: Loops and arcs **2016**(Nov 5) (2012). http://www.lostgarden.com/2012/04/loops-and-arcs.html
17. for Disaster Risk Reduction (UNDRR), U.N.O.: Hazard definition and classification review: Technical report. Technical report (2020)
18. Elsden, C., et al.: On speculative enactments. Proc. CHI **2017**, 5386–5399 (2017)
19. Fleming, K., et al.: The use of serious games in engaging stakeholders for disaster risk reduction, management and climate change adaption information elicitation. Int. J. Disaster Risk Reduct. **49**, 101669 (2020)
20. Fry, T.: Design futuring, pp. 71–77. University of New South Wales Press, Sydney (2009)
21. Gall, T., Vallet, F., Yannou, B.: How to visualise futures studies concepts: revision of the futures cone. Futures **143**, 103024 (2022)
22. Gemino, A., Wand, Y.: A framework for empirical evaluation of conceptual modeling techniques. Requir. Eng. **9**(4), 248–260 (2004)
23. Gill, J.C., Malamud, B.D.: Hazard interactions and interaction networks (cascades) within multi-hazard methodologies. Earth Syst. Dyn. **7**(3), 659–679 (2016)
24. Glette-Iversen, I., Flage, R., Aven, T.: Extending and improving current frameworks for risk management and decision-making: a new approach for incorporating dynamic aspects of risk and uncertainty. Saf. Sci. **168**, 106317 (2023)
25. Guizzardi, G., Botti Benevides, A., Fonseca, C.M., Porello, D., Almeida, J.P.A., Prince Sales, T.: UFO: unified foundational ontology. Appl. Ontol. **17**, 1–44 (2021)
26. IPCC: Ipcc fifth assessment report-synthesis report. Technical report (2014)
27. Irwin, A.: Risk, science and public communication: third-order thinking about scientific culture. In: Routledge Handbook of Public Communication of Science and Technology, pp. 147–162. Routledge (2021)
28. Janssen, M.A., Falk, T., Meinzen-Dick, R., Vollan, B.: Using games for social learning to promote self-governance. Curr. Opin. Environ. Sustain. **62**, 101289 (2023)
29. Lee, H., et al.: Synthesis report of the ipcc sixth assessment report (ar6), longer report. ipcc. (2023)
30. McGinnis, M.D., Ostrom, E.: Social-ecological system framework: initial changes and continuing challenges. Ecol. Soc. **19**(2) (2014)
31. Mossoux, S., Delcamp, A., Poppe, S., Michellier, C., Canters, F., Kervyn, M.: Hazagora: will you survive the next disaster?-A serious game to raise awareness about geohazards and disaster risk reduction. Nat. Hazard. **16**(1), 135–147 (2016)
32. Mylopoulos, J.: Conceptual modelling and telos. Conceptual modelling, databases, and CASE: an integrated view of information system development, pp. 49–68 (1992)
33. Nealen, A., Saltsman, A., Boxerman, E.: Towards minimalist game design. In: Proceedings of the 6th International Conference on Foundations of Digital Games, pp. 38–45 (2011)
34. Nelson, H.J., Poels, G., Genero, M., Piattini, M.: A conceptual modeling quality framework. Softw. Qual. J. **20**(1), 201–228 (2012)

35. Perez, P., et al.: Assessment and monitoring of the effects of the commod approach. Companion modelling: a participatory approach to support sustainable development, pp. 155–187 (2014)
36. Potschin, M., Haines-Young, R., et al.: Defining and measuring ecosystem services. Routledge handbook of ecosystem services, pp. 25–44 (2016)
37. Rosa, E.A.: Metatheoretical foundations for post-normal risk. J. Risk Res. **1**(1), 15–44 (1998)
38. Sales, T.P., Baião, F., Guizzardi, G., Almeida, J.P.A., Guarino, N., Mylopoulos, J.: The common ontology of value and risk. In: Conceptual Modeling - 37th International Conference, ER 2018, Xi'an, China, 22–25 October 2018, Proceedings. LNCS, vol. 11157, pp. 121–135. Springer (2018)
39. Shi, P., et al.: Disaster risk science: a geographical perspective and a research framework. Int. J. Disaster Risk Sci. **11**, 426–440 (2020)
40. Simpson, N.P., et al.: A framework for complex climate change risk assessment. One Earth **4**(4), 489–501 (2021)
41. Solinska-Nowak, A., et al.: An overview of serious games for disaster risk management-prospects and limitations for informing actions to arrest increasing risk. Int. J. Disaster Risk Reduct. **31**, 1013–1029 (2018)
42. Steelman, T.A., McCaffrey, S.: Best practices in risk and crisis communication: implications for natural hazards management. Nat. Hazards **65**, 683–705 (2013)
43. Stewart, I.: Advancing disaster risk communications. Earth-Sci. Rev. 104677 (2024)
44. Terti, G., et al.: Anycare: a role-playing game to investigate crisis decision-making and communication challenges in weather-related hazards. Nat. Hazards Earth Syst. Sci. **19**(3), 507–533 (2019)
45. UNDRR: Hazard information profiles: Supplement to undrr-isc hazard definition & classification review. Technical report, United Nations Office for DisasterRisk Reduction (UNDRR) (2021)
46. UNESCO: Transforming the future: Anticipation in the 21st century. Taylor & Francis (2018)
47. Voros, J.: Big history and anticipation: using big history as a framework for global foresight. Handbook of anticipation: theoretical and applied aspects of the use of future in decision making, pp. 1–40 (2017)
48. Willis, M., Adamo, G.: Ontology-based interaction design for social-ecological systems research. In: International Conference on Research Challenges in Information Science, pp. 74–82. Springer (2024)
49. Willis, M., De Angeli, A., Zancanaro, M.: Experience probes: immersion and reflection between reality and virtuality. In: Human-Computer Interaction–INTERACT 2017: 16th IFIP TC 13 International Conference, pp. 253–262. Springer (2017)
50. Yamori, K.: Disaster Risk Communication: A Challenge from a Social Psychological Perspective. Springer Publishing Company, Incorporated, 1st edn. (2020)

Modeling Methods and Requirements Engineering

Modelling Hierarchies of Organisational Rules

Jöran Lindeberg[1(✉)], Martin Henkel[1], and Katarina Fast Lappalainen[2]

[1] Department of Computer and Systems Sciences, Stockholm University, Stockholm, Sweden
{joran,martinh}@dsv.su.se

[2] Department of International and Operational Law, Swedish Defence University, Stockholm, Sweden
Katarina.FastLappalainen@fhs.se

Abstract. The vast and growing web of rules and regulations that govern modern organisations is increasingly difficult to comprehend and manage. One technique for understanding and managing rules among stakeholders is enterprise modelling of organisational rules. However, rule models are usually flat, lacking hierarchical structure that allows for effective abstraction and reduced complexity. The aim of this paper is to examine the different aspects of hierarchy among organisational rules. A model pattern is proposed, based on the main constructs rule description, norm hierarchy, scope, and citation. The model pattern is demonstrated by instantiating it in a concrete case, the European Union's General Data Protection Regulation (GDPR), using Wikibase Cloud to structure the data and Kumu.io to visualise it.

Keywords: Organisational Rule System · Enterprise Modelling · Rule Hierarchy · Legal Design · Privacy Regulation · OntoUML

1 Introduction

The rules and regulations that govern modern organisations form an increasingly complex web of demands to manage. Despite the best efforts of law makers, problems with vagueness, imprecision, and ambiguity remain [11]. Moreover, the sheer volume of legislation at different levels, industry standards, and an organisation's internal rules (which main purpose often is to comply with externally imposed rules) makes it difficult for legal experts, and even more so for IT people and other stakeholders, to navigate and comprehend these systems. This is problematic when building regulatory-compliant software systems in heavily regulated sectors, such as healthcare [8]. Another area is when trying to understand the rules that govern a certain sector.

Without any means to structure rules, an effort to map out rules governing a certain domain may become unwieldy. The authors are involved in a project, Health Data Sweden, where a part of the work is to create a legal toolbox

J. Grabis et al. (Eds.): RCIS 2025, LNBIP 547, pp. 175–191, 2025.
https://doi.org/10.1007/978-3-031-92474-3_11

- an easy to use guide for understanding regulations in the health care area. The experience of creating the legal toolbox for the increased use of health data touched on the complexity of rules handling. One part of the toolbox was a map, not yet published, of the health data ecosystem, including actors, regulations, and health data sets. The map built with the Kumu visualisation tool, pulling data from Wikidata, over time became very large, including around 700 elements and over 1,300 connections. This example shows how important it is to build hierarchies that allow users to manage complexity by, for example, regulating the granularity of the map.

One technique for handling complexity and creating a common understanding among stakeholders is enterprise modelling [20]. Enterprise modelling, as a form of knowledge creation in organisations, helps individuals internalise the explicit information of the organisation into tacit knowledge, a process called sense-making [12]. As argued by Liu et al. [14], there is a highly fluid dynamic between mental models and external visualisations. Mental models are affected by visualisations, and effective visualisations are normally based on existing mental models. With time, they converge. Pure mental models have two major weaknesses: lack of precision and lack of working memory (The Magical Number Seven, plus or minus two [17]). Fortunately, a mental model can be augmented by an external visualisation. Improved sense-making means better comprehension of problems, leading to better solutions. For example, an IT-project in highly regulated environment benefits if all stakeholders develop a common understanding of basic legal conditions.

Enterprise modelling deals with many aspects of an organisation, including organisational rules. However, rule models are usually flat, lacking the hierarchical structure that allows for abstraction and reduced complexity. As expressed by Krogstie [12, pp. 130–131] in 2012:

> Whereas the traditional operational models (e.g. process models) have decomposition and modularisation facilities that make it possible to view a system at various levels of abstraction and to navigate in a hierarchical structure, rule models are usually flat. With many rules, such a model soon becomes difficult to grasp, even if each rule should be understandable in itself.

As far as we have been able to establish, this observation still holds.

Organisational rules form intricate, complex, webs: organisational rule systems. Our definition of an *organisational rule* is broad. It includes all the rules and regulations that affect the action space of an organisation. This covers both legislation, contracts, and internal rules established by the organisation itself. Since law is a more mature field than the study of internal organisational rules, most of our examples in this paper are from the legal area. Moreover, legislation is publicly available and usually well-structured.

These rule systems have substantial hierarchical aspects that could be modelled to facilitate common understanding among stakeholders. Modelling can be seen as "a type of hierarchical abstraction" [12, p. 89]. Hierarchical abstraction is an effective strategy for humans trying to make sense of complex systems.

Depending on the question at hand, the involved people need to find a suitable level for information search and discussion. Although some decisions require a high level of detail, others can be taken based on a more abstract understanding [15].

The research question is thus the following: How can hierarchies in organisational rule systems be modelled to facilitate sense-making?

When talking about hierarchies and sense-making, what we concretely have in mind is a digital map analogy that allows you to zoom in and zoom out. The first situation is when you start from above and gradually go downward to higher granularity. As more details become visible, you adjust your direction. The second situation is when you start from below and look upward to understand what is around or above you.

Since the main purpose of the study is sense-making, the pattern should be easy to use for different kinds of stakeholders. Thus, as for non-functional requirements, we prioritise ease of use over correctness and completeness.

The remainder of this paper is organised as follows. Section 2.1 expands on hierarchies. Section 2 presents related research, mostly from legal frameworks. Section 3 proposes four hierarchical dimensions to be used in organisational rule systems. Section 4 presents the proposed pattern. Section 5 demonstrates the pattern by instantiating it as a model of the General Data Protection Regulation (GDPR) and its related rules. Section 6 discusses the strengths and weaknesses of the pattern. Section 7 concludes.

2 Related Research

2.1 Modelling Hierarchies

By *hierarchy* we subscribe to the view of Krogstie that this concept also includes weakly hierarchical graphs, not just strict ones [12, p. 90]. A strictly hierarchical graph has only one top node, each subnode has exactly one superior node, and has no lateral relations. In contrast, a weakly hierarchical graph largely looks like a strict one, but deviations are tolerated. There in no clear definition of when a weakly hierarchical graph ceases to be hierarchical, but Krogstie emphasises that there the term should be used only for graphs that are close to strictly hierarchical graphs.

According to Krogstie [12, p. 93], there are four standard hierarchical relations.

- *Classification.* E.g. Harry Potter is a human.
- *Aggregation.* E.g. A book is composed by pages. (Also called composition.)
- *Generalisation.* E.g. Cats are mammals.
- *Association.* E.g. RCIS 2025 is a member of the set RCIS conferences.

In addition, Krogstie considers an additional standard hierarchical relation: *supervision*, which means a phenomenon that controls other phenomena. For example, a regional healthcare system may be controlled by a ministry of health.

Another aspect of hierarchy and abstraction is discussed by Vermier [22], observing that conceptual models are helpful for common understanding between stakeholder, but that real-life conceptual models are often too large to be easily understood. Different algorithms and other techniques have been proposed for abstraction. One approach is to make a hierarchy of increasingly detailed entity relationship models, in which a lower-level model is represented as a single entity by a higher-level model [6].

There are also downsides to abstraction. As explained by Zugal et al. [25, p. 126], "the introduction of submodels may force the reader to switch her attention between the submodels, leading to the so-called split-attention effect". Consequently, small models are less likely to benefit from abstraction, and it is beneficial if the submodel can be expanded close to its representation in the higher-level model.

2.2 Existing Models of Rule Hierarchies

Although, to the best of our knowledge, there are no studies that directly propose how to model hierarchical structures of rules, there is work that deals with rules and their application. Some of these contributions include aggregation relations as well as supervision relations.

The European Legislation Identifier (ELI) [3] "is a framework to make legislation metadata available online in a standardised format". ELI represents that a legal resource is part of another legal resource and also that a work can be a member of another work. Another legal meta-data framework, SALI (Legal Industry Standards) [4], offers a taxonomy that, for example, states that "privacy law" is part of "information security law". Moreover, Hamdaqa & Hamou-Lhadj [10], present a model in which a legal act is composed of one or more provisions, which in turn can be composed by other provisions.

Other contributions have focused on cross-referencing, that is, *citations* [10, 16]. Some types of citation convey hierarchical relations. Moreover, cross-referencing needs precise references, down to the lowest-level clause of a legal text, which connects this area with that of aggregation, discussed above. Thus, several works included hierarchical decomposition of legal texts. However, none of these contributions proposes how citations can be used for abstraction.

Some meta-models can express that a rule exists to provide detail within the limits of a higher-level rule. Kholkar et al. [11] present a meta-model of a generic set of regulation concepts that represent that a Directive is supported by a Regulation. A similar pattern is found in the meta-model of Semantics Of Business Vocabulary And Business Rules (SBVR) [18], which has Directives made operational by Business Rules. Also the taxonomy of SALI shows a hierarchy of norms. However, in our view, it is strangely flat, with, for example, "legislation" and "constitution" at the same level.

Other meta-models can express that rules can have different scope. A rule can have a scope along different dimensions. The most obvious and manageable is the territorial scope, also known as jurisdiction. The Akoma Ntoso (AKN) ontology [7], LegalRuleML [5], SALI, and ELI also have some support for this. ELI also has the property "coverage" that can be not only spatial (as in jurisdiction), but also temporal. However, the scope can also be material. The legislation could, for example, only apply to large enterprises, but not to small and medium sized enterprises. Here, SALI includes, for example, a taxonomy of different types of industries. However, none of these models addresses how to use spatial, temporal, or thematic scope to construct hierarchies.

In summary, earlier contributions have created hierarchies but have not much addressed how to leverage them for abstraction.

3 Hierarchies in Organisational Rule Systems

In the meta-models presented in Sect. 2, we identified four rule hierarchy dimensions that can be useful for modelling: norm hierarchy, scope, rule description decomposition, and citations. While the first two are about the rules themselves, the latter two originate from the structure of the texts that express the rules, as shown in Table 1.

Table 1. Two types of rule hierarchies

Type	Dimension
A. Rule	1. Norm hierarchy
	2. Scope
B. Rule description	3. Rule description decomposition
	4. Citations

A concept that is related to rule hierarchy, but different, is *rule precedence*. While the former is about how to systematically build layers of abstraction, the latter is about determining which rules win over the other in case of conflict in a particular situation. This paper is about rule hierarchies, but here also follows a brief discussion about rule precedence.

In the legal field, there are three general principles for conflict resolution, *lex superior*, *lex posterior*, and *lex specialis* [23] According to the *lex superior* principle, hierarchically superior norms have precedence over hierarchically inferior norms. The principle *lex posterior* states that later norms supersedes earlier norms. The principle *lex specialis* implies that the more specific norm prevails over the more general one. These principles may in turn be in conflict with each other. A law can, for example, be very specific and perfectly subsume a case but be found to be unconstitutional.

However, we found only one of these principles, lex superior, useful for building layers of abstraction. In other words, rule precedence is only partially overlapping with hierarchical abstraction of rules.

3.1 Norm Hierarchy

Lex superior is a type of supervision hierarchical relation, presented in Sect. 2.1. It is useful because lower-level rules may only add more granularity to higher-level rules. In most if not all legal systems, there is a *norm hierarchy* [15] that prohibits ordinary law from violating the constitution of a country. Going down the ladder, the statutes of an organisation must respect the law of the land. Descending even further, in most organisations the executive director cannot make rules that conflict with those decided by an annual assembly.

For example, in healthcare, the hierarchy of rules can include international health regulations, national health policies, hospital guidelines, and specific departmental protocols.

3.2 Scope

A rule is only applicable within a certain *scope* [11]. In Sect. 2 three types of scope were presented: spatial, temporal, and material. A rule that covers a wide geographical area for a long time, and is applicable for many types of actors, processes, industries, etc., is in general more important than in the opposite case. Going back to the digital map analogy, this is the type of rule you want to see already from a high altitude. Continuing with the same analogy, if you are on the ground looking upward, you probably first want to see the rules that have the most specific scope for your situation.

The *spatial scope* is what, in a legal context, is usually called jurisdiction. For example, Swedish law is normally only applicable in Sweden. There are also cases of competing jurisdictions, but that is beyond the scope (no pun intended) of this paper. There are plenty of taxonomies that decompose the world into administrative areas, from continent to neighbourhood. For example, ELI recommends the *Getty Thesaurus of Geographic Names*.

The *temporal scope* may seem obvious. Most often, you would first and foremost be interested in current legislation. It could also be argued that laws that have existed for a long period of time generally have had more time to have an impact and affect other laws. There is also the *Lindy effect* [21], suggesting that the longer something non-material has existed, and withstood the test of time, the longer it is likely to continue existing.

The *material scope* is the messy scope. It is usually not difficult to establish to which region a city belongs to or which year a month is part of. However, only imagination sets the limit for how many different material scopes one can invent. One dimension could be the actors who must comply with a rule. For example, only organisations or also individuals? Another dimension could be activity. For example, healthcare or construction? If you are building a hospital, you are probably doing both. However, in spite of the messiness, material scopes could be useful. All other things alike; a regulation that applies to all farming activities should be higher up in a hierarchy than one that is only about growing tomatoes.

3.3 Rule Description Decomposition

Most laws are published as self-contained documents. These documents are divided into sections of different levels, such as chapters, clauses, paragraphs, etc. Other types of rules than legislation, such as intra-organisational rules, or industry standards, have similar structures, even if often less rigid.

Self-contained pieces of legislation can be grouped into sets, such as privacy law or commercial law. A set can be part of other sets. For example, international privacy law is part of both privacy law and international law. These are loose categorisations, but some useful taxonomies exist, such as the SALI standard. We refer to this as *rule description decomposition.*

3.4 Citations

Rule descriptions can be related by citing [19]. These *citations* can be made for a number of purposes, and some types of citation convey something about the influence of a particular rule description. One type of citation states that the citing rule description must comply with the cited rule description, establishing a supervision relation between the two described rules. We call this type *compliance citation.* A second type of citation delegates from the citing rule to the cited rules to decide about the details. This also creates a supervision relation, but in the opposite direction than a compliance citation. We denominate this type as *delegation citation.*

A difference between this relationship and the relationship that arises between a rule of lesser norm hierarchy than another rule, is that a compliance citation established a supervision relation between particular rules instead of types of rule.

4 Proposed Pattern

In this section, we turn the discussion from Sect. 3 into an OntoUML model. The OntoUMl modelling language [1] is grounded in the Unified Foundational Ontology (UFO) [2], and using foundational ontologies generally results in better models. OntoUML has also been used previously for legal modelling [9,24]. However, we did not completely follow UFO's multiplicity rules. There was a conflict between the rigour of UFO and our intention to keep the pattern simple for the sake of usability. For example, in UFO the characterization relation stereotype only allows for a multiplicity of exactly 1 at the source end.

Figure 1 shows our proposed OntoUML pattern for rule hierarchies.

In the centre of the pattern is a Self-contained Rule Description, which can be recursively composed of Rule Description Sections. For example, a law can be composed of parts, which in turn are composed by chapters, which are composed by sections, etc. Both Rule Description Section and Self-contained Rule Description are subclasses of Rule Description.

We concure with Griffo [9] and later Weigand et al. [24] in that norms are different from the texts that express them. Some norms are unwritten, such as

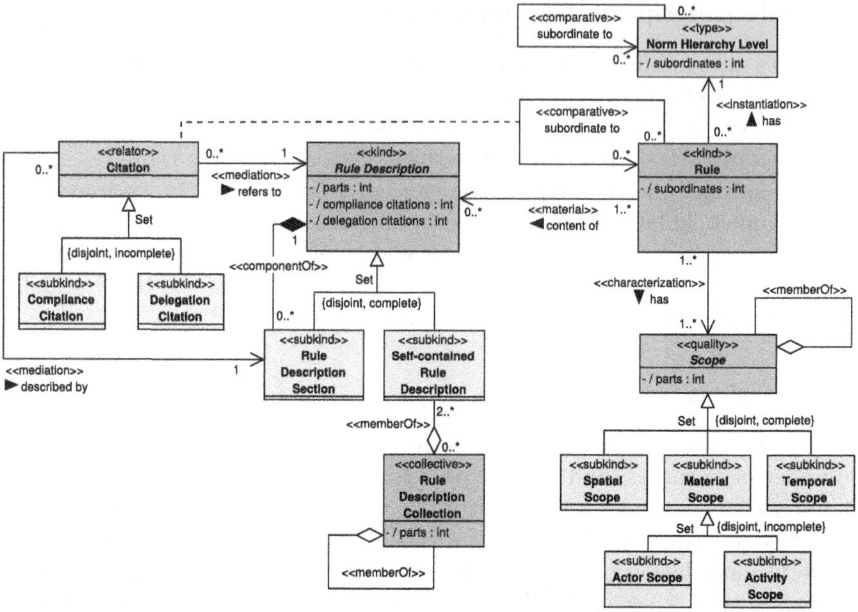

Fig. 1. Rule Hierarchies Model Pattern. Note that in OntoUML, arrows at the end of an association represent target end, rather than navigability.

customary law, but still have Scope and Norm Hierarchy, grounded in their social acceptance instead of a formal declaration. Thus, from a strictly ontological point of view, there should be one class for Rule Description and another one for Rules. We did consider, however, to merge these two classes for pragmatic reasons. As mentioned in Sect. 1, this study prioritises usability over correctness. When instantiating the pattern, it became clear that separating Rule Description from Rule would make the process cumbersome. Moreover, it is our understanding that people in general do not make this separation in their mental models of rules and regulations. Rather, they think about the content of particular rule sections. Since the purpose of the pattern is sense-making, in the instantiation we considered it preferable to align with what we believe is the common mental model. In other words, the instantiation does not completely align with the proposed pattern. The hierarchies could be demonstrated also with this solution.

A Rule has exactly one Norm Hierarchy Level. One type of norm will often be superior to some types and inferior to other types. For example, a constitutional norm beats a norm in ordinary legislation, which in turn beats a company's own rules.

Norm Hierarchy Level has an attribute: subordinates. Its value is calculated as the number of instances that an instance is recursively supreme of, according to the -supreme to- relation. See also Sect. 5 to better understand how this works in practice.

A Rule also has a Scope. There are three subclasses: Spatial Scope, Temporal Scope, and Material Scope. These classes need to be integrated with external taxonomies. For example, a taxonomy of administrative areas or a taxonomy about types of legal persons. In an organisation, the organisational units could be captured in an internal taxonomy, and an example value could be a company division.

Scope has a recursive part-of-relation with itself, showing that, for example, the state of Brandenburg is composed of several districts, or that a company division includes lower-level organisational units.

Scope has an attribute: parts. This value is calculated as the number of instances that an instance recursively has as parts, according to the -part of-relation.

A Self-contained Rule Description is commonly organised into parts, sections, paragraphs, etc. [10]. In the model, Rule Description Section is used as a generic construct for any type of these divisions. In some legal documents, these hierarchies can be several levels deep.

Rule Description and a Rule Description Collection have an attribute: parts. This value is calculated as the number of instances that an instance of these classes recursively has as parts, according to the -part of- relation (including both aggregations and compositions).

Rule Description Section can cite a Rule Description, either a complete Self-contained Rule Description or a Rule Description Section. As explained in Sect. 3.4, we have identified two types of citations that create hierarchical relationships: compliance and delegation. Compliance occurs when a rule states that it should be interpreted within the framework of another rule, which establishes a supervision relationship. Delegation can be considered the opposite of compliance. Delegation is when a rule declares that it will be followed by other more detailed rules. Like compliance, delegation establishes a supervision relation, but with the citing rule as the superior party. Citations are between Rule Descriptions but are *truth-makers* [1, Relator] to subordinate associations between rules. The Rule class has an attribute: subordinates. This value is calculated as the number of subordinate citation associations that an instance of this classes recursively has pointing to it.

The Rule Description Class has an attribute: compliance citations. This value is calculated as the number of compliance citation relations that an instance of these classes recursively has pointing to it. The delegation citation attribute works the opposite way. Its value is calculated as the number of delegation citation relations that an instance of these classes recursively has pointing from it.

5 Demonstration

Having presented the pattern, we will now demonstrate how its four hierarchies can be used. First, we present the chosen case for the instantiation: GDPR. Second, we present used tools: Wikibase for data management and Kumu for visualisation. Third, we present the four hierarchies that could be built.

5.1 The Case of GDPR

We choose to instantiate the pattern as a model of the GDPR regulation, including its surrounding rules and regulations, all the way down to the internal rules of a Swedish public authority. The case was chosen for two reasons. First, it includes all elements of the proposed pattern. Second, since GDPR is well known, the case will be relatively easy for the reader to follow.

The Data Protection Convention 108 of the Council of Europe (not the same as the Council of the European Union) is an international treaty that is applicable in all 46 member states of the Council of Europe, including Sweden.

GDPR is an EU legislation that followed the Data Protection Directive. GDPR is divided into chapters, which are divided into Sections. Section 94 or Chap. 11 includes the citations that repeal the Data Protection Directive. Section 6 of Chap. 2 has the citation that delegates to each EU member state to create their won, complementary, legislation. In the case of Sweden, this became The Data Protection Act (2018:218).

The Data Protection Act (2018:218) is a Swedish ordinary law. It applies in the Swedish jurisdiction. Swedish ordinary law is a rule type that is subordinate to EU legislation. The act is divided into Chapters and Sections. Section 1 of Chap. 1 has a compliance citation to GDPR. There is also Sect. 2 of Chap. 3, which covers personal data in the context of employment.

The Data Protection Ordinance (2018:219) is a Swedish government ordinance. It applies in Sweden. A Swedish government ordinance is a rule type that is subordinate to the Swedish ordinary law. The ordinance is divided into sections (without chapters). Section 1 contains three compliance citations: the Data Protection Act (2018: 218), GDPR, and the Data Protection Convention 108. Section 3 of this ordinance designates the Swedish Authority for Privacy Protection (Integritetsskyddsmyndigheten, IMY) as the supervisory authority with respect to GDPR and the Data Protection Act (2018:218).

We do not know the internal rules of IMY, but it seems plausible that it had to make its staff policy in light of Sect. 2 of Chap. 3 of the Data Protection Act 2018:218, which, as mentioned, covers personal data in the context of employment. It also seems plausible that IMY had to rewrite its Supervision Policy according to Sect. 3 of the Data Protection Ordinance 2018:219, since this section designates IMY as the supervisory authority. All rules created by IMY, being a Swedish public authority, have a rule type subordinate to Swedish government ordinance. They apply only to the "jurisdiction" of IMY. This part of the pattern illustrates how the cascading effects of legislation such as GDPR continue within organisations. IMY has, moreover, probably defined its own internal, hierarchical, rule structure stating that rules created by the General Director are superior to those created by the HR Director.

5.2 Used Tools: Wikibase and Kumu

The instantiation was implemented using the Wikibase Cloud platform, which is the tool of Wikidata.org. Wikibase offers a simple way to build knowledge graphs while simultaneously making them publicly available. Another reason for choosing Wikibase is that, as seen in Sect. 2, some of the most common legal modelling frameworks are knowledge graphs, and for future work it may prove useful to align with that environment. Thirdly, managing large number of rules is clearly more scalable with a database such as Wikibase than Kumu, which is essentially a visualisation tool.

A downside of Wikibase compared to a relational database is that it has less support for automatic calculations, which is needed to make the proposed pattern scalable. (These operations are not included in Fig. 1.) For the demonstration, these values were calculated and updated manually.

To visualise the Wikibase model, we exported a csv file through a SPARQL query and then imported it into Kumu. This tool was also used to create Figs. 2, 3, and 4. The Wikibase instantiation, the Kumu map, the SPARQL query, the class diagram project file, and the figures are available in an online supplement [13].

5.3 Norm Hierarchy

As explained in Sect. 4, in the demonstration, the Rule class in merged into the Rule Description class.

The recursive -subordinate to- relation of Norm Hierarchy Level allows a hierarchy to be built, shown in Fig. 2. The size of each Rule Description item is decided by the number of -subordinate to- relations that its respective Norm Hierarchy Level item recursively has pointing to it.

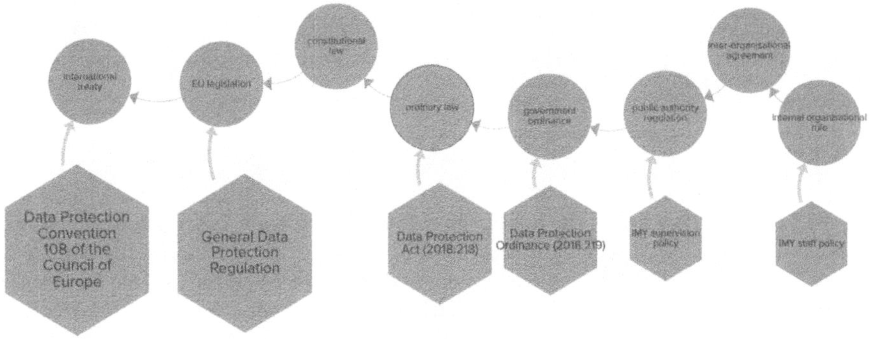

Fig. 2. The higher the norm hierarchy the larger the Self-contained Rule Description. Norm Hierarchy is calculated as the number of instances that a Norm Hierarchy Level instance (green circle) is recursively supreme of. The value is propagated to the Self-contained Rule Description instances (blue hexagon) that has a relationship to the Norm Hierarchy instance. (Color figure online)

The instantiated model is of modest size and complexity, which may make this hierarchy seem obvious. That said, it shows how all Rule Descriptions will have a link to exactly one Norm Hierarchy Level instance, and that these levels form a hierarchy that segments the Rule Descriptions. For example, all ordinary laws will have the same hierarchy level.

Returning to the digital map analogy of Sect. 1, a user can first see international treaties, such as the Data Protection Convention 108, and then drill all the way down to internal organisational rules, such as the IMY staff policy.

5.4 Scope

The Scope hierarchy, shown in Fig. 3, is built by the recursive -part of- relation. This demonstration will cover only the spatial scope.

The size of each Rule Description item is decided by the number of -part of- relations that its respective Scope item recursively has below it.

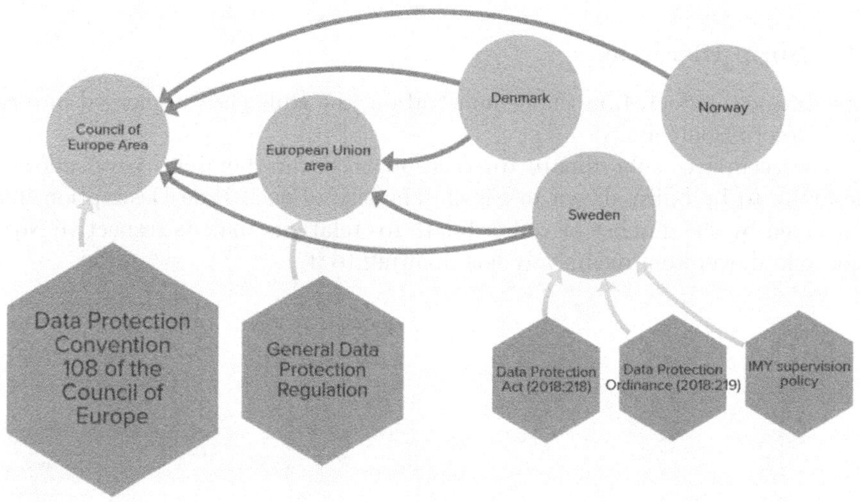

Fig. 3. The larger the Scope the larger the Self-contained Rule Description. The hierarchy value is calculated as the number of instances that a Scope instance (orange circle) recursively has as parts. The value is propagated to the Self-contained Rule Description instances (blue hexagon) that have a relationship to a Scope instance. (Color figure online)

The Council of Europe Area is on the top of the Spatial Scope hierarchy because it has the highest number of elements as parts. Therefore, GDPR will be hierarchically lower than the Data Protection Convention 108, but higher than Sweden's Data Protection Act.

As can be seen, there is a relationship between the European Union area and the Council of Europe area. This choice is debatable. Even if all EU member states are currently also members of the Council of Europe, that does not always have to be the case. However, we decided to focus on the geographical space and not the institutional relations.

Returning to the analogy of the digital map in Sect. 1, a user can first see international treaties, such as the Data Protection Convention 108, and then drill down to reach internal organisational rules, such as the IMY staff policy.

5.5 Rule Description

The Rule Description hierarchy, shown in Fig. 4, is built by recursively composing sections of the Rule Description Section into Self-contained Rule Descriptions, which in turn are aggregated into Rule Description Collections. As shown, in GDPR, Sect. 6 is part of Chap. 2, which is part of the regulation, which is part of the EU privacy regulation, which is part of the privacy regulation in general.

The size of each Rule Description or Rule Description Collection instance is decided by the number of -part of- relations that it recursively has below it.

Using the digital map analogy, a user could start from the top, only seeing "privacy regulation" as an instance among other rule collections on the highest level, such as commercial regulation or maritime regulation. In the next step, the user can see the "EU privacy regulation" among, for example, the African Union Privacy Regulation. The next step is to choose zooming in on either "GDPR" (a Self-contained Rule Description) or complementary EU member state privacy regulation (a rule collection). If instead starting from the bottom, a user that has been studying Chap. 2 of GDPR may benefit from looking upward through the layers of, first, a Self-contained Rule Description, and, further above, different levels of Rule Description Collections.

5.6 Citation

A Citation compares two Rule Descriptions directly, creating a weakly hierarchical graph. The two types of citations (compliance and delegation) presented in Sect. 3.4, will make some Rule Descriptions emerge as more influential than others. GDPR is arguably a prime example: (a) it explicitly states that other legislation may complement it, and (b) a multitude of laws, ordinances, regulations, and internal company rules state that they will comply with it.

Figure 5, for simplicity, only shows compliance citations, not delegation citations. The figure depicts the web of compliance citations that is formed by the Rule Description Self-Contained items and their sections.

The size, showing its hierarchy values, of each Self-Contained Rule Description instance is decided by the number of -comply with- relations that it recursively has pointing at itself or its Rule Description Sections. For example, GDPR has a value of 5, and Data Protection Convention 108 has a value of 2.

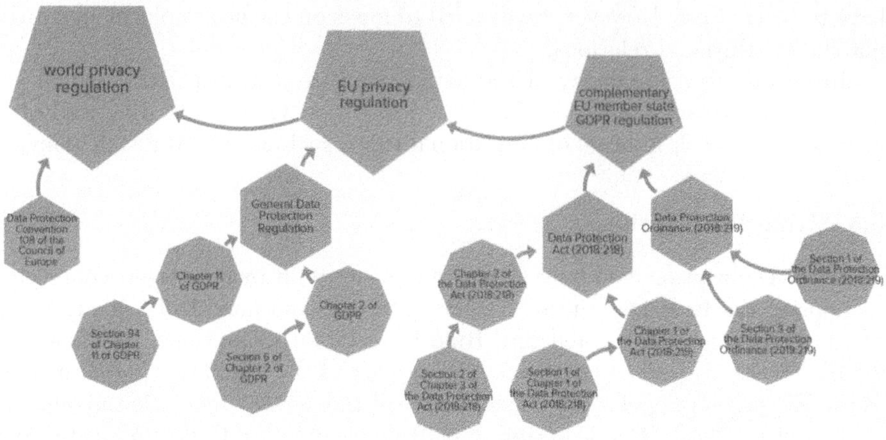

Fig. 4. The higher number of parts the larger Self-contained Rule Description. The hierarchy value is calculated as the number of instances that a Rule Description (hexagon and octagon) or Rule Description Collection instance (pentagon) recursively has as parts.

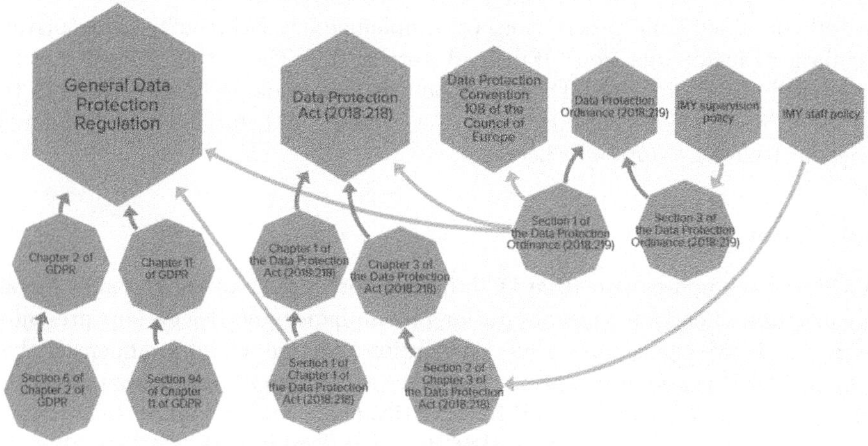

Fig. 5. The higher number of compliance citations the larger Self-contained Rule Description. The hierarchy value is calculated as the number citations (green arrow) that a Self-contained Rule Description (hexagon) recursively has pointing to it. (Color figure online)

6 Discussion

In the previous sections, we have used OntoUML to design a pattern for modelling hierarchies of organisational rules, and made a modest instantiation of GDPR, using the tools Wikibase and Kumu.

The pattern included four hierarchical dimensions. Additionally, the Citation dimension has two subclasses and the Scope has three subclasses. One of these is the Material Scope class, which can be used for a multitude of sub-dimensions. As can be seen, several competing, or alternative, hierarchies may emerge from one rule system. The advantage is flexibility. The user can choose among various possible perspectives. In addition, the hierarchy values for each dimension can be merged into one, creating a more general hierarchy. If the used tool allow, the user could assign different weights to each dimension to test what makes the most sense. The disadvantage is that multiple perspectives can be confusing, and users might disagree on the "best" general hierarchy.

As mentioned, Material Scope can be used for any number of sub-dimensions. Modelling several Material Scope simultaneously might lead to a degree of complexity that is not helpful for sense-making. Users should probably prioritise the scope sub-dimension most relevant to their needs.

Both the Scope and Norm Hierarchy dimensions rely on the use of well-founded taxonomies. Which norm is hierarchically superior to which is not always obvious. For example, in our instantiation, we have represented International Treaty as superior to EU Legislation. But this is only true for treaties that the EU has ratified. Some treaties are ratified by some, but not all, of the member states, and other treaties are not ratified no by the EU nor by a particular member state. Representing all these types of issues in a hierarchical taxonomy can be complex.

The pattern has weaknesses. As discussed in Sect. 3, the four hierarchies are built by two different principles. While the hierarchy values for Rule Description decomposition and Citation, depend by the relations between Rule Descriptions themselves, the hierarchy values for Norm Hierarchy and Scope are transferred to the Rule Descriptions. This can make a model harder to comprehend.

It could happen that sufficiently clear hierarchies do not emerge. As discussed in Sect. 2.1, there is no exact definition of when a weakly hierarchical graph ceases to be a hierarchy at all [12]. A user evaluation may be needed to understand the extent of this problem.

The proposed pattern is rather generic and can be used for varying use-cases. An example could be the legal toolbox project described in Sect. 1, which entails trying to comprehend and get an overview of a complex rule system, such as health data.

A limitation of this study is that we have only demonstrated the model on a rule system of a very modest size. Future research could use a dataset from, for example, ELI, with hundreds or thousands of items. In case of some of the relations in the proposed pattern are missing, perhaps a Large Language Model (LLM), could be used to add them, based on the textual content of the rules. More visualisation tools could then be tested in addition to Kumu.

7 Conclusion

This study aimed to investigate how the hierarchies of organisational rules can be modelled for the purpose of abstraction and sense-making among stakeholders.

The proposed pattern allows for building hierarchies along four different aspects of organisational rules: norm hierarchy, scope, rule description decomposition, and citations.

The study demonstrates how weakly hierarchical graph can emerge from each of these four aspects. The extent these hierarchies are sufficiently clear to substantially facilitate sense-making needs to be evaluated with much a larger organisational rule systems data set.

The proposed pattern not only aids in simplifying complex rule sets but also facilitates regulatory compliance and improves stakeholder communication. Its intended application in workshops and data warehouses suggests it has versatility and practicality in various organisational contexts.

References

1. OntoUML specification. https://ontouml.readthedocs.io/
2. Unified Foundational Ontology (UFO) specification. https://dev.nemo.inf.ufes.br/seon/UFO.html
3. European Legislation Identifier (ELI) (2012). https://op.europa.eu/en/web/eu-vocabularies/eli
4. SALI (Legal Industry Standards) (2020). https://www.sali.org/
5. LegalRuleML Core Specification Version 1.0 (2021). https://docs.oasis-open.org/legalruleml/legalruleml-core-spec/v1.0/os/legalruleml-core-spec-v1.0-os.html
6. Campbell, L.J., Halpin, T.A., Proper, H.A.: Conceptual schemas with abstractions making flat conceptual schemas more comprehensible. Data Knowl. Eng. **20**(1), 39–85 (1996). https://doi.org/10.1016/0169-023X(96)00005-5, https://www.sciencedirect.com/science/article/pii/0169023X96000055
7. Cervone, L., Palmirani, M., Vitali, F.: Akoma Ntoso | Akoma Ntoso Site (2008). http://akomantoso.info/
8. Cook, P.D., Mengal, S., Parameswaran, S.: SHAMROQ: towards semantic models of regulations. In: SEKE, pp. 93–100 (2020). https://doi.org/10.18293/SEKE2020-144
9. Griffo, C.: UFO-L: A Core Ontology of Legal Aspects Built on Legal Relations Perspective. Ph.D. thesis, Technological Center, Computer Science Department, Federal University of Espírito Santo State (2018)
10. Hamdaqa, M., Hamou-Lhadj, A.: An approach based on citation analysis to support effective handling of regulatory compliance. Futur. Gener. Comput. Syst. **27**(4), 395–410 (2011). https://doi.org/10.1016/j.future.2010.09.007
11. Kholkar, D., Sunkle, S., Kulkarni, V.: Semi-automated creation of regulation rule bases using generic template-driven rule extraction. In: Proceedings of the Second Workshop on Automated Semantic Analysis of Information in Legal Text (ASAIL 2017). London (2017). https://ceur-ws.org/Vol-2143/paper1.pdf
12. Krogstie, J.: Model-Based Development and Evolution of Information Systems. Springer, London (2012). https://doi.org/10.1007/978-1-4471-2936-3
13. Lindeberg, J.: Online supplements for the article "Modelling Hierarchies of Organisational Rules" for RCIS 2025 (2025). https://github.com/JoranL/rcis2025/
14. Liu, Z., Stasko, J.: Mental models, visual reasoning and interaction in information visualization: a top-down perspective. IEEE Trans. Visual Comput. Graph. **16**(6), 999–1008 (2010). https://doi.org/10.1109/TVCG.2010.177

15. Lyalin, D., Williams, W.: Business rules hierarchy: the world of business rules is not flat. Bus. Rules J. **17**(2) (2016). http://www.brcommunity.com/a2016/b850.html
16. Maxwell, J.C., Antón, A.I., Swire, P., Riaz, M., McCraw, C.M.: A legal cross-references taxonomy for reasoning about compliance requirements. Requirements Eng. **17**(2), 99–115 (2012). https://doi.org/10.1007/s00766-012-0152-5
17. Miller, G.A.: The magical number seven, plus or minus two: some limits on our capacity for processing information. Psychol. Rev. **101**(2), 343–352 (1994). https://doi.org/10.1037/0033-295X.101.2.343
18. Object Management Group (OMG): semantics of business vocabulary and business rules. Version 1.4. Technical report (2017)
19. Sannier, N., Adedjouma, M., Sabetzadeh, M., Briand, L.: Automated classification of legal cross references based on semantic intent. In: Daneva, M., Pastor, O. (eds.) REFSQ 2016. LNCS, vol. 9619, pp. 119–134. Springer, Cham (2016). https://doi.org/10.1007/978-3-319-30282-9_8
20. Stirna, J., Persson, A.: Enterprise Modeling: Facilitating the Process and the People. Springer, Cham (2018). https://doi.org/10.1007/978-3-319-94857-7
21. Taleb, N.N.: Antifragile: Things That Gain from Disorder. Random House Publishing Group (2012)
22. Vermeir, D.: Semantic hierarchies and abstractions in conceptual schemata. Inf. Syst. **8**(2), 117–124 (1983). https://doi.org/10.1016/0306-4379(83)90036-4
23. Vranes, E.: The principles of conflict resolution. In: Vranes, E. (ed.) Trade and the Environment: Fundamental Issues in International Law, WTO Law, and Legal Theory. Oxford University Press (2009). https://doi.org/10.1093/acprof:oso/9780199562787.003.0003
24. Weigand, H., Johannesson, P., Andersson, B.: Ontological analysis of policy-based decision making. In: 17th International Workshop on Value Modelling and Business Ontologies. Hertogenbosch (2024). https://ceur-ws.org/Vol-3821/paper7.pdf
25. Zugal, S., Pinggera, J., Weber, B., Mendling, J., Reijers, H.A.: Assessing the impact of hierarchy on model understandability – a cognitive perspective. In: Kienzle, J. (ed.) MODELS 2011. LNCS, vol. 7167, pp. 123–133. Springer, Heidelberg (2012). https://doi.org/10.1007/978-3-642-29645-1_14

If Complexity Is the Problem, Collaboration Is the Solution: Drivers and Trust-Building Interventions in DSR

Xabier Garmendia(✉) and Oscar Díaz

University of the Basque Country (UPV/EHU), San Sebastián, Spain
{xabier.garmendiad,oscar.diaz}@ehu.eus

Abstract. Design Science Research (DSR) increasingly addresses complex socio-technical challenges that require effective collaboration among diverse stakeholders. While trust is recognized as fundamental to successful collaboration, there is limited understanding of how trust mechanisms operate specifically within DSR contexts. This study addresses this gap through a two-phase investigation. First, we examine collaboration drivers in DSR through a survey of practitioners (n = 21), revealing that knowledge exchange predominates over artifact development as a primary motivation. Second, we conduct a systematic mapping study identifying 32 distinct trust-building mechanisms. The framework is empirically validated with the DESRIST community, demonstrating that interpersonal factors (such as honesty and consistent communication) are more crucial than institutional mechanisms in building trust within DSR. This research contributes to both the theoretical understanding of trust development in DSR and practical guidance for researchers managing DSR projects.

Keywords: Design Science Research · Collaboration · Trust-building · Trust-building mechanisms

1 Introduction

"Advancing Information Science and Information Systems *Quality* in the Era of *Complexity*" is both a technical and social challenge[1]. This complexity manifests in multiple dimensions, including **technical complexity**, characterized by increasingly sophisticated systems and technologies; **social complexity**, stemming from the growing number of stakeholders and user interactions; **organizational complexity**, driven by intricate business processes and evolving requirements; and **environmental complexity**, shaped by rapidly changing external conditions [14]. This calls for inter-disciplinary teams. Additionally, socio-technical issues are often "wicked problems," where boundaries are unclear and information can be incomplete or conflicting, necessitating continuous reassessment and adaptation [28]. In summary, quality and complexity necessitate ongoing collaboration among interdisciplinary teams.

[1] The quote stands for the theme of RCIS'25.

J. Grabis et al. (Eds.): RCIS 2025, LNBIP 547, pp. 192–207, 2025.
https://doi.org/10.1007/978-3-031-92474-3_12

In complex socio-technical projects, team members from different disciplines must rely on each other's expertise while often working outside their comfort zones [25]. Trust allows researchers to acknowledge the validity of different methodological approaches and epistemological perspectives, even when they differ from their own disciplinary traditions [5]. This is particularly important when dealing with quality aspects, as different disciplines may have varying definitions and metrics of quality that need to be reconciled. Furthermore, the iterative nature of addressing complex problems requires team members to maintain confidence in the process and in each other, even when immediate results are uncertain or when initial approaches need revision [6]. Without trust, interdisciplinary teams risk fragmenting into siloed approaches that fail to leverage the full potential of their diverse expertise [8]. This underscores the importance of dedicated study on trust-building in IS research teams.

However, trust-building is highly contextual: interventions (such as providing timely feedback, for example) that have been proven effective in specific contexts (like software agile teams [33]) may not produce the same outcomes in other settings (e.g., research teams). The effectiveness of trust-building mechanisms can vary significantly across different settings. This work focuses on *research* teams within the practice of *Design Science Research* (DSR). On these grounds, this work tackles two research questions:

- RQ1: What are the main drivers for collaboration in DSR projects?
- RQ2: What interventions can be devised to promote trust in DSR projects?

We contribute to RQ1 by gathering evidence from DSR practitioners through positioning collaboration in a two-dimensional space: *"the what"* and *"the what for"* (Sect. 3). On the other hand, we address RQ2 by conducting a Systematic Mapping Study (Sect. 4), and validating the findings with the DSR community (n = 21) (Sect. 5). We begin by highlighting the specifics of collaboration in DSR projects.

2 The Practice: Collaboration in DSR

In Design Science Research (DSR), solutions must consider both the human and organizational impacts, along with technological efficiency and effectiveness [18]. This multifaceted nature necessitates the collaboration of a diverse group of individuals . The question arises about how DSR collaboration differ from collaboration in other settings.

DSR projects possess several distinctive characteristics that may affect collaboration dynamics, and ultimately, trust-building.

1. The dual imperative of theoretical contribution and practical utility creates unique tensions. Unlike explanatory research or standard product development, DSR teams must balance academic rigor with practical relevance, requiring trust across different professional cultures. Practitioners must trust that academic rigor will yield practical value, while researchers must value practitioners' insights [31].
2. The iterative and emergent nature of DSR creates higher uncertainty compared to more linear research or development processes [4]. While agile software development shares some of these characteristics, DSR's combination of artifact development with theory building introduces additional complexity [4]. Team members

must maintain trust despite frequent pivots, failed iterations, and evolving project goals - situations that might strain traditional trust mechanisms based on predictability and consistency [19].

3. The multidisciplinary nature of DSR teams extends beyond typical cross-functional collaboration. DSR often requires integration of diverse epistemological perspectives, research traditions, and evaluation criteria [12]. This goes deeper than typical interdisciplinary projects because team members must not only understand each other's domains but also trust in fundamentally different ways of knowing and validating knowledge [12].
4. Unlike explanatory research where the methodology might be well-established, or product development where market requirements provide direction, DSR teams often operate in uncharted territory. This requires trust in the process itself, even when immediate outcomes are uncertain or when traditional metrics of progress may not apply [15].

These differences in collaboration dynamics are partially explained by the motivations that researchers have for collaboration. This raises our next question.

3 RQ1: What Are the Main Drivers for Collaboration?

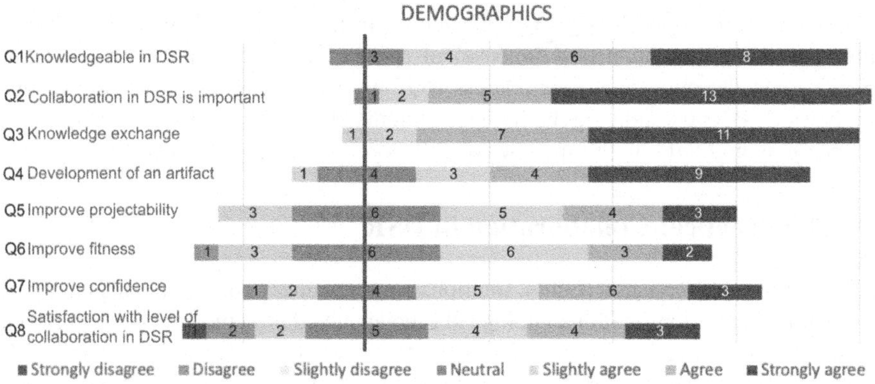

Fig. 1. What is the purpose of collaboration in DSR? Answers from DESRIST participants (n = 21).

This section abounds on the relevance of collaboration by gathering evidence from DSR practitioners. First, we introduce the goal in terms of a framework of collaboration drivers. Next, we transform this framework into a questionnaire, and finally, we distribute this questionnaire to DSR practitioners.

The Goal. For our purposes, we position collaboration in a two-dimensional space: "the what" and "the what for". First dimension captures *the object of the collaboration*, specifically, "knowledge" and "artifact". Collaboration centered on knowledge is primarily driven by the desire to share and transfer information, insights, and expertise between individuals or groups. The outcome of knowledge-based collaboration is typically increased knowledge, skills, and expertise among the participants. Alternatively, a collaboration focused on artifact development is centered on creating tangible products, tools, or solutions. The primary goal is to produce a concrete intervention that can be utilized.

As for the second dimension, "the what for" captures *the aim of the collaboration*. Here, we tap into the three venues for knowledge accumulation where acollaboration might be driven by applying somewhere else's solution to a different domain (*projectability*), improving the solution with additional features (*fitness*), or providing more reliable evaluations about the problem-solution pair (*confidence*) ([34]).

The Subjects. Since trust-building is highly contextual, we must ensure that subjects share the same context. Since this context is DSR, we should strive for all subjects to be well aware not only of the theory of DSR but also knowledgeable about the practice of DSR so that they ground their opinions in vivid experiences of conducting DSR in teams. To this end, we considered to tap into attendees of distinct conferences including, CAiSE, RCIS, ECIS and DESRIST. For convenience (attending the conference in person for paper presentation), we resorted to the opinion of the DESRIST 2022's and DESRIST 2023's participants. DESRIST (International Conference on Design Science Research in Information Systems and Technology) is a main venue for DSR practitioners.

The Questionnaire. We collect evidence of the relevance of collaboration in DSR along these two dimensions using a Likert questions with six questions (see Fig. 1).

- Q1 collects information about the participants' DSR experience,
- Q2 collects data about the importance of collaboration in DSR practices,
- Q3-Q4 gather data on collaboration goals, let this be for knowledge exchange or artefact development,
- Q5-Q7 examine the reasons for collaboration in terms of projectability, fitness, and confidence [34].
- Q8 evaluates satisfaction with current level of collaboration in DSR projects.

The Process. For quality sake, a pilot study was conducted with the members of our research group to clarify the values for the two dimensions under stud. The questionnaire was first delivered to DESRIST22 attendees who were contacted via email in February 2023. A total of 13 individuals completed the questionnaire. We considered this number low, and hence we waited a year to tap into DESRIST23 attendees. This time, a different approach was employed, where we attended in person and used a poster showcasing the work along with a QR code linked to the questionnaire. This strategy yielded an additional eight responses, bringing the total number of answers to 21.

Results. Figure 1 depicts the results with the help of a divergent bar chart diagram[2]. The strong knowledge levels in DSR (Q1) suggest that respondents have likely experienced enough projects to recognize collaboration gaps, increasing the reliability of the results.

"The what". *Knowledge exchange* was deemed more prevalent than *artifact*. Knowledge exchange (Q3) is viewed as a crucial collaboration goal, with 20 out of 21 participants expressing some level of agreement, and notably, 11 strongly agreeing.

"The what for". Projectability, fitness, and confidence were all highly valued as key drivers of collaboration. However, *projectability* received a higher level of agreement with no disagreement. The prevailing scenario is for DSR teams to engage in collaboration to seek 'knowledge' while working towards projecting the solution into various domains.

In general, we can observe a common belief regarding the crucial role of collaboration in DSR, as demonstrated by the vertical bar, where agreement values extend to the right and disagreement values to the left. However, the moderate responses regarding practical benefits (Q5-Q7) indicate some friction or challenges in current collaborative practices. Indeed, Q8 reflects a common phenomenon in academia, where the recognized importance of a practice (collaboration) often surpasses satisfaction with its current implementation, particularly when considering real-world constraints and challenges in research projects. These results suggest insufficient levels of collaboration despite agreement about its importance. This makes it relevant to explore the enablers of collaboration in DSR. In other domains, "trust-building" emerges as a major enabler for collaboration [6]. This leads us to our second research question: *What trust-building interventions can be devised to promote collaboration in DSR projects (RQ2)?*

However, interventions (such as providing timely feedback, for example) that have proven effective in specific contexts (like agile software teams) may not yield the same outcomes in other settings (e.g., DSR teams). Hence, RQ2 is divided into two questions:

- RQ2.1: *What trust-building mechanisms have been proposed for DSR-like scenarios?*
- RQ2.2: *To what extent can these interventions be applied to DSR teams?*

4 RQ2.1: What Trust-Building Mechanisms Have Been Proposed for DSR-Like Scenarios?

We tackle this question through a Systematic Mapping Study (SMS). We first introduce Lewicki and Bunker's trust development model ([24]), which provides the framework for the review. Next, we organize primary studies according to this framework to compile a list of trust mechanisms. Finally, we concentrate on those mechanisms that are suggested for a domain resembling DSR.

[2] Divergent-bar charts are recommended for Likert scales because they effectively display the distribution of responses across multiple categories, allowing viewers to see both the overall pattern and the distribution within each category [13].

4.1 A Model for Trust Building

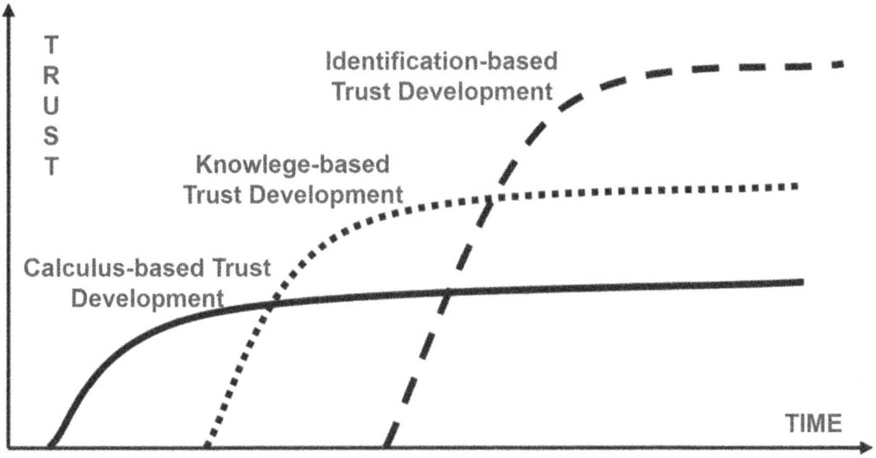

Fig. 2. Lewicki and Bunker's trust development model (adapted from [24]).

Trust is understood as *"a psychological state comprising the intention to accept vulnerability based upon positive expectations of the intentions or behaviours of another"* ([29]). Acknowledging the notion of trust as residing within the individual, Lewicki and Bunker draw on the work of Shapiro to argue for trust development to be an iterative process that "takes on a different character in the early, developing and mature stages of a relationship" as knowledge of the other person grows ([24]). They introduced a staged model[3] with three levels of trust that are linked in a sequential iteration (see Fig. 2). Firstly, ***calculation-based trust*** is based on an individual's subjective assessment of the costs and benefits of establishing and maintaining a relationship. Trust emerges when one party believes the other will contribute beneficially to the collaboration. Secondly, ***knowledge-based trust*** is developed over time through repeated interactions and observations. It is based on predictability developed by knowing the other party well enough to anticipate their behavior. This trust relies on information. Finally, ***identification-Based trust*** is based on a deep sense of shared identity between individuals or groups. At this level, trust exists because the parties have a deep understanding and appreciation of each other's desires, and this mutual understanding is developed to the point where they can effectively act on behalf of one another [32].

Shapiro's tree-stage model offers a way to position each mechanism within a stage. However, the utility of this mechanism very much depends on the organization. Trust-building is highly contextual. For instance, in organizational contexts where opportunism is possible, implementing mechanisms such as establishing norms or ensuring

[3] This model assumes that the parties are entering into a new relationship with no prior history to overcome, feeling vulnerable if disclosure comes too quickly, and uncertain about the future of the relationship.

confidentiality can foster greater trust than in more transparent settings, such as those found in educational environments.

Certain trust-building mechanisms are considered to be more appropriate and therefore more likely to be used during certain stages ([32]). Calculus-based trust can be developed through mechanisms that support initial interactions, laying the foundation to start collaboration. Communication is an effective mechanism to improve understanding and predictability, thereby enhancing knowledge-based trust. Finally, at the point of identification-based trust, trust mechanisms can help strengthen the relationship by allowing each party to act effectively for the other. In our research, we use Shapiro et al.'s classification to categorize the identified trust-building mechanisms into these three levels of trust. To explore trust-building mechanisms for DSR collaborations, we conduct a literature review.

4.2 Systematic Mapping Review

Worth noting, our use of SMR aims for replicability rather than completeness. We are interested in the systematicity that SMR offers to enable others to replicate the study. We adhere to the recommendations presented by [22] (see Fig. 3).

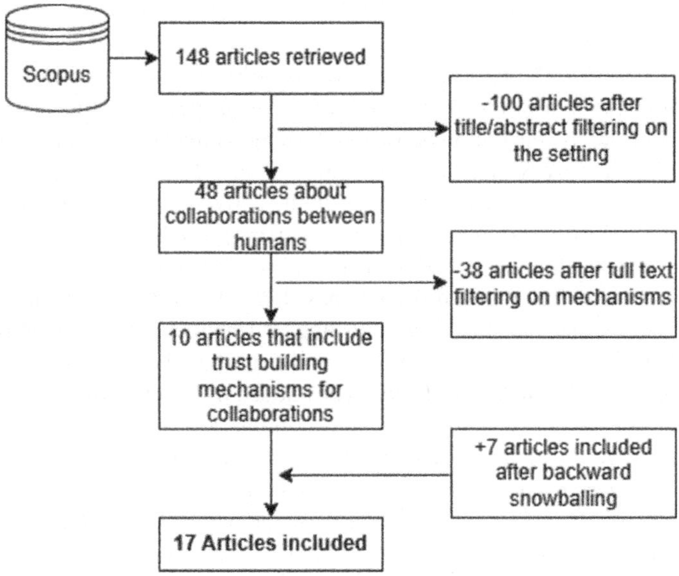

Fig. 3. Article selection process for trust-building mechanisms in collaborations.

Search String. We broke down the question into the PICO facets: population (any organization), intervention (trust-building), comparison (not applicable), and outcome (mechanisms). Next, we searched for synonyms. We observed that the development of trust can be described in various ways, such as "trust-building", "building trust", or "trust development". Similarly, mechanisms can also be suggested by models, frameworks, processes, or interventions. This led to the following search string:

("trust-building" OR "building trust" OR "trust development") AND ("model" OR "framework" OR "mechanisms" OR "process" OR "intervention")

Procedure. We conducted the query in the Scopus databases. Reviewing papers based on titles not only streamlines the review process but may also promote consistency in the filtering process. Titles are less susceptible to subjective interpretations compared to abstracts or full texts, thus reducing potential biases in paper selection and promoting replicability. Moreover, the limited impact of false negatives (i.e., not identifying relevant studies) is acceptable as our goal is to offer a representative list of trust-building mechanisms rather than an exhaustive one. By focusing on the titles, we identified 148 primary studies (PSs). The filtering process was twofold (see Table 1):

Table 1. Exclusion and inclusion criteria

Type	Criteria
Inclusion criteria - Setting	- The study focuses on trust between humans
	- The study focuses on trust as a means for collaboration
	- The collaboration aims to creating knowledge and artifacts
Inclusion criteria - Mechanisms	- The study proposes some mechanism for trust-building
Exclusion criteria - Type of paper	- The study is in a language other than English
	- The study is a previous version of another study in the review
	- The work is a secondary study such as a literature review
	- The study is not peer-reviewed

- Focus on the setting. The PS should concentrate on trust between humans to promote collaboration. Trust in platforms, services, or systems like buy-sell environments should not be included. Moreover, trust as a factor for achieving other objectives like customer relationships or social media is also excluded. The filtering process was carried out using the abstracts of the PSs, leading to 48 selected papers.
- Filter by mechanisms. The research paper should suggest mechanisms for building trust. Research papers that only examine the phenomenon without suggesting practical measures (interventions, mechanisms) to improve trust are not considered. Unfortunately, it was common for research papers not to reveal the mechanisms in the abstract but rather later in the manuscript. This required reading all 48 research papers, of which 10 went further than analyzing the phenomena to test actionable propositions.

These 10 PSs were then subjected to backward snowballing. This resulted in 7 additional PSs. In total, 17 PSs were identified, all of which introduce mechanisms for building trust among individuals with the aim of collaborating to create knowledge and artifacts.[4]. The list of PSs can be found in the References[5]. Main threat to validity comes from internal validity, specifically the potential for not capturing all relevant publications. To mitigate this issue, we utilized snowballing ([22]). Additionally, to enhance the filtering process, we involved extra reviewers for cases of uncertainty and strive for consensus among them.

Data Extraction. These 17 PSs were then moved to the coding stage. We extracted the trust-building mechanisms proposed in the included articles using a hierarchical coding approach where the top themes were those of Lewicki and Bunker's model: calculus-based trust, knowledge-based trust, and identification-based trust. Excerpts were extracted by one researcher and open-coded iteratively by the two authors. Operationally, we resorted to a mind map where leaf nodes stand for excerpts and ancestor nodes correspond to themes and sub-themes[6]. The resulting code-book which gathers all the trust-building mechanisms, is described in this URL[7] together with excerpts that illustrate the realization of the distinct codes in the PSs.

4.3 Results: Trust Mechanism from the Literature

We have identified a total of 32 mechanisms, categorized as 11 for calculus-based trust, 9 for knowledge-based trust, and 12 for identification-based trust (see Fig. 5). Next, we provide an overview.

Calculus-Based Trust. It builds upon an individual's subjective assessment of the costs and benefits of establishing and maintaining a relationship. The resulting sub-themes include:

- **Initial interaction**. Positive initial interactions between team members can enhance trust, and increasing the frequency and richness of these interactions can lead to higher levels of initial trust. This sub-theme includes three codes: *StrongReputation*, *GettingAcquainted* & *ThirdPartyInvolvement*.
- **Attitude**. Attitude is a critical component in building trust, as it helps to establish initial impressions and sets the tone for the relationship. This sub-theme includes four codes: *Honesty*, *Willingness*, *RiskTaking* & *TeamWorking*.
- **Predictability**. Projects can have unpredictable costs and schedules. Although many of the factors that drive this unpredictability are systemic (e.g. rapidly evolving technological markets, scope creep, and poorly defined requirements), others are linked to individual attributes. This sub-theme includes four codes: *Confidence*, *Consistency*, *Complementarity*, & *Norms*.

[4] The filtering data is available at https://labur.eus/filteringProcess.

[5] Primary Sources (PSs): P1: [1], P2: [2], P3: [3], P4: [5], P5: [7], P6: [9], P7: [10], P8: [11], P9: [16], P10: [17], P11: [21], P12: [23], P13: [26], P14: [27], P15: [30], P16: [33], P17: [35].

[6] Available at: https://mm.tt/app/map/3178582539?t=Q2Y7KDAkqA.

[7] https://rebrand.ly/mechanismsCodebook.

Knowledge-Based Trust. It builds upon the predictability that the other party will act in a trustworthy manner. The resulting sub-themes include:

- **Communication.** Research indicates that frequent and face-to-face interactions are essential for establishing a shared understanding and reducing the risk of misinterpretation. This sub-theme includes seven codes: *EncourageCommunication*, *TimelyFeedback*, *SharingKnowledge*, *SharedVision*, *HandlingExpectations*, *OpenCommunication*, & *Receptivity*.
- **Adaptation.** Adaptation helps overcome challenges arising from cultural differences, work styles, and processes. The formation of a relationship requires mutual adjustment, resulting in a more harmonious and effective collaboration. This sub-theme includes two codes: *CulturalBlending* & *OrganisationBlending*.

Identification-based Trust. It builds upon gains on the understanding of each other's needs, choices, and preferences. The resulting sub-themes include:

- **Equity preservation.** Equity is related to the perception of control and influence in decision-making processes that grounds one's ability to protect their own interests. This sub-theme includes two codes: *ResponsibilityEquity* & *RewardsEquity*.
- **Commitment.** Shared commitment can strengthen trust among participants. Research shows that commitment is instrumental in ensuring that team members accept decisions and work cooperatively to carry out decisions. Codes include *CareForOthers*, *GoodFollower*, *Familiarisation*, *LongTermInterests*, *AchieveEarlyResults*, *Confidentiality*, *LooseSchedules* and *Performance*.
- **Management.** Project management might play a key role to foster trust amongst participants. Such opportunities are captured through these two codes: *ManagementStyle*, and *IS-Support*.

The upcoming section evaluates the effectiveness of the aforementioned trust-building mechanisms in DSR collaborations as perceived by the DSR community.

5 RQ2.2: To What Extent Can These Interventions Be Projected Onto DSR Teams?

When considering projectability, the question arises regarding the extent to which the intervention (e.g., trust-building mechanisms) that have been successfully applied in one setting (e.g., outsourced information systems development projects) can be applicable in other settings (e.g., DSR projects). We tackle this issue by directly prompting the practitioners. We then tap into the same participants (n = 21), and use the same protocol as in the survey in Sect. 3.

When designing the questionnaire, a 7-point Likert scale was utilized. Additionally, an open-ended question was included to gather qualitative comments. The questionnaire was structured based on three trust levels as shown in Fig. 4. Each trust-building mechanism identified in the literature was assessed for its anticipated effectiveness in fostering trust in DSR collaborations. The format of the questions followed this structure: *"In my experience, [mechanism][example] fosters trust-building in DSR collaborations."* Figure 4 depicts the results that will be discussed next.

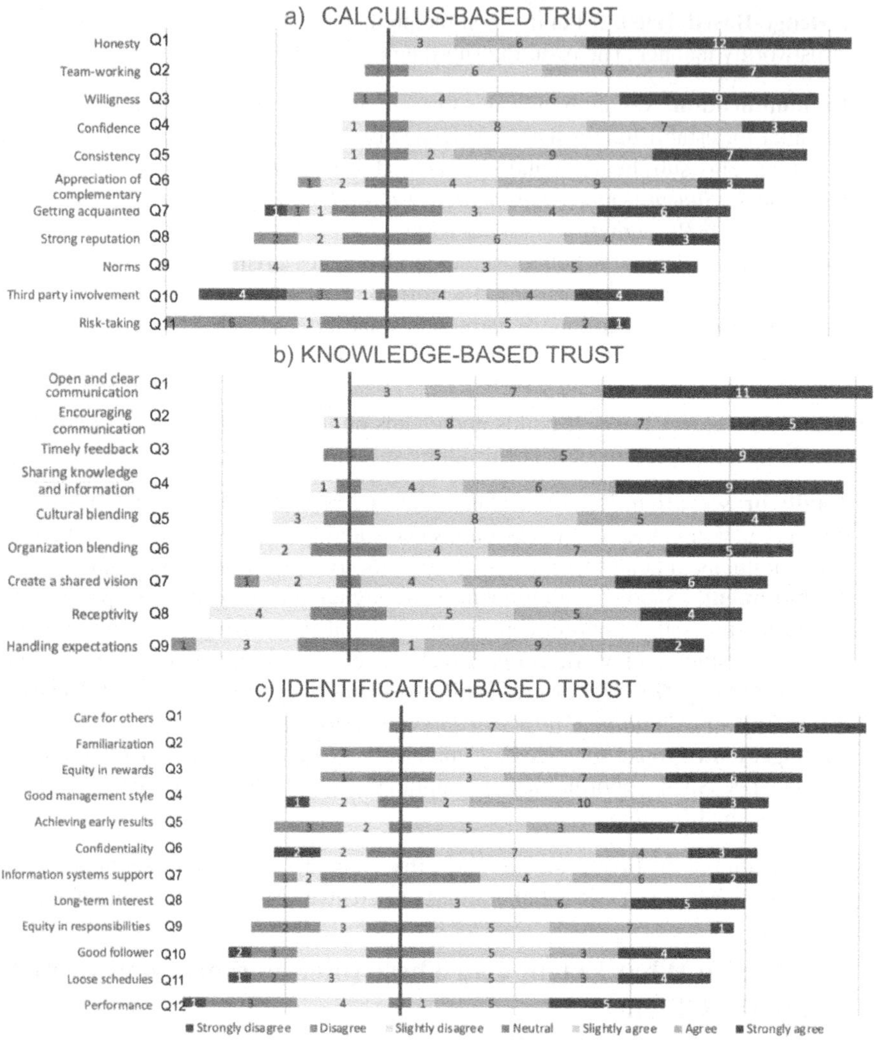

Fig. 4. Trust-building mechanisms questionnaire: *In my experience, Q fosters trust-building in DSR collaborations.*

5.1 Results: Calculus-Based Trust Mechanisms

Interpersonal factors emerge as the most potent trust-building elements. Honesty stands out as the most universally accepted mechanism, receiving unanimous positive agreement with an impressive 12 participants strongly agreeing. This is closely followed by team-working and willingness, both of which garnered overwhelming positive responses and particularly strong agreement levels. On the other hand, the more institutionalized or external mechanisms for trust-building received notably more mixed responses. Strong reputation and established norms showed larger neutral segments, indicating less certainty about their effectiveness in building trust. Third-party involvement emerged as particularly controversial, with a clear split between those who see it as beneficial and those who do not. Most strikingly, risk-taking was viewed with significant skepticism, receiving the highest number of negative responses and lowest levels of strong agreement. This pattern of responses suggests that trust in DSR collaborations is built primarily through direct interaction and demonstrated behavior rather than through external validation or formal mechanisms. The strong preference for interpersonal factors over institutional ones indicates that trust-building in DSR is perceived as a predominantly relationship-based process, where personal qualities and direct interactions carry more weight than formal credentials or structural arrangements.

5.2 Results: Knowledge-Based Trust Mechanisms

Communication emerges as the cornerstone of knowledge-based trust. Open and clear communication (Q1) receives nearly unanimous positive endorsement, with 11 participants strongly agreeing and no negative responses. This is reinforced by similarly strong positive responses to encouraging communication (Q2) and timely feedback (Q3), suggesting that the quality and regularity of communication are fundamental to building knowledge-based trust. Cultural and organizational blending (Q5, Q6) show more moderate but still generally positive responses. This suggests that while organizational and cultural alignment contributes to trust-building, they are seen as supporting rather than primary factors. Creating a shared vision (Q7) receives similar moderate positive endorsement, though with some disagreement. The more nuanced aspects of knowledge-based trust - receptivity (Q8) and handling expectations (Q9) - show more varied responses with larger neutral segments and some disagreement. This pattern suggests that these aspects, while important, may be more challenging to implement effectively or may depend more heavily on individual circumstances and relationships.

5.3 Results: Identification-Based Trust Mechanisms

Care for others (Q1) emerges as the most positively received mechanism, with 20 participants expressing agreement and only one neutral response. This strong endorsement suggests that demonstrating genuine concern for collaborators' well-being is fundamental to building identification-based trust. Familiarization (Q2) and equity in rewards (Q3) also receive strong positive responses. The more operational aspects show mixed responses. Information systems support (Q7), confidentiality (Q6), and equity in responsibilities (Q9) receive moderately positive endorsements but with significant neutral segments. Long-term interest (Q8) shows stronger positive agreement, suggesting that commitment to sustained collaboration matters more than specific operational mechanisms. Finally, the lower-rated factors - good follower (Q10), loose schedules (Q11), and performance (Q12) - show the most divided opinions, with significant disagreement alongside positive responses. This suggests these factors might be more context-dependent or may not be universally applicable to trust-building.

5.4 Threats to Validity

Construct Validity relies on respondents' interpretation and understanding of the survey items. To address this challenge, an initial pilot study was conducted. The first version of the questionnaire was sent to members of the research group to improve comprehension of the questions and detect any possible erroneous interpretations. The survey also includes a detailed description of what is meant by each code (e.g., honesty), along with examples of how it can be operationalized (e.g., fulfilling promises, forthrightness, frankness). In this case, **internal validity** is concerned with the sample size. For DESRIST22, we received 13 responses out of over 140 DESRIST authors contacted. We attempted to create a brief and non-intrusive questionnaire, but we understand that researchers receive numerous emails daily and may not have had the opportunity to read or respond to our survey. To get more responses, we decided to reach out to DESRIST23, in this case using a poster. This effort resulted in 8 more answers. Regarding **external validity**, we believe the findings could also apply to other interdisciplinary research scenarios ([12]). **Conclusion validity** refers to the extent to which the conclusions are drawn from the data that were collected and analysed. We opted for a 7-point Likert scale to increase the sensitivity of responses and provide a wider range of options for participants to choose from [20]. On the downside, Likert scales have been criticized as being arbitrary, with no fixed intervals between response options. While participants may interpret the relative distance between options differently, the scale assumes equal intervals. Therefore, the results should be interpreted in qualitative terms rather than quantitative (A is being favored more than B rather than A is twice more favoured than B). We address this last point in the conclusive section.

6 Discussion and Conclusions

We investigated trust-building in DSR projects using a projectability approach. This resulted in 32 mechanisms that were then shared with the DESRIST community. We collected feedback through a Likert-based survey (n = 21). However, Likert scales have been criticized for being arbitrary, with no fixed intervals between response options. Participants may interpret the relative distance between options differently, while the scale assumes equal intervals. In order to enhance this indication and better profile each mechanism, we can normalize an agreement factor. The agreement factor is calculated as (#Agree + #Strongly Agree) / #Participants. Figure 5 depicts the results.

We can now conclude with an informed conjecture about the "musts" for trust-building in DSR (greenish background in Fig. 5). At the beginning of the project, *honesty, consistency*, and *willingness* are prioritized over reputation, establishing norms, or accepting risks. In trust based on calculations, these factors are considered better indicators of proactive engagement and a collaborative spirit than the reputation or established safety norms of a potential partner. Operational practices aimed at achieving these goals include keeping promises, being transparent, conveying a willingness to contribute useful knowledge, and being consistent in both your words and actions.

Throughout the project, *open communication* and *knowledge sharing* take precedence over adaptation. Operationally, team members are actively encouraged to express

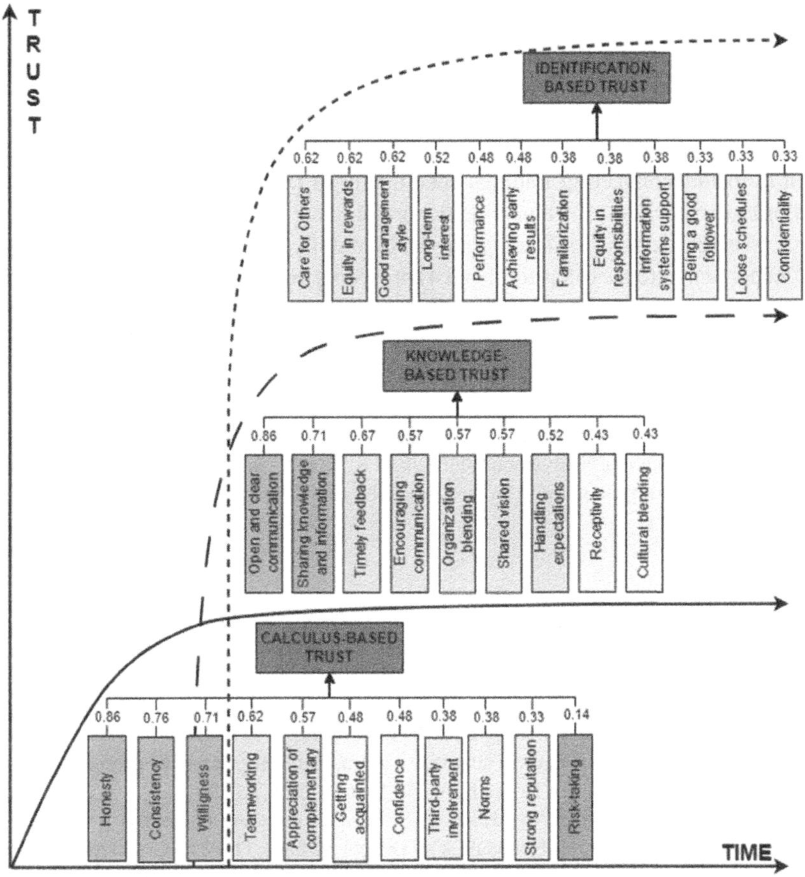

Fig. 5. Three-stage trust-building model with mechanisms weighted by DSR researchers' agreement.

their thoughts and ideas openly, fostering a culture of transparency and trust. Regular team meetings and discussions provide ample opportunities for sharing expertise and insights. Sharing promotes maintaining transparent information flow and making resources easily accessible to all. Utilizing collaboration tools and platforms enables seamless information exchange, facilitating efficient communication and collaboration. Emphasizing constructive and timely feedback in response to messages and information requests streamlines collaboration and minimizes delays. Additionally, responding promptly to emails and other information requests shows that you value the other person's time and effort. Beyond the project, *equity in rewards* and *good management style* are favored over performance or equity in responsibilities. Establishing a culture of fairness ensures that rewards, recognition, and growth opportunities are distributed based on merit. Regarding management style, good practices include active listening and task delegation. To conclude, we employ a two-phase mixed-method approach: (1) A Systematic Mapping Study examining 148 papers (filtered to 17 relevant studies) and (2) Empirical validation through a survey of DESRIST conference participants (n = 21). Data analysis utilizes hierarchical coding based on Lewicki and Bunker's trust model, with data collection spanning two DESRIST conferences (2022–2023).

Improvement areas for this research include: 1) strengthening theoretical contribution by better connecting findings specifically to DSR contexts rather than general collaboration principles; 2) enhancing methodological rigor through larger samples and clearer research design; 3) developing more analytical rather than descriptive results with deeper contextual exploration; and 4) providing concrete, practical implementation guidance with real-world examples for DSR practitioners. Without strong trust relationships, the collective learning and knowledge building that are essential to advancing the field of Information Systems cannot fully develop.

Acknowledgements. Research supported by MCIN/AEI/10.13039/501100011033/FEDER, UE and the "European Union NextGenerationEU/PRTR" under contract PID2021-125438OB-I00. Xabier Garmendia enjoys a grant from the University of the Basque Country - PIF20/236.

References

1. de Araujo, D.V.B., Franco, M.: Trust-building mechanisms in a coopetition relationship: a case study design. IJOA (2017)
2. Babar, M.A., Verner, J.M., et al.: Establishing and maintaining trust in software outsourcing relationships: an empirical investigation. JSS **80**(9), 1438–1449 (2007)
3. Barnett, M., Anderson, J., et al.: The process of trust building between university researchers and urban school personnel. Urban Educ. **45**(5), 630–660 (2010)
4. Baskerville, R., Kaul, M., Pries-Heje, J., Storey, V.: Inducing creativity in design science research. In: Extending the Boundaries of Design Science Theory and Practice: 14th International Conference on Design Science Research in Information Systems and Technology, DESRIST 2019, Worcester, MA, USA, 4–6 June 2019, Proceedings 14, pp. 3–17. Springer (2019)
5. Blomqvist, K., Ståhle, P.: Building organizational trust. In: 16th Annual IMP Conference, Bath, UK, pp. 7–9. Citeseer (2000)
6. Bond-Barnard, T.J., Fletcher, L., Steyn, H.: Linking trust and collaboration in project teams to project management success. IJMPB **11**(2), 432–457 (2018)
7. Chang, C.M., Yen, C.H., Cheng, H.L.: Trust-building mechanisms and knowledge sharing in virtual communities. In: Proceedings of the 9th ICEB, pp. 1070–1079 (2009)
8. Chen, Y., Yu, C., Yuan, Y., Lu, F., Shen, W.: The influence of trust on creativity: a review. Front. Psychol. **12**, 706234 (2021)
9. Chow, P.T., Cheung, S.O., Chan, K.Y.: Trust-building in construction contracting: mechanism and expectation. IJPM **30**(8), 927–937 (2012)
10. Czakon, W., Czernek, K.: The role of trust-building mechanisms in entering into network coopetition: the case of tourism networks in Poland. Ind. Mark. Manag. **57**, 64–74 (2016)
11. Das, T.K., Teng, B.S.: Between trust and control: developing confidence in partner cooperation in alliances. Acad. Manag. Rev. **23**(3), 491–512 (1998)
12. Dickhaut, E., Janson, A., Hevner, A., Leimeister, J.M.: Sharing design knowledge through codification in interdisciplinary dsr collaborations. In: HICSS 2023 (2022)
13. Heiberger, R., Robbins, N.: Design of diverging stacked bar charts for likert scales and other applications. J. Stat. Softw. **57**, 1–32 (2014)

14. Henson, V.R., et al.: A practical guide for managing interdisciplinary teams: lessons learned from coupled natural and human systems research. Soc. Sci. **9**(7), 119 (2020)
15. Hevner, A.R., et al.: Transparency in design science research. Decis. Support Syst. **182**, 114236 (2024)
16. Jarvenpaa, S.L., Leidner, D.E.: Communication and trust in global virtual teams. Organ. Sci. **10**(6), 791–815 (1999)
17. Järvinen, M.: Trust-building in a coopetitive multi-supplier network. J. Public Procure. (2019)
18. Johannesson, P., Perjons, E.: An Introduction to Design Science, vol. 9783319106. Springer, Cham (2014). https://doi.org/10.1007/978-3-319-10632-8, https://link.springer.com/book/10.1007/978-3-319-10632-8
19. Jørgensen, L., Åsgård, T.: Trust and control in project management. Procedia Comput. Sci. **164**, 397–406 (2019)
20. Joshi, A., Kale, S., Chandel, S., Pal, D.K.: Likert scale: explored and explained. Br. J. Appl. Sci. Technol. **7**(4), 396 (2015)
21. Khalfan, M.M., McDermott, P., Swan, W.: Building trust in construction projects. Supply Chain Manag. Int. J. (2007)
22. Kitchenham, B.A., Budgen, D., Brereton, P.: Evidence-Based Software Engineering and Systematic Reviews, vol. 4. CRC press, Boca Raton (2015)
23. Lander, M.C., Purvis, R.L., et al.: Trust-building mechanisms utilized in outsourced is development projects: a case study. Inf. Manag. **41**(4), 509–528 (2004)
24. Lewicki, R.J., Bunker, B.B., et al.: Developing and maintaining trust in work relationships. Trust Organ. Front. Theory Res. **114**, 139 (1996)
25. Lorenzoni, G., Lipparini, A.: The leveraging of interfirm relationships as a distinctive organizational capability: a longitudinal study. SMJ **20**(4), 317–338 (1999)
26. Meier, M., Lütkewitte, M., Mellewigt, T., Decker, C.: How managers can build trust in strategic alliances: a meta-analysis on the central trust-building mechanisms. J. Bus. Econ. **86**(3), 229–257 (2016)
27. Oza, N.V., Hall, T., Rainer, A., Grey, S.: Trust in software outsourcing relationships: an empirical investigation of Indian software companies. IST **48**(5), 345–354 (2006)
28. Rittel, H.W., Webber, M.M.: Dilemmas in a general theory of planning. Policy Sci. **4**(2), 155–169 (1973)
29. Rousseau, D.M., Sitkin, S.B., Burt, R.S., Camerer, C.: Not so different after all: a cross-discipline view of trust. Acad. Manag. Rev. **23**(3), 393–404 (1998)
30. Schilke, O., Cook, K.S.: A cross-level process theory of trust development in interorganizational relationships. Strateg. Organ. **11**(3), 281–303 (2013)
31. Sein, M.K., Henfridsson, O., Purao, S., Rossi, M., Lindgren, R.: Action design research. MIS q. 37–56 (2011)
32. Shapiro, D.L., Sheppard, B.H., Cheraskin, L.: Business on a handshake. Negot. J. **8**(4), 365–377 (1992). https://doi.org/10.1111/j.1571-9979.1992.tb00679.x
33. Tyagi, S., Sibal, R., Suri, B.: Empirically developed framework for building trust in distributed agile teams. IST 106828 (2022)
34. Vom Brocke, J., Winter, R., Hevner, A., Maedche, A.: Accumulation and evolution of design knowledge in design science research: a journey through time and space. JAIS **21**(3), 9 (2020). https://doi.org/10.17705/1jais.00611
35. Wildman, J.L., Shuffler, M.L., et al.: Trust development in swift starting action teams: a multilevel framework. Group Organ. Manag. **37**(2), 137–170 (2012)

Synthline: A Product Line Approach for Synthetic Requirements Engineering Data Generation Using Large Language Models

Abdelkarim El-Hajjami(✉) and Camille Salinesi

Paris 1 Panthéon-Sorbonne University, Paris, France
{abdelkarim.el-hajjami,camille.salinesi}@univ-paris1.fr

Abstract. While modern Requirements Engineering (RE) heavily relies on natural language processing and Machine Learning (ML) techniques, their effectiveness is limited by the scarcity of high-quality datasets. This paper introduces Synthline, a Product Line (PL) approach that leverages Large Language Models to systematically generate synthetic RE data for classification-based use cases. Through an empirical evaluation conducted in the context of using ML for the identification of requirements specification defects, we investigated both the diversity of the generated data and its utility for training downstream models. Our analysis reveals that while synthetic datasets exhibit less diversity than real data, they are good enough to serve as viable training resources. Moreover, our evaluation shows that combining synthetic and real data leads to substantial performance improvements. Specifically, hybrid approaches achieve up to 85% improvement in precision and a 2× increase in recall compared to models trained exclusively on real data. These findings demonstrate the potential of PL-based synthetic data generation to address data scarcity in RE. We make both our implementation and generated datasets publicly available to support reproducibility and advancement in the field.

Keywords: Requirements Engineering · Synthetic Data · Large Language Models · Product Line

1 Introduction

Requirements Engineering (RE) largely operates through Natural Language (NL), which is the primary medium for capturing, specifying, and validating all types of requirements [1]. Although this reliance on NL has enabled significant advancements in the use of Natural Language Processing (NLP) techniques and Artificial Intelligence (AI) models in RE [9,10], their effectiveness remains tied to the availability of diverse high-quality datasets. Even with the paradigm shift introduced by in-context learning in Large Language Models (LLMs) [7], achieving optimal performance still requires sufficient high-quality training data and additional fine-tuning [8]. Therefore, the shortage of freely available RE-specific

J. Grabis et al. (Eds.): RCIS 2025, LNBIP 547, pp. 208–225, 2025.
https://doi.org/10.1007/978-3-031-92474-3_13

datasets remains a major bottleneck [9,17], slowing progress and limiting the effective use of AI models in RE.

However, the acquisition of high-quality data presents significant challenges. It is both costly and time-consuming, with the data annotation phase often being laborious and error-prone [12]. In addition, companies are often reluctant to share sensitive and private data, restricting the availability and use of such data for research purposes outside the organizations that produce it. Consequently, synthetic data emerges as a promising solution to address these issues.

Data synthesis may take several forms: anonymization of sensitive data [16]; annotation for supervised learning [12]; augmentation through text transformations (e.g., word deletion or synonym replacement) [4,6]; and the generation of entirely new samples from scratch [19]. Our research follows this last approach, leveraging LLMs' advanced text generation and multilingual capabilities [2] to address data scarcity in RE. This approach not only enables cost-effective, large-scale data generation but also provides systematic control over data properties, supporting varied RE use cases and addressing specific challenges such as data imbalance [15] and the scarcity of non-English resources [9].

To enable systematic controllability, we propose a generative approach that formalizes the variability of RE use cases and their specific requirements using a Feature Model (FM). This study represents the first iteration of a broader design science cycle [3], where we design and evaluate a minimal Product Line (PL) of synthetic datasets designed for classification-based RE use cases. Within this scope, we examine the following research questions:

- **RQ1.** How does the diversity of the synthetic data compare to that of real data?
- **RQ2.** How does training on synthetic data affect model performance relative to training exclusively on real data?

Contributions. ① We propose *Synthline*, a minimal and configurable approach for systematically controlling LLM-based data generation. ② We evaluate our approach on the use case of identifying requirements specification defects. The evaluation was conducted along two dimensions: data diversity and downstream model performance. ③ We provide both the generated datasets and our implementation code as open-access resources to support broader adoption and reproducibility in the RE community[1].

The paper is organized as follows: Sect. 2 reviews the related work relevant to our study. Section 3 introduces the methodology, detailing our PL-based approach and its implementation through Synthline. Section 4 presents our evaluation approach for assessing data diversity and utility. Section 5 discusses the results of our experiments. Section 6 examines threats to validity. Finally, Sect. 7 concludes the paper and outlines future research directions.

[1] https://github.com/abdelkarim-elhajjami/synthline/tree/v0.0.0.

2 Related Work

LLMs have motivated numerous studies to explore different approaches for synthetic data generation. The first methods relied on simple context-informed prompting, either in a zero-shot setting using only task specification [13] or in a few-shot setting, guided by a small number of examples [11]. Recent works have explored attributed prompting, incorporating explicit attributes to control generation and address issues of low informativeness and redundancy [19]. Furthermore, multi-step generation approaches decompose the workflow into sequential steps, generating multiple data series to ensure greater diversity [18]. These studies underscore the importance of controlled generation for both data diversity and quality.

In the RE domain, synthetic data research has primarily focused on data augmentation. Malik et al. [4] use techniques like back translation, paraphrasing, and RE-specific substitutions (e.g., Noun-Verb or Actor-Action) to improve conflict and duplicate detection in imbalanced datasets. Similarly, Majidzadeh et al. [5] introduce code-based augmentation methods such as variable renaming and operand swapping, to enhance requirements traceability, yielding improved precision, recall, and F1-score. To the best of our knowledge, Cheng et al. [6] are the only researchers to employ LLMs (GPT-J) for RE data augmentation; they demonstrated performance gains in classification across three industry case studies. However, these augmentation approaches have several key limitations. First, they rely on existing datasets, making them ineffective in cases of severe data scarcity. Second, their transformations often produce data that remains close to the original domain and distribution, restricting the potential for genuinely novel data. Third, they typically offer limited control over data attributes such as linguistic, or domain variability. By contrast, our research focuses on generating entirely new synthetic datasets with LLMs, guided by an FM to offer fine-grained controllability.

3 Overview of the Synthline Approach

This study constitutes the first iteration of a broader design science cycle, where design problems and knowledge questions are iteratively addressed through cycles of improvement design and empirical investigation [3]. We chose the PL approach to systematically control and configure the generation of synthetic datasets according to various RE use cases. The PL approach provides a form of systematic customization of a common platform from which multiple variants can be derived [21]. This paradigm offers two important benefits in our synthetic data generation context. First, it enables systematic customization, ensuring each variant aligns with the specific needs and constraints of its RE use case. Second, it helps improve the quality of the generated data as the platform commonalities can be systematically checked and tested across multiple datasets. This approach comprises two main phases: *Domain Engineering* and *Application Engineering*. In Domain Engineering, we begin with *domain analysis*, where we define the scope and variability of classification-based RE use cases, resulting in an FM. We then carry out *domain implementation* by developing a configurable generative workflow called *Synthline* as our shared platform. During Application Engineering, we select and configure features from the FM to meet the requirements of a specific RE use case and then use Synthline to generate the corresponding data.

3.1 Domain Engineering

Feature Model. The first step in the PL approach is domain analysis, which typically results in an FM [22]. An FM is a widely used notation to describe a PL's variability [22]. The standard FM notation can represent both concrete and abstract features, marking them as either optional or mandatory. The notation also defines constraints between features to prevent invalid or conflicting configurations. It also supports *or*- and *xor*- decomposition of features, to either allow selection of multiple subfeatures (requiring at least one) from a given feature set, or, restrict selection to exactly one subfeature.

To build the FM, we followed a reactive approach that starts from one or more concrete variants and then generalizes them into an FM. This approach can be done incrementally, gradually refining the model and introducing new features over time, lowering initial costs but potentially taking longer to achieve a stable platform [23]. Although our ultimate goal is to cover all RE use cases that can be automated with ML, achieving such broad coverage at once is impractical. Therefore, our incremental reactive approach starts in this study with a subset of RE use cases only. We intend to expand the coverage to additional use cases in future work.

To identify a suitable starting point, we relied on a recent systematic literature review on ML for RE [10], which highlighted that classification is the predominant ML task used in the RE domain, significantly surpassing other tasks like clustering and regression. It is important to distinguish between classification as an ML task and requirements classification as an RE use case. Requirements classification, as an RE use case, refers specifically to categorizing requirements into predefined types (e.g., functional vs. non-functional requirements). Classification as an ML task, on the other hand, supports not only requirements classification but also other RE use cases that can be formulated as classification problems, such as identifying requirements specification defects, where requirements are classified based on defect-related labels. Another key insight from [10] is that textual requirements artifacts are the primary data source in most RE-related ML studies, compared to other data types such as design documents, domain knowledge, user feedback, or policies and regulations. Based on these findings, we defined the scope of our PL in this study to focus on classification-based RE use cases, with textual requirements as the main artifact. We structured our FM around four core features: `Generator`, `Artifact`, `MLTask`, and `Output`.

The first feature of the model is `Generator`, which defines the configuration of the LLM generator. It includes the `LLM`, `Temperature`, and `TopP` features. The `LLM` subfeature supports two options: GPT-4o and DeepSeek-V3 models. `Temperature` regulates randomness in the generation process, with lower values producing more deterministic outputs and higher values allowing for greater creativity. `TopP` enables nucleus sampling, providing fine-grained control over output diversity while maintaining coherence.

The second core feature `Artifact` defines the characteristics of the generated data through three mandatory subfeatures. The `Requirement` subfeature, currently the only supported artifact type aligned with our initial scope, contains the core specifications for the requirements content. The `Domain` specifies the context(s) and `Language` controls the output language(s). Within the current `Requirement` feature, four key aspects are defined. The `RequirementType` subfeature follows the categorization defined in IEEE 830, including `ExternalInterfaces` (e.g., user interfaces, hardware interfaces), `Functions`, `Performance` requirements, `LogicalDatabase` specifications, `DesignConstraints`, and `SystemAttributes` (e.g., reliability, security, maintainability). The `SpecificationLevel` allows distinguishing between `HighLevelSpecification` and `DetailedSpecification`, enabling generation at different granularity levels. The `RequirementSource` identifies the stakeholder origin of requirements, such as `EndUsers`, `BusinessManagers`, `DevelopmentTeam`, or `RegulatoryBodies`. Finally, the `SpecificationFormat` supports different notation standards including natural language (`NL`), constrained NL (`ConstrainedNL`), use cases (`UseCase`), and user stories (`UserStory`).

The third core feature `MLTask` specifies the ML task characteristics that will be applied to the generated data. Within our current scope, we focus on the `Classification` subfeature, which requires two mandatory components: `Label` that defines the target classification categories, and `LabelDescription` that provides detailed specifications of these categories.

The last core feature `Output` defines the format and size of the generated data. It consists of two distinct subfeatures: `OutputFormat`, structured as an alternative group that requires choosing between `CSV` or `JSON` format, and `SubsetSize`, enabling control over the volume of data generated in each run. The complete graphical representation of the FM is provided in the Appendix 7.

Synthline Architecture. Our application of the design science methodology leads us to design Synthline iteratively. Synthline consists of a PL-based synthetic data generation workflow for RE use cases that can be framed as classification tasks in AI terms. The architecture of Synthline, illustrated in Fig. 1, transforms a configuration of the FM we designed for this purpose into synthetic data through a series of components.

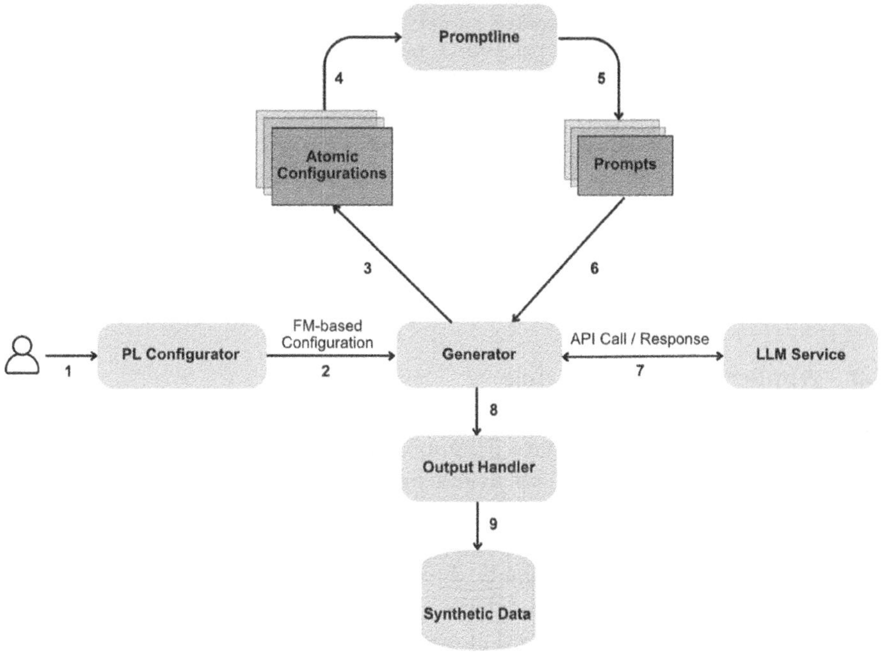

Fig. 1. High-level architecture of Synthline.

The workflow begins with the PL Configurator, which implements our FM as a graphical interface for feature selection. The interface directly mirrors the hierarchical structure of our FM, organizing controls into four distinct sections corresponding to our core features. For feature selections that require choosing from predefined options, such as `OutputFormat` between `CSV` and `JSON`, the interface provides dropdown menus. For numerical parameters like `Temperature` or `SubsetSize`, text input fields are implemented with appropriate validation. This design ensures that users can only create valid configurations while maintaining flexibility in their choices.

The Generator transforms this FM-based configuration into Atomic Configurations, where an Atomic Configuration represents a unique combination of feature values.

The Promptline component converts each Atomic Configuration into a Prompt using a standardized template, based on the structured prompt pattern from [34], which has proven effective:

```
Generate a requirement that:
1. Is classified as {label} (Description: {label_description})
2. Is of type {requirement_type}
3. Is written in {language}
4. Is for a {domain} system
5. Is from {requirement_source} perspective
```

```
6. Follows {specification_format} format
7. Is written at {specification_level} level
Important: Generate ONLY the requirement text. No additional context.
```

The Generator manages the distribution of the requested number of samples evenly across all atomic configurations, adjusting for cases where perfect division is not possible. It then interfaces with the LLM Service through asynchronous API calls for text generation. The generated content flows to the Output Handler, which persists the Synthetic Data with its configuration metadata in the specified format.

3.2 Application Engineering

Identification of Requirements Specification Defects. For demonstrating and evaluating our synthetic data PL approach, we selected the identification of requirements specification defects as our primary use case.

Requirements specification defects present a critical challenge in software development, as they can lead to significant issues in later development stages when problems become increasingly difficult and costly to resolve [24]. Automated methods to identify these defects, have shown promise in reducing human effort and improving defect detection accuracy [25–27].

The identification of requirements defects can be effectively framed as a classification problem, where each requirement is analyzed to determine the presence of specific types of defects. Several researchers in the RE community have identified different classes of requirements specification defects. Berry and Kamsties [28] highlighted ambiguity as a primary concern, while other research has emphasized other key defect categories such as non-measurable, non-atomic, and uncertain requirements [25].

Table 1 details the common requirements specification defects as defined by Fazelnia et al. [25]. We focus on these defects because we have access to their publicly available dataset[2] in which the requirements are labeled with these six categories.

The dataset combines requirements from real software systems, including Electronic Health Records Systems from the U.S. Department of Health and Human Services, and high-quality requirements from student projects' SysRS documents. Two experienced subject matter experts with over 20 and 10 years of experience conducted a rigorous three-stage review process, including grammatical analysis, defect labeling, and final peer review, to ensure high-quality annotations [25].

The dataset contains 131 total samples across 6 defect classes, with Ambiguous requirements being the most common (34 samples) and Directive requirements being the least represented (4 samples). This dataset presents two significant limitations: the substantial class imbalance and its relatively small size, which constrains the effective training or fine-tuning of AI models without risking overfitting (Table 2).

[2] Dataset available at https://zenodo.org/records/11000349.

Table 1. Categories of requirements defects and their definitions [25]

Category	Definition
Ambiguous Requirements	The requirements specifications are unclear, imprecise, and open to multiple interpretations
Directive Requirements	Direct the developer to look at additional sources beyond the requirement (e.g., referring to a figure or table)
Non-Measurable Requirements	The requirement is not measurable, or testable, it contain a qualitative value that is unable to be quantified or measured (e.g., "some," "large," "long")
Optional Requirements	Contain terms that may be interpreted in multiple ways and leave the choice of how to implement the requirement to the developer (e.g., "can," "may," "optionally," "as desired," "at last," "either," "eventually," "if appropriate," "in case of," "if necessary")
Uncertain Requirements	Contain a qualitative value that is insufficiently defined (e.g., "good," "bad," "strong," "easy"), or they may contain under-referenced elements (e.g., "compliant with standards (which ones?)")
Non-Atomic Requirements	Describe more than one action to be taken and typically have a conjunction, such as "and"

Table 2. Distribution of Requirements Specification Defects

Class	# of samples
Ambiguous	34
Directive	4
Non-Measurable	18
Optional	31
Uncertain	16
Non-Atomic	28

Data Generation. Our synthetic data generation methodology employs each LLM to produce six class-specific subsets aligned with the defect categories defined in Table 1, using their textual definitions as label descriptors. This aligns with the multi-step generation approach that creates multiple data subsets in separate runs, providing more control [18]. The configuration settings for our generation process are detailed in Table 3.

We selected GPT-4o and DeepSeek-V3 as our primary LLM options. The inclusion of GPT-4o builds on the demonstrated efficacy of its predecessors in generating high-quality synthetic data [12,19,29]. For open-source alternatives, DeepSeek-V3 was selected as the state-of-the-art model as of our experimentation date (5 January 2025), with empirical evaluations indicating it surpasses leading proprietary models, including GPT-4o and Claude 3.5 Sonnet, across most benchmarks [2]. Following [19] and [29], a Temperature of 1.0 and TopP of 1.0 were chosen, as these settings have been shown to produce high-quality synthetic data. The Healthcare domain was selected to directly align with the real dataset's focus on Electronic Health Records Systems from the U.S. Department of Health and Human Services, while Restaurant Operations Management was chosen as the second domain to reflect the requirements related to restaurant employee registration, menu management, and operational functionalities present in the dataset. Similarly, we chose Constrained NL format and English language to ensure our synthetic requirements mirror the linguistic characteristics of the real dataset. The CSV output format was chosen for its simplicity and compatibility with the real dataset's format. We deliberately aligned our synthetic data features with those of the real dataset, as the evaluation of synthetic models was to be performed on a holdout subset of real data, ensuring a fair assessment of their performance. Additionally, we included all available subfeatures for requirement types, specification levels, and requirement sources. This comprehensive selection ensures diversity and avoids systematic exclusions that could bias the dataset.

Table 3. Configuration parameters for controlled multi-step generation runs, applied per LLM and defect class in synthetic data subset production.

Feature	Values	# Atomic Configurations
Temperature	1	1
TopP	1	1
Requirement Type	External Interfaces, Functions, Performance, Logical Database, Design Constraints, System Attributes	7
Specification Level	High-Level Specification, Detailed Specification	2
Requirement Source	End Users, Business Managers, Development Team, Regulatory Bodies	4
Specification Format	Constrained NL	1
Language	English	1
Domain	Healthcare, Restaurant Operations Management	2
Output Format	CSV	1
Subset Size	1120	1
Total # Atomic Configurations		**112**

For each LLM (GPT-4o and DeepSeek-V3), we generated 1,120 requirements per defect category, resulting in 6,720 synthetic samples per model and 13,440 samples in total.

4 Evaluation

As highlighted in a recent survey [14], evaluation methods for synthetic data can be broadly classified into direct and indirect approaches. Direct evaluation focuses on the properties of the generated data itself. A commonly studied aspect is data diversity, which is typically evaluated using vocabulary-based statistics (e.g., vocabulary size and N-gram frequency) or semantic variation metrics such as cosine similarity [19]. On the other hand, indirect evaluation primarily assesses the utility of synthetic data with respect to the performance of downstream models trained on it, commonly applying a train-on-synthetic, test-on-real paradigm [19]. In our research, we focus on these two complementary dimensions to comprehensively assess our generated data.

Diversity Evaluation. To comprehensively assess the diversity of our synthetic datasets compared to real data, we performed this evaluation after deduplicating[3] the synthetic datasets (both GPT-4o and DeepSeek). We used several complementary metrics to capture different aspects of textual diversity. We began with two vocabulary-based measurements: absolute vocabulary size, defined as the number of unique terms in each dataset, and normalized vocabulary size per data point, which accounts for varying dataset sizes. These metrics provide a basic indication of lexical richness, where a larger vocabulary suggests greater linguistic variety and potentially richer expression in the generated samples.

For semantic diversity assessment, we computed cosine similarities between text pairs using the embedding of Sentence-BERT [31]. The Average Pairwise Similarity (APS) measures the overall semantic similarity across all samples in the dataset. A lower APS value indicates greater semantic diversity overall. The Intra-Class APS, on the other hand, measures the semantic similarity between samples within the same defect class. Lower Intra-Class APS values suggest more diverse examples of the same defect type, indicating less redundancy within classes.

To evaluate phrase-level diversity, we calculated the Inter-sample N-gram Frequency (INGF), which measures repetitive patterns in the generated text. This metric computes the average frequency of n-grams across all samples in a dataset, where lower values indicate less repetition and thus more diverse phrase construction. Together, these metrics provide a multi-faceted view of dataset diversity, from lexical richness to semantic variation and phrase-level uniqueness.

The synthetic datasets' statistics are summarized in Table 4, showing the total number of samples and their distribution across defect classes after deduplication.

[3] Deduplication consists of removing identical requirements that were occasionally generated multiple times by the LLMs.

Table 4. Synthetic Datasets' Statistics After Deduplication.

Defect Class	DeepSeek Samples	GPT-4o Samples
Ambiguous	1,007	1,117
Directive	1,015	1,120
Non-Atomic	828	1,118
Non-Measurable	701	1,117
Optional	934	1,119
Uncertain	828	1,112
Total Samples	**5,313**	**6,703**

Utility Evaluation. To enable a direct comparison between synthetic and real data performance, we fine-tuned BERT-base-uncased [20] as the backbone and used the standard cross-entropy loss by default. For hyperparameter selection, we adhered to the recommendations in [30] and did not use the validation set for model selection. Table 5 details the hyperparameters used in our experiments.

Table 5. Training hyperparameters.

Hyperparameter	Value
Learning Rate	5e-5
Batch Size	32
Training Epochs	6
Weight Decay	1e-4
Warmup Ratio	6%

Our evaluation methodology follows a train-on-synthetic, test-on-real paradigm [19]. We maintained a holdout test set composed of 30% of the real data, which remains constant across all experiments to ensure consistent evaluation. The remaining 70% of real data is used only for training the real-data baseline model. For synthetic data experiments, we trained models on either purely synthetic data or combinations of synthetic and real training data. This approach allows us to directly compare the effectiveness of synthetic data against, and in combination with, real data.

To ensure robust evaluation, we conducted multiple training runs for each configuration, accounting for the inherent variability in model training. We evaluated model performance using macro-averaged precision and recall metrics.

5 Results

5.1 RQ1: How Does the Diversity of the Generated Data Compare to that of Real Data?

As shown in Table 6, while both GPT-4o and DeepSeek generate datasets with larger absolute vocabularies (4,874 and 3,659 terms respectively) compared to the real dataset (566 terms), this is primarily due to their larger sample sizes. When normalized by sample count, the synthetic datasets demonstrate significantly lower lexical density, with 0.73 and 0.69 words per sample for GPT-4o and DeepSeek respectively, compared to 4.32 words per sample in the real data. This 6.2–6.3× reduction in lexical density suggests that synthetic data tends toward a more constrained vocabulary despite having access to a broader lexical range.

Table 6. Vocabulary size analysis across datasets.

Dataset	Vocab. Size	Normalized Vocab. Size
Real Data	566	4.32
DeepSeek	3,659	0.69
GPT-4o	4,874	0.73

Examining semantic diversity through Table 7, we observe that synthetic datasets produce more semantically homogeneous content. Both synthetic datasets show higher overall APS values (DeepSeek: 0.649, GPT-4o: 0.612) compared to real data (0.544). This pattern is particularly pronounced in the intra-class analysis, where synthetic data shows notably higher similarity scores (DeepSeek: 0.729, GPT-4o: 0.689) compared to real data (0.587). As illustrated in Fig. 2, the distribution of cosine similarities for synthetic data is shifted toward higher values and exhibits a narrower spread.

Table 7. Semantic similarity and phrase repetition metrics.

Dataset	APS	Intra-class APS	INGF
Real Data	0.544	0.587	1.214
DeepSeek	0.649	0.729	5.086
GPT-4o	0.612	0.689	2.472

Table 7 shows that, at the phrase level, synthetic data exhibit significantly higher n-gram repetition, with INGF values 2–4 times higher than real data (DeepSeek: 5.086, GPT-4o: 2.472, vs. Real: 1.214). This increased repetition could be attributed to our choice of Constrained NL Specification format in the generation process, which encourages more standardized phrase construction.

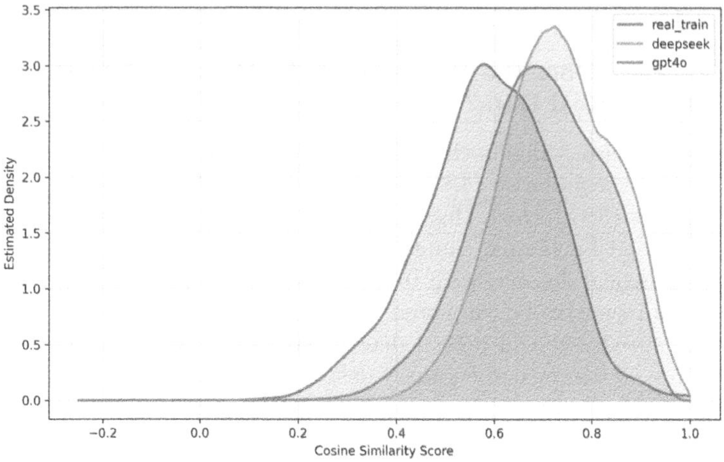

Fig. 2. Distribution of cosine similarity of text pairs sampled from the same class.

While synthetic datasets offer broader absolute vocabulary coverage, they show lower diversity than real data across all measured dimensions - exhibiting 6× lower lexical density, higher semantic similarity (0.612-0.649 vs 0.544), and increased phrase repetition (2-4× higher INGF values).

5.2 RQ2: How Does Training on Synthetic Data Affect Model Performance Relative to Training Exclusively on Real Data?

We examine model performance across three key training approaches: synthetic-only data, hybrid training combining real and synthetic data, and combined synthetic datasets. First, models trained exclusively on synthetic data show improved performance over the real-data baseline. As shown in Table 8, the DeepSeek-based model achieved a 37% improvement in precision (0.425 vs 0.310), while the GPT-4o-based model showed a 20% improvement (0.372 vs 0.310). This suggests that synthetic data alone can provide a promising viable training alternative in scenarios where real data is scarce.

The most notable performance gains, however, emerge from hybrid training approaches combining real and synthetic data. The combination of real and GPT-4o synthetic data yielded the best overall performance, achieving 85% improvement in precision (0.575 vs. 0.310) compared to the real-data baseline. This substantial improvement suggests strong complementarity between real and synthetic training data. Interestingly, we observed significant variations in performance stability across different synthetic data sources. Models trained with DeepSeek synthetic data showed more consistent performance (standard deviation: ±0.027 in precision) compared to those trained with GPT-4o data (±0.229). This difference in stability could have important implications for practical applications where performance consistency is important. A particularly

Table 8. Model performance with different training data compositions. All experiments use identical model architecture, varying only in training data.

Training Data	Precision	Recall
Real data only (baseline)	0.310 ± 0.050	0.256 ± 0.043
DeepSeek only	0.425 ± 0.083	0.334 ± 0.025
GPT-4o only	0.372 ± 0.060	0.329 ± 0.080
GPT-4o + DeepSeek	0.308 ± 0.037	0.297 ± 0.015
Real + DeepSeek	0.456 ± 0.027	0.431 ± 0.042
Real + GPT-4o	0.575 ± 0.229	0.512 ± 0.166
Real + GPT-4o + DeepSeek	0.469 ± 0.121	0.427 ± 0.089

noteworthy finding is that combining multiple synthetic datasets (GPT-4o + DeepSeek) actually degraded performance below that achieved by individual synthetic datasets. This negative synergy persisted even when real data was included in the combination (Real + GPT-4o + DeepSeek: 0.469 vs. Real + GPT-4o: 0.575), suggesting that careful curation of synthetic data sources may be more important than maximizing the volume of synthetic data.

The relatively low performance metrics across all configurations can be attributed to our methodological choice of using recommended hyperparameters [30] without dataset-specific tuning. While hyperparameter optimization might improve absolute performance, using a fixed configuration across all experiments ensures fair comparisons between different training data compositions, making any relative improvements directly attributable to the data quality.

> Synthetic data can not only match but significantly exceed the performance of real data in training downstream models for classifying requirements defects, with hybrid approaches showing particular promise. However, the success appears highly dependent on the specific combination and curation of synthetic data sources rather than merely increasing data volume.

6 Threats To Validity

We identified several threats to validity across the four standard dimensions: internal, external, construct, and conclusion validity. Internally, our models' performance could be influenced by the specific hyperparameter settings chosen for training. While we followed established recommendations from previous research [30], different hyperparameter configurations might yield varying results. Our evaluation methodology's reliance on BERT-base-uncased as the backbone model may introduce bias, as other pre-trained models might perform differently with synthetic data. Additionally, while we followed the structured prompt pattern from [19], the synthetic data quality inherently depends on prompt design choices, where even subtle variations in template structure or phrasing could systematically influence the characteristics of generated samples.

Regarding construct validity, our diversity metrics (vocabulary size, APS, INGF) may not fully capture all aspects of diversity in NL generation outputs. While these metrics are widely used in synthetic data evaluation [14,19] and provide useful quantitative measures, they may have inherent limitations in representing the nuanced characteristics that define high-quality diverse text. Our choice to use constrained NL specification format across all synthetic data generation inherently limits linguistic variation and diversity potential, which may affect our assessment of the approach's ability to generate diverse outputs. Moreover, our approach relies on Fazelnia et al.'s [25] defect category definitions as label descriptions to guide LLM generation, where categories like "Non-Measurable" may have subjective interpretation boundaries, potentially affecting how the LLM interprets and generates examples for each defect type. Externally, our findings may not generalize beyond the specific context of identifying requirements specification defects to other classification-based RE use cases. The effectiveness of our approach was demonstrated primarily in the healthcare and restaurant operations management domains, and different application domains might present unique challenges not addressed in our current evaluation. Additionally, our results are based on GPT-4o and DeepSeek-V3 models, and the performance characteristics might vary with different LLM versions or alternative models. Finally, conclusion validity might be affected by the small size of our test set, which could impact the statistical significance of our performance comparisons. Additionally, the variability in model training due to random initialization and stochastic optimization processes introduces potential instability in our results, although we attempted to mitigate this by conducting multiple training runs for each configuration.

7 Conclusion

This paper presented Synthline, a synthetic data generation approach that leverages LLMs within a PL strategy to control the generation of synthetic data for classification-based RE use cases. Through an empirical evaluation focused on identifying requirements specification defects, we found that synthetic data, particularly when combined with real data, can significantly enhance downstream classifier performance. Specifically, our experiments showed improvements of up to 85% in precision and a 2$\times$ increase in recall compared to a baseline using only real data. Furthermore, although synthetic datasets exhibited less diversity than real data, the results confirm that carefully curated synthetic samples serve as a viable-and, in many cases, superior-training resource when real data is limited.

In the future, we plan to broaden Synthline's scope to support additional RE use cases and refine generation techniques to further enhance synthetic data quality. We will also develop a more comprehensive evaluation framework that goes beyond diversity and utility, enabling a deeper understanding of how synthetic data contributes to robust, reliable AI solutions in RE.

Appendix: Graphical Representation of the FM

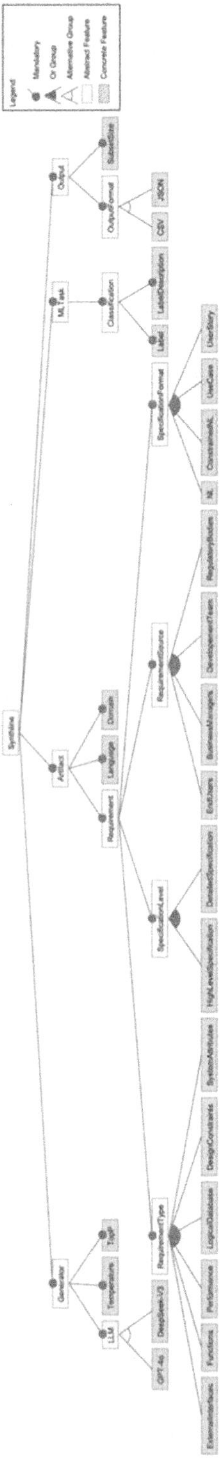

References

1. Franch, X., Palomares, C., Quer, C., Chatzipetrou, P., Gorschek, T.: The state-of-practice in requirements specification: an extended interview study at 12 companies. Requir. Eng. 1–33 (2023). https://doi.org/10.1007/s00766-023-00399-7
2. DeepSeek-AI, DeepSeek-V3 Technical Report, arXiv:2412.19437 [cs.CL], 2024
3. Wieringa, R.: Design science as nested problem solving. In: Proceedings of the 2009 ACM International Conference, New York, NY, USA, 2009. https://doi.org/10.1145/1555619.1555630
4. Malik, G., Cevik, M., Başar, A.: Data Augmentation for Conflict and Duplicate Detection in Software Engineering Sentence Pairs, arXiv:2305.09608 [cs.SE], 2023
5. Majidzadeh, A., Ashtiani, M., Zakeri-Nasrabadi, M.: Multi-type requirements traceability prediction by code data augmentation and fine-tuning MS-CodeBERT. Comput. Stand. Interfaces. **90**, Article 103850 (2024). https://doi.org/10.1016/j.csi.2024.103850
6. Gräßler, I., Preuß, D., Brandt, L., Mohr, M.: Efficient extraction of technical requirements applying data augmentation. In: Proceedings of the IEEE International Symposium on Systems Engineering (ISSE). Vienna, Austria 2022, pp. 1–8 (2022). https://doi.org/10.1109/ISSE54508.2022.10005452
7. Brown, T.B., et al.: Language Models are Few-Shot Learners, arXiv:2005.14165 [cs.CL], 2020
8. Mosbach, M., Pimentel, T., Ravfogel, S., Klakow, D., Elazar, Y.: Few-shot fine-tuning vs. in-context learning: a fair comparison and evaluation. In: Findings of the Association for Computational Linguistics: ACL 2023, Toronto, Canada, pp. 12284–12314, 2023. Association for Computational Linguistics (2023)
9. Zhao, L., et al.: Natural language processing for requirements engineering: a systematic mapping study. ACM Comput. Surv. **54**(3), Article 55, 1–41 (2022)
10. Li, T., Zhang, X., Wang, Y., Zhou, Q., Wang, Y., Dong, F.: Machine learning for requirements engineering (ML4RE): a systematic literature review complemented by practitioners' voices from stack overflow. Inf. Softw. Technol. **172**, Article 107477 (2024)
11. Li, Z., Zhu, H., Lu, Z., Yin, M.: Synthetic data generation with large language models for text classification: potential and limitations. In: Proceedings of the 2023 Conference on Empirical Methods in Natural Language Processing (EMNLP), 2023
12. Gilardi, F., Alizadeh, M., Kubli, M.: ChatGPT outperforms crowd workers for text-annotation tasks. Proc. Natl. Acad. Sci. **120**(30), Article e2305016120 (2023)
13. Ye, J., et al.: ZeroGen: Efficient Zero-shot Learning via Dataset Generation, arXiv:2202.07922 [cs.CL], 2022
14. Long, L., et al.: On LLMs-Driven Synthetic Data Generation, Curation, and Evaluation: A Survey, arXiv:2406.15126 [cs.CL], 2024
15. El-Hajjami, A., Fafin, N., Salinesi, C.: Which AI Technique Is Better to Classify Requirements? An Experiment with SVM, LSTM, and ChatGPT, arXiv:2311.11547 [cs.AI], 2024
16. Yoon, J. Jordon, J., van der Schaar, M.: PATE-GAN: generating synthetic data with differential privacy guarantees. In: Proceedings of the International Conference on Learning Representations (ICLR), 2019
17. Abualhaija, S., et al.: Replication in Requirements Engineering: the NLP for RE Case, arXiv:2304.10265 [cs.SE], 2024
18. Shao, Z., Gong, Y. Shen, Y., Huang, M., Duan, N., Chen, W.: Synthetic Prompting: Generating Chain-of-Thought Demonstrations for Large Language Models, arXiv:2302.00618 [cs.CL], 2023

19. Yu, Y., et al.: Large language model as attributed training data generator: a tale of diversity and bias. In: Proceedings of the Thirty-seventh Conference on Neural Information Processing Systems (NeurIPS) Datasets and Benchmarks Track, 2023
20. Devlin, J., Chang, M.-W., Lee, K., Toutanova, K.: BERT: pre-training of deep bidirectional transformers for language understanding. In: Proceedings of the 2019 Conference of the North American Chapter of the Association for Computational Linguistics: Human Language Technologies, Volume 1 (Long and Short Papers). In: Burstein, J., Doran, C., Solorio, T. (eds.), Minneapolis, Minnesota: Association for Computational Linguistics, pp. 4171–4186, 2019. https://doi.org/10.18653/v1/N19-1423
21. Pohl, K., Böckle, G., Van Der Linden, F.: Software Product Line Engineering: Foundations, Principles, and Techniques, vol. 1. Springer, Berlin, Germany (2005)
22. Apel, S., Batory, D., Kästner, C., Saake, G.: Feature-Oriented Software Product Lines: Concepts and Implementation. Springer, Berlin, Heidelberg (2013)
23. Northrop, L.: Software Product Lines Essentials. SEI Carnegie Mellon University, Pittsburgh, 2008
24. Lauesen, S., Vinter, O.: Preventing requirement defects: an experiment in process improvement. Requir. Eng. J. **6**(1), 37–50 (2001)
25. Fazelnia, M., Koscinski, V., Herzog, S., Mirakhorli, M.: Lessons from the use of natural language inference (NLI) in requirements engineering tasks. In: Proceedings of the 32nd IEEE International Requirements Engineering Conference (RE), 2024
26. Goyal, S., et al.: SDP-BB: a software defect prediction model using BiLSTM and BERT-based semantic features. IEEE Trans. Softw. Eng. (2022)
27. Dautovic, A., Plösch, R., Saft, M.: Automated quality defect detection in software development documents. In: Proceedings of the 5th International Conference on Software Quality Management, 2011
28. Berry, D.M., Kamsties, E.: Ambiguity in requirements specification. In: Leite, M., Doorn, J. (eds.), Requirements Engineering: State of the Practice, pp. 7–44. Springer US, Boston, MA, 2004
29. Peng, B., Li, C., He, P., Galley, M., Gao, J.: Instruction Tuning with GPT-4, arXiv:2304.03277 [cs.CL], 2023
30. Perez, E., Kiela, D., Cho, K.: True few-shot learning with language models. In: Beygelzimer, A., Dauphin, Y., Liang, P., Wortman Vaughan, J. (eds.), Proceedings of the Advances in Neural Information Processing Systems, 2021
31. Reimers, N., Gurevych, I.: Sentence-BERT: sentence embeddings using Siamese BERT-networks. In: Inui, K., Jiang, J., Ng, V., Wan, X. (eds.), Proceedings of the 2019 Conference on Empirical Methods in Natural Language Processing and the 9th International Joint Conference on Natural Language Processing (EMNLP-IJCNLP), Hong Kong, China: Association for Computational Linguistics, 2019, pp. 3982–3992, 2019. https://doi.org/10.18653/v1/D19-1410

The Role of Ethics in Requirements Engineering for Developing Information Systems

Christopher Julian Kern[1(✉)], Karin Hübner[1], Leo Poss[2], Stefan Schönig[2], and Julia Kroenung[1]

[1] University of Hagen, Hagen, Germany
{christopher.kern,julia.kroenung}@fernuni-hagen.de
[2] University of Regensburg, Regensburg, Germany
{leo.poss,stefan.schoenig}@ur.de

Abstract. This systematic review examines the integration of ethical considerations in Requirements Engineering (RE), analyzing methodological approaches and implementation challenges across software development practices. Through comprehensive literature analysis, we identify five primary frameworks: data governance, expert systems, stakeholder engagement, ethical oversight, and participatory methods.

Our findings reveal critical success factors, including stakeholder communication protocols, transparent data handling practices, and human-centered governance structures. While established frameworks like ETHICS and IEEE 7000-2021 standard offer promising directions, their practical implementation remains challenging. Key obstacles include responsibility diffusion among stakeholders, ethical-functional requirement conflicts, and resource-intensive processes. The study exposes a significant research gap in practitioner guidance for ethical value trade-offs. We propose recommendations encompassing structured ethical frameworks, enhanced stakeholder participation mechanisms, and formal oversight protocols.

Keywords: Ethical Values · Requirements Engineering · Responsible Software Development · Artificial Intelligence

1 Introduction

Information systems rapidly reshape societal paradigms and interpersonal interactions in the contemporary technological landscape, raising ethical concerns that demand careful consideration. Increasing integration of information systems, especially those based on artificial intelligence (AI), into critical domains such as healthcare, finance, and transportation amplifies the need to ensure these systems are developed and deployed responsibly [13,23,35]. *Requirements Engineering* (RE) shapes these systems, translating user needs and societal values into concrete specifications that guide the development of these systems. RE

J. Grabis et al. (Eds.): RCIS 2025, LNBIP 547, pp. 226–242, 2025.
https://doi.org/10.1007/978-3-031-92474-3_14

directly influences ethical considerations such as the preservation of privacy, the mitigation of algorithmic bias, and the overall impact on human well-being [7].

Existing literature suggests that traditional software development processes often fail to adequately address ethical dimensions like ethical value requirements, responsibilities, and social impact [5,7,46]. This lack of specific guidance creates ambiguity regarding deriving, documenting, and managing ethical requirements, hindering the development of ethical information systems [6,46]. To systematically address this issue, we will answer the following research questions:

RQ1. How can the role of ethical values in requirements engineering for information systems be categorized?

RQ2. What methods and technologies are recommended to facilitate the development of ethical information systems?

This study addresses the gap by systematically reviewing relevant literature to identify methods and techniques for integrating ethical values into software development. Through a structured literature analysis of top academic journals, the study identifies key themes and recommendations for ethical software development, focusing on RE and human-centered design. The findings highlight the essential building blocks of stakeholder participation, extraction of ethical requirements, and addressing biases in algorithms and data. We categorize existing approaches considering data governance, expert systems, stakeholders, and ethical oversight. Our findings contribute to existing knowledge by proposing a framework for integrating ethical considerations into RE for developing information systems at the beginning of the software development lifecycle, emphasizing a proactive and collaborative approach while motivating future research and potential future tool support toward ethical information systems.

The paper is structured as follows: After the motivation section elaborates on the need for this research, Sect. 2 describes the necessary background on ethics and RE as part of software development and agile practices. The following Sect. 3 describes how we conducted our study adhering to best practices like the PRISMA statement [41]. Section 4 summarizes our findings on ethical values in RE and is divided into two main sections: *(i)* describing the results categorized in the respective groups and *(ii)* the topology of the identified recommended actions. This is followed by a discussion of the findings and their implications for the diffusion of responsibility, including limitations of this manuscript in Sect. 5, and a conclusion including future work in Sect. 6.

2 Background

2.1 Ethics

The concept of value, derived from the Latin *valere* meaning "to be strong" or "to be worthy", covers both subjective and objective perspectives. Subjectively, values represent personal or societal principles that guide decision-making and define what is considered important or good in life. Objectively, values are the targets of emotional and conscientious acts [46]. In the context of ethical system design, values become actionable through structured frameworks that translate abstract ethical principles into concrete system requirements.

The IEEE 7000-2021 standard [21] defines ethical values as culturally grounded principles that help individuals distinguish right from wrong. It adopts a three-layered value framework: *core value – value qualities – value demonstrator*. A *core value* lies at the center of a cluster comprising related values and demonstrators. Core values like privacy or beauty are positive principles frequently referenced by user groups [48]. *Value qualities* specify different aspects of a core value, while *value demonstrators* refer to observable characteristics or practices that promote or undermine the core value. For example, within the core value of privacy, demonstrators might include data security, prohibition of unauthorized recordings, or restrictions on knowledge-sharing about the user by the system [48]. These demonstrators closely align with stakeholder needs and preferences, emphasizing their relevance in system design.

In value-based engineering, *ethical value requirements* (EVRs) are derived from identified values to ensure that the *system of interest* embodies positive value dispositions. The IEEE 7000-2021 standard defines these value dispositions as design features that fulfill ethical requirements ("created in systems when designs are created that fulfill the ethical requirements" [21]). Sometimes, value dispositions and value demonstrators are used interchangeably as part of a value cluster [47]. This structured approach ensures that values are systematically incorporated into the system's design, fostering ethical outcomes that align with stakeholder expectations.

Ethical considerations in software development arise from the need to safeguard fundamental human rights and promote well-being beyond monetary aspects [54]. In contexts involving *artificial intelligence* (AI) and autonomous systems, transparency regarding algorithms and decision-making processes is critical to ensure accountability and enable human oversight during critical decisions [5,12]. One approach by [54] identified 31 overarching ethical values, ranging from accountability to usability, and has been employed in [26] to map the ethical values of business process management. Similarly, in a comprehensive review of global policy documents on AI ethics, [23] extracted five primary ethical principles frequently referenced across various frameworks relevant to broader information systems and not just limited to AI:

- **Transparency:** Transparency encompasses the goals of clarity, explainability, and interpretability. It ensures that decision-making processes and underlying data are understandable, fostering trust, mitigating harm, and supporting accountability in legal contexts.
- **Justice and Fairness:** This principle emphasizes inclusivity and fairness by advocating for mechanisms that reduce biases and discrimination in algorithmic systems. It seeks to ensure fair access to the technologies and their benefits to create just outcomes.
- **Non-maleficence:** Non-maleficence focuses on preventing harm caused by technology use. This includes minimizing risks to physical, psychological, and

economic well-being and applying risk management strategies when harm is unavoidable.

- **Responsibility:** Responsibility refers to the clear assignment of accountability for developing, deploying, and overseeing AI systems. It encompasses organizational and individual responsibility, though whether AI systems can bear responsibility remains debated.
- **Privacy:** Privacy involves the ethical management of personal data, ensuring that data collection, storage, and usage are conducted responsibly. It calls for robust data protection measures and the adaptation of regulatory frameworks to AI-driven technologies.

Beyond these core principles, the authors identify additional values, such as *beneficence*, *autonomy*, *trust*, and *sustainability*. Nevertheless, they stress significant ambiguities in interpreting and operationalizing these principles and emphasize transitioning from high-level ethical principles to concrete implementation strategies, particularly as AI intensifies the ethical challenges faced in system design and usage [23].

2.2 Requirements Engineering and Agile Approaches

Requirements engineering (RE) focuses on identifying, documenting, and managing a system's functional and non-functional requirements. It ensures that software products meet stakeholder needs and expectations. The process includes several key stages: requirement elicitation, analysis, specification, validation, and management. Like this, RE bridges stakeholders and developers to mitigate misunderstandings that could lead to project failures [42]. Various methods, including interviews, brainstorming, and prototyping, are used to gather and refine requirements effectively [18].

Agile methods prioritize working software over comprehensive documentation, individuals over processes, and collaboration over contractual negotiations [1]. They promise faster and more flexible development than traditional software development models by focusing on iterative progress and close collaboration with customers. The *Agile Manifesto* offers a mindset for team and project organization while not explicitly addressing process structures, including RE. This absence does not mean that RE is irrelevant; instead, agile approaches require RE to be adapted into flexible processes and structures aligned with the agile mindset while maintaining the primacy of people over processes [28].

Requirements gathering and management remain essential elements of software development to ensure deliverables align with customer expectations [28]. Agile frameworks provide an ideal environment for continuous specification and refinement of requirements through iterative development, close customer collaboration, and enhanced team communication tools. In this context, RE should be integrated as an ongoing activity where requirements evolve throughout development. However, fundamental quality requirements and constraints must be defined early, following identifying goals and stakeholders, as they significantly influence the specifications, architecture, and design. These foundational requirements also form a basis for later testing [28].

3 Methodology

Our systematic literature analysis (SLA) addresses the research questions, adhering to the methodology proposed in [4]. We opted to focus our analysis on high-quality publications and identified them similar to [4] by using scholarly rankings, such as the *VHB Journal Rating* [14] and *SJR (SCImago Journal Rank indicator)*, that serve as a reliable basis for the selection of qualitative sound sources. The literature search targeted scientific journals with *A+* or *A* ratings in the fields of Information Systems (IS) and Business Ethics, according to the *VHB Journal Rating* as well as journals in the domains of Computer Science and Information Systems ranked *Q1* with an *SJR (SCImago Journal Rank indicator)* of at least *5.0*.[1]

The search was conducted on the *EBSCOhost* meta-database in Q4 2024, followed by a backward search. Only peer-reviewed academic articles available online were considered, with no restrictions on the publication date, as older approaches to implementation were deemed equally relevant to the research questions as more recent methodologies. The selection criteria focus on articles that mention methods and techniques related to the research questions and provide detailed descriptions and evaluations. These methods can involve engineering tool requirements or other disciplines' relevant techniques.

To identify our journals, we used the following string: *[Journal Name] AND "ethic*"* in Abstract and *("value" OR "requirement*") AND ("information system*" OR "software")* in Full Text. This search string was designed to comprehensively identify articles discussing values and requirements in system design while minimizing overly restrictive criteria. The asterisk refers to wildcards.

Preliminary search iterations using various strings revealed that restrictive search strategies like requiring all search terms to appear in the abstract or keywords yielded relatively few results. Consequently, the search string was intentionally kept broad and general to maximize the initial pool of articles. This broad dataset was subsequently refined based on thematic titles, keywords, and relevance. A backward search was followed using the references cited in the selected articles. The article selection process is documented in Fig. 1, presented according to the PRISMA framework [41].

During the initial database search, 2,226 records were identified. Duplicates, editorials, book reviews, and similar entries were excluded during the first screening, retaining only peer-reviewed articles, case studies, and reports. After a detailed review, 23 articles remained in the sample. An additional article was identified and extracted through the following backward search. Resulting in a final sample of 24 publications. Many of the excluded articles focused on ethical guidelines at the corporate level or examined ethical attitudes and decision-making on an individual level without linking these aspects to implementing ethics in information systems.

[1] A comprehensive list of all included journals is available from the authors upon request.

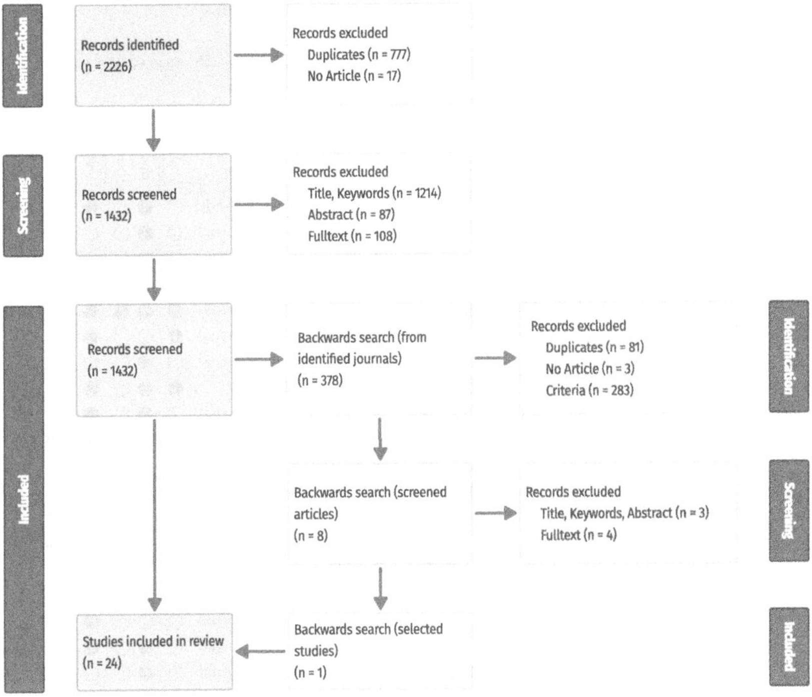

Fig. 1. PRISMA diagram following [41].

The final sample includes articles on methods and techniques for integrating ethical values into information systems design or software development, particularly within requirements engineering and ethical frameworks guiding development processes. Articles have been excluded when they primarily address corporate-level ethical guidelines without linking them to specific methods or techniques for system design, individual-level ethical attitudes, and decision-making. For instance, discussions on personal or professional ethics in IT are not considered to establish a direct connection to implementing ethics in information systems. Furthermore, ethical education and training for raising moral awareness in IT professionals, where ethics are addressed indirectly through the developer rather than integrated into system design methodologies, have also been excluded.

4 Findings

Based on the literature review results, we identified several key themes and non-exclusive categories that help structure the role of RE in ethical information systems shown in the concept matrix (see Table 1): *Data Governance and Ethical Information Systems* (DATA-GOV) that focus on data as the underlying resource for decision-making in information technology and *Expert Systems and*

Table 1. Concept matrix following [53]

	Category							Ethical Values [23]				
Author	DATA-GOV	DATA-XPSYS	STAKE	STAKE-ETH	STAKE-OV	Context	Methodology	Trans-parency	Justice	Harm-lessness	Respon-sibility	Privacy
Adman & Warren (2000) [2]	○	○	●	●	○	IS-Design	Case Study	○	●	○	●	○
Alder (1998) [3]	○	○	●	○	●	IS-Design	Conceptual	○	●	○	○	○
Chatterjee *et al.* (2009) [6]	○	○	●	○	●	IS-Design	Conceptual	○	●	○	○	○
Culnan & Williams (2009) [9]	○	○	○	○	●	Corporate	Case Study	○	○	○	●	●
Díaz-Rodríguez *et al.* (2023) [11]	●	○	○	○	○	IS-Design	Conceptual	●	●	●	●	●
Dillard & Yuthas (2001) [10]	○	○	●	○	○	IS-Design	Conceptual	●	○	○	●	○
Gonçalves *et al.* (2012) [15]	○	○	●	○	○	Corporate	Conceptual	○	●	○	○	○
Hunkenschroer & Luetge (2022) [20]	●	○	○	○	○	Corporate	Review	●	●	○	●	●
John-Mathews *et al.* (2022) [24]	●	○	●	○	●	IS-Design	Conceptual	○	●	○	●	○
Khalil (1993) [27]	●	●	○	○	●	IS-Design	Conceptual	○	○	○	●	●
Lobschat *et al.* (2021) [31]	●	○	●	○	●	Corporate	Conceptual	●	○	○	●	●
Martin (2018) [33]	●	○	○	○	●	IS-Design	Conceptual	●	●	●	●	○
Mingers & Walsham (2010) [34]	○	○	●	○	○	IS-Design	Conceptual	○	●	○	○	●
Mittelstadt (2019) [35]	○	○	●	○	●	IS-Design	Conceptual	○	●	○	●	●
Nussbaumer *et al.* (2021) [39]	○	●	●	○	●	IS-Design	Case Study	●	●	○	●	●
Olphert & Damodaran (2007)[40]	○	○	●	●	○	IS-Design	Case Study	○	●	○	○	○
Saravanamuthu (2002) [43]	○	○	●	●	○	Corporate	Conceptual	○	○	○	●	○
Smith & Hasnas (1999)[44]	○	○	○	○	●	General	Conceptual	○	●	○	●	●
Som *et al.* (2009) [45]	●	○	●	○	●	IS-Design	Conceptual	○	●	○	○	●
Stahl (2007) [49]	○	○	●	●	○	General	Conceptual	○	○	○	●	○
Telkamp & Anderson (2022) [50]	○	○	●	●	●	General	Review	●	○	●	●	●
Tóth *et al.* (2022) [52]	●	○	●	●	○	IS-Design	Conceptual	●	●	○	●	●
Wood *et al.* (2020) [55]	○	●	●	○	○	IS-Design	Review	●	○	○	○	●
Yuthas & Dillard (1999) [56]	○	●	●	○	●	IS-Design	Conceptual	○	○	○	●	○

Ethical Considerations (DATA-XPSYS) which adds another layer to this process by providing the necessary system expertise. The category of *Stakeholder Engagement* (STAKE) focuses on communication and participation of stakeholders as the key to developing ethical systems. This category can further be divided into two subcategories, namely *ETHICS Methodology* (STAKE-ETH) based on the ETHICS concept[2] of [36] and *Ethical Oversight in Software Development* (STAKE-OV) that includes all authors calling for a 'guardian of ethics' as the person responsible for ethical development. Finally, we categorized all publications by their context and methodology and mentioned the ethical values of the values presented in [23].

DATA-GOV: Data Governance and Ethical Information Systems. As foundational components of data-driven systems, algorithms analyze datasets and identify patterns, which they then use to provide decision-making recommendations or make decisions autonomously [51]. Since algorithms are developed

[2] The ETHICS methodology by Enid Mumford [36] is a participatory socio-technical framework focusing on user involvement and ethical considerations in system design in distinction to the philosophical concept of ethics.

within a non-neutral world, they inherently reflect societal biases [27,33,52]. Ethical concerns in data governance can arise at various stages, particularly regarding the selection and handling of training data [20,24,33]. Algorithms cannot independently exhibit ethical behavior, as ethical decision-making involves complex human values and subjective judgments [50].

The presence of value-laden[3] algorithms implies legal accountability for organizations regarding the ethical implications of their systems [33]. This accountability is linked to both the degree of decision-making power delegated to algorithms versus humans [33,52] and the societal impact of the decisions being made, often referred to as the *moral intensity* of a situation. In response to these ethical challenges, three types of actions are recommended: internal organizational measures, internal technical measures, and external regulatory approaches. Participatory processes and inclusive discussions during the development phase are crucial for ethical data governance, ensuring diverse moral perspectives are considered [10,24]. Developers are urged to evaluate training data and implement transparent decision-making processes to empower users [11,33]. Externally, dynamic regulatory frameworks are proposed to provide adaptable legal boundaries while fostering public education on AI ethics and data governance [11].

DATA-XPSYS: Expert Systems and Ethical Considerations. Expert systems and decision support systems aim to select, analyze, and sequentially present facts to facilitate decision-making by a human expert [30]. These systems support human decision-makers by organizing relevant information or autonomously providing recommendations in well-structured problem settings where expert knowledge is traditionally required [10]. The authors of [10] propose a responsibility framework to integrate ethical considerations into system development, aiming for continuous stakeholder engagement to identify potential system tasks and characteristics while exploring possible ethical dilemmas. This process fosters a sense of group solidarity, enabling decision-makers to understand stakeholders' collective interests and values better. The framework encourages decision-makers to internalize a sense of *responsible self* [37], ensuring that design choices are aligned with stakeholder needs and ethical principles. The ultimate goal is to create expert systems capable of making moral decisions in line with the values of the community they serve. Building on this framework, [39] introduce an ethics-by-technology model in emergency management. Their approach ensures that ethical values are systematically integrated throughout the entire system development life-cycle, from research and design to deployment.

The following Table 2 contains inductively defined and combined typology of recommended actions split into three categories: internal organizational actions, internal technological actions, and external measures for ethical value development.

---

[3] The term *value-laden* refers to concepts, theories, language, or inquiries that are not purely descriptive or objective but rather implicitly or explicitly embody, reflect, promote, or presuppose specific normative commitments, evaluative stances, or prescriptive ideals (cf. [32]).

Table 2. Typology of recommended actions for DATA-GOV and DATA-XPSYS

		Recommended action	Sources
Internal: Organizational Measures	Establishment of Ethical Frameworks & Comprehensive Ethics Training	Implementation of formal ethics guidelines supported by comprehensive employee training covering guidelines themselves, practical AI development applications (including bias detection and mitigation), and case studies	[20, 39, 50]
	Stakeholder Engagement for Ethical Data Practices	Diverse stakeholder input on data selection, preparation, management, and ethical implications actively addressed through structured communication and participation	[10, 24, 39, 50]
	Continuous Ethical Evaluation of AI Systems	Continuous ethical evaluation throughout the development lifecycle for ethically informed (conscious) data management, rigorous diverse output review, documented ethical decisions, and clear output evaluation criteria	[24, 33, 50, 52]
	Transparency towards Affected Parties	Ensuring affected individuals are informed about the use of information systems and their impact	[20]
Internal: Technical Measures	Human-Centered IS Governance	Auditability, transparency, and human-in/on-the-loop control ensured in information systems through design principles incorporating clear protocols for decision-making authority and defined roles and responsibilities	[11, 20, 27, 33, 39, 52]
	Interactive Machine Learning	Implementing interactive machine learning paradigms that allow continuous human input and feedback during the learning process	[10]
	AI Quality Assurance and Ethical Development	AI quality and ethical development through established testing methods (validation, verification, robustness testing against adversarial attacks) and comprehensive training on bias identification, mitigation, and prevention, incorporating algorithmic bias reduction techniques (data preprocessing, algorithmic adjustments, and careful training data curation)	[10, 20, 24, 52, 55]
	Data Protection through Encryption	Implementing robust data protection measures, including advanced encryption techniques, to ensure data confidentiality and privacy throughout the lifecycle	[55]
External Measures	Legal Framework and Control Mechanisms	Legal framework defining system responsibilities and implementing control mechanisms will ensure ethical and legal compliance	[11, 35, 52]
	Ethical Education and Competency Development	Integrating ethics into computer science curricula and providing AI ethics training alongside public digital literacy initiatives to foster a broad understanding of ethical implications	[11, 31, 45]

STAKE: Stakeholder Engagement. Integrating ethical values in software development is diverse, with ongoing debates about whether information systems are inherently value-laden [31,33] or value-neutral [3]. Despite this discussion, there is broad agreement that communication and stakeholder participation is essential in creating ethical systems [8]. Equal and consensus-driven collaboration with stakeholders enhances the inclusion of ethical perspectives [10,34]. In particular, involving marginalized groups such as individuals with disabilities helps promote social inclusion in system design [3,15].

In AI development, value embedding requires participatory communication across both high-level and algorithmic design stages [24,35]. Translating ethical principles into specific requirements faces methodological challenges due to contextual dependencies and the absence of universal principle prioritization [35]. To address this, the *Corporate Digital Responsibility* (CDR) framework proposes a layered approach to integrating values, standards, and behaviors throughout the software lifecycle [31]. Ethics-by-design offers a practical approach to value-driven system development, from goal setting and value selection to RE, implementation, and user training [39]. Additionally, adopting the multiple-perspectives approach [43] ensures a holistic view of system development by considering technical, organizational, and personal dimensions. Through active stakeholder participation, software projects become more sustainable and gain user acceptance [45,56].

STAKE-ETH: ETHICS Methodology [36,49]. The ETHICS framework (Effective Technical & Human Implementation of Computer-based Systems), developed by Mumford [36], emerges as a prominent approach within the STAKE

Table 3. Typology of recommended actions for STAKE-ETH

		Recommended action	Sources
Internal: Organizational Measures	Comprehensive Stakeholder Engagement Framework	Systematic stakeholder selection based on demographic heterogeneity and fair discourse frameworks to enable participatory value assessment and algorithmic parameter calibration	[3,6,10,15,24, 34,35,39,40,56]
	Integrated Methodological Framework	Synthesis of existing frameworks like ETHICS methodology, Viable System Model, QuickEthics, Linston's Multiple Perspectives, Ten Point Plan, CATWOE, and Joint Application Development protocols for systematic ethical implementation	[2,34,43,49]
	Development Accountability and Capacity Framework	Integration of ethical evaluation protocols, development team capacity enhancement, and systematic attribution of developer responsibility across software architecture, systems implementation, and process reengineering	[6,39]
	Organizational Documentation and Transparency Framework	Systematic implementation of ethical guidelines, decision justification protocols, and algorithmic training data documentation incorporating stakeholder moral frameworks	[31,35,39,50]
Internal: Technical Measures	Comprehensive Technical Capability Development	Integration of specialized developer training protocols focusing on ethical value implementation methods and technical execution frameworks.	[15]
	Data Minimization Architecture	Implementation of systematic data processing protocols emphasizing personal data minimization through architectural constraints and technical safeguards.	[39]
	Value Integration in General	Systematic technical assessment and implementation methodology for integrating ethical values into system architecture and development processes	[34]
External Measures	Integrated Sociopolitical Discourse Framework	Discourse in politics, business and society on digital ethics	[2]

cluster, focusing on human and organizational aspects of system design [36]. In e-government contexts, early implementation of ETHICS demonstrates enhanced technology adoption, acceptance, and sustainability through citizen participation [40].

Furthermore, participatory development fosters social inclusion and strengthens democracy. However, to enable comprehensive citizen involvement throughout all stages of software development, more support from commissioning authorities, enhanced training for public employees, and citizen empowerment are essential. Within the STAKE-ETH subcategory, the primary values addressed are justice, fairness, and responsibility, while other values are emphasized less. The recommended strategies are outlined in Table 3.

While some authors in the STAKE cluster support ETHICS' cooperative approach, they also highlight its limitations. Scholars have raised concerns about its embeddedness within existing capitalist frameworks [43,44], while others have noted fundamental limitations in its theoretical foundations [49]. In particular, the ETHICS framework is ambiguous regarding whether ethical values are essential to system development or merely instrumental for better implementation. This distinction is crucial for determining whether ethical or functional requirements should take priority in specific cases [49]. To address these shortcomings, scholars have suggested supplementing the framework with principles from information ethics to provide a stronger theoretical foundation [49] or enhancing its

Table 4. Typology of recommended actions for STAKE-OV

		Recommended action	Sources
Internal: Organizational Measures	Systematic Integration of Ethical Values	Ethical values and considerations systematically integrated throughout the software development lifecycle	[44]
	Ethical Accountability within the Development Team	Ethical conduct through dedicated ethics representatives, individual developer responsibility for stakeholder considerations, and ongoing ethical training	[6,31,35,39]
	Diversity in Stakeholder Engagement	Diverse stakeholder representation for incorporating varied perspectives and values into the development process.	[24,50]
	Privacy Impact Assessment	Mandatory execution of Privacy Impact Assessments (PIAs) in all software development projects to ensure early identification and mitigation of potential impacts on data protection	[9]
External Measures	Ethical Licensing and Certification	Regulatory frameworks for ethical AI licensing and certification, including defined assessment criteria and licensing/certification processes	[27,33]

methodological toolkit with approaches like Quick-Ethics or the Viable System Model [2].

STAKE-OV: Ethical Oversight in Software Development. Designated ethical overseers in the software development process can also provide a solution for more ethical software development – while technical and organizational structures alone are insufficient for the incorporation of ethical values, the inclusion of certified or licensed individuals can aid in sensitive decisions, particularly in algorithm design [27,33]. These ethical overseers, or *moral agents*, ensure that ethical principles are actively integrated throughout the development process, maintaining a balance between stakeholder interests and ethical imperatives [6].

Beyond individual oversight, teams of diverse ethical perspectives can play a significant role in ensuring ethical balance [24,50]. [44] argue that professional or corporate codes alone cannot safeguard ethical behavior. Instead, they propose a structured three-step approach: first, engaging management in discussions about the organization's ethical foundations; second, embedding an ethical oversight routine as part of the development process; and third, appointing an ethics officer or responsible for handling ethical concerns raised by IS professionals as shown in Table 4.

5 Discussion

Our literature review identifies four key building blocks for integrating ethical values into software development: *stakeholder communication*, *data and algorithm governance*, *decision-making authority*, and *ethical representation*. Our distilled insights show that, while technical solutions are necessary, they are insufficient without participatory processes. For example, in the category STAKE-OV, involving diverse stakeholder groups can ensure values such as fairness, inclusion, and transparency. Ethical challenges in data handling, algorithm design, and decision-making can be addressed through collaborative frameworks and oversight mechanisms, emphasizing that high-stakes decisions must remain under human control. Dedicated ethical representatives – external reviewers or internal advisors – are essential in safeguarding ethical standards throughout the system lifecycle.

5.1 Discussion of Categories and Recommended Actions

Ethical practices in information systems provide different approaches to ensure ethical data governance (DATA-GOV) and expert system design (DATA-XPSYS). Likewise, they address ethical concerns in stakeholder engagement (STAKE-ETH), focusing on inclusivity, accountability, and transparency throughout the software development lifecycle. They also propose ethical oversight in software development (STAKE-OV), emphasizing embedding ethical values consistently across organizational processes and ensuring accountability at multiple levels.

These perspectives, including diverse stakeholder perspectives and transparent communication, are essential for reducing ethical risks during data handling and algorithmic decision-making. On the technical side, human-centered IS governance and AI quality assurance play can aid in preventing bias and ensuring accountability through rigorous testing, robustness measures, and transparent feedback mechanisms. External constraints, such as legal frameworks and educational initiatives, support long-term compliance and foster ethical competency among developers. On the other hand, the actions for addressing ethical concerns in stakeholder engagement (STAKE-ETH) focus on fostering inclusivity, accountability, and transparency throughout the software development lifecycle. Ethical oversight in software development (STAKE-OV) aids in embedding ethical values consistently across organizational processes and ensuring accountability at multiple levels. Likewise, this also includes participatory design and stakeholder engagement, which reduces the gap between perceived and actual stakeholder interests, preventing resistance and ensuring more effective implementations [16].

5.2 Diffusion of Responsibility

Integrating ethical values into developing (information) systems is a complex task. While various measures across the software development lifecycle are proposed, no consensus on a transparent methodology or responsibility for embedding ethics has emerged. Beyond software development processes, external factors such as corporate ethics, personal morality, professional codes, and legal frameworks are frequently seen as key elements for ensuring ethical information systems [9,25]. Several authors argue that codes of ethics should be systematically enforced and integrated into corporate culture through management leadership [31,44]. Calls for ethical application, particularly in AI systems, have shifted responsibility toward the personal development of developers, with proposals to integrate ethical training into software engineering curricula [6,9].

Despite the growing recognition of the need for stakeholder engagement and discourse, significant gaps remain in defining processes and best practices for value-based system design. While involving stakeholders offers recognized benefits such as increased user acceptance, sustainability, and faster product launches,

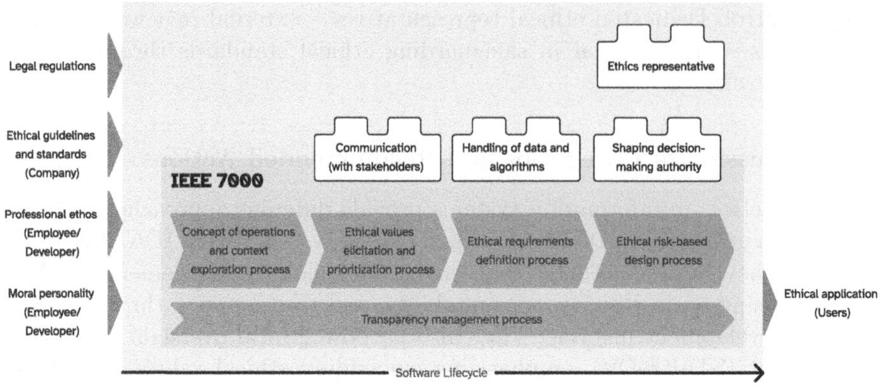

Fig. 2. Diffusion of responsibilities – Building blocks of ethical software development with contextualization.

it also imposes considerable time and cost burdens during early development. This raises concerns about the economic viability of such approaches, with questions about whether "ethical" companies or those employing ethically certified developers might gain a competitive advantage in the market [35].

Our findings suggest a diffusion of accountability, where multiple actors and processes could assume responsibility for embedding ethical values (Fig. 2). This mirrors the phenomenon of "diffusion of responsibility", where individual action diminishes as the number of potential actors increases [29]. A possible way to counter this phenomenon is to utilize standardized frameworks and regulations. IEEE 7000-2021 [21], for example, offers a promising step toward clarifying responsibilities and methods for value-based system design, integrating ethical value elicitation and ethical information systems.

Another approach to addressing this issue could be the installation of ethics and compliance officers to ensure that organizations uphold their ethical standards and compliance frameworks by monitoring, advising, and enforcing ethical guidelines [19]. Their effectiveness, however, depends on their ability to operate independently of company management, as conflicts of interest can arise when ethics officers report to executives rather than directly to the board of directors, potentially limiting their capacity to challenge unethical behavior [38]. Likewise, ethical awareness [22] in general could counter the diffusion of responsibility through universal accountability. Further research into these methods is required within their respective contexts, as their practicality has yet to be demonstrated.

5.3 Limitations

The literature review focuses on methods within RE for identifying and implementing ethical requirements, which cannot directly ensure transferability and applicability to broader ethical concerns in software development. Instead, it examines a specific part of software development and can serve as a use case and first step for further research. To this end, the sample includes only high-ranking information systems and business ethics journals. Expanding the review to include interdisciplinary journals and conference proceedings could provide additional insights. Despite using a systematic methodology, some relevant studies may not have been included due to publication bias or limited database coverage. One limitation of this meta-study is its predominantly theoretical foundation, which requires further qualitative or quantitative methodological validation to enable robust transfer to practical applications.

6 Conclusion and Future Work

This meta-study presents a structured review of existing research on ethics in RE and methods for identifying and integrating ethical values in software development. The findings illustrate different ways of embedding ethical considerations into RE processes, showcasing the lack of a unified methodology or clearly defined responsibilities. Many publications highlight the role of RE in conceptualizing and specifying ethical values. The identified methods span the entire software development life-cycle, involving technical processes and external governance structures (e.g., [24,50]).

A significant contribution of this study is categorizing approaches into four key building blocks for ethical software development, including *stakeholder communication*, *data and algorithm management*, *decision-making authority*, and *ethical oversight*. In providing these categories, we effectively answer our first research question. These elements include different aspects, such as participatory processes, continuous ethical evaluation, and transparency in RE. One key problem we identified was the diffusion of responsibility among various stakeholders – from developers to external ethics officers – indicating the absence of clear accountability frameworks. Despite references to codes of ethics and professional standards, current practices seem insufficient to systematically ensure the integration of ethical values, opening up a valuable avenue for further research.

We identified a need for further standardization and methodological clarity in ethical RE. Frameworks like IEEE 7000-2021 [21] offer an essential first step by embedding ethical value considerations directly into RE processes; however, their adoption and practical application are still uncertain. In identifying these frameworks, we answer our second research question. Future research could explore empirical case studies, regional variations in ethical value prioritization, and the effectiveness of existing frameworks in real-world RE settings. Another avenue for further research could be the inclusion of ethical user stories in different contexts. Ethical user stories can counter the diffusion of responsibility by assigning clear ownership of ethical decisions to specific roles within the development process [17].

All these possibilities showcase the need for actionable guidelines and standardized methodologies to bridge the gap between ethical theory and practical application in RE, ultimately setting the foundation for developing truly responsible and human-centered software systems.

References

1. Abrahamsson, P., Salo, O., Ronkainen, J., Warsta, J.: Agile software development methods: review and analysis (2017)
2. Adman, P., Warren, L.: Participatory sociotechnical design of organizations and information systems - an adaptation of ETHICS methodology. J. Inform. Technol. **15**(1), 39–51 (2000)
3. Alder, G.S.: Ethical issues in electronic performance monitoring: a consideration of deontological and teleological perspectives. J. Bus. Ethics **17**(7), 729–743 (1998)
4. vom Brocke, J., Simons, A., Niehaves, B., Reimer, K., Plattfaut, R., et al.: Reconstructing the giant: on the importance of rigour in documenting the literature search process. In: ECIS 2009 Proceedings (2009)
5. Chatila, R., Havens, J.C.: The IEEE global initiative on ethics of autonomous and intelligent systems, pp. 11–16. Springer (2019)
6. Chatterjee, S., Sarker, S., Fuller, M.: Ethical information systems development: a Baumanian postmodernist perspective. J. Assoc. Inf. Syst. **10**(11) (2009)
7. Cheng, B.H., Atlee, J.M.: Research directions in requirements engineering. In: Future of Software Engineering (FOSE 2007), pp. 285–303. IEEE (2007)
8. Cludts, S.: Organisation theory and the ethics of participation. J. Bus. Ethics **21**(2/3), 157–171 (1999)
9. Culnan, W.: How ethics can enhance organizational privacy: lessons from the choicepoint and TJX data breaches. MIS Q. **33**(4), 673 (2009)
10. Dillard, J.F., Yuthas, K.: A responsibility ethic for audit expert systems. J. Bus. Ethics **30**(4), 337–359 (2001)
11. Díaz-Rodríguez, N., Del Ser, J., Coeckelbergh, M., López de Prado, M., Herrera-Viedma, E., et al.: Connecting the dots in trustworthy artificial intelligence: from AI principles, ethics, and key requirements to responsible AI systems and regulation. Inform. Fusion **99**, 101896 (2023)
12. European Union: EU Artificial Intelligence Act (2024)
13. Floridi, L.: Establishing the rules for building trustworthy AI. Nat. Mach. Intell. **1**(6), 261–262 (2019)
14. German Academic Association for Business Research: VHB Publication Media Rating 2024 (2024). https://vhbonline.org/en/service/vhb-rating-2024/
15. Gonçalves, R., Martins, J., Pereira, J., Oliveira, M., Ferreira, J.: Enterprise web accessibility levels amongst the forbes 250: where art thou O virtuous leader? J. Bus. Ethics **113**(2), 363–375 (2012)
16. de Gooyert, V., Rouwette, E., van Kranenburg, H., Freeman, E.: Reviewing the role of stakeholders in operational research: a stakeholder theory perspective. Eur. J. Oper. Res. **262**(2), 402–410 (2017)
17. Halme, E., et al.: How to write ethical user stories? Impacts of the ECCOLA method. In: Gregory, P., Lassenius, C., Wang, X., Kruchten, P. (eds.) Agile Processes in Software Engineering and Extreme Programming, pp. 36–52. Springer (2021)
18. Hegade, P.: Design for requirements engineering. J. Eng. Educ. Transform. **33**, 219 (2020)
19. Hoffman, W.M., Neill, J.D., Stovall, O.S.: An investigation of ethics officer independence. J. Bus. Ethics **78**(1–2), 87–95 (2007)
20. Hunkenschroer, A.L., Luetge, C.: Ethics of AI-enabled recruiting and selection: a review and research agenda. J. Bus. Ethics **178**(4), 977–1007 (2022)

21. Institute of Electrical and Electronics Engineers: IEEE 7000:2021 IEEE Standard Model Process for Addressing Ethical Concerns during System Design (2021)
22. Jaganjac, B., Abrahamsen, L.M., Olsen, T.S., Hunnes, J.A.: Is it time to reclaim the "ethics" in business ethics education? J. Bus. Ethics **190**(1), 1–22 (2023)
23. Jobin, A., Ienca, M., Vayena, E.: The global landscape of AI ethics guidelines. Nat. Mach. Intell. **1**(9), 389–399 (2019)
24. John-Mathews, J.M., Cardon, D., Balagué, C.: From reality to world. A critical perspective on AI fairness. J. Bus. Ethics **178**(4), 945–959 (2022)
25. Kamoun, F., Almourad, M.B.: Accessibility as an integral factor in e-government web site evaluation. Inform. Technol. Peopl. **27**(2), 208–227 (2014)
26. Kern, C.J., Poss, L., Kroenung, J., Schönig, S.: Navigating the moral maze: a literature review of ethical values in business process management. Bus. Process. Manag. J. **30**(8), 343–370 (2024)
27. Khalil, O.: Artificial decision-making and artificial ethics: a management concern. J. Bus. Ethics **12**(4), 313–321 (1993)
28. Laplante, P.A.: Requirements Engineering for Software and Systems. Applied Software Engineering, 3rd edn. CRC Press (2018)
29. Latané, B., Darley, J.M.: The Unresponsive Bystander: Why Doesn't He Help? Appleton-Century-Crofts, New York (1970)
30. Ligęza, A.: Expert systems approach to decision support. Eur. J. Oper. Res. **37**(1), 100–110 (1988)
31. Lobschat, L., Mueller, B., Eggers, F., Brandimarte, L., Diefenbach, S., et al.: Corporate digital responsibility. J. Bus. Res. **122**, 875–888 (2021)
32. Longino, H.E.: Science as Social Knowledge: Values and Objectivity in Scientific Inquiry. Princeton University Press (1990)
33. Martin, K.: Ethical implications and accountability of algorithms. J. Bus. Ethics **160**(4), 835–850 (2018).
34. Mingers, W.: Toward ethical information systems: the contribution of discourse ethics. MIS Q. **34**(4), 833 (2010)
35. Mittelstadt, B.: Principles alone cannot guarantee ethical AI. Nat. Mach. Intell. **1**(11), 501–507 (2019)
36. Mumford, E.: Effective Systems Design and Requirements Analysis. Macmillan Education, UK (1995)
37. Niebuhr, H.R.: The Responsible Self: An Essay in Christian Moral Philosophy. Harper & Row, New York (1963)
38. den Nieuwenboer, N.A., Treviño, L.K., Bishop, D., Kreiner, G.E., Murphy, C.: Growth through ethical role identity work: the case of ethics and compliance officers. J. Bus. Ethics (2024)
39. Nussbaumer, A., Pope, A., Neville, K.: A framework for applying ethics-by-design to decision support systems for emergency management. Inform. Syst. J. **33**(1), 34–55 (2021)
40. Olphert, W., Damodaran, L.: Citizen participation and engagement in the design of e-government services: the missing link in effective ICT design and delivery. J. Assoc. Inf. Syst. **8**(9) (2007)
41. Page, M.J., McKenzie, J.E., Bossuyt, P.M., et al.: The PRISMA 2020 statement: an updated guideline for reporting systematic reviews. BMJ n71 (2021)
42. Pilat, L., Kaindl, H.: A knowledge management perspective of requirements engineering. In: 2011 Fifth International Conference on Research Challenges in Information Science, pp. 1–12. IEEE (2011)
43. Saravanamuthu, K.: The political lacuna in participatory systems design. J. Inform. Technol. **17**(4), 185–198 (2002)

44. Smith, H.J., Hasnas, J.: Ethics and information systems: the corporate domain. MIS Q. **23**(1), 109 (1999)
45. Som, C., Hilty, L.M., Köhler, A.R.: The precautionary principle as a framework for a sustainable information society. J. Bus. Ethics **85**(S3), 493–505 (2009)
46. Spiekermann, S.: Value-Based Engineering: Building Ethical Technology for Humanity. De Gruyter, Berlin (2023)
47. Spiekermann, S., Krasnova, H., Hinz, O., et al.: Values and ethics in information systems: a state-of-the-art analysis and avenues for future research. Bus. Inform. Syst. Eng+.+ **64**(2), 247–264 (2022)
48. Spiekermann, S., Winkler, T.: Value-based engineering with IEEE 7000. IEEE Technol. Soc. Mag. **41**(3), 71–80 (2022)
49. Stahl, B.C.: ETHICS, morality and critique: an essay on Enid Mumford's socio-technical approach. J. Assoc. Inf. Syst. **8**(9) (2008)
50. Telkamp, J.B., Anderson, M.H.: The implications of diverse human moral foundations for assessing the ethicality of artificial intelligence. J. Bus. Ethics **178**(4), 961–976 (2022)
51. de la Torre Gutierrez, H., Pham, D.: Estimation and generation of training patterns for control chart pattern recognition. Comput. Ind. Eng. **95**, 72–82 (2016)
52. Tóth, Z., Zsófia, R., Gruber, T., Loebbecke, C.: The dawn of the AI robots: towards a new framework of AI robot accountability. J. Bus. Ethics **178**(4), 895–916 (2022)
53. Webster, J., Watson, R.T.: Analyzing the past to prepare for the future: writing a literature review. MIS Q. **26**(2), xiii–xxiii (2002)
54. Winkler, T., Spiekermann, S.: Human values as the basis for sustainable information system design. IEEE Technol. Soc. Mag. **38**(3), 34–43 (2019)
55. Wood, A., Najarian, K., Kahrobaei, D.: Homomorphic encryption for machine learning in medicine and bioinformatics. ACM Comput. Surv. **53**(4), 1–35 (2020)
56. Yuthas, K., Dillard, J.F.: Ethical development of advanced technology: a postmodern stakeholder perspective. J. Bus. Ethics **19**(1), 35–49 (1999)

Databases and Information Management

Model-Agnostic Evolution Management

Pavel Koupil(✉), Jáchym Bártík, and Irena Holubová

Department of Software Engineering, Charles University, Prague, Czech Republic
{pavel.koupil,jachym.bartik,irena.holubova}@matfyz.cuni.cz

Abstract. This paper tackles a critical challenge in data management: adapting to evolving user requirements, a concept we term *evolution management*. Our approach specifically addresses the evolution management of multi-model data. By leveraging category theory, we provide an abstract representation of the combined models. We extend this categorical framework by introducing schema-modification operations of varying types and complexities. And, we propose corresponding propagation strategies to adapt queries, with a particular focus on mitigating the impact of *decreasing information capacity*. Through experimental verification, we demonstrate significant reductions in effort across different strategies and their effects on querying.

Keywords: Evolution · Multi-Model Data · Query Synchronization

1 Introduction

More than two-thirds of the 50 most widely used database management systems (DBMSs)[1] fall into the category of *multi-model*, i.e., supporting multiple models as first-class citizens. However, due to the absence of established standards governing which models should be combined and how or how to query the multi-model data, the features of existing multi-model databases exhibit significant variations in the supported models, the way they are combined, the multi-model query languages, etc. [16]. In addition, combining different models, such as aggregate-oriented and aggregate-ignorant, flat and structured, or schema-full and schema-less, naturally introduces conflicting features requiring specific handling.

This situation presents numerous challenges across all aspects of data management. In this paper, we focus on the situation when user requirements change, resulting in the need to adapt the data structures, so-called *evolution management*, i.e., propagation of the user-required changes in the schema to all affected parts of the system while preserving its integrity and efficiency. We must also consider the cross-model redundancy and the contradictory features of the combined

[1] https://db-engines.com/en/ranking.

Supported by the GAČR grant no. 23-07781S and GAUK grant no. 292323.

J. Grabis et al. (Eds.): RCIS 2025, LNBIP 547, pp. 245–261, 2025.
https://doi.org/10.1007/978-3-031-92474-3_15

models/DBMSs. Our approach is based on our unifying abstract categorical representation of multi-model data [11], which covers all popular data models and their combinations, including cross-model redundancy. In addition, its natural graph representation shields the user from the technical details of the underlying models and DBMSs.

The key and, to the best of our knowledge, genuine contributions of this paper to the evolution management of multi-model data are as follows:

- For the categorical model of multi-model data, we formally define three types of schema-modification operations (SMOs) – *basic*, *composite*, and *complex* – enabling natural and user-friendly model modifications. In addition, our set is among the most comprehensive among related solutions, while model abstraction eliminates redundant operations for analogous concepts across underlying logical models, ensuring both expressiveness and conciseness.
- We extend our multi-model evolution-management approach by synchronizing multi-model queries, i.e., ensuring that the queries provide the same results after applying SMOs. We primarily focus on the challenging aspect of *decreasing information capacity*, i.e., the possible loss of information. We utilize several approaches to the multi-model context to solve the problem.
- Through a proof-of-concept implementation, we demonstrate how our approach reduces manual, error-prone work. And, we experimentally evaluate the efficiency of the distinct synchronization strategies.

Outline. Section 2 introduces the categorical representation of multi-model data. Section 3 describes the evolution management strategy, the SMOs, and their propagation. Their implementation and experiments are described in Sect. 4. Section 5 overviews related work, and Sect. 6 concludes.

2 Categorical Representation of Multi-model Data

First, let us introduce our categorical representation of multi-model data. To unify the terminology of distinct models, we use the following terms: A *kind* corresponds to a class of items (e.g., a relational table or a collection of documents). A *record* corresponds to an item of a kind (e.g., a table row or a document). A record consists of simple/complex *properties* having *domains*.

Let us also remember the basic notions of category theory. A *category* $\mathbf{C} = (\mathcal{O}, \mathcal{M}, \circ)$ consists of a set of objects $\mathcal{O}$, set of morphisms $\mathcal{M}$, and a composition operation $\circ$ over the morphisms ensuring transitivity and associativity. Each morphism is modeled as an arrow $f : A \rightarrow B$, where $A, B \in \mathcal{O}$, $A = dom(f), B = cod(f)$. And there is an *identity* morphism $1_A \in \mathcal{M}$ for each object A. Apparently, a category can be visualized as a multigraph, with objects represented as vertices and morphisms as directed edges.

The core and the integration point of all our solutions is an abstract representation of multi-model data called *schema category* [11]. It is defined as a tuple $\mathbf{S} = (\mathcal{O}_\mathbf{S}, \mathcal{M}_\mathbf{S}, \circ_\mathbf{S})$. Objects in $\mathcal{O}_\mathbf{S}$ correspond to the domains of the ER model's entity types, attributes, and relationship types. Morphisms in $\mathcal{M}_\mathbf{S}$ (of semantics "has a property", "has an identifier", "has a role", or "is a" and respectively labels) connect appropriate pairs of objects.

The decomposition of a schema category **S**, eventually partial or overlapping, is defined via a set of *mappings*. For each kind, the mapping specifies where and how its records are stored in a selected single-/multi-model DBMS utilizing a concept of *functor*, which mutually maps corresponding objects of logical and conceptual schemas and preserves the hierarchical structure of a kind.

Example 1. Consider a part of the sample multi-model scenario in Fig. 1(c). In the categorical schema, kinds *User* and *Business* (whose root objects are highlighted with colors) are mapped to the respective relational table (purple) and document collection (green). The kind *User* has only simple properties, e.g., *name* or *fans* (we omit the most common morphism label ***#property*** for morphisms of semantics "has a property"). In the curly braces, we can see the identifiers of the kinds, e.g., *user_id* for kind *User*. The kind *Business* also has a complex property *address* consisting of several simple properties (e.g., *city* and *street*). There is also kind *Tip*, which represents the relationship of kinds denoted using the respective morphisms with label ***#role***, of semantics "has a role" (i.e., *User* and *Business*). Kind *Friends* represents an array of pairs of objects denoted using morphisms with label ***#role***, whose position is given by the simple property *_INDEX*. Last but not least, complex property *attributes* of kind *Business* represents a map of *(key, value)* pairs associated with kind *Business*. □

2.1 Conceptual Querying of Multi-Model Data

The categorical representation can be viewed as a graph. Thus, it can be queried using a graph query language. For this purpose, we introduced the *Multi-Model Query Language* (MMQL) [9] utilizing the well-known SPARQL notation.

An MMQL expression consists of a projection (`SELECT`) that defines the structure of the result and a selection (`WHERE`), where graph patterns can be defined along with filter conditions (`FILTER`). We also support result modifiers such as `ORDER BY`, `LIMIT`, or aggregation functions. Graph patterns are expressed using triples $p : s \rightarrow o$, where o, s (e.g., variables or constants) correspond to objects from $\mathcal{O}_\mathbf{S}$ and p maps to morphisms in $\mathcal{M}_\mathbf{S}$. In other words, instead of RDF graph data, we query the schema category **S**. (Sample MMQL queries are provided in Fig. 2(a) and explained in Example 3).

3 Evolution Management Approach

A schema category **S** represents the conceptual model of a given multi-model use case, abstracting from the technical details of underlying logical models. In addition to their specific features, the logical schemas can overlap, i.e., parts of the data can be represented redundantly using distinct logical models. Users interact with **S** represented as a graph while mapping to logical schemas ensures information preservation. This mapping can be inferred from sample data instances [12] or created manually for selected parts of **S** using the operations *createMapping()*, *editMapping()*, and *deleteMapping()*.

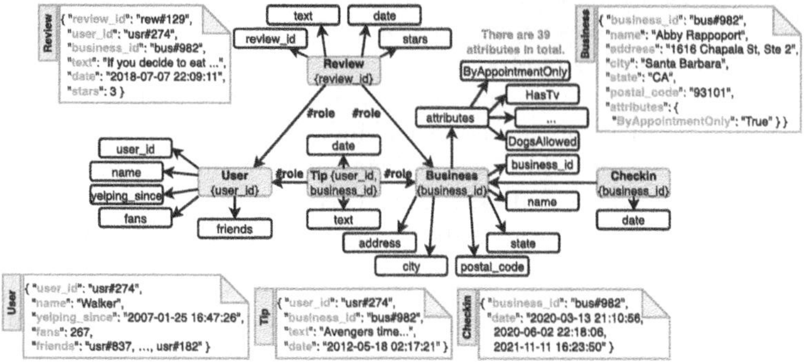

(a) Version 1: Semi-automatically inferred unnatural schema.

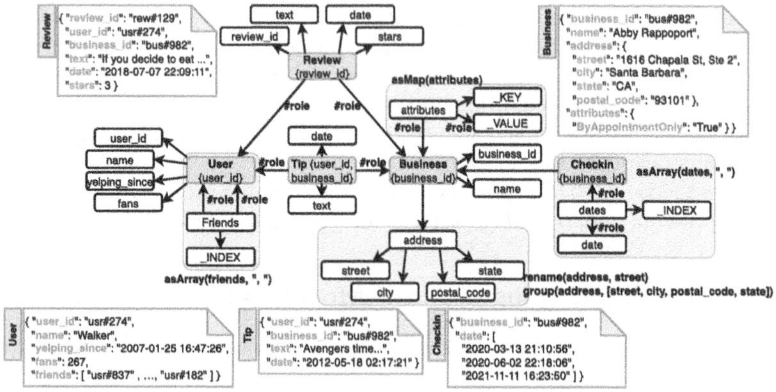

(b) Version 2: More realistic schema: Attributes forming an *address* are grouped together. A large varying set of *attributes* is transformed into a map.

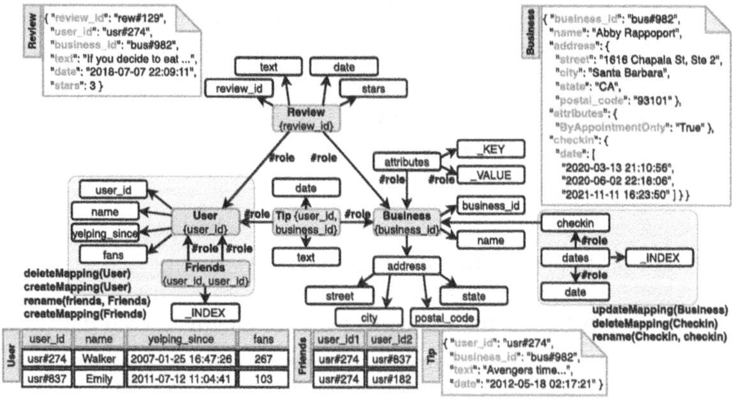

(c) Version 3: The transformation of the single-model schema and data to multi-model.

Fig. 1. Schema evolution and its propagation to data.

3.1 Schema Modification Operations

For the specification of user-required changes, we define the *Multi-Model Schema Evolution Language*, i.e., a set of schema modification operations (SMOs) that enables creating and deleting any allowed categorical conceptual schema.

We divide the SMOs for the conceptual layer into three groups: *Basic* SMOs (see Table 1) are atomic and low-level, working with a single object or morphism. They may, however, affect multiple objects or morphisms (and even use other basic operations). To guarantee correct propagation, basic SMOs must satisfy the prerequisites (as indicated in Table 1). The *composite* SMOs (see Table 2) and the *complex* SMOs (see Table 3) involve multiple objects or morphisms and usually require more parameters than the basic ones. For example, the operations *split()/join()* require a custom user input, which none of the basic SMOs can provide. Composite SMOs enable working with the categorical model using higher-level constructs, i.e., creating/deleting common structures, such as simple/complex properties, relationships, arrays, sets, or maps. As such, these operations correspond to the composition of basic SMOs. Complex SMOs aim to mediate transformations on existing parts of the schema category **S**.

Example 2. Consider the examples from Fig. 1 illustrating categorical schema automatically inferred [12] from the *Yelp Academic Dataset*[2], subsequently modified using SMOs and changes propagated to data. SMO-targeted areas are highlighted in yellow.

Version 1 in Fig. 1(a) maps the inferred schema solely to the input document (green) model. There are kinds *User*, *Tip*, *Business*, *Checkin*, and *Review*. Relationship objects *Review* and *Tip* include morphisms labeled `#role` indicating participants, while *Checkin* is a weak-entity-like kind identified by *business_ id* of kind *Business*.

Table 1. Basic SMOs

SMO	Behavior	Prerequisite
$create() \rightarrow O$	Creates object O with a unique identifier and default settings	–
$create(D, C) \rightarrow m$	Creates morphism $m : D \rightarrow C$ with a unique signature and default settings	$D, C \in \mathcal{O}_{\mathbf{S}}$
$delete(O)$	Deletes object O	$O \in \mathcal{O}_{\mathbf{S}}$; O can participate only in $1_O \in \mathcal{M}_{\mathbf{S}}$
$delete(m)$	Deletes morphism m	$m \in \mathcal{M}_{\mathbf{S}}$
$rename(O, n)$	Changes the name of object O to n	$O \in \mathcal{O}_{\mathbf{S}}$
$relabel(\mathcal{M}, l)$	Changes the label of morphisms in $\mathcal{M}$ to l	$\mathcal{M} \subseteq \mathcal{M}_{\mathbf{S}}$; If $l =$ `#role`, then $\lvert\mathcal{M}\rvert > 1$ and $m \in \mathcal{M}$ share *dom*
$move(m, p)$	Sets $m.dom$ to $p[last].cod$, i.e. to the end of a path p	$m \in \mathcal{M}_{\mathbf{S}}$; $p[first].dom = m.dom$
$copy(O) \rightarrow O'$	Creates a copy O' of object O (with a new unique identifier)	$O \in \mathcal{O}_{\mathbf{S}}$
$copy(m, D, C) \rightarrow m'$	Creates a copy $m' : D \rightarrow C$ of morphism m (with a new unique signature)	$D, C \in \mathcal{O}_{\mathbf{S}}$; $D = copy(m.dom)$; $C = copy(m.cod)$

[2] https://www.yelp.com/dataset.

Table 2. Composite SMOs

SMO	Behavior
$createProperty(l, P) \rightarrow (O)$	Creates object O and morphism $m : P \rightarrow O$ with label `#property`
$deleteProperty(O)$	Removes property O, i.e., deletes object O and all incoming morphisms. If O is part of an identifier of an object P, it is removed from the respective set $P.ids$
$deleteComplexProperty(O)$	(1) Deletes all nested properties without path from another complex property. Otherwise, it deletes the respective morphism (2) Deletes all incoming morphisms. (3) If O participates in a relationship which would be broken, the relationship is deleted too (4) Deletes O
$createIdent$ $(O, \{I_1, ..., I_n\}) \rightarrow T$	Creates an identifier T from properties $I_1, ..., I_n$ of a complex property O
$deleteIdent(T)$	Removes identifier T of an object $O \in \mathcal{O}_S$
$createRelationship$ $(l, \{O_1, \ldots, O_n\}) \rightarrow R$	(1) Creates object R representing a *relationship* object with label l (2) For each O_i creates morphism $m_i : R \rightarrow O_i$ with label `#role`
$deleteRelationship(R)$	(1) Deletes all outgoing morphisms (2) Deletes R
$createArray(P, l) \rightarrow A$	For object P, creates an *array* complex property A with label l and properties index I and element E
$deleteArray(A)$	(1) Deletes all outgoing morphisms (2) Deletes A
$createSet(P, l) \rightarrow S$	For object P, creates a *set* complex property S with label l and a property E
$deleteSet(S)$	(1) Deletes all outgoing morphisms (2) Deletes S
$createMap(P, l) \rightarrow M$	For object P, creates a *map* complex property M with label l and properties key K and value V
$deleteMap(M)$	(1) Deletes all outgoing morphisms (2) Deletes M

Table 3. Complex SMOs

SMO	Behaviour
$group(l, \{O_1, \ldots, O_n\}) \rightarrow G$	(1) Creates object G representing a *group* with label l (2) For each O_i creates morphism $g_i : G \rightarrow O_i$ (3) For $P \in \mathcal{O}_S$, a parent of objects O_i, creates $g : P \rightarrow G$
$ungroup(\{O_1, \ldots, O_n\})$	Inverse operation to group
$join(\{O_1, \ldots, O_n\}, \Phi) \rightarrow O'$	(1) Creates object O' such that its active domain is formed of results of transformation function Φ applied to values from the active domains of $O_1, \ldots, O_n$ (2) For $P \in \mathcal{O}_S$, a parent of objects O_i, creates morphism $m : P \rightarrow O'$ (3) Deletes morphisms $m_i : P \rightarrow O_i$
$split(O', \Phi') \rightarrow \{O_1, \ldots, O_n\}$	Inverse operation to join
$asArray(P)$	Converts a complex property P to an *array*
$asArray(P, \Psi)$	Converts a simple property P to an *array* so that the values of the active domain of P are separated using separator ψ and become the values of the active domain of the array
$asMap(P)$	Converts a complex property P to a *map*

As we can see, the automatic schema inference led to unnatural structures. For instance, kind *Business* includes a complex property *attributes*, comprising 39 optional properties, each found in 19 to 119,765 source documents. Hence, we transformed property *attributes* into a map (using SMO *asMap()*). Also, kind *Business* contains several properties forming an address (e.g., *state*, *city*, etc.). So, we renamed *address* to *street* (using SMO *rename()*) and grouped related properties into a complex property *address* (using SMO *group()*). Furthermore, we converted properties *friend* and *date* into arrays (using SMO *asArray()*), resulting in **Version 2** in Fig. 1(b).

Subsequently, we applied SMOs *deleteMapping()* and *createMapping()* to change the logical representation of the data, i.e., we restructured kinds *User* and *Friends* into separate relational tables (violet). Finally, we embedded kind *Checkin* into kind *Business* via *updateMapping()* and deleted the original kind *Checkin*. These SMOs resulted in the categorical schema **Version 3** in Fig. 1(c). □

3.2 Propagation to Data

Regarding propagation to data, introduced in [8], the SMOs can be further divided into heavy and light. A *heavy operation* is an SMO propagated from the conceptual level to the mapping, the logical level, and the data instances. Conversely, a *light operation* only triggers propagation to the mapping. Each SMO can be heavy. Operations that do not consist of *delete(O)* or *delete(m)* can be also light, e.g., SMOs *createProperty()*, *asArray()*, or *asMap()*.

There are multiple strategies for heavy SMO propagation, namely *eager*, *lazy*, and *hybrid*. In the first case, we update all the data instances at once. In the second case, we update them one by one only when we need them. The last option is a combination of the first two [11,21]. Our approach [8] is compatible with all of them; however, we focus only on the eager strategy at the moment.

3.3 Propagation to Queries

The propagation of changes to queries depends on how we handle the schema and data changes. The first option involves versioning the schema and the data, allowing the original queries to continue working with the original schema version. For new queries and applications, we use the most recent schema version. The second option we focus on involves propagating changes to both the data [8] and queries [7] and maintaining only the latest version. Generally, the propagation strategies follow the SMOs and the respective mappings when possible. However, the core problem is resolving cases that cannot be propagated straightforwardly.

More formally, let I be a database instance consisting of data D structurally corresponding to categorical schema $\mathbf{S}$ (represented as a functor $I : D \to \mathbf{S}$). The impact of SMO $O : \mathbf{S} \to \mathbf{S}'$ propagated to instances as $P_O : Inst(S) \to Inst(S')$ can be characterized using the concept *information capacity* [1,5]:

- If I' is structurally equivalent to I, then it is *information capacity-preserving* operation. An example is SMO *rename()*.
- If for each $I \in Inst(S)$ there is at most one instance $I' \in Inst(S')$, i.e., the functor P_O is surjective on objects, the operation is *decreasing information capacity*. An example is SMO *deleteProperty()*.
- If for each $I \in Inst(S)$ there is at least one instance $I' \in Inst(S')$, i.e., the functor P_O is injective on objects, the operation is *increasing information capacity*. An example is SMO *createProperty()*.

In this paper, we primarily focus on applying strategies for synchronizing queries in the context of information capacity-decreasing SMOs. These operations are the most critical as they cause information loss, fundamentally altering the query. When reconstructing a query, several strategies [2,15,17] can be employed. Strategy-selection criteria include:

1. Whether the query can be reconstructed using the remaining data.
2. The increased complexity of the new version of the query, including the expression complexity and the time/memory complexity of its evaluation.
3. The metadata we maintain about the data (e.g., a *catalog*) and dependencies, including the size and memory intensity of this metadata and how efficiently it can be extracted, maintained, and updated as data changes.
4. The fidelity of the new query version, or how much the new query results differ from the original, such as returning different, fewer, or more records.

Additionally, changes to integrity constraints can impact query performance. For example, modifying the primary or secondary key (i.e., corresponding indices) impacts queries using the old index (e.g., within a join/filtering). Without optimal propagation, query performance deteriorates.

Example 3. Let us introduce an example of the impact of capacity-decreasing SMOs on representative queries. Figure 2(a) illustrates a fragment of the Version 3 schema from Fig. 1(c), where modifications are performed using the SMOs *deleteProperty()*, *createProperty()*, and *rename()*. Additionally, Fig. 2(a) presents three specific queries: Q_1v: "Retrieve the *text* and *stars* from *Reviews* where *review_id* exceeds 100", Q_2: "Retrieve pairs of (*state, countOfBusinesses*) representing the number of businesses in each state", and Q_3: "Retrieve all *Reviews* ordered by *review_id*".

The capacity-decreasing SMO *deleteProperty(review_id)* affects queries Q_1 and Q_3, while SMO *deleteProperty(state)* impacts Q_2. □

A common basic strategy is to utilize an *impact catalog* for each column, recording the impact of its removal on queries and allowing users to assess column importance. Such information can be variously complex. For example, paper [19] classifies attributes as *replaceable* or *dispensable* upon removal, based on metadata annotations specifying whether an attribute, condition, or relation can be replaced in specific query contexts (e.g., selection predicates, projection lists, join conditions). These metadata elements, represented as Boolean values, indicate replacement feasibility and associated constraints. The subsequent EVE project [18], which serves as an inspiration for our work, employs an extended metadata for query synchronization. When a column or table is deleted, the system identifies affected queries and views, determines their dependencies, and selects the best replacement strategy using recorded metadata (e.g., to preserve query result, minimize query complexity and execution time). The impact catalog further characterizes columns using factors such as distinct values, value ranges, default values, functional dependencies, containment relationships, and integrity constraints. These attributes inform replacement selection to maintain query correctness. Metadata can be manually specified by database administrators or inferred from schema constraints and query patterns.

Generally, such metadata facilitates the selection of appropriate replacement methods. We introduce and utilize several representatives:

A. Phantom Strategy. The first propagation option involves replacing the removed column with a *phantom*, wherein the original column values are substituted with those from the minimal domain (smallest number of values and memory size) to preserve query functionality. This process entails analyzing queries, utilizing metadata, and substituting original values with new ones.

The representation of phantom values varies based on the query Q where property p is removed. If p appears only in filtering conditions, the phantom can be a bitmap indicating whether each row was part of the original result by modifying the filter condition to $p =$ `True`. For clauses such as `DISTINCT`, aggregation functions, and `JOIN`, the original values are replaced with enumerated values from a less memory-intensive domain (e.g., replacing strings with integers). For the `ORDER BY` clause, the order is maintained using ordered unsigned integers while retaining original values for projection. A minimal memory *footprint* is sought since the phantom size matches the number of original records. For each query, a phantom of the deleted column can be created (e.g., using multiple bitmaps for greater memory efficiency). For some queries, phantoms can be consolidated into fewer representations.

Example 4. Consider capacity-decreasing operations from Fig. 2(a) that are applied to *review_id* in kind *Review* and *state* in the address of kind *Business*, impacting queries Q_1–Q_3. Let us analyze them to determine the optimal phantom representation based on query composition.

For example, removing *review_id* impacts both selection $reviews > rew\#100$ in Q_1 and sorting in Q_3. For selection, a binary phantom (True/False) suffices to indicate document inclusion. However, the phantom must also capture sequence for ordering, requiring string substitution with a more memory-efficient type.

Aggregation in Q_2 adds additional complexity. If the aggregated value is excluded from the result, replacing strings with primitive enumeration values conserves memory (as a resulting phantom representation, as illustrated in Fig. 2(b)). Conversely, if included, the original values must be retained, resulting in phantoms as memory intensive as the original data.

In all cases, the modified queries still yield results identical to their originals. □

B. Functional-Dependency Strategies Let us first recall the basic definition of a functional dependency. Consider a schema $R(A_1, A_2, \ldots, A_n)$ with attributes $A_1, A_2, \ldots, A_n$. A *functional dependency* (FD) $X \rightarrow Y$, where $X, Y \subseteq R$, holds in R if for any tuples $t_1, t_2 \in R$, $t_1[X] = t_2[X]$ implies $t_1[Y] = t_2[Y]$, i.e., the attributes in X uniquely determine those in Y. A *relaxed functional dependency* (RFD) generalizes an FD by introducing flexibility through a parameter ϵ, allowing the dependency to hold with some tolerance, as a strict FD may not be applicable across the entire database.

Inspired by the approach in [2], we can synchronize queries using (relaxed) functional dependencies to maintain the original query semantics. Furthermore, we categorize (R)FDs as surjective or injective. A surjective FD enables the

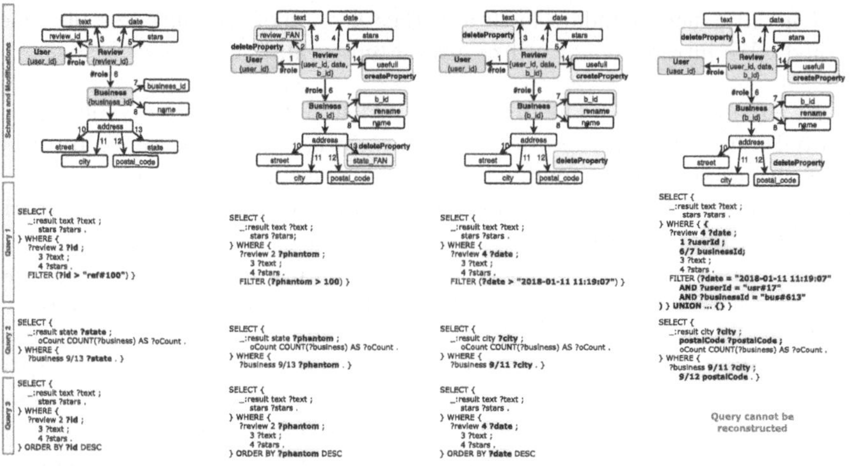

(a) Original data (version 3) (b) Phantom strategy (c) FDs strategy (d) Relaxed FDs strategy

Fig. 2. An example of application of various strategies to propagate structural changes to the queries

replacement of all values in the deleted column, while an injective FD ensures that the replacement values are unambiguous.

Example 5. To assess the substitutability of a removed property, valid (relaxed) FDs are utilized (see Fig. 2(c) and 2(d)). For example, removing *review_id* from kind *Review* is governed by FD $user\_id, business\_id, date \rightarrow review\_id$, allowing the replacement with $(user\_id, business\_id, date)$. This substitution preserves review identification but complicates queries, making filtering operation inefficient (see Q_1 in Fig. 2(c)) and significantly altering ordering in result (i.e., Q_3 cannot be reconstructed). A more efficient alternative uses relaxed FD $date \rightarrow review\_id$, improving filtering (see Q_1 in Fig. 2(d)) but possibly including additional documents when multiple reviews share the same date and causing slight discrepancies in ordering (see Q_3 in Fig. 2(d)).

For aggregation queries, employing FD $city, postalCode \rightarrow state$ changes the aggregation root but permits result recovery through transformation and external knowledge (see Q_2 in Fig. 2(c)). However, the relaxed FD $city \rightarrow state$ may not accurately reproduce results when cities share names between states (see Q_2 in Fig. 2(d)). □

Additional Considerations. Beyond the outlined strategies, several factors must be addressed. First, different sequences of identical operations can yield varying query propagation effects [19]. For instance, executing *delete(A)* followed by *create(A)* may negate each other, rendering the sequence ineffective.

Additionally, deciding whether to synchronize queries with the initial or the latest version is crucial. As proposed in [6], referencing the first query version requires backtracking a history of queries and changes to restore deleted columns, which is memory-intensive due to storing original-sized column phantoms.

Moreover, information capacity-decreasing operations can be classified as no longer needed (e.g., remove query) or destructive (i.e., requiring replacement). A preliminary what-if analysis, derived from the metadata catalog and visually indicated to the user, can determine the impact of propagating the SMO.

If query synchronization is unfeasible due to capacity-decreasing operations without an efficient replacement, decisions must be made to drop queries or restrict schema evolution [1]. Typically, three policies are adopted: propagate (i.e., apply changes and synchronize queries), block (i.e., prohibit schema changes), or prompt (i.e., consult the database administrator). The application of individual strategies may result in:

- The query is fully reconstructed to return identical results, though it may require more execution time or increased memory due to complex queries or extensive metadata. Nonetheless, the metadata size remains smaller than the original deleted data, reducing the overall memory footprint.
- The query is partially reconstructed to approximate original results, potentially causing time inefficiency with non-injective or non-surjective FDs.
- The query cannot be reconstructed, necessitating a decision to either propagate the SMO and delete the query or block the SMO propagation.

Utilization of AI. AI-based approaches can be used to select an appropriate strategy, e.g., considering the frequency of execution of that query (for example, frequently executed and important queries will lead to a phantom strategy, even if it is memory intensive) while considering the available system resources. In other words, it is an optimization problem where the AI agent operates in a state space defined by the system's current resource availability (e.g., memory, CPU usage, etc.) and the historical performance of previously used strategies. The actions then correspond to selecting specific propagation strategies.

A promising solution is to use reinforcement learning because we do not know which actions to take, but we can easily measure their outcome (e.g., consumption of resources during and after the propagation). Another option is to use simpler methods (e.g., heuristics). The advantages of this approach include a lower need for training data, less training time, and better explainability of the resulting policy.

4 Implementation and Evaluation

Our solution maintains versions of a schema category **S** and a chain of applied SMOs. We also maintain the mapping between queries and schemas. So, the propagation is straightforward since each such mapped query targets a subset or a view of the data. The benefits of our approach lie in the use of MMQL abstraction and its translation to Domain Specific Languages of the underlying DBMSs. We do not have to deal with propagation to individual languages and their specific constructs. Depending on the queries an SMO affects, we can always use the most appropriate strategy for the given property. Currently, we heuristically calculate the price per change and operation of new versions of queries. In the future, we will utilize a suitable statistical model.

The experimental implementation of the described idea is based on a family of tools systematically developed in our research group for managing multi-model data based on the category theory. Around this representation, we have already built the following tools:

- *MM-cat* [14] enables multi-model modeling, i.e., the manual creation of a categorical model and its mapping to any combination of the supported logical models. It also supports the respective data transformations between the logical and categorical models.
- *MM-infer* [12] enables automatic inference of the categorical schema from sample multi-model instances. In addition, as the data set can be large, the data is processed in parallel.
- *MM-evocat* [8] extends *MM-cat* with the propagation of changes in the categorical schema to the data, i.e., it extends the SMOs with the respective propagation strategies.
- *MM-quecat* [10] adds support for querying over the categorical schema category **S** using MMQL. The queries are then decomposed according to the mapping. The subqueries are evaluated in the underlying DBMSs, and the partial results are combined to produce the final result.
- *MM-evoque* [7] extends *MM-quecat* for propagation of changes in the categorical schema to queries in MMQL.

4.1 Propagation to Data Instances

The toolset was implemented as a proof of concept to demonstrate that the proposed strategy enables (1) the model-agnostic conceptualization of any multi-model logical schema, (2) mapping to specific data logical schemas, (3) querying using MMQL, and (4) schema evolution based on user requirements (SMOs) with correct and complete propagation of changes to data instances and queries.

Experimental evaluations of various schema-to-data propagation strategies were conducted on a 14" MacBook Pro (2021) featuring an Apple M1 Pro processor (8 cores), 16 GB LPDDR5 memory, and a 512 GB SSD. To mitigate potential biases from resource fluctuations and initial overhead, each experiment configuration was executed 30 times, excluding initial runs and averaging the remaining measurements. The experiments utilized five *Yelp Academic Dataset* collections in *MongoDB*: *Review* (6,990,247 records), *User* (1,987,897 records), *Business* (150,346 records), *Tip* (908,915 records), and *Checkin* (131,930 records) – for sample data see Fig. 1(a).

The modifications correspond to the SMOs depicted in Fig. 1(b) and Fig. 1(c). Three eager strategies were utilized for substantial SMO propagation:

- *insertMany*: Drops integrity constraints, deletes records with schema **S**, performs a bulk insert of records with schema **S′**, and restores constraints.
- *insertOne*: Similar to *insertMany*, but inserts records individually.

- *update*: Retains integrity constraints and updates affected documents using update or set operations based on the target system.

Moreover, the *insertOne* and *insertMany* strategies are implemented as *standard*, i.e., proactively preventing database driver exceptions such as integrity constraint violations, and *alter*, i.e., disabling integrity constraints before insertion and re-enabling them afterwards.

Table 4 illustrates the time requirements of the three SMO propagation strategies applied to kinds (*Business*, *CheckIn*, *User*, and *Friends*) during schema evolution from Version 1 to 2 and 2 to 3, as illustrated in Fig. 1. The times are divided into the *preparation phase PreP* (i.e., creating tables/collections or disabling constraints), *transformation phase TraP* (i.e., updating data to match the target structure and representation), *propagation phase ProP* (i.e., inserting data into the DBMS), and the *alter IC phase AltP* (i.e., re-enabling constraints).

As observed, solely updating data is the slowest strategy by an order of magnitude between systems. The fastest approach involves creating a new collection or deleting data and bulk inserting a new version, achieving up to eight times greater performance than individual inserts or updates when propagating SMOs to the entire collection. Performance is additionally enhanced by disabling integrity constraints before insertion despite the increased effort in the preparation and transformation phases.

Table 4. Time [*s*] needed for propagation of SMOs

Change	Strategy	PreP	TraP	ProP	AltP	Total Time
Business Address SMOs	updateOne	0.000	1.841	80.018	–	81.858
	insertOne	0.000	1.746	73.764	–	75.511
	insertMany	0.000	1.865	3.246	–	5.111
Checkin Dates SMOs	updateOne	0.000	3.446	88.863	–	92.310
	insertOne	0.000	3.706	86.362	–	90.068
	insertMany	0.000	3.976	9.201	–	13.178
User Friends SMOs	updateOne	0.001	101.064	1,443.758	–	1,544.823
	insertOne	0.004	88.367	1,378.334	–	1,466.705
	insertMany	0.002	82.045	129.150	–	211.197
Business Checkin SMOs	updateOne	0.000	6.369	104.918	–	111.286
	insertOne	0.000	7.531	102.470	–	110.001
	insertMany	0.000	8.070	14.712	–	22.782
User Relational SMOs	insertOne	0.001	75.102	130.275	–	205.378
	insertOneAlter	0.025	74.468	116.480	2.159	193.131
	insertMany	0.001	54.723	18.762	–	73.486
	insertManyAlter	0.023	54.315	9.466	1.942	65.745
Friends Relational SMOs	insertOne	2.023	304.041	1,605.221	–	1,911.285
	insertOneAlter	2.121	305.847	1,355.017	27.931	1,690.916
	insertMany	2.053	161.603	381.701	–	545.357
	insertManyAlter	2.094	162.392	55.954	28.212	248.652

4.2 Propagation to Operations

Table 5 illustrates the effectiveness of phantom and (relaxed) FD strategies on high-impact filtering, aggregation, and ordering queries from Fig. 2. As we can see, in the phantom strategy, queries can be fully reconstructed with minimal alterations to the original MMQL expression, thus preserving execution time and memory usage. However, this strategy effectively substitutes the removed property with a memory-efficient placeholder, which may be seen as a disadvantage. In contrast, the FD and relaxed FD strategies remove properties entirely but often increase the complexity of the MMQL expression (e.g., the expression contains numerous `UNION` clauses), potentially hindering performance and resource efficiency. The fidelity of query outcomes in these approaches depends on the properties of the FD, such as injectivity, surjectivity, and degree of relaxation. When a dependency is both injective and surjective, the resulting dataset remains identical to the original query. One notable benefit of these methods is their minimal metadata requirements, relying solely on query synchronization with SMO propagation.

As for data changes, they affect all strategies differently. In the phantom strategy, newly introduced data remain excluded as if the corresponding rows were not part of the original query. Conversely, FD-based approaches may include such data, thereby risking the invalidation of the original functional dependency. Once a property is removed, its validity can no longer be verified.

5 Related Work

The evolution management is generally challenging, even for single-model data. In a multi-model world, we have to cope with additional problems such as a missing unifying interface for both data and operations, distinct ways of combining the models, contradictory features of the models, different propagation

Table 5. An experimental comparison of query synchronization strategies

Query	Strategy	QR	QF	SC	ET	EM	MM	ME	MU
Q_1	Phantom	Yes	Good	Good	Good	Good	Average	Good	Good
	FD	Yes	Good	Worse	Worse	Worse	Good	Worse	Good
	Relaxed FD	Yes	Average	Worse	Good	Worse	Good	Worse	Good
Q_2	Phantom	Yes	Good	Good	Good	Good	Worse	Good	Good
	FD	Yes	Good	Worse	Worse	Worse	Good	Worse	Average
	Relaxed FD	Yes	Average	Average	Good	Good	Good	Worse	Average
Q_3	Phantom	Yes	Good	Good	Good	Good	Worse	Good	Good
	FD	No	n/a	n/a	n/a	n/a	n/a	n/a	n/a
	Relaxed FD	Yes	Average	Average	Good	Good	Good	Worse	Average
General case	Phantom	Yes	Good	Good	Good	Good	Worse	Good	Good
	FD	Yes	Good	Worse	Worse	Worse	Good	Worse	Average
	Relaxed FD	Yes	Average	Average	Average	Average	Good	Worse	Average

[QR] Query reconstruction [QF] Query fidelity [SC] Statement complexity [ET] Execution time [EM] Execution memory [MM] Metadata memory [ME] Metadata extraction [MU] Metadata updating

Table 6. A comparison of features of selected evolution management approaches

	CGOOD [22]	Spivak et al. [20]	*Darwin* [21]	*MM-evolver* [4]	*Orion* [3]	*MM-evocat* [8]
Schema modification	Yes	Yes	Yes	Yes	Yes	Yes
Data models	Relational, object, object-relational	Relational, RDF	Graph, key/value, document, columnar	Relational, graph, key/value, document, columnar	Relational, graph, key/value, document, columnar	Relational, array, graph, RDF, key/value, document, columnar
Multi-model	No	No	No	Yes	No	Yes
Propagation to IC	No	No	No	Partial	Partial	Yes
Data migration	Eager	Eager	Eager, lazy, hybrid	Eager	Eager	Eager
Schema inference	No	No	Tracked, untracked	No	Untracked	Yes (using *MM-infer* [13])
Version jumps	No	No	Yes	No	No	No
Query rewriting	No	No	Yes	No	No	Yes

strategies for distinct models, cross-model data migrations, etc. Several evolution management approaches currently consider at least two models [3,4,20–22]. However, they have various limitations. Table 6 summarizes the comparison of our approach with them.

The main common limitations of existing approaches are targeting only a small fraction of data models and/or (partially) ignoring integrity constraints and/or queries over the data and other features arising from combining multiple data models. Our approach genuinely supports all popular data models in a multi-model way. The set of SMOs is the richest, and the changes are propagated eagerly to data and ensure query synchronization, even for the capacity-decreasing SMOs. The approaches we adopt and utilize/extend [2,19] were originally proposed only for single-model use cases. To our knowledge, no multi-model solution deals with this challenge and offers such a range of strategies. Only system *Darwin* [21] allows for other data migration strategies, namely lazy and hybrid, and supports optimization using version jumps. However, it focuses only on propagation to data instances.

6 Conclusion

Evolution management is one of the core and most complex data management tasks, as most applications need to evolve according to changing user requirements. The situation is much more complex in the context of multi-model data.

This paper presents our strategy based on the idea of capturing multi-model data in a unified, model-agnostic manner using category theory, treating the data naturally as a categorical graph. Having such a representation, we introduce a set of schema modification operations that can modify the categorical graph without dealing with the technical details of the underlying DBMSs and models. These operations are classified based on their complexity, and we define the respective propagation strategies that adapt the data instances and ensure query synchronization. We especially focus on capacity-decreasing SMOs, where the information loss needs to be reflected respectively. Finally, we empirically evaluate and compare the strategies that were introduced.

Even though the main aim of the introduced toolset is to simplify the critical tasks of data management in the world of multi-model data, the user is still required to make complex decisions. Our near future aim is to exploit suitable AI strategies and extend the toolset towards self-adaptability. The system can learn from the given data, given or extracted metadata, operations, change logs, and user decisions and automatically or proactively make these complex decisions.

References

1. Caruccio, L., Polese, G., Tortora, G.: Synchronization of queries and views upon schema evolutions: a survey. ACM Trans. Database Syst. **41**(2) (2016)
2. Caruccio, L., Polese, G., Tortora, G.: Dependency-based query/view synchronization upon schema evolutions. In: Proceedings of ER 2018 Workshops. LNCS, vol. 11158, pp. 91–105. Springer (2018)
3. Chillón, A.H., Klettke, M., Ruiz, D.S., Molina, J.G.: A taxonomy of schema changes for NoSQL databases. CoRR abs/2205.11660 (2022)
4. Holubová, I., Vavrek, M., Scherzinger, S.: Evolution management in multi-model databases. Data Knowl. Eng. **136**, 101932 (2021)
5. Hull, R.: Relative information capacity of simple relational database schemata. SIAM J. Comput. **15**(3), 856–886 (1986)
6. Koeller, A., Rundensteiner, E.A.: A history-driven approach at evolving views under meta data changes. Knowl. Inf. Syst. **8**(1), 34–67 (2005)
7. Koupil, P., Bártík, J., Holubová, I.: MM-evoque: query synchronisation in multi-model databases. In: Proceedings of EDBT 2024, pp. 818–821. OpenProceedings.org (2024)
8. Koupil, P., Bártík, J., Holubová, I.: MM-evocat: a tool for modelling and evolution management of multi-model data. In: Proceeding of CIKM 2022, pp. 4892–4896. ACM (2022)
9. Koupil, P., Crha, D., Holubová, I.: A universal approach for simplified redundancy-aware cross-model querying. Inf. Syst. **127**, 102456 (2025)
10. Koupil, P., Crha, D., Holubová, I.: MM-quecat: a tool for unified querying of multi-model data. In: Proceedings of EDBT 2023, pp. 831–834. OpenProceedings.org (2023)
11. Koupil, P., Holubová, I.: A unified representation and transformation of multi-model data using category theory. J. Big Data **9**(1), 61 (2022)
12. Koupil, P., Hricko, S., Holubová, I.: A universal approach for multi-model schema inference. J. Big Data **9**(1), 97 (2022)
13. Koupil, P., Hricko, S., Holubová, I.: MM-infer: a tool for inference of multi-model schemas. In: Proceedings of EDBT 2022. pp. 2:566–2:569. OpenProceedings.org (2022)
14. Koupil, P., Svoboda, M., Holubová, I.: MM-cat: A tool for modeling and transformation of multi-model data using category theory. In: Proceedings of MODELS 2021, pp. 635–639. IEEE, New York (2021)

15. Liu, J., Mozafari, B.: Query rewriting via large language models. CoRR abs/2403.09060 (2024)
16. Lu, J., Holubová, I.: Multi-model databases: a new journey to handle the variety of data. ACM Comput. Surv. **52**(3) (2019)
17. Ravve, E.V.: Maintenance of queries under database changes: a unified logic based approach. In: Proceedings of FoIKS 2016. LNCS, vol. 9616, pp. 191–208. Springer (2016)
18. Rundensteiner, E.A., et al.: Evolvable view environment (EVE): non-equivalent view maintenance under schema changes. In: Proceedings of SIGMOD 1999, pp. 553–555. ACM Press (1999)
19. Rundensteiner, E.A., Lee, A.J., Nica, A.: On preserving views in evolving environments. In: Proceedings of Workshop KRDB 1997. CEUR Workshop Proceedings, vol. 8, pp. 13.1–13.11. CEUR-WS.org (1997)
20. Spivak, D.I., Wisnesky, R.: Relational foundations for functorial data migration. In: Proceedings of DBPL 2015, pp. 21–28. ACM, New York (2015)
21. Störl, U., Klettke, M.: Darwin: a data platform for NoSQL schema evolution management and data migration. In: Proceedings of the EDBT/ICDT 2022 Workshops. CEUR-WS.org, Edinburgh (2022)
22. Tuijn, C., Gyssens, M.: CGOOD, a categorical graph-oriented object data model. Theoret. Comput. Sci. **160**(1), 217–239 (1996)

A Flexible Framework for Transposition-Aware Querying of a Musical Score Database

Adel Aly(✉), Olivier Pivert, and Virginie Thion

Univ. Rennes, CNRS, IRISA, Lannion, France
{adel.aly,olivier.pivert,virginie.thion}@irisa.fr

Abstract. Retrieving music pieces based on their content is an emerging topic in the database management and information retrieval fields. This paper tackles the challenge of retrieving musical scores containing a given melodic pattern while allowing flexibility in the matching. It provides two contributions: (i) a fuzzy-logic-based theoretical framework that makes it possible to support customizable and explainable retrieval of approximate answers, including transposition-aware searches; (ii) an implementation of this framework within a real-world digital score library, along with some experimental performance results.

1 Introduction

Music Information Retrieval (MIR) has emerged as a prominent field of research, driven by advancements in data management and the increasing availability of music data. This multidisciplinary domain spans areas such as information science, musicology, and computer science. While metadata-based retrieval is common in platforms like digital libraries, content-based retrieval, such as searching by melody, is crucial for specific use cases, including exploratory music discovery, musicological analysis, and plagiarism detection.

The work presented in this paper focuses on the task of *melodic pattern searching* within databases of musical scores (encoded in Western classical notation). Unlike metadata-based retrieval, melodic pattern searching involves finding musical works that contain a specific sequence of musical notes, each characterized by attributes such as pitch and duration. The complexity of this task lies in addressing the variability inherent in music, as exact matches are often too restrictive and fail to reflect the needs of musicians and researchers who seek flexibility. *Flexible search* methods aim to address this task by allowing approximate matches, ranking the results based on their closeness to the input pattern. But MIR systems still grapple with this challenge [1], even more for recognizing patterns when they are shifted uniformly in pitch, referred to as *transposition-agnostic search* in the following, which is particularly relevant when considering folk and traditional music querying.

In this paper, we propose a **framework for approximate melodic pattern search tolerant to transposition**, capable of retrieving musical scores

J. Grabis et al. (Eds.): RCIS 2025, LNBIP 547, pp. 262–278, 2025.
https://doi.org/10.1007/978-3-031-92474-3_16

that contain a given pattern, even if transposed on the pitch axis. Our approach leverages fuzzy logic, offering parameterizable tolerance across dimensions and explainable results. Additionally, we present an **implementation** of this theoretical framework, and some **experimental performance results**.

The paper is structured as follows: Sect. 2 reviews related work on melodic pattern matching. Section 3 introduces basic concepts of musical notation. Section 4 details the proposed approximate pattern matching technique for retrieving musical scores containing user-defined patterns, including transposed forms. Section 5 describes the implementation within a digital score library for folk music from the Brittany French region. Section 6 provides experimental results that concern both the overhead induced by making the matching flexible and the way the query execution time evolves with the size of the database. Finally, Sect. 7 discusses future work.

2 State of the Art

The issue of searching for melodic patterns in musical scores has been addressed through different techniques ranging from exact match methods to more flexible approaches that account for variations in pitch, duration, or sequencing.

Most melodic search systems rely on exact matching of note sequences. Text-based formats, such as ABC notation and the Humdrum Toolkit's **kern notation [2,3], encode pitch using letters and duration using numbers. Associated exact search techniques via string search algorithms have been used to query such data. Semi-structured representations, like MusicXML [4] and MEI [5], lead to expressing queries through XML-based languages such as XQuery [6]. Graph-based models also exist, allowing musical scores to be queried with SPARQL over RDF [7] or Cypher over property graphs [8]. While the latter approaches provide effective query mechanisms, they only support exact matches.

To introduce flexibility, methods relying on edit distances have been proposed [9]. *Tunepal* [10], another approach developed for Irish traditional music and using audio input to perform melody recognition, normalizes pitch—removing octave information—and standardizes duration to improve tolerance of variations [11]. However, the kind of flexibility obtained is rather limited and these methods do not handle transpositions, which are essential for many music retrieval tasks, specifically in the context of querying folk music databases.

Few systems address transposition challenges by focusing on interval-based or contour matching techniques, allowing melodic patterns to be retrieved regardless of their starting pitch. *Themefinder* [3] is a web platform based on the Humdrum toolkit that supports interval-based searches. Users input a sequence of symbols representing contours or intervals, and the system retrieves matching patterns. *Musipedia* employs contour-based searches using Parson code, which encodes melodic direction (up, down, repeat) instead of absolute pitches, offering a way to identify melodies from their contour. While these methods demonstrate the importance of transposition-aware search in music retrieval, they are limited in their ability to retrieve patterns that are similar but not identical to the query.

Fig. 1. A simple musical score (beginning of *Belle nous irons dans tes verts prés* [12])

Our work addresses this problem by introducing customizable tolerance levels for pitch, duration, and sequencing, and enabling more flexible melodic pattern searches while preserving transposition awareness.

3 Musical Notation

Let us first introduce some basic notions of Western music notation, kept to the bare minimum for the understanding of the approach. Even though the model we rely on (cf. Sect. 4) can also deal with polyphonic music, we restrict the presentation to monophonic music for the sake of simplicity. We use the musical score from Fig. 1 for illustrating the introduced concepts[1]. We applied colors to certain notes to simplify their identification in the examples in the following of the paper.

In Western music notation, a finite set of frequencies is used to refer to the sounds that may appear in a musical piece. Each frequency is associated with a pitch class (a letter C, D, E, F, G, A, and B) and an index known as the octave (in the set $\{1, \ldots, 7\}$). On a music sheet, *musical events* are graphically represented to define the sounds that compose the music piece. The symbols are written all along a group of horizontal lines called *staves*. A main symbol that appears in a musical score is the *note* (e.g. a symbol among , , , , , etc.), which denotes a sound to be played. A note symbol encodes both the information of the frequency and the duration of the denoted sound. The vertical position of the note on the staff determines the frequency (e.g. if we consider the first staff of Fig. 1, the first note is a $G4$, then follows an $A4$, and a $B4$, etc.), while the shape of a note (including the head, stem, and flags) determines the duration (for instance, a note (resp.) is two times shorter than a note (resp.)). Time is divided into frames called *measures*, synchronized over the staves, and visually delineated by vertical *bars*. In Fig. 1, we can see five measures.

[1] In the corpus this musical score belongs to, time signature is voluntary not imposed (See [12] for the explanation by musicologists who collected and encoded the corpus). Also, measures do not necessary have the same length.

Additional symbols exist, including the rest symbols, like that appears near the end of the second measure. The whole musical content of a score is composed of several synchronized sequences of musical events associated with instruments. The digitized content of a musical score also contains additional information in the form of metadata (the title, the composer, the collector, the composition date, etc.), as well as rendering information (margins for the music sheet, spacing between the elements, direction of the stems, fonts, etc.).

4 Theoretical Framework

This section presents a framework for retrieving musical content that matches a given melodic pattern, possibly approximately, authorizing or not musical transposition. The framework relies on a data model presented in Subsect. 4.1. The notion of a melodic pattern and the general principles of the selection operators are given in Subsect. 4.2. In Subsect. 4.3, we describe crisp selection operators, and extend them in Subsect. 4.4 so as to achieve flexible matching.

4.1 Data Model

The data model the theoretical framework relies on represents the musical content, centered on the concept of a voice as the fundamental element of a musical score. A voice encapsulates the sequence of notes played by an instrument and serves as the primary unit for content retrieval. The model adopts an event-based approach, where notes are viewed as temporally bound objects characterized by their pitch and duration. This perspective draws inspiration from the work in [13], which conceptualizes voices as temporal functions, mapping sounds to their corresponding time intervals.

Definition 1 (Musical event). *A musical event, denoted by $s_{t_1}^{t_2}$, is defined by its pitch s, its onset time t_1, and its ending time t_2, where $t_2 > t_1$. Rests are denoted by $\perp_{t_1}^{t_2}$.*

In Fig. 1, the first musical event is the note event $G4_0^1$, then follows a $A4_1^2$, itself followed by a $B4_2^3$, etc. The last event of the first line is the rest event $\perp_{32}^{34}$.

Definition 2 (Event properties). *Let $e = s_t^{t'}$ be an event. We use the notations $start(e) = t$, $end(e) = t'$ and $sound(e) = s$. We also denote the duration of e by $dur(e) = end(e) - start(e)$. By definition, $dur(e) > 0$ for any e.*

Definition 3 (Interval). *Let e_1 and e_2 be two events that are both not rests. The interval between e_1 and e_2 is defined as $interval(e_1, e_2) = sound(e_2) - sound(e_1)$. Classically, we consider that an interval value of ± 1 corresponds to a tone*[2].

[2] A tone is a standard interval between two pitches, equivalent to two semitones. For example, the interval between $C4$ and $D4$ is $+\ 1$ tone.

We can now turn to the notion of musical content of a *voice.*

Definition 4 (Voice). *A* voice V *is a non-empty set* $\bigcup_{j=1}^{n} \{s_{j\,t_j}^{\;t_{j+1}}\}$ *of musical events (with* $t_1 = 0$*) satisfying the three following conditions: (i)* $\exists e \in V$ *s.t.* $start(e) = 0$*, (ii)* $\forall e \in V$*, either* $start(e) = 0$ *or* $\exists e' \in V$ *s.t.* $e \neq e'$ *and* $start(e) = end(e')$*, i.e. there are no "holes" in a voice, and (iii)* $\forall (e, e') \in V_{end}$*,* $end(e) = end(e')$ *where* $V_{end} = \{e \in V \mid \nexists\, e'$ *s.t.* $e \neq e'$ *and* $start(e') = end(e)\}$*, i.e. the "end events" all end at the same time.*

Example 1. The voice of Fig. 1 can be written as follows (where a time step of 1 is equal to the duration of a [3]), denoted by V in the following.

$$\{G4_0^1,\ A4_1^2,\ B4_2^3,\ A4_3^4, G4_4^6,\ A4_6^8,\ B4_8^{10}, C5_{10}^{12},\ D5_{12}^{16},$$
$$D5_{16}^{17}, D5_{17}^{18},\ D5_{18}^{19},\ E5_{19}^{20}, D5_{20}^{22},\ C5_{22}^{23},\ B4_{23}^{24}, A4_{24}^{26},\ D5_{26}^{28},\ G4_{28}^{32}, \perp_{32}^{34},$$
$$G4_{34}^{36}, B4_{36}^{37},\ C5_{37}^{38},\ D5_{38}^{40}, C5_{40}^{42},\ B4_{42}^{44},\ A4_{44}^{46}, G4_{46}^{48},\ A4_{48}^{50},\ \cdots\}$$

The nine first musical events model the notes that appear in the first measure of Fig. 1, the eleven following events model the notes that appear in the second measure, the next events model the notes that appear in the third measure, etc.

Remark 1. This set-oriented model allows for a straightforward representation of overlapping musical events, which is particularly useful in the context of polyphonic music. Furthermore, this type of modeling aligns closely with existing data formats for music, leading to an intuitive implementation of the framework (see Sect. 5). Hereafter, for the sake of simplicity, we only consider monodic voices in the examples.

By extension, the musical content of a musical score is composed of a set of voices, for instance $\{V_{violin}, V_{lute}\}$, where each voice is defined as in Definition 4.

4.2 Melodic Pattern and General Principles

A melodic pattern is a sequence of notes that a user wants to retrieve in a musical score database. Figure 2 gives two examples of musical pattern. As an example, if a user defines Pattern P_2, then he would search for a sequence of five notes: a sixteenth $D4$, and then the eighth notes $C4$, $D4$, $E4$, and $F4$. Formally, a melodic pattern is defined as bellow.

Definition 5 (Melodic pattern). *A melodic pattern* P *has the form* $\langle N_1, ..., N_k\rangle$ *(with* $k \geq 1$*) where each* $N_i \in \{N_1, \cdots, N_k\}$ *is a pair* (s_i, d_i) *composed of a sound and a duration (of the* i^{th} *event).*

Example 2. See[4] Fig. 2, P_2 is $\langle (D4, 1), (C4, 2), (D4, 2), (E4, 2), (F4, 2)\rangle$.

We now expose the general principles underlying the selection operators.

[3] The smallest duration found in the score, see [13] for details.

[4] We recall that 1 is the duration of a sixteenth note (see Definition 4 and Example 1).

Fig. 2. Two examples of melodic patterns

The Three Aspects of a Musical Pattern. A selection relies on checking *three aspects* of the pattern: the pitch of the notes (if approximated, the pitch of the notes may vary in the answer w.r.t. the pitches declared in the pattern), the duration of the notes (if approximated, the duration of the notes may vary), and their sequencing (if approximated, some notes may be inserted in the answer).

Tone-Stable vs Translation-Agnostic Selection. A *"tone-stable" selection* is used when a user seeks a sequence of specific notes, for instance to practice on a musical instrument (the pitches of the notes correspond to a gesture meant to be practiced on the instrument), for studying in a specific tonality, or for similar purposes. Such a selection considers the pitches of the notes, their duration and sequencing. On the other hand, a *translation-agnostic selection* caters to scenarios where the user wants to identify a melody based on "how it sounds", regardless of its tonality. It focuses on the musical intervals modelled in the pattern rather than the absolute pitches. The operators defined in the following capture these two semantics.

4.3 Melodic Pattern Selections

In this subsection, we define the process of selecting the occurrences of a melodic pattern P (without approximation), first in the tone-stable way, denoted by σ_P in the following, and then in the transposition-agnostic way, denoted by σ_P^t. In the definitions, we consider a melodic pattern $P = \langle N_1, ..., N_k \rangle$, and a voice V with its power set $\mathcal{P}(V)$.

Definition 6 (Selection over a voice (σ)). *The selection of P over V, i.e. set of matches of P in V, is defined by $\sigma_P(V) = \{A \mid A \in \mathcal{P}(V)$ and A "matches" $P\}$ where A "matches" P iff there is an isomorphism h from P to A s.t. for all N_i in $\{N_1, ..., N_k\}$, one has:*

- $(Cond_{sound})$ $sound(N_i) = sound(h(N_i))$
- $(Cond_{dur})$ $dur(N_i) = dur(h(N_i))$
- $(Cond_{seq})$ *if* $i > 1$, $start(h(N_i)) = end(h(N_{i-1}))$

For each image (in A), the first condition states that the image has the same sound as its preimage (in P), the second condition states that the image has the same duration as its preimage, and the third condition states that each image follows the next one, in the correct order, with no inserted event between them. While $(Cond_{seq})$ is always necessary, $(Cond_{dur})$ or $(Cond_{sound})$ can be omitted if the user is interested only in the pitches or the rhythm.

Example 3. Let $P_1 = \langle (C5, 1), (D5, 2), (C5, 2), (B4, 2) \rangle$ be the pattern illustrated in Fig. 2. The third to sixth notes (in cyan) in the third measure of Fig. 1 match this pattern as there is an isomorphism h that satisfies Definition 6, from P to $A = \{C5_{37}^{38}, D5_{38}^{40}, C5_{40}^{42}, B4_{42}^{44}\}$ defined as $\{h((C5, 1)) = C5_{37}^{38}, h((D5, 2)) = D5_{38}^{40}, h((C5, 2)) = C5_{40}^{42}, h((B4, 2)) = B4_{42}^{44}\}$. Another answer can be seen (in cyan) in the third to sixth notes of the fourth measure.

Let us note that there is no answer for P_2 in Fig. 1.

Definition 7 (Transposition-agnostic selection over a voice (σ^t)). *The transposition-agnostic selection of P over V, i.e. set of matches of P in V, is defined by $\sigma_P^t(V) = \{A \mid A \in \mathcal{P}(V)$ and A "matches" $P\}$ where A "matches" P iff there is an isomorphism h from P to A s.t. for all $i \in [1..k]$, one has:*

- $(Cond_{interval})$ *if* $i > 1$, $interval(N_{i-1}, N_i) = interval(h(N_{i-1}), h(N_i))$
- $(Cond_{dur})$ $dur(N_i) = dur(h(N_i))$
- $(Cond_{seq})$ *if* $i > 1$, $start(h(N_i)) = end(h(N_{i-1}))$

The first condition states that the sequence of intervals in A is the same as in the melodic pattern. The second and third conditions are the same as in Definition 6. Rests are not taken into account in this definition, and in the following. Indeed, dealing with rests requires more intricate definitions. Due to space limitation, we cannot go into the detail here.

Example 4. Consider the pattern $P_2 = \langle (D4, 1), (C4, 2), (D4, 2), (E4, 2), (F4, 2) \rangle$ from Fig. 2. There is no exact match for this pattern in Fig. 1, but we can find a match if we use the transposition-agnostic selection. Indeed, the sequence of intervals between the notes of P_2 is -1 tone (from $D4$ to $C4$), $+1$ tone (from $C4$ to $D4$), $+1$ tone ($D4$ to $E4$), and $+0.5$ tone ($E4$ to $F4$)). The isomorphism h from P_2 to $A = \{A4_3^4, G4_4^6, A4_6^8, B4_8^{10}, C5_{10}^{12}\}$ defined as $\{h((D4, 1)) = A4_3^4, h((C4, 2)) = G4_4^6, h((D4, 2)) = A4_6^8, h((E4, 2)) = B4_8^{10}, h((F4, 2)) = C5_{10}^{12}\}$ satisfies Definition 7. It corresponds to the answer in green in Fig. 1.

Remark 2. A selection operator over several voices (typically (a subset of) the voices of a musical score) can be defined as follows: if $\mathcal{V}$ is a set of voices and σ_P is a selection operator, then the selection of P in $\mathcal{V}$ is defined by $\cup_{V \in \mathcal{V}} \sigma_P(V)$. This remains true for the selection operators defined in the next section.

4.4 Fuzzy Melodic Pattern Selections

We now consider the notion of an *approximate matching* and generalize the previous selection operators using fuzzy logic. Let us recall that fuzzy logic [14] extends Boolean logic by considering truth values in the interval [0, 1] instead of the pair {0, 1}, which makes it possible to interpret gradual predicates, i.e., predicates whose satisfaction is a matter of *degree* (such as *tall*, *young*, etc.).

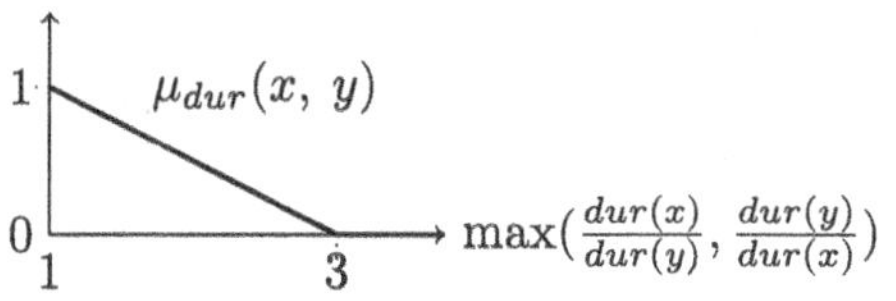

Fig. 3. Membership function μ_{dur}

An example of a fuzzy membership function is given in Fig. 3. It assesses the extent to which two notes x and y of a musical score are similar in terms of duration. The idea is that the satisfaction of the fuzzy criterion (the ordinates in Fig. 3, which go from 0 to 1) is a function of the ratio between the durations of the two notes (that must be smaller than 3 here). This function will be used in Example 5. Let us recall that fuzzy logic also provides aggregation functions for combining multiple satisfaction degrees into a single one, in the case where compound fuzzy conditions are considered. Common aggregation functions include the minimum (for interpreting conjunction), maximum (for disjunction), and averaging operators (for expressing different types of trade-offs between criteria). For instance, if we consider two fuzzy sets, "tall" and "young", and we want to find the degree to which an individual x is both tall and young, we may use the minimum function (or any other triangular norm): $\mu_{tall\ and\ young}(x)\ \min(\mu_{tall}(x), \mu_{young}(x))$.

Before extending the previous selection operators, we need to introduce a few notions.

Definition 8 (order-preserving mapping). *Let $P = \langle N_1, \cdots, N_k \rangle$ be a pattern, and $A = \{e_1, \cdots, e_k\}$ be a set of events. Let h be an isomorphism from P to A. The isomorphism h is said to be* order-preserving *iff for all $i \in [1..k-1]$, $start(h(N_{i+1})) > start(h(N_i))$.*

Definition 9 (Membership functions and aggregators). *In order to assess how close a pattern element is to its matching element in an answer, three fuzzy membership functions are needed (one over each aspect considered in the matching process, see Sect. 4.2):*

- *$\mu_{sound}(N, a)$, where N is a pattern element (see Definition 5) and a is an event, assesses how close N is to a in terms of pitch;*
- *$\mu_{dur}(N, a)$ assesses how close N is to a in terms of duration;*
- *$\mu_{seq}(a_1, a_2)$, where a_1 and a_2 are musical events with $start(a_2) > start(a_1)$, assesses how close—in terms of timing—$start(a_2)$ is to $end(a_1)$.*

For transposition-agnostic matching, instead of $\mu_{sound}(N, a)$, we use:

- *$\mu_{interval}(N, N', a, a')$, which assesses the closeness between $interval(N, N')$ and $interval(a, a')$.*

Two aggregation functions are also needed: agg_{note} for aggregating the satisfaction degrees over the musical aspects for a note, and agg for aggregating the satisfaction degrees of the notes in an answer.

In practice, the membership functions and aggregates depend on the retrieval context. They can be defined by experts for a group of users, or be inferred from parameters set by the user. The latter case is illustrated in the following where the membership functions are derived from user-defined tolerance thresholds (in Fig. 3, function μ_{dur} corresponds to a tolerance value of 3).

Remark 3. Straightforwardly, crisp selection is recovered using Boolean membership functions (corresponding to no flexibility) for pitch/interval, duration and sequencing.

Definition 10 (Fuzzy tone-stable selection (σ_P^f)). *Let V be a voice and $P = \langle N_1, \cdots, N_k \rangle$ a pattern. The fuzzy set of matches of P in V, denoted by $\sigma_P^f(V)$, is defined as $\sigma_P^f(V) = \{(\mu_P(A)/A) \mid A \in \mathcal{P}(V)$ and there exists an order-preserving isomorphim h from P to $A\}$, where $\mu_P(A) = agg(\{\mu_{note}(h(N_i)) \mid h(N_i) \in A\})$, meaning that $\mu_P(A)$ is an aggregation of the degrees associated with the musical events from A. μ_{note} is defined below.*
In order to lighten the notation, let us denote $h(N_i)$ by a_i. Then, the "tone-stable" satisfaction degree of a_i relatively to P, denoted by $\mu_{note}(a_i)$, is defined as:

$$\mu_{note}(a_i) = \begin{cases} agg_{note}(\mu_{sound}(N_i,\, a_i),\, \mu_{dur}(N_i,\, a_i)) & if\ i = 1 \\ agg_{note}(\mu_{sound}(N_i,\, a_i),\, \mu_{dur}(N_i,\, a_i), \mu_{seq}(a_{i-1},\, a_i)) & if\ i > 1 \end{cases}$$

Definition 11 (Fuzzy translation-agnostic selection ($\sigma_P^{f,t}$)). *Let V be a voice and P be a musical pattern. The fuzzy set of translation-agnostic matches of P in V, denoted by $\sigma_P^{f,t}(V)$, is almost the same as in Definition 10, except that the closeness of the musical intervals is used instead of the closeness between pitches. Then, the definition of $\mu_{note}(a_i)$ becomes:*

$$\begin{cases} agg_{note}(\mu_{dur}(N_i, a_i)) & if\ i = 1 \\ agg_{note}(\mu_{interval}(N_{i-1},\, N_i,\, a_{i-1},\, a_i), \mu_{dur}(N_i, a_i), \mu_{seq}(a_{i-1},\, a_i)) & if\ i > 1 \end{cases}$$

Example 5. Let us consider again the voice V from Example 1 and the pattern $P_2 = \langle (D4, 1), (C4, 2), (D4, 2), (E4, 2), (F4, 2) \rangle$ from Fig. 2. We consider a user searching for the occurrences of P_2 in V and wanting all translation-agnostic approximate occurrences of the pattern. The three flexibility aspects—pitch, duration, and sequencing—are considered.

The user allows flexibility in the melodic intervals between consecutive notes, accepting entries where the interval in the match is close to the interval in the pattern, within a tolerance of 2 whole tones. For example, consider the first two notes in the pattern: $(D4, 1)$ and $(C4, 2)$. The interval between D and C is 1 tone. With a tolerance of 2, we accept any pair of notes where the interval between them lies between -0.5 and $+2.5$ tones. This can be translated by the membership function (where the parameter appears in the denominator):
$\mu_{interval}(e_1, e_2, a_1, a_2) = \max\left(1 - \frac{|interval(e_1,e_2) - interval(a_1,a_2)|}{2}, 0\right)$.

The user also allows variability in note durations, accepting matches between one-third and three times the pattern's notes duration, leading to the following function for μ_{dur} if we choose the triangular membership function, which is illustrated in Fig. 7, that can be formalized (taking $z = \max(\frac{dur(x)}{dur(y)}, \frac{dur(y)}{dur(x)})$) as:

$$\mu_{dur}(e_1, e_2) = \begin{cases} 0 & \text{if } z \geq 3 \\ \frac{3-z}{2} & \text{otherwise.} \end{cases}$$

The shape of the membership functions and their parameter values (2 for pitch, 3 for duration, and 2 for sequencing) are chosen specifically for this example. In practice, these values are use-case specific and depend on the user's retrieval needs.

$A = \{A4_{76}^{78}, G4_{78}^{80}, A4_{80}^{82}, C5_{82}^{84}, B5_{84}^{86}\}$ belongs to $\sigma_{P_2}^{f,t}(V)$ as there is an order-preserving isomorphism h defined as $\{h((D4,\ 1)) = A4_{76}^{78}, h((C4,\ 2)) = G4_{78}^{80}, \ldots, h((F4,\ 2)) = B5_{84}^{86}\}$. We now consider the calculation of $\mu_{P_2}(A)$. For each event a_i of A, $\mu_{note}(a_i)$ is calculated according to Definition 11. We take two events of A as example: the first one ($A4_{76}^{78}$) and the fourth one ($C5_{82}^{84}$). For the first event ($A4_{76}^{78}$), the duration only is taken into account for μ_{note} according to Definition 11.[5] We have $\max(\frac{dur(D4,\ 1)}{dur(A4_{76}^{78})}, \frac{dur(A4_{76}^{78})}{dur(D4,\ 1)}) = \max(0.5, 2) = 2$ Thus, $\mu_{dur}((D4,\ 1), A4_{76}^{78}) = \frac{3-2}{2} = 0.5$. For $C5_{82}^{84}$, the fourth note in A, the three aspects have to be considered (see Definition 11). The duration and the sequencing are exact but the interval (involving the previous event $C5_{82}^{84}$) is not:

- $\mu_{interval}((D4,\ 2), (E4,\ 2), h((D4,\ 2)), h((E4,\ 2)))$
  $= \mu_{interval}((D4,\ 2), (E4,\ 2), A4_{80}^{82}, C5_{82}^{84})$
  $= \max(1 - \frac{|(sound((E4,\ 2)) - sound((D4,\ 2))) - (sound(C5_{82}^{84}) - sound(A4_{80}^{82}))|}{2}, 0)$
  $= \max(1 - \frac{|1-(1.5)|}{2}, 0) = 0.75$
- $\mu_{dur}((E4,\ 2), h((E4,\ 2))) = \mu_{dur}((E4,\ 2), C5_{82}^{84}) = 1$
- $\mu_{seq}(h((D4,\ 2)), h((E4,\ 2)))$
  $= \mu_{seq}(A4_{80}^{82}, C5_{82}^{84}) = \max(1 - \frac{start(C5_{82}^{84}) - end(A4_{80}^{82})}{2}, 0)$
  $= \max(1 - \frac{82-82}{2}, 0) = 1$

To aggregate the three degrees associated with each note, we assume the use of the minimum operator (agg_{note} aggregator in Definition 10), which results in an aggregated degree of 0.25 for the first note and 0.5 for the last note in the match A. The choice of the aggregation operator is important, as it impacts evaluation and result filtering and ranking. In this case, the minimum operator imposes a penalty on notes with low degrees in any aspect, even if the other aspects do well. This reflects a conservative approach, often used in fuzzy systems to prioritize the weakest link in a multi-criteria evaluation.

When all the degrees associated to each note of A are computed, the overall degree of the match is computed by aggregation (agg aggregator in Definition 10) of the individual notes' degrees. For instance, assuming the note degrees to be 0.25, 1.0, 1.0, 0.75 and 0.5, using the arithmetic mean as the aggregation operator yields an overall degree of 0.7 for the match. This operator balances

[5] Recall that interval and sequencing criteria require two consecutive pattern elements.

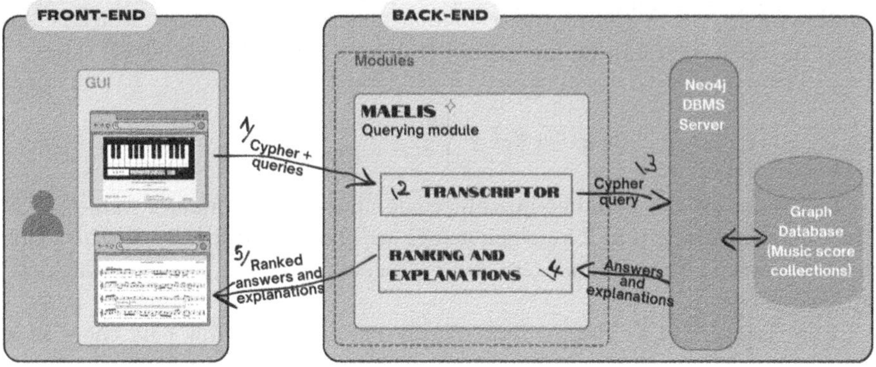

Fig. 4. System architecture and content-based querying (Color figure online)

the contributions of all notes, ensuring the retrieved sequence is, on average, close to the query pattern P, while tolerating some deviation for individual notes.

Ranking of the Answers. The overall degree assigned to a match makes it possible to rank the matches according to their closeness to the pattern. For example, if another match B has an overall degree of 0.4, this indicates that B is less satisfactory than A under the chosen aggregation strategy.

Explainability of the Answers. According to the process used to define the satisfaction degree of an answer, a explanation of the answer can be provided, taking the form of the detail of the satisfaction degrees obtained by each note (over the aspects pitch/interval, duration and sequencing).

Parametrization. As seen above, three aspects—pitch/interval, duration, and sequencing—may be taken into account in a selection. However, not all three aspects are equally relevant in every context. Indeed, use cases vary: a rhythm-centric search might prioritize duration and sequencing, while a melody-focused search would consider all three. The definition of the membership functions allows adjusting to specific needs by more or less relaxing a tolerance threshold.

5 Implementation in the Skrid Digital Score Library

Skrid [15] is a digital score library dedicated to folk musical scores from the Brittany French region. It stems from a collaborative effort with Dastum [16], an association dedicated to preserving the region's oral folk heritage. Skrid's architecture (see Fig. 4) is composed of two parts.

A *front-end* (in blue in Fig. 4) contains the modules the users interact with. Given that the users are musicians and musicologists, we developed an interface that allows users to define a pattern using a piano keyboard and sliders for setting the flexibility parameters[6]. For the same reason, melodic patterns and results are

[6] This GUI interface is available at https://shaman.enssat.fr/skrid/searchInterface.

displayed in Western music notation. In the *back-end* of the application (in green in Fig. 4), data is stored, managed, and processed. SKRID relies on a storage of the musical scores in a property graph database (Neo4J DBMS [17]). The data model enabling this storage is beyond the scope of this paper; the reader may refer to [8] for details.

The implementation of the framework described in Sect. 4 takes the form of CYPHER+, an extension of the CYPHER language [18]. A CYPHER+ query makes it possible to define a melodic pattern and set flexible parameters. Listing 1 is an example of CYPHER+ query. In Listing 1, lines 3, 4 and 5 are specific to the augmented syntax and are not part of native CYPHER. They serve to define the flexibility parameters. Then follows native-cypher: lines 7 to 12 define the melodic pattern sought for (here, it is pattern P from Fig. 2). The `RETURN` clause specifies the output of the query, which in this case includes the source of the first event (`e1.source`), representing the identifier of the musical score, and the start time of the answer (`e1.start`).

```
1   MATCH
2       // Non-native Cypher: to define the flexibility parameters :
3       ALLOW_TRANSPOSITION // transposition-awareness
4       TOLERANT interval=1.0, dur=3, seq=0.0
5       ALPHA 0.5 // threshold for the minimum acceptable matching degree
6       // Native Cypher: to define the melodic pattern to search for :
7       (e1:Event)-[:NEXT]->(e2:Event)-[:NEXT]->(e3:Event)-[:NEXT]->(e4:Event)-[:NEXT]->(e5:
            Event), // Chain of 5 consecutive events
8       (e1)--(f1{class:'d', octave:4, dur:16}), // e1 is a sixteenth (dur:16) D4 note
9       (e2)--(f2{class:'c', octave:4, dur:8}), // e2 is a eighth note (dur:8) C4 note
10      (e3)--(f3{class:'d', octave:4, dur:8}), // e3 is a eighth note D4
11      (e4)--(f4{class:'e', octave:4, dur:8}), // e4 is a eighth note E4
12      (e5)--(f5{class:'f', octave:4, dur:8}), // e4 is a eighth note F4
13  RETURN e1.source AS source, e1.start AS start // Information returned in the answer
```

Listing 1. CYPHER+ query based on P

A software tool called MAELIS[7] (yellow portion in Fig. 4) was developed, whose purpose is to execute CYPHER+ queries. In SKRID, MAELIS serves as a software layer positioned between the graph DBMS and the front-end that sends CYPHER+ queries on the basis of the information provided by the user through the GUI querying module (piano keyboard and sliders). MAELIS is composed of two modules: a *Transcriptor* module and a *Ranking and explanations* one. The *Transcriptor* module takes as input a CYPHER$^+$ query (stage 1 in Fig. 4) and, based on the membership functions defined in the system, converts the CYPHER$^+$ query into a (classical) CYPHER query that is used to retrieve the answers as well as the information needed to compute the satisfaction degrees. The transcription step corresponds to stage 2 in Fig. 4. The transcripted query is then sent to the Neo4j DBMS that executes it over the musical score collections (stage 3 in Fig. 4). From the data returned by Neo4J, the *Ranking and explanations* module computes the satisfaction degrees for the retrieved answers (stage 4 in Fig. 4). The overall satisfaction degree makes it possible to rank-order the answers before presenting them to the user. As to the partial satisfaction degrees, they make it possible to provide a detailed explanation for each answer (stage 5 in Fig. 4).

[7] MAELIS is an open-source software. Its code (in Python) is available at https://github.com/aa196883/compilation_requete_fuzzy.

Let us consider again pattern P_2 (see Fig. 2), along with the following flexibility parameters: pitch tolerance of at most one tone in either direction, duration tolerance corresponding to a multiplicative factor of at most 3, with transposition allowed. These requirements correspond to the CYPHER+ query of Listing 1. Figure 5 is a screenshot of the user interface of the platform displaying two (among more) of the answers to the aforementioned query. In the first match, all notes are highlighted in green, indicating that each note exactly matches the user's query. This answer corresponds to Example 4. This match has a matching degree of 1. The second match is approximate, it corresponds to the matching explained in Example 5. The satisfaction degrees for each note (on each of its aspects) are displayed when the user hovers over it (cf. the beige box in Fig. 5).

Fig. 5. Two matches for P, one exact and another approximate

6 Performance Results : The Cost of Flexibility

In this section, we present the results of performance tests aimed to assess the execution time of fuzzy queries for the different flexibility aspects (pitch/interval, duration, sequencing) and associated parameter values. The goal of these experiments is threefold: (1) to show how increasing tolerance impacts query runtimes; (2) to analyze how query execution times vary with the length of the pattern; and (3) to assess the cost of the transcription and ranking steps. The goal here is not to compare this work with the existing literature. The capabilities and expressiveness of our framework do not align with those of existing systems, making a comparison in terms of response time moot. Instead, the focus is on understanding the behavior and performance characteristics of our system in the context of its unique features.

Figures 6, 7, and 8 show the variation of the query execution time with respect to the pitch/interval, duration, and sequencing flexibility aspects, respectively with respect to pattern length. Each figure contains multiple graphs corresponding to different parameter values (e.g., different tolerance thresholds for the parameter considered). For a given figure, the matching over the two other criteria is crisp (non-approximate). Each point represents the average execution time

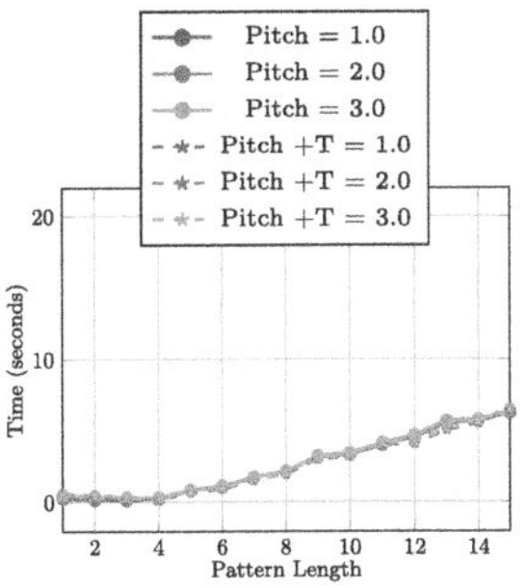

Fig. 6. Runtimes over pattern length for different pitch flexibility

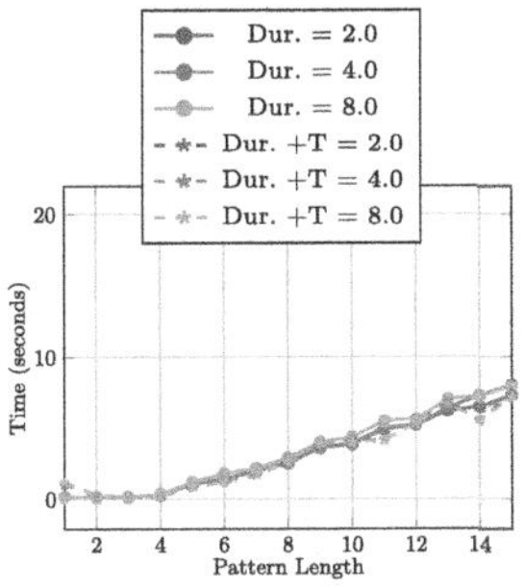

Fig. 7. Runtimes over pattern length for different duration flexibility

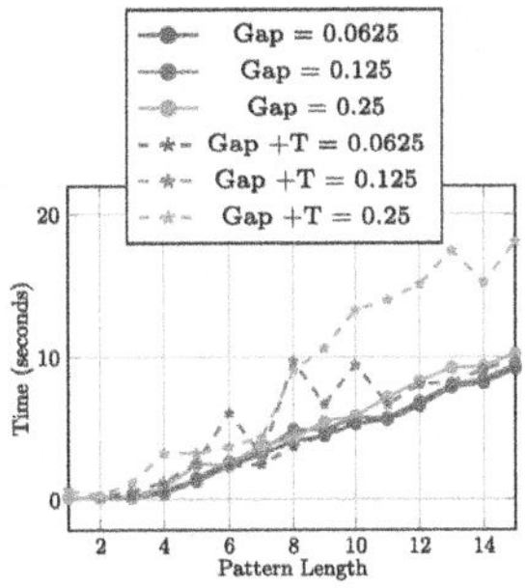

Fig. 8. Runtimes over pattern length for different gap flexibility

of 15 different queries with the same flexibility parameters. For each query, we ensured that there was at least one exact (i.e., non-fuzzy) match in the database. The database used for these tests consists of 658 real-life musical scores, each being a traditional piece from Brittany. This corresponds to a total of 81,032 nodes and 124,326 relationships (stemming from the storage of the scores in the graph-based model of [8]). All the tests were conducted on a PC with an Intel Core i7-13700H 13th generation processor and 16 GB of RAM. Each figure includes two sets of graphs: one set shows query execution times without transposition, while the other set includes transposition-aware queries (denoted by +T in the figures).

From Figs. 6 and 7, we observe that all the curves largely overlap, regardless of the parameter values. For pitch/interval, parameter values of 1, 2, and 3 yield nearly identical query execution times. Similarly, for duration, parameter values of 2, 4, and 8 show no noticeable differences in runtime. This also extends to transposition: curves representing queries with and without transposition overlap significantly. A likely explanation for this behavior is that while increasing flexibility (via higher tolerance values) increases the number of matches, the growth in the number of answers remains reasonable. As a result, the query execution time does not increase substantially. Additionally, the impact of pattern length appears to be linear for both pitch/interval and duration.

From Fig. 8, we observe that, unlike for pitch/interval and duration, two specific curves stand out: those representing transposition-aware queries with gap parameter values of 0.125 and 0.25. While the number of results for these queries is not remarkable compared to other tests, the query evaluation times are noticeably higher and exhibit significant variability across queries, even when pattern length, parameter value, and transposition settings remain constant. The reason for this unpredictability is unclear and represents a starting point for future work focused on optimizing the system. Importantly, this variability only occurs for certain patterns under a specific context: high gap parameter values combined

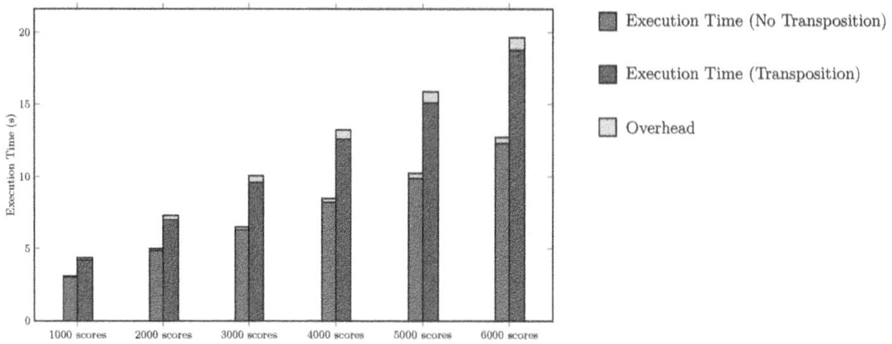

Fig. 9. Runtimes for a 6-element fuzzy query over different database sizes (Color figure online)

with transposition-aware queries. This suggests that the observed behavior is a niche issue rather than a systemic limitation. Its resolution is nonetheless important in order to ensure the performance consistency of the system.

Figure 9 presents the runtimes for a six-element fuzzy query (with parameters set to 1, 2 and 0.0625 for pitch/interval, duration and sequencing respectively) across databases of increasing size, measured both with and without transposition. Each database is generated by cloning an initial set of 500 distinct scores, making the size of the result scale proportionally with database size. The figure also includes the overhead, which accounts for the time required for query transcription, degree computation, and ranking (see yellow portion in Fig. 4). Several key observations can be made from this figure. First, the runtime scales linearly with database size, whether transposition is enabled or not, which shows the scalability of the framework. Second, while queries with transposition (blue bars) consistently take longer to run than those without transposition (red bars), this is no surprise due to the significantly larger number of possible matches when considering all transpositions. The runtime ratio between transposition-aware and no-transposition queries remains around 1.5, which is a reasonable trade-off given that every possible transposition is accounted for in the results. Finally, the overhead (yellow bars) is very small compared to the overall query execution time. This overhead, primarily due to score computation across all retrieved results, scales with the volume of results and is linear with respect to the number of total notes accross all results. These findings highlight the efficiency of the system under scenarios involving transposition and larger datasets.

7 Conclusion and Perspectives

In this paper, we have presented a theoretical framework based on fuzzy logic for retrieving a melodic pattern in a musical score. Three fuzzy criteria—on pitch, duration, and sequencing—can be specified in order to make the matching process tolerant. Additionally, the selection operator can be configured in such a way that it accounts for transposition. We implemented this framework as part of a digital score library, offering a concrete tool for performing flexible searches. The salient properties of this framework are: *i)* its high parameterizability and *ii)* its capability to provide ranked and explainable answers thanks to the use of fuzzy logic. Performance tests show the tractability of the approach, as well as the impact of flexibility and transposition awareness on query execution times. These results indicate that the system scales well with database size and query length, and guarantees a reasonable overhead cost even for flexible and transposition-aware queries. Looking ahead, several extensions to the framework are planned. First, we aim to generalize the approach in order to support *polyphonic queries* (where the pattern itself is polyphonic). Second, we plan to collaborate with musicians so as to conduct an *in-depth user study*, as no existing benchmark for musical score retrieval in a database context currently exists. This feedback will be invaluable for assessing/refining our approach and carry on its development. Lastly, *optimization* efforts will focus on improving performance, particularly for queries involving flexibility on the sequencing dimension.

Acknowledgements. We would like to thank DASTUM for their collaboration, with special thanks to Anne-Marie Nicol and Gwenaël Piel for their involvement. We also extend our gratitude to Vincent Barreaud and Tommaso Padovano for their work on the first version of SKRID. For the current version of the platform, and the development of the fuzzy engine, we thank Louis Thomas-Girardey for his contributions.

References

1. Aly, A., Pivert, O., Thion, V.: Database approaches to the modelling and querying of musical scores: a survey. In: Proceedings of the International Conference on Theory and Practice of Digital Libraries (TPDL). LNCS, vol. 15177, pp. 292–309 (2024)
2. Walshaw, C.: ABC notation (1993). https://abcnotation.com/
3. Huron, D.: Music information processing using the humdrum toolkit: concepts, examples, and lessons. Comput. Music. J. **26**(2), 11–26 (2002)
4. Good, M.: MusicXML for notation and analysis. In: The Virtual Score: Representation, Retrieval, Restoration, pp. 113–124. MIT Press (2001)
5. Rolland, P.: The music encoding initiative (MEI). In: Proceedings of the International Conference on Musical Applications Using XML, pp. 55–59 (2002)
6. Chamberlin, D.D., et al.: XQuery from the Experts: A Guide to the W3C XML Query Language. Addison Wesley, Boston (2003)
7. Jones, J., de Siqueira Braga, D., Tertuliano, K., Kauppinen, T.: MusicOWL: the music score ontology. In: Proceedings of the International Conference on Web Intelligence (WI 2017), pp. 1222–1229. Association for Computing Machinery (2017)
8. Rigaux, P., Thion, V.: Topological querying of music scores. Data Knowl. Eng. (DKE) **153** (2024)
9. Sapp, C., Liu, Y.-W., Selfridge-Field, E.: Search effectiveness measures for symbolic music queries in very large databases. In: Proceedings of the International Conference on Music Information Retrieval (ISMIR) (2004)

10. Duggan, B., O'Shea, B.: Tunepal: searching a digital library of traditional music scores. OCLC Syst. Serv. **27**(4), 284–297 (2011)
11. Duggan, B., O'Shea, B., Gainza, M., Cunningham, P.: Compensating for expressiveness in queries to a content based music information retrieval system. In: Proceedings the International Computer Music Conference (ICMC) (2009)
12. Poulain, A.: Carnets de route – Chansons traditionnelles de Haute-Bretagne. Presses Universitaires de Rennes, co-édition Dastum (2011)
13. Fournier-S'niehotta, R., Rigaux, P., Travers, N.: Modeling music as synchronized time series: application to music score collections. Inf. Syst. **73**, 35–49 (2018)
14. Zadeh, L.A.: Fuzzy sets. Inf. Control **8**(3), 338–353 (1965)
15. The SKRID platform (2024). https://shaman.enssat.fr/skrid/
16. Dastum association (1993). https://www.dastum.bzh/association/
17. Neo Technology. Neo4j web site (2025). www.neo4j.org
18. Neo Technology. The Cypher Manual (2025). https://neo4j.com/docs/cypher-manual/current

Implementing Digital Twin Query Views

Emilio Carrión[1(✉)] and Pedro Valderas[2]

[1] Mercadona Tech, Mercadona, Barcelona, Spain
emcarrio@mercadona.es

[2] PROS - VRAIN, Universitat Politècnica de València, Valencia, Spain
pvalderas@dsic.upv.es

Abstract. Digital twins (DTs) have become increasingly complex as they integrate data from diverse heterogeneous sources, which presents a major challenge in querying and aggregating information in a unified and secure way. The Entity-Relationship Digital Twin (ERDT) model provides a basis for representing both physical entities and their digital counterparts at the conceptual level. This paper extends this modeling solution by proposing a Domain Specific Language to define queries on the DT through ERDT views. This DSL is supported by a code generation strategy that translates the defined queries into executable Python specifications. In addition, we also present a reference architecture for deploying the software artifacts generated from models to obtain a fully operative DT. We validate our approach through the development of a DT for the management of the truck fleet in a real industrial scenario located within the logistics operations of Mercadona, one of Spain's largest supermarket chains. The results demonstrate that our solution covers data heterogeneity and security constraints while also providing enhanced query capabilities in a production environment.

Keywords: Digital Twin · Modelling · Query

1 Introduction

Digital twins (DTs) have emerged as a game-changing technology in a wide range of fields, from manufacturing and urban infrastructure to logistics and healthcare [27]. DTs are virtual representations of real-world entities synchronised at a specific frequency and fidelity [13]. These virtual counterparts of physical objects and systems enable real-time monitoring, simulation and optimisation, providing advanced insights into operational performance [19].

However, as the complexity and scale of DT implementations increase, so does the need for software engineering solutions to ensure quality development. One problem when managing DTs is the fact that they often make use of a wide range of data formats and sources to represent their physical counterparts. These sources include 3D models, building information models (BIM), proprietary software systems, physics engines and other domain-specific data formats.

Pedro Valderas' work is supported by the project PID2023-146224OB-I00 founded by MICIU/AEI/10.13039/501100011033, FEDER, and the UE.

J. Grabis et al. (Eds.): RCIS 2025, LNBIP 547, pp. 279–294, 2025.
https://doi.org/10.1007/978-3-031-92474-3_17

This heterogeneity presents a major challenge in querying and aggregating information in a unified way. Without a standardised approach, extracting meaningful information is problematic, as each data source may require specialised management. In addition, DTs are increasingly used in industries with stringent security requirements. One example is the healthcare field where private personal data must be protected. Ensuring secure access to this data is key, as any unauthorised access could lead to privacy leaks.

Model-Driven Development (MDD) seems to be one of the best options to face the development of DT [6]. However, although there are some MDD solutions that face this challenge [18,20,24], most of them fail when considering queries.

In previous research [10], we introduced the Entity-Relationship Digital Twin (ERDT) model to conceptually define DTs. This modelling solution is an extension of the traditional entity-relationship (ER) model that allows representing both the physical entities and the digital counterparts that define a DT. In [11], we introduced a preliminary version of a DSL that allows us to define queries on the DT represented by views over an ERDT model. By abstracting and structuring different types of data within a conceptual modelling framework, we created a standardised query interface that integrates these various data sources and applies security constraints.

This paper extends and evolves our previous work by incorporating semantics to the DSL by transforming DSL descriptions into a language with well-defined semantics. In particular, we present a transformation strategy that automatically generates executable Python specifications from the queries defined by using the DSL. In addition, we also present a reference architecture for deploying the software artifacts generated from models to obtain a fully operative DT by following the principles of MDD. To demonstrate the practical application of this solution, we also present as a case study based on the development of a DT for truck fleet management in a real industrial scenario located within the logistics operations of Mercadona, one of Spain's largest supermarket chains.

In summary, this work contributes to improving the state-of-the-art in DT development by:

- Proving a complete DSL based on ERDT views to create flexible query mechanisms that combine data from multiple entities while respecting security constraints.
- Demonstrate the practical application of these innovations through a real industrial case study of a truck fleet management system.
- By improving the query capabilities of ERDT models, this work aims to provide a more robust framework for managing the complex data generated by DTs in various domains.

The paper is structured as follows. Section 2 introduces the related work. Section 3 briefly summarises the ERDT model and presents the real-world case study of a truck fleet management system. Section 4 introduces the DSL to define ERDT query views. Section 5 presents a reference architecture to execute ERDT query views and a code generation strategy to automatically obtain executable code from conceptual ERDT query view models. Section 6 introduces a proof of concept validation based on the implementation of the real case study presented in the paper. Section 7 discusses the main contributions of this work. Finally, Sect. 8 provides conclusions and potential future research directions.

2 Related Work

Querying complex models is a well-established concern in the Model-Driven Engineering (MDE) community [22]. Thus, over time, different approaches to address this problem have emerged, in particular in the context of UML and OCL (Object Constraint Language).

Early work [16] used OCL as a query language and highlighted the similarity between OCL syntax and traditional database query languages. They also compared OCL and ER model query languages showing that OCL had potential to be used as an expressive query mechanism in object-oriented systems. However, they noted several problems in the definition of OCL that need to be addressed.

Building on this foundation, the work of [5] showed that the combination of UML and OCL could achieve the expressiveness of relational algebra, making it ideal for complex query tasks. However, they concluded that specific extensions to OCL were needed for it to be used in practice as a query language, especially when working with UML. [25] proposed a novel modelling notation to represent queries using UML. This proposal introduced dedicated symbols for common selection tasks with OCL semantics. It also stressed the importance of making queries more accessible by visual modelling in a language-agnostic manner.

More recently, models have grown in size and complexity, shifting the focus to making persistence and queries scalable. Morsa, a model repository that uses No-SQL databases to efficiently store models [15], is an example of this trend. It shows the need for new approaches to managing large-scale models.

Regarding DTs, we find even more varied problems. [6] highlights that DTs face three major challenges: managing heterogeneous models from different disciplines, maintaining bidirectional synchronization between DTs and physical systems, and supporting collaborative development throughout the entire lifecycle. These challenges directly influence the approach to queries in DT environments.

The recent advances in DT modelling discuses by [26] show the importance of data-driven and model-driven approaches to enable important DT functionalities such as monitoring, simulation, and prediction. This analysis of the state-of-the-art in DT modelling demonstrates the need for comprehensive modelling solutions that can cover diverse fields and functionalities.

Our work builds on these foundations, but addresses the main challenges posed by DTs. Previous work focused on querying static data, however DTs need real-time access to dynamic data from multiple heterogeneous sources. Our view-based approach extends traditional model query concepts to fit DT-specific needs such as real-time synchronization.

Furthermore, unlike traditional OCL-based approaches that are primarily used as constraint languages, our query views are intended to handle the complex nature of DT data while maintaining security and performance, aligning with [17] observations on the limitations of OCL in data processing but providing a solution more tailored to the domain of DTs.

3 ERDT Model Overview

The Entity-Relationship Digital Twin (ERDT) model [10] extends the well-known Entity-Relationship (ER) [12] to model DTs of physical systems. Its main contributions are the definition of historical values (as DTs often rely on historical data for operational functionalities) and the introduction of interfaces and data flows, key elements in the ERDT model that ensure real-time synchronisation between the physical and virtual components of the DT.

Figure 1 introduces a partial view of the ERDT model presented as a case study in a previous work [10], which will be used as running example throughout the rest of the paper. This model represents the last-mile fleet of delivery trucks that we have at Mercadona and its logistics warehouse operations (called hives).

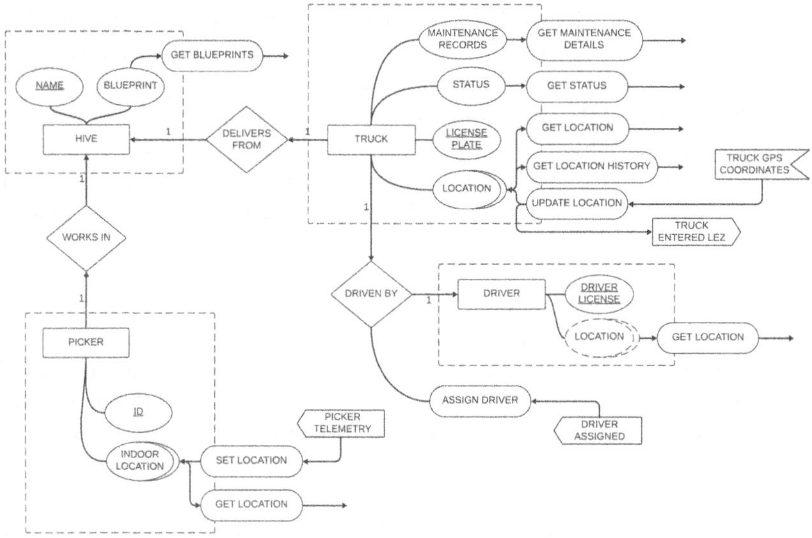

Fig. 1. ERDT model of Mercadona's logistics operations.

Entity Set and Relationships. Entities represent the physical objects we are digitising in our DT. Relationships show how entities are related between them. In the running example, several entities and relationships are defined for the DT, as it shown in the model in Fig. 1. For instance, the *Truck* and *Driver* entities represent the different trucks of the fleet and the drivers of the trucks, respectively. In the same way, there is a relationship between both that represents that each Truck is assigned to a Driver.

Entities, as in the base ER model, have attributes. However, our modelling approach introduces support for features characteristic to DTs, such as attributes that need to keep a historical record of their values. For example, the Truck entity includes a *Location* attribute that tracks the past locations of the truck. This property is key for the right representation of the temporal aspects of a DT and enables more complex queries and analyses over time.

Interfaces act as controlled access points to entities and their attributes, hiding internal complexity and facilitating interaction with the DT. They are specially useful in systems that manage heterogeneous data sources, as they offer a single access point for fetching or updating information related to an entity, abstracting away the complexity of interacting with multiple systems.

Defined as functions that receive values and update or return the attributes or relationships of an entity, interfaces vary from simple updates to more complex operations. They also allow for the implementation of security and privacy restrictions, controlling which users can access and modify information, providing a first layer of security for data, as they explicitly define what can be done with the data. Any operations not supported or defined by the interface are not allowed, ensuring secured access to sensitive information.

Each entity is equipped with interfaces to handle its queries and updates. For example, in our DT case study, location data for trucks is managed by external proprietary systems. However, instead of accessing these systems directly, the Truck entity's interfaces provide a unified access point to retrieve this data, such as the *getLocationHistory()* interface. This abstraction simplifies interaction and shields users from the complexity of the underlying systems, allowing seamless querying and updates.

Value Sets are used in ER models to define a set of values that can be assigned to the attribute of one entity. We use them for the same purpose in the context of ERDT. Although they are note graphically represented in the model in Fig. 1, some examples defined in the case study are: (1) *GPSReading[]*, which is an historical value set of coordinate readings. It has an *unique()* method that returns a new collection with non-repeating coordinate records, and a *count()* method that returns the collection size. And (2) *Polygon*, which is a set of values that represents coordinate polygons. It has a *contains(coordinate)* method that checks if a given coordinate is inside the polygon boundaries.

Data Flows represent the bidirectional communications between the DT and the physical components. There are three main types:

- Incoming Events: Sent by the physical entities to notify changes in their state. Examples: the *Driver Assigned* event shown in Fig. 1.
- Data Requests: Sent from the DT to the physical entities to query their current status. They are executed periodically or on demand. Examples: the *Truck GPS Coordinates* request.
- Outgoing Events: Sent from the DT to the physical components to execute actions or send commands. They let the DT to influence the physical system. Examples: the event *Truck Entered LEZ* that notifies that a truck has entered a low emissions zone.

In short, interfaces and data flows enable secure and structured information management in the ERDT model. They facilitate the representation of the bidirectional communication between the DT and the physical components, enabling real-time synchronization and the analysis of the collected data.

3.1 Querying Limitations

Although interfaces in the ERDT model act as a controlled access point to entity attributes and encapsulate internal complexity, they also have some limitations. Mainly, these interfaces work at the entity level, meaning that they are designed to ease interactions with individual entities rather than enabling more complex, system-wide operations that span multiple entities or relationships. This is not useful to define queries or operations that need to aggregate data from different entities in the DT, limiting the flexibility of the model in more complex scenarios.

Besides, the security policies applied through these interfaces are often limited to simple operations, such as controlling access to updating or querying specific fields of an entity. For example, if we wanted to retrieve all trucks that are less than 20 km away from a given location in the running example (See Fig. 1), we would need to query each of the trucks through the *getLocation()* interface and apply a transformation to get the distance to the specified point. Then, we would need to filter out the trucks that do not comply to the given condition. Querying all the trucks one by one and managing the needed calculations and comparisons becomes difficult as the complexity of the DT grows, making this kind of query a challenging task.

These limitations are covered by ERDT query views, a more flexible querying mechanism that operates beyond the scope of individual entities.

4 ERDT Query Views

In the context of DTs, querying data efficiently and securely across multiple entities is essential for gaining insights into complex systems. However, as DTs grow in complexity, relying only on entity-level interfaces can lead to simple, repetitive queries that require users to query each entity's attributes and relationships independently. To address this, a preliminary version of a DSL for defining views over ERDT models was introduced in [11]. In this work, we improve this DSL in such a way we not only propose a syntax to define queries in an abstract way, but also incorporate semantics that become these queries into executable artifacts. This DSL serves as a mechanism to simplify querying and offer higher-level abstractions that enable the reuse of software artifacts [23] in the ERDT model.

A DSL is generally composed of three main components [21]: abstract syntax, concrete syntax, and semantics. Next subsections explain them in detail.

4.1 Abstract Syntax

The abstract syntax defines the main concepts of the DSL and their relationships including the rules that define how models can be built. In this section, we present a formal definition of the concepts proposed to create ERDT viers.

A *view* in an ERDT model is defined as a high-level query abstraction that interacts with multiple entities and their relationships by invoking operations from their respective entity interfaces. The view is a function that aggregates, filters, or transforms the data fetched from these entities, while applying the security constraints defined by the entity interfaces. Let:

- $E = \{e_1, e_2, ..., e_n\}$ be the set of all entities in the ERDT model.
- $I(e) = \{o_1, o_2, ..., o_m\}$ be the set of interface operations defined for entity $e \in E$, where each operation o_i is a function $o : P \rightarrow A$, where P is the set of parameters, and A is the set of attributes or results for the operation.
- $S(o)$ is the set of security constraints for operation o_i, restricting which users $u \in U$ can invoke o_i.

A view $v \in V$ in the ERDT model is then a function that performs system-wide queries by linking the results of multiple entity operations while applying security and filtering conditions. The formal definition of a view v is as follows:

$$v = f(\{e_k | e_k \in E, o_i \in I(e_k), C(o_i(p_j)), u \in S(o_i)\}) \tag{1}$$

where:

- $e_k \in E$ represents an entity involved in the view.
- $o_i \in I(e_k)$ is an interface operation of the entity e_k, and $p_j \in P$ are the parameters passed to that operation.
- $C(o_i(p_j))$ represents filtering conditions applied to the result of the operation $o_i(p_j)$, such as specific status or thresholds.
- $S(o_i)$ is the set of security restrictions and $u \in S(o_i)$ ensures that the user u has permission to access the operation o_i.
- f is an aggregation or transformation function that operates on the results.

4.2 Concrete Syntax

To facilitate the definition of ERDT views based on the abtract sintax presented above, we proposed a concrete syntax based on Yaml, which provides a structured format that allows users to specify permissions, parameters, and query logic without needing in-depth knowledge of the underlying entity interfaces.

The definition language follows a straightforward format that emphasises clarity. The general structure can be described as shown in Listing 1.1.

```
<View Name>:
  entity: <EntitySet>
  permissions:
    - <Permission Condition>
  parameters:
    - name: <Parameter Name>
      type: <Parameter Type>
  query: <Filtering Conditions>
```

Code Listing 1.1. View syntax structure.

This DSL starts with the name given to the view. Then, it specifies which Entity Set the view acts on. Following this definition, we find several attributes:

- *Permissions*: Declares the list of permissions the user invoking the query must have in order to retrieve the results. These permissions can be specified at interface level (the user must have permission to execute the interface, e.g. *Truck::get_status*) or using a role (the user must have assigned a specific role, e.g. *Role::FleetManager*).
- *Parameters*: Specifies the entry-point parameters that the user can pass to the view to dynamically limit the results through the following defined query.
- *Query*: Specifies the query that will be applied over the Entity Set instances. An entity will be returned if this query is evaluated as true for that instance. Interfaces can be used to access entity attributes; only interfaces existing in the Entity Set definition specified in the view can be used. Value Sets instances and parameters specified in the view can also be used to be compared with the entity attributes through comparison operators.

An example of a view that filters trucks by status can be seen in Listing 1.2. It selects entities of type Truck, and users require access permission to their location to run the query. It receives a zone parameter of the Polygon value set and the query is defined as the entities (of the selected type Truck) which get_location() interface returns a Location value contained by the zone.

```
get_trucks_inside_zone:
  entity: Truck
  permissions:
    - Truck::get_location
  parameters:
    - name: zone
      type: Polygon
  query: zone.contains(entity.get_location())
```

Code Listing 1.2. Truck inside a zone view.

4.3 Semantics

The semantics of the DSL are defined by creating a transformation from DSL-based specifications to descriptions defined with a well-defined semantic language such as Python. We have chosen Python due to its readability, rich ecosystem of libraries for data processing, and seamless integration with diverse systems.

This allows us to ensure that ERDT views are not only high-level abstractions of queries but also efficient executable artifacts that can be seamlessly integrated into DT systems. The generated code interacts with the respective entity interfaces by invoking the necessary operations, ensures that security constraints are applied, and applies any specified filtering conditions.

The transformation process uses text templates to convert high-level ERDT views into executable queries based on python. These templates are introduced in the next section since we need to know the architecture proposed to execute the generated queries in order to properly understand the templates.

5 Execution Architecture and Code Generation

This section introduces how ERDT views are transformed into executable Python-based queries in detail. However, to properly understand this we need to previously know the architecture in which Python-based queries are deployed and executed, which is shown in Fig. 2.

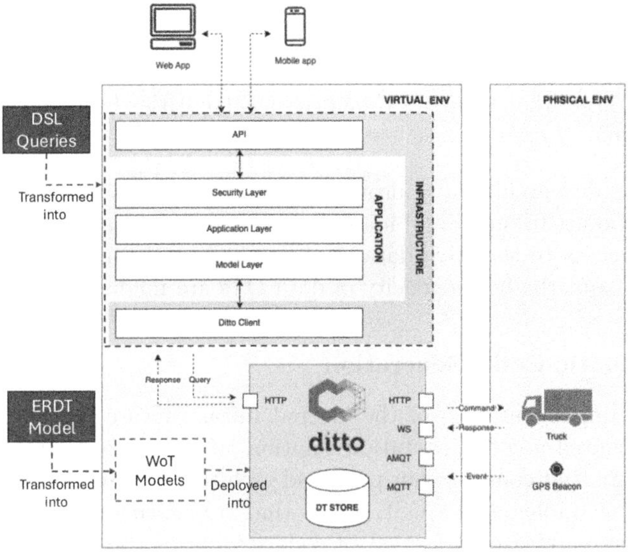

Fig. 2. Architecture.

First, we need to know that ERDT models are transformed into Web of Things (WoT) models [4]. Although this transformation is out of the scope of this paper, some details can be found in [8]. For the purpose of this paper, we just need to know that WoT is a standard defined by the W3C in order to represent physical things in the context of the Internet of Things. In addition, WoT models are directly interpretable by DT platforms such as Eclipse Ditto [1] or the Onesite platform [3]. These platforms allow us to instantiate DTs from WoT modes into runnable environments.

Our proposal leverages on Ditto to act as a data warehouse of our DTs entities and to manage the real-time synchronized of these entities with their physical counterparts. Ditto also exposes interfaces to update and query DT entities by their attributes, which we use to connect our DT system with it.

However, platforms like Ditto lack essential features for DT management such as advanced security policies, complex query mechanisms, and enterprise authentication methods. To solve this problems, we complement Ditto with a layered infrastructure that is generated from the views defined with the DSL presented in this paper. The process of code generation is presented in the next subsection. The generated code implements the following layers:

- API: A HTTP interface to allow users to access the DT views. This layer includes an authentication system to identify each user request.
- Security: A layer that verifies that the user that is invoking the view has the necessary permissions to access it as defined in the ERDT views created with the proposed DSL.
- Application: A layer that orchestrates the querying of entities and runs the querying conditions to obtain a subset of entities.
- Model: A layer that represents the different entities that we have defined for our DT. It implements their attributes and interfaces to be executed by the queries.
- Client: A layer that retrieves the current DT entities from the runtime platform Ditto.

This proposed architecture provides all the functionality required to query DTs: access points to query and manage data, a security layer that prevents not authorized access to sensitive data, and an application and model layers that take into account the heterogeneity of data DTs are made of.

5.1 Automatic Code Generation

To ease the implementation of the layered infrastructure presented above, we have implemented a code generation solution that automatically generates the different layers that comprise our proposed architecture from ERDT views. This code generator implements a DSL parser that transforms the ERDT views into Python objects and executes a set of code templates through Jinja [2], a fast, expressive, extensible templating engine for Python. The source code of the complete code generator can be accessed at a Github repository [14].

As a representative example, Listing 1.3 shows the code template used to generate the application layer, which includes the code that access Ditto in order to retrieve the entities that satisfy a query.

```
from client import DittoClient
{% for entity in entities %}
from model import {{ entity }}
{% endfor %}
ditto_client = DittoClient()
{% for name , view in views . items () %}
def {{ name }}({{ view.get_parameters() }}) -> list[{{ view
                                        .entity }}]:
    entities = ditto_client.query({{ view.entity }})
    matching_entities = []
    for entity in entities:
        if {{ view.query }}:
            matching_entities.append(entity)
    return matching_entities
{% endfor %}
```

Code Listing 1.3. Template for generating application layer code.

Listing 1.4 shows the code generated by this template when it is applied to the view presented in Listing 1.2. This code interacts with the Ditto client in order to obtain the Trucks of the system and filter those that satisfy the condition defined in the query.

```
from client import DittoClient
from model import Truck, Polygon
ditto_client = DittoClient()
def get_trucks_inside_zone(zone: Polygon) -> list[Truck]:
    entities = ditto_client.query(Truck)
    matching_entities = []
    for entity in entities:
        if zone.contains(entity.get_location()):
            matching_entities.append(entity)
    return matching_entities
```

Code Listing 1.4. Application layer code generated from the *get_ trucks_ inside_ zone* query.

Manual Code Completion. It is worth to remark that some implementation details of the Model and Security layers cannot be generated from EDRT views and need to be manually defined. On the one hand, note that the Model layer includes classes that implement ERDT entities from the data stored by Ditto. Ditto stores this data in structures that highly depend on the physical devices that provide the data. For instance, depending on the device used to detect the location of a physical thing, different fields can be used to provide it (e.g., lat, lon, latitude, longitude, lat.degrees, lat.minutes, etc.). This information cannot be automatically generated from ERDT views. Thus, mappings between Ditto data structures and ERDT entities must be manually defined. As a representative example, Listing 1.5 shows the code of the Model layer that is generated to implement the Truck entity. It is completed with some code added manually (highlighted as comments), which maps the internal structure used by Ditto to store locations with the Location of a Truck.

```
@dataclass
class Truck:
  thing_id: str
  location: Location  # Manually Defined
  @classmethod
  def from_ditto(cls, data) -> "Truck":
    return cls(
        thing_id=data["thingId"],
        location=Location(  # Manually Defined
            latitude=data["attributes"]["latitude"],
            longitude=data["attributes"]["longitude"])
    )
```

Code Listing 1.5. Complete Truck model class.

On the other hand, the data security layer is in charge of checking whether the user executing a query has the correct permissions. This can be done by using different authentication systems such as ADFS (Active Directory Federation Services) or SAML (Security Assertion Markup Language). Thus, the interaction with these systems must be manually implemented in the security layer.

6 Proof of Concept Validation

As a proof of concept, we have developed a DT for the real case study used as a running example, which is based on the logistics operations of Mercadona. This company has a production-ready fleet management system that controls a large fleet of delivery trucks. This system is built on a combination of data sources including a traditional relational database management system (RDBMS) and different proprietary third-party software solutions that manage real-time data feeds, such as GPS tracking and maintenance logs. These heterogeneous data sources make the architecture complex and depends on interconnected tables representing logistics entities such as trucks, drivers, routes, and sensor data from deliveries.

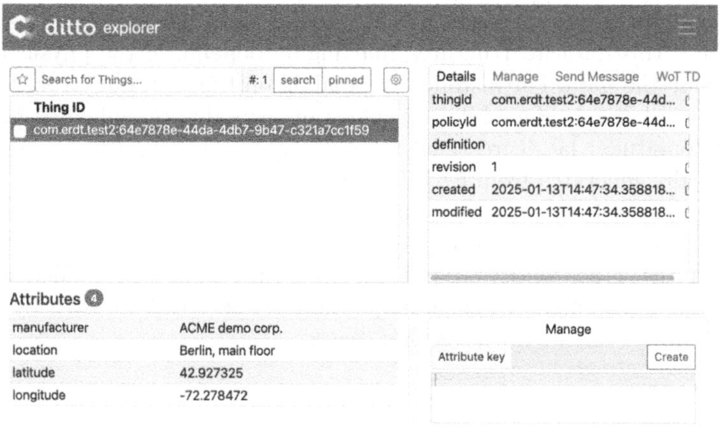

Fig. 3. Ditto Dashboard.

To develop a DT that improves the management and monitoring of the heterogeneous data sources we created the corresponding ERDT model (a partial view of it is presented in Fig. 1), generated the equivalent WoT models, and deployed them into Ditto. Figure 3 shows a snapshot of the Ditto dashboard after deploying these models. In particular, it shows the data obtained in real time for a specific Truck including attributes such as latitude and longitude.

Note that Ditto plays the role of middleware between the DT and both the physical things and third-party systems. However, it lacks query mechanisms to execute the complex queries that were required by the fleet management system. Some examples of these queries are the following:

- Identifying trucks that are inside specified coordinate polygons.
- Retrieving the location of active trucks.
- Listing trucks that are due for maintenance.
- Listing trucks that have been idle at the same location for more than 15 min.

Thus, we used the proposed DSL and code generation strategy to define and implement these queries. For example, the first query was defined in Listing 1.2. The previous sections show some of the generated code to implement the application and model layers that support this query. Figure 4 shows the Open API interface that is generated as an implementation of the API layer. As we can see, it exposes the query through an HTTP API. It includes an end-point that receives a list of coordinate points to calculate and return the list of trucks that are inside the polygon they form.

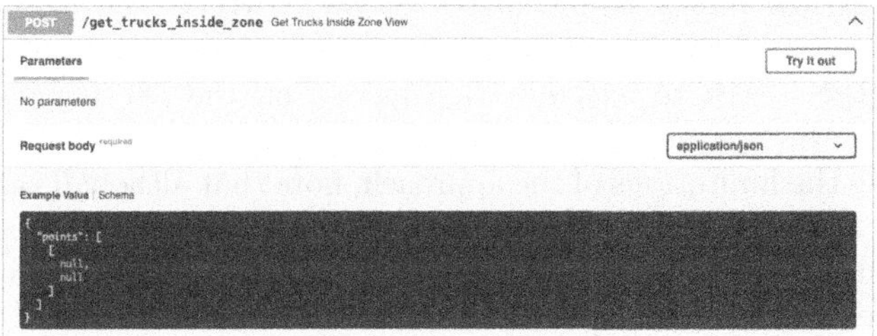

Fig. 4. OpenAPI Dashboard.

All the software artifacts created to develop this DT, as well as the resulting implementation, are available in a Github repository [7]. A video showcasing this process is also available [9].

Finally, it should be noted that this proof-of-concept validation demonstrates that our proposal can effectively handle real use cases in production environments. It also demonstrates from a pragmatic point of view that the proposed architecture and model transformation allow: (1) transforming conceptual view definitions into executable queries; (2) improving current DT platforms with advanced functionality; (3) supporting the application of security restrictions; and (4) exposing functionality through a documented REST API. However, it is important to acknowledge that this evaluation only demonstrates the feasibility of the proposed method and does not provide a thorough comparison or assessment of its effectiveness against alternative approaches. This improved validation will be carried out in future work.

7 Discussion

The ERDT model together with both the DSL for defining queries and the proposed code generation strategy provide a solid model-driven development framework for developing DTs that improves the state of the art.

On the one hand, this solution provides greater abstraction and modularity. ERDT models encapsulate complex interactions and data through entities and interfaces. They allow developers to work with high-level concepts without the need to manage complex details, simplifying development and maintenance and offering a distinct advantage in environments involving complex data interactions.

On the other hand, it facilitates the implementation of DTs. The proposed automatic code generation strategy allows practitioners and researchers to easily implement and deploy DT systems, abstracting away technical complexities. First, although this is not the scope of this paper, ERDT models can be automatically transformed into WoT models that can be directly deployed into platforms such as Ditto, obtaining a fully operative infrastructure to synchronize physical things with the digital counterparts. Second, ERDT views are transformed into an executable Python implementation based on a layered architecture that provides a REST API to executes the queries on the DT from any external application.

Regarding the limitations of the approach, note that although the majority of the software artifacts required for the implementation of a DT are automatically generated, there are certain aspects of the process that require manual intervention. This need for manual code completion introduces potential limitations to the overall approach.

First, manually defining parts of the Model layer, such as the mapping between Ditto data structures and ERDT entities, can be time-consuming and error-prone. Second, the manual implementation of the security layer to integrate with authentication systems such as ADFS or SAML also poses challenges. Incorporating such systems requires detailed knowledge of the specific protocols and standards used by each system. Moreover, implementing the security layer by hand increases the risk of introducing security vulnerabilities due to human error.

The requirement to write custom code reduces the potential efficiency gains provided by the model-driven approach and introduces variability in the quality and consistency of the generated artifacts. In future work, we will enhance the generation process to better support the automatic handling of heterogeneous data structures and security requirements.

8 Conclusions and Further Work

In this article, we have presented a reference architecture and an automatic code generation strategy to implement and deploy ERDT query views in operational environments. We have also shown how this solution addresses the challenges of complex DT implementations by showing a real case study at Mercadona.

Our ongoing work focusses on the development of a visual tool that provides a complete model transformation framework to automate the generation of code from ERDT models and views to operational systems, with special attention to aspects such as security and scalability.

The ERDT model together with the DSL for defining queries represent an important step forward in the development of DTs. It provides a solid basis for building complex and secure systems. As DTs continue to evolve and become more widespread in different sectors, the need for efficient query mechanisms will grow. This work contributes to meeting that need and opens up new opportunities for research and practical application.

References

1. Eclipse ditto - open source framework for digital twins in the IoT. https://eclipse.dev/ditto/
2. Jinja. https://palletsprojects.com/projects/jinja/
3. Onesite platform. https://www.onesait.com/technology/platform
4. Web of Things (WoT)—w3.org. https://www.w3.org/WoT/
5. Akehurst, D.H., Bordbar, B.: On querying UML data models with OCL. In: Gogolla, M., Kobryn, C. (eds.) UML 2001. LNCS, vol. 2185, pp. 91–103. Springer, Heidelberg (2001). https://doi.org/10.1007/3-540-45441-1_8
6. Bordeleau, F., Combemale, B., Eramo, R., van den Brand, M., Wimmer, M.: Towards model-driven digital twin engineering: current opportunities and future challenges. In: Babur, Ö., Denil, J., Vogel-Heuser, B. (eds.) ICSMM 2020. CCIS, vol. 1262, pp. 43–54. Springer, Cham (2020). https://doi.org/10.1007/978-3-030-58167-1_4
7. Carrión, E., Pastor, O., Valderas, P.: Demo: Erdt views. https://github.com/EmilioCarrion/digital-twins/tree/main/Demos/ERDT%20Views
8. Carrión, E., Pastor, O., Valderas, P.: Research report: model-driven development for digital twins. https://github.com/EmilioCarrion/digital-twins/blob/main/Research%20reports/Model%20Driven%20Development%20for%20Digital%20Twins%20Report.pdf
9. Carrión, E., Pastor, O., Valderas, P.: Video: Erdt views. https://media.upv.es/player/?id=f4ca0d40-d707-11ef-b31b-0ff72988f975
10. Carrión, E., Pastor, O., Valderas, P.: Conceptual modelling method for digital twins. In: Maass, W., Han, H., Yasar, H., Multari, N. (eds.) ER 2024. LNCS, vol. 15238, pp. 417–435. Springer, Cham (2025). https://doi.org/10.1007/978-3-031-75872-0_22
11. Carrión, E., Pastor, O., Valderas, P.: Querying digital twin models. In: 20h International Conference on Evaluation of Novel Approaches to Software Engineering (Publication Pending) (2025)
12. Chen, P.: The entity-relationship model-toward a unified view of data. ACM Trans. Database Syst. **1**(1), 9–36 (1976). https://doi.org/10.1145/320434.320440
13. Digital twin consortium: digital twin definition (2024). https://www.digitaltwinconsortium.org/glossary/glossary/#digital-twin
14. EmilioCarrion: GitHub - emiliocarrion/erdttowot. https://github.com/EmilioCarrion/ERDTtoWoT
15. Espinazo Pagán, J., García Molina, J.: Querying large models efficiently. Inf. Softw. Technol. **56**(6), 586–622 (2014). https://doi.org/10.1016/j.infsof.2014.01.005

16. Gogolla, M., Richters, M.: On constraints and queries in UML. In: Schader, M., Korthaus, A. (eds.) The Unified Modeling Language, pp. 109–121. Physica-Verlag, Heidelberg (1998)
17. Habela, P., Kaczmarski, K., Stencel, K., Subieta, K.: OCL as the query language for UML model execution. In: Bubak, M., van Albada, G.D., Dongarra, J., Sloot, P. (eds.) ICCS 2008. LNCS, vol. 5103, pp. 311–320. Springer, Heidelberg (2008). https://doi.org/10.1007/978-3-540-69389-5_36
18. Jia, W., Wang, W., Zhang, Z.: From simple digital twin to complex digital twin part I: a novel modeling method for multi-scale and multi-scenario digital twin. Adv. Eng. Inform. **53**, 101706 (2022). https://doi.org/10.1016/j.aei.2022.101706
19. Jones, D., Snider, C., Nassehi, A., Yon, J., Hicks, B.: Characterising the digital twin: a systematic literature review. CIRP J. Manuf. Sci. Technol. **29**, 36–52 (2020)
20. Kirchhof, J.C., Michael, J., Rumpe, B., Varga, S., Wortmann, A.: Model-driven digital twin construction: synthesizing the integration of cyber-physical systems with their information systems. In: Proceedings of the 23rd ACM/IEEE International Conference on Model Driven Engineering Languages and Systems, MODELS 2020, pp. 90–101. Association for Computing Machinery, New York (2020). https://doi.org/10.1145/3365438.3410941
21. Kleppe, A.: Software Language Engineering: Creating Domain-Specific Languages Using Metamodels. Pearson Education (2008)
22. Kolovos, D.S., et al.: A research roadmap towards achieving scalability in model driven engineering. In: Proceedings of the Workshop on Scalability in Model Driven Engineering, BigMDE 2013, pp. 1–10. Association for Computing Machinery, New York (2013). https://doi.org/10.1145/2487766.2487768
23. Mernik, M., Heering, J., Sloane, A.M.: When and how to develop domain-specific languages. ACM Comput. Surv. **37**(4), 316–344 (2005). https://doi.org/10.1145/1118890.1118892
24. Schroeder, G.N., Steinmetz, C., Rodrigues, R.N., Henriques, R.V.B., Rettberg, A., Pereira, C.E.: A methodology for digital twin modeling and deployment for industry 4.0. Proc. IEEE **109**(4), 556–567 (2021).https://doi.org/10.1109/JPROC.2020.3032444
25. Stein, D., Hanenberg, S., Unland, R.: Query models. In: Baar, T., Strohmeier, A., Moreira, A., Mellor, S.J. (eds.) UML 2004. LNCS, vol. 3273, pp. 98–112. Springer, Heidelberg (2004). https://doi.org/10.1007/978-3-540-30187-5_8
26. Tao, F., Xiao, B., Qi, Q., Cheng, J., Ji, P.: Digital twin modeling. J. Manuf. Syst. **64**, 372–389 (2022)
27. Tao, F., Zhang, H., Liu, A., Nee, A.: Digital twin in industry: state-of-the-art. IEEE Trans. Industr. Inf. **15**(4), 2405–2415 (2019). https://doi.org/10.1109/TII.2018.2873186

Framework for the Design and Development of an Automated Virtual Data Integration System for Genomic Data

Ana León Palacio(✉)

Valencian Research Institute for Artificial Intelligence (VRAIN), Universitat Politècnica de València, Valencia, Spain
aleon@vrain.upv.es

Abstract. Designing correct data integration systems is crucial, especially in AI projects where the datasets for analysis must be consistent, homogeneous, up-to-date, representative, and free from significant errors or biases. This is especially true in complex domains such as genomics, where integrating and querying many data sources with different data schemas, quality, and formats is required. This work presents a framework for creating an Automated Virtual Data Integration System combining different solutions presented in the literature and adapted to the particularities of genomic data.

Keywords: AVDIS · Genomics · ETL · Virtual Data Integration

1 Introduction

Using genomic data in Precision Medicine (PM) facilitates elucidating complex disease mechanisms at the molecular level. The analysis of genomic data can lead to the development of targeted therapeutic approaches, the design of individualized treatment plans, and the anticipation of possible undesired effects. Nevertheless, genomic data management poses significant challenges that can be summarized as [18]:

- Genomic data is spread in hundreds of repositories with different formats. Each time a new source is added to the system, its structure must be transformed to match the global schema. Due to the complexity of the data stored in some repositories, this can be a time-consuming task that requires expert knowledge to map the source to the global schemas. Furthermore, any change in the source schema requires adapting the integration processes to include the changes in the mappings to the global schema, which requires a huge maintenance effort and may compromise the consistency of the system.
- Genomic data sources are volatile, which means that new data sources are created while others disappear. This increases the effort of adding and removing sources to the system, as well as the task of maintaining the integrated datasets updated.

J. Grabis et al. (Eds.): RCIS 2025, LNBIP 547, pp. 295–310, 2025.
https://doi.org/10.1007/978-3-031-92474-3_18

- Genomic data is characterized by the huge size of the datasets and the number of sources to integrate. Therefore, tasks such as identifying duplicated records become a time-consuming task that constitutes a bottleneck in complex systems.
- Identifying connections and relationships between many different datasets requires performing queries over the integrated data in many different ways. These queries must be flexible enough to adapt to the changing nature of data and their schemas.

Each time a new PM project starts, the same question arises: How can these heterogeneous data be integrated, queried, and analyzed to obtain meaningful insights? Answering this question is especially relevant when designing projects that use Artificial Intelligence (AI) because the datasets must be consistent, homogeneous, up-to-date, representative, and free from significant errors or biases.

Data integration *is the problem of combining data residing at different sources and providing the user with a unified view of these data* [16]. To ease the addition of new sources, improve query flexibility, and reduce the maintainability efforts of the Extract-Transform-Load (ETL) processes related to genomic data, a Virtual Data Integration System (VDIS) combined with an incremental approach can be helpful. A VDIS relies on data as they are in their original sources, builds an intermediate infrastructure to provide virtual data integration, and provides means to retrieve data at query time [10]. Furthermore, following an incremental approach can allow adding new sources without affecting the already existing ones and reduce the number of entities to be compared when measuring similarity. Finally, data fusion techniques can be used to evaluate the accuracy, the freshness, and the dependence of the sources.

Despite the advantages that a VDIS can provide to genomic data integration, the main problem of these systems is the high degree of manual work required to perform integration tasks such as the mappings between the source and the global schemas. Therefore, a key aspect of any VDIS is the number of tasks that can be automated to reduce the integration effort as much as possible.

In this work, the previously mentioned aspects are considered, and a theoretical proposal for creating an Automated Virtual Data Integration System (AVDIS) is proposed, based on joining different and current solutions found in the literature. The proposed AVDIS serves two purposes: first, to address how heterogeneous data can be integrated, queried, and analyzed for meaningful insights; second, to offer a solution that enhances automation in the ETL process.

The methodology used to develop this work is the Design Science Research Methodology (DSRM) for Information Systems [22]. The selected DSRM involves a rigorous process to design artifacts to solve observed problems, make research contributions, evaluate the designs, and communicate the results to appropriate audiences. It consists of six activities: Problem identification and motivation, Definition of the objectives for a solution, Design and development, Demonstration, Evaluation, and Communication. Since the result of this research work is to propose a theoretical approach, this document describes the following activities: problem identification and motivation, the definition of the objectives for a solution, and the design of the solution. The development, demonstration, and evaluation tasks are considered in future work.

Therefore, after this introduction, Sect. 2 reviews related works that address data integration challenges and define the objectives of the proposed approach, aligning with the first and second activities of the DSRM. Section 3 aligns with the third task of the DSRM by describing the fundamental constructs to design a VDIS according to the tasks required for managing genomic data and proposing a framework for automating these tasks. The AVDIS framework represents the artifact and the main outcome of this research. The document ends with the conclusions and future work.

2 Related Works

Data integration encompasses three main tasks: schema integration, record linkage, and data fusion. Much literature and solutions have been produced for each of them, but in an isolated manner.

For schema integration, the most up-to-date work is the one proposed by [10]. In this work, the authors propose a VDIS based on a bottom-up approach, where the structure of the source schemas is extracted from the datasets and generates a globally integrated schema. The authors also provide a fully automated and open-source tool. This approach avoids the possibility of having a conceptual model of the domain as the target schema, which is interesting to ensure semantic integrity and represent data in a standardized way. This is important since works such as the ones developed by the GA4GH[1] are putting efforts into defining standards for representing genomic data in a structured way. Other works also propose conceptual models for representing different aspects of the domain to better capture the semantics and structure of the genomic data [2,11,17]. The proposal presented in this work is based on the approach proposed by [10] but extended to provide the possibility of considering domain conceptual models to ensure the semantic integrity and standardization of the data.

Record linkage and data fusion solutions are mainly focused on providing algorithms such as the ones compared by [21] and [20]. These algorithms are not fed with domain information, which means that it is quite complicated to resolve data conflicts considering the characteristics of the data sources such as currency, dependency, or reliability. The system proposed in this paper complements the schema integration with the record linkage and data fusion tasks to provide an already treated dataset to the user as an answer to any question performed over the system. It also includes domain knowledge information about the data to improve performance and results during record linkage and data fusion.

Based on the gaps identified in the related works, we address the second task of the DSRM by defining the objective of our proposal: to design and AVDIS as an artifact that addresses two key challenges: first, designing a framework for

[1] https://www.ga4gh.org/.

integrating, querying, and analyzing heterogeneous data to extract meaningful insights; and second, enhancing automation within the ETL process to improve efficiency and scalability.

3 Virtual Data Integration System

A VDIS integrates the data sources by implementing three levels of tasks [8,10], as shown in (Fig. 1).

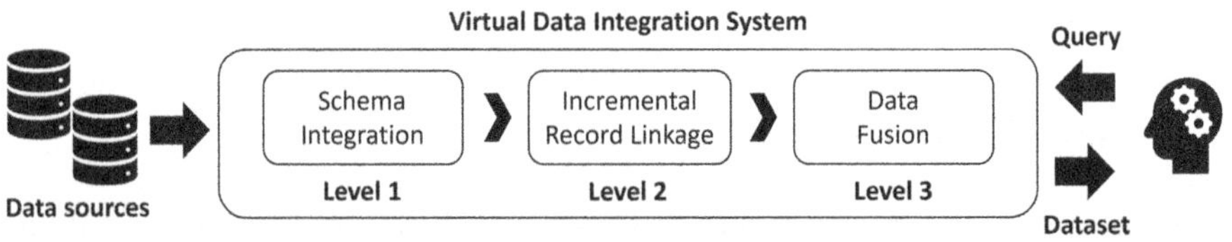

Fig. 1. The three levels of tasks implemented by a VDIS.

In Level 1, the VDIS needs to resolve heterogeneity at the schema level by establishing semantic mappings between the contents of disparate data sources and a global schema that represents the domain knowledge (schema integration). In Level 2, the VDIS needs to resolve heterogeneity at the instance level by detecting records that refer to the same real-world entity (incremental record linkage). In Level 3, the VDIS needs to combine records that refer to the same real-world entity by fusing them into a single representation (data fusion). As a result of a query performed over the system, the user obtains an integrated dataset.

3.1 Level 1: Schema Integration

The schema integration in a VDIS must support flexible and on-demand extraction of data from the source schemas and their homogenization and integration into a global schema [10]. The general workflow of the schema integration level when the global schema is unknown covers the following main tasks [20]:

- Pre-processing: Since schemas are commonly expressed in different languages, a normalization step is necessary to translate them from one language or formalism into a uniform representation without changing their semantics.
- Schema matching: It generates correspondences between schemas, i.e. relationships between one or more elements of one schema and one or more elements of the other [3].
- Schema merging: It merges the input schemas into a new integrated global schema.
- Post-processing: It evaluates, repairs, and refines the resulting schema by checking its consistency, and coherence, resolving its cycles, and pruning its redundancies [20].

Having a conceptual model representing the knowledge to be integrated reduces the tasks required during the schema integration. The schema merging and the post-processing tasks are no longer required because a new data schema is not needed and the global schema is already consistent and coherent with the domain knowledge. Only the mappings between the source schemas and the global schema are required.

3.2 Level 2: Incremental Record Linkage

Data sources may contain different, complementary, or additional information, and a combination of these provides more complete information. The goal of the incremental record linkage (also known as entity resolution) is to identify records that refer to the same logical entity across different data sources when they cannot be linked via a unique identification number [9].

Record linkage can be seen essentially as a clustering problem, where each cluster contains records that correspond to a single distinct real-world entity [12] and which general workflow has three main steps:

- Blocking: This step aims to reduce the number of record pairs to be compared by building an index or block key for each record [6,21,24].
- Similarity computation: This step compares records of the same block according to a set of matching rules and creates a similarity graph where each node represents a record and each edge represents the similarity between two records.
- Graph clustering: This step clusters the nodes according to the results of the similarity graph.

3.3 Level 3: Data Fusion

Data fusion refers to resolving conflicts from different sources and finding the truth that reflects the real world increasing the correctness of the integrated data [9]. There are two types of data conflicts: uncertainty and contradictions [4,8]. Uncertainty is caused by missing information (null values or missing attributes in a source) and contradiction is caused by different sources providing different values for the same attribute of a real-world entity. Conflicts can be ignored, avoided, or resolved.

4 A Framework for Automating a VDIS

According to the three levels of tasks described in the previous section, Fig. 2 presents the proposed framework to complete the design of an AVDIS to automatically integrate and query multiple and diverse data sources. Additionally, the AVDIS must have a Rewriting System to translate the user queries over the global schema into queries over the corresponding data sources.

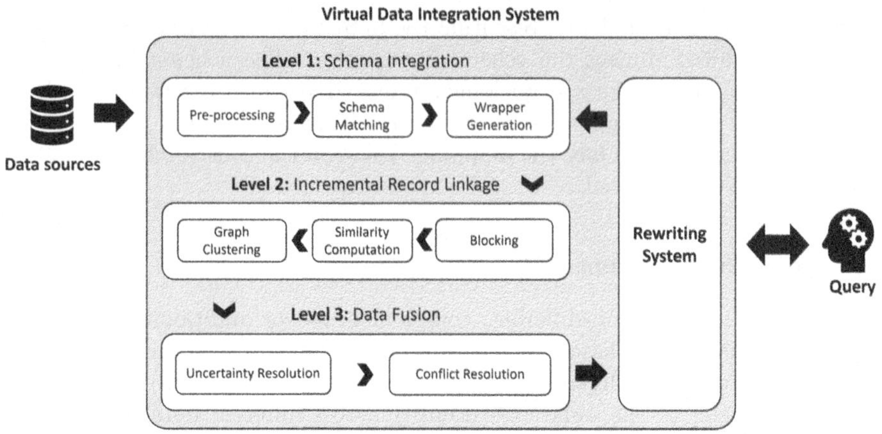

Fig. 2. AVDIS workflow.

The tasks of the Schema Integration level are covered by the Pre-processing, the Schema Matching, and the Wrapper Generation modules. The tasks of the Incremental Record Linkage level are covered by the Graph Clustering, the Similarity Computation, and the Blocking modules. Finally, the tasks of the Data Fusion level are covered by the Uncertainty Resolution and Conflict Resolution tasks.

4.1 Automating the Schema Integration

The schema integration pipeline depicted in Fig. 3 is based on the approach proposed by [10], where each module is divided into a set of consecutive steps, from the inference of the data source schema to the generation of the wrappers that allow retrieving the data from the sources.

The schema integration process starts with the pre-processing of the data and/or meta-data associated with the data sources to be queried. Source data are commonly expressed in different formats (e.g., XML, JSON, CSV). Therefore, Bootstrapping and Normalization tasks are necessary to infer the schemas from the data sources and translate them from one language or formalism to a uniform representation (or canonical data model) without changing their semantics.

To such an aim, [10] suggests representing the source schemas as typed graphs according to a set of metamodels, one metamodel for each type of data source format to be translated. The canonical schema can be represented as a Resource Description Framework Schema (RDFS) according to another metamodel, as depicted in Fig. 4.

The transformation of the typed graph into the canonical model is done by applying a set of transformation rules, which details can be found in [10]. At the end of the pre-processing task, the data from the data sources have been normalized and a canonical schema from each data source has been generated. Fig 5 shows an example of transforming a JSON schema into a canonical schema represented as an RDFS.

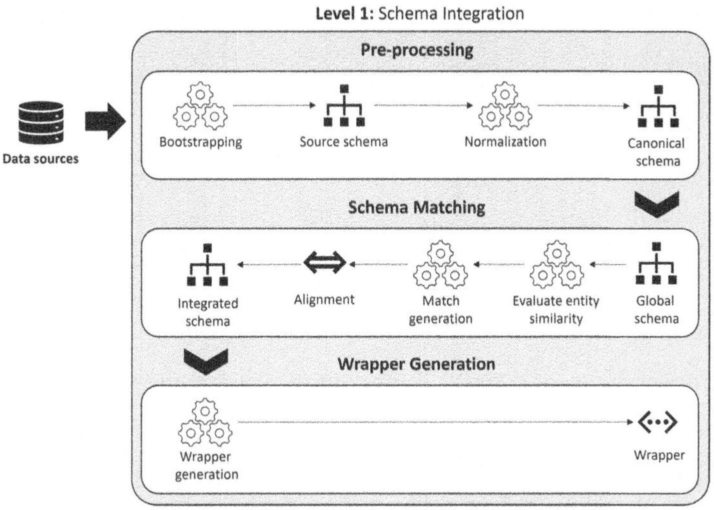

Fig. 3. Schema integration pipeline.

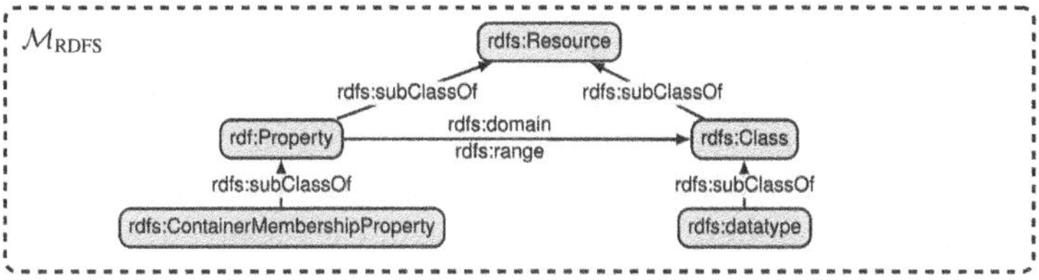

Fig. 4. RDFS metamodel used to represent a canonical schema [10].

Once the canonical schemas are generated, the next stage involves the identification of semantic correspondences between the elements of each canonical schema and the elements of the unified representation of the domain (e.g., the Conceptual Schema of the Genome). The schema matching process takes as input the set of schemas from the pre-processing task, and the specification of the global schema, and returns a set of correspondences between entities (classes and attributes) which constitutes an alignment.

The matching task measures the similarity between entities by combining three different metrics related to syntactic, semantic, and structural similarity [5]. Each correspondence commonly has as components the unique identifier of the correspondence, the members of the correspondence, the expressions required to express complex relations and transformations, and a confidence measure assigning a degree of trust in the identified relation [20]. Correspondences can be described using EDOAL[2] (Expressive and Declarative Ontology Alignment Language). Finally, a threshold is applied to the confidence measure to determine

[2] https://moex.gitlabpages.inria.fr/alignapi/.

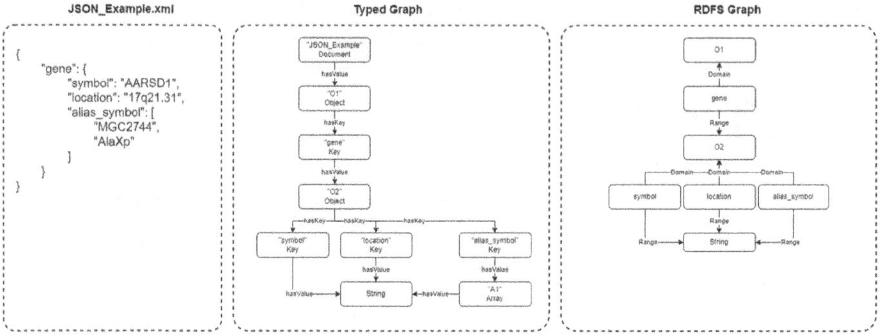

Fig. 5. Transformation of a JSON example into a typed graph and a canonical model.

the correspondences that will be included in the alignment. Figure 6 shows the specification of a global schema that will be used as a running example.

Once the alignment is finished, an integration graph containing the mappings between the source schemas and the global schema is created. Figure 7 shows the integration graph resulting from the matching between three different schemas and the global schema shown in Fig. 6.

The Wrapper Generation task consists of creating programs that allow retrieving data from the sources. Following a Local as View (LAV) strategy, each local schema is described as a function over the global schema, describing which data of the global database are present in each source [15]. This strategy allows adding sources to the system independently of other existing sources. The wrappers translate queries over the global schema into requests understood by specific data sources. The role of the wrappers is to allow the global schema to see each data source as relational, no matter which actual format it uses [1]. Figure 8 shows how three wrappers cover different parts of the global schema in the integrated graph. More details about the implementation of the wrappers can be found in [1,15,19].

4.2 Automating the Incremental Record Linkage

The Incremental Record Linkage is executed after a user executes a query on the VDIS. The result of this query is a dataset containing the integrated data coming from the sources and represented according to the global schema. The Incremental Record Linkage pipeline depicted in Fig. 9 is divided into the Blocking, the Similarity Computation, and the Graph Clustering tasks.

The record linkage process starts with an indexing technique commonly called blocking, which splits the datasets into non-overlapping blocks, such that only records within each block are compared with each other [6]. To such an aim, a classical blocking approach can be used since the description of the domain knowledge is provided by the global schema (i.e., the CSG).

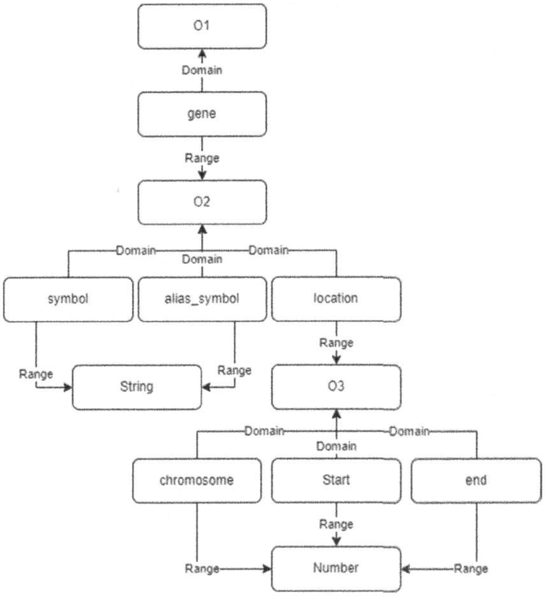

Fig. 6. Example of global schema in RDFS format that represents genes with the string attributes *symbol* and *alias_ symbol*, and the *location* attribute represented as an object with *chromosome*, *start*, and *end* as number attributes.

To build the blocks, a set of rules must be defined so an index or blocking key is assigned to each record. This step requires domain knowledge to select the most appropriate blocking keys. In this case, the CSG may contain the definition of which attributes are required to identify an entity univocally, or they can be defined during the specification of the CSG by the domain expert. For example, the set of rules to define the blocking keys for each entity of the global schema shown in Fig. 6 could be $R1 : gene \rightarrow symbol$ and $R2 : location \rightarrow chromosome \cup start \cup end$. $R1$ defines *symbol* as the attribute used as the blocking key for each gene entity and *R2* defines the union of attributes *chromosome*, *start*, and *end* as the blocking key to identify each location entity.

The blocking key can be based on a single record field (attribute), or the concatenation of values from several fields. To overcome errors and increase accuracy, several blocking keys based on different record fields can be generated (i.e., blocking schema). Matching records have at least one blocking key in common and will be inserted into the same block as shown in Fig. 10. It is crucial that this calculation be scalable since it must be checked for all pairs of records. When all records are indexed with their corresponding blocking keys, they are inserted into appropriate index data structures.

After blocking, all entities of a block are pairwise compared with each other. For each entity pair, the similarity of their attribute values is calculated according to a set of matching rules that specify the required minimal similarity for the

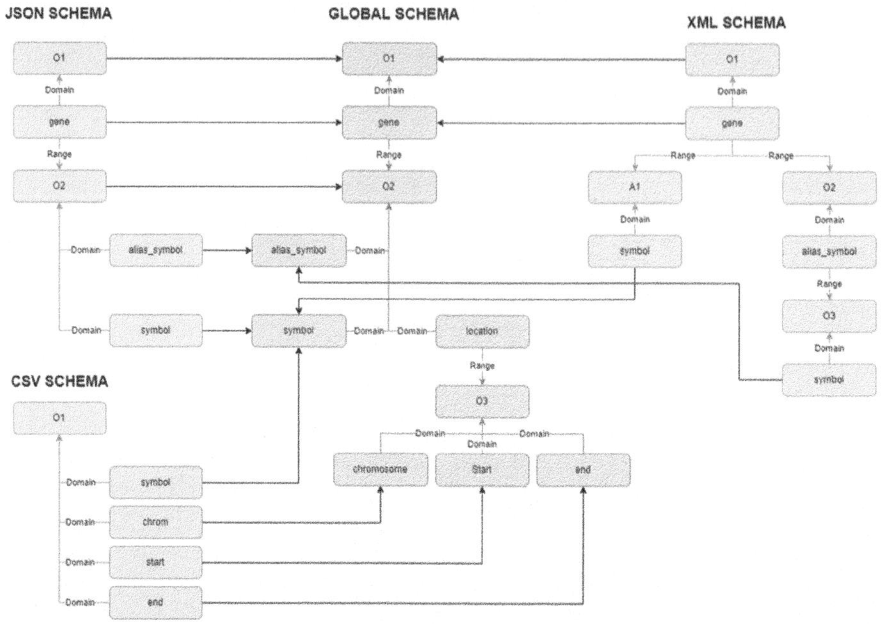

Fig. 7. Example of integration graph.

considered attributes. For example, using the blocks obtained in the example of Fig. 10, for each pair of records r_i and r_j, some similarity rules could be $SR1$: $r_i.symbol == r_j.symbol$, $SR2$: $r_i.chromosome == r_j.chromosome$, $SR3$: $r_i.start == r_j.start$, and $SR4$: $r_i.end == r_j.end$. Therefore, the similarity measure of each pair of records could be the sum of the similarity rules. This step requires domain knowledge to define the similarity rules. The similarity values are used in the following clustering step to decide whether or not a pair of entities match.

The output of this task is the set of matching entity pairs together with a similarity value per link. This output is stored as a similarity graph (see Fig. 11) where entities are represented as vertices and match links as edges [23].

Finally, clustering algorithms typically try to group entities such that the similarity between entities within a cluster is maximized while the similarity between entities of different clusters is minimized [23]. Each cluster is represented by a cluster graph with the clustered entities and intra-cluster similarity links. In this work, we propose using the CLIP algorithm [23] for the graph clustering stage. An example of the clusters created for block 1 is shown in Fig. 11.

4.3 Automating the Data Fusion

Once the integrated data is clustered, the next level is to solve the conflicts between the entities of each cluster to get a single merged entity. As shown in Fig. 12, the Data Fusion pipeline is divided into the Uncertainty Resolution and

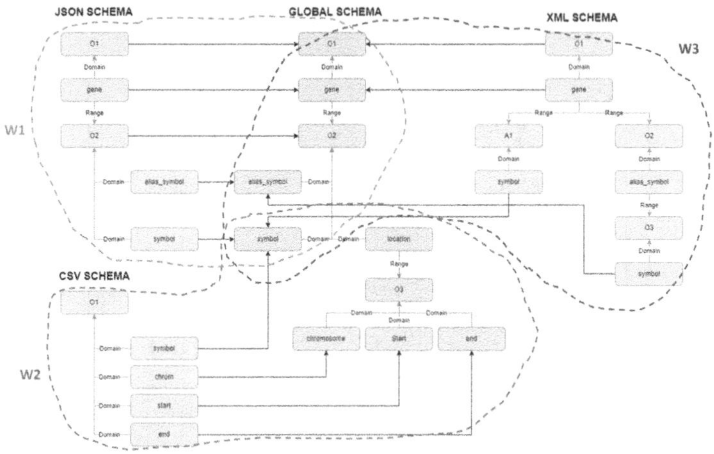

Fig. 8. Example of how wrappers (W1, W2, and W3) cover different parts of the global schema in an integration graph.

the Conflict Resolution tasks. Once each cluster is processed, the results are joined to obtain the final dataset.

Uncertainty resolution is based on solving the presence of null values or missing attributes in one source. At this stage, the entities are represented according to the global schema, which means that the missing attributes in a source are represented as null values. Therefore, the strategy to solve uncertainty can focus on solving the following scenarios [4]:

- Presence of exact tuples which requires identifying those entities having the same values in all of the attributes.
- Presence of subsumed tuples, i.e. tuples having more null values.
- Presence of complement tuples, i.e. tuples with null values and not contradictions.

The challenge in solving uncertainty is how to resolve these conflicts efficiently and in a meaningful way. To such an aim, relational attributes such as *join* and *union* have been traditionally used. The presence of exact tuples is basically a redundancy-solving problem. For example, the *full disjunction* operator is a variation of the *join* operator that maximally combines tuples from connected relations while preserving all information in the relations [7].

The result of applying *full disjunction* to Alice records in Table 1 would be the first record. This operator solves the presence of subsumed and complemented tuples while preserving uniqueness.

Conflicts or contradictions are caused by different sources providing different values for the same attribute of a real-world entity. Solving contradictions requires considering the accuracy, the freshness, and the dependence of the sources. This information is domain-dependent which means that requires expert knowledge to define the conflict rules and information about the source of each record.

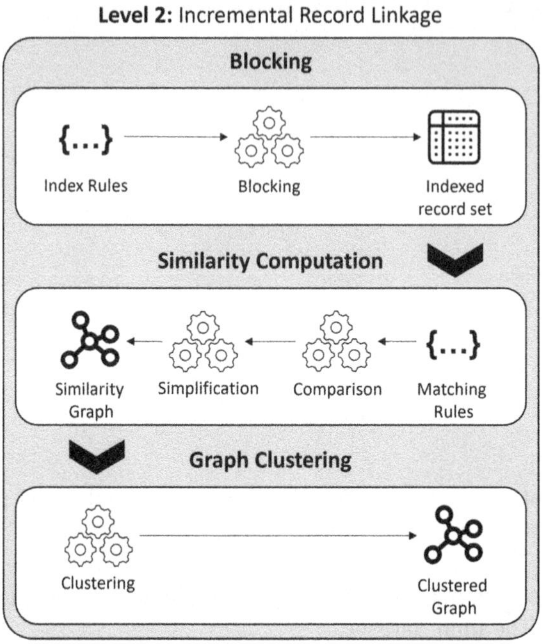

Fig. 9. Incremental record linkage pipeline.

Table 1. Example dataset with duplicated records.

	Name	Age	Room
1	Alice	23	A764
2	Charles	32	S432
3	Alice	23	
4	Alice	23	A764

The conflict rules define the actions to be performed in the following scenarios:

- The conflict requires computing a new value (e.g., max, min, merge, or join). In this case, the formula to compute the new value is required. For example, when merging Alice records in Table 2, a rule for the value of the *online_ time* column could be *online_ time = MAX(online_ time)*, and the result would be record 4.
- The conflict requires selecting one of the provided values according to the priority of the sources. For example, when merging Charles records in Table 2, if source 3 has more accurate data for a user department a rule for the *Department* column could be *Department = Department WHERE Source == "Source_ 3"* and the result would be record 3.

	Record Fields					Blocking Schema	
ID	symbol	alias_symbol	chromosome	start	end	R1	R2
ID1	AARSD1	MGC2744, AlaXp				AARSD1	
ID2	PSEN1		14	73136417	73223691	PSEN1	14-73136417-73223691
ID3	AARSD1	MGC2744, AlaXp				AARSD1	
ID4	AARSD1		17	42950525	42964453	AARSD1	17-42950525-42964453

Block 1		
ID	R1	R2
ID1	AARSD1	
ID3	AARSD1	
ID4	AARSD1	17-42950525-42964453

Block 2		
ID	R1	R2
ID2	PSEN1	14-73136417-73223691

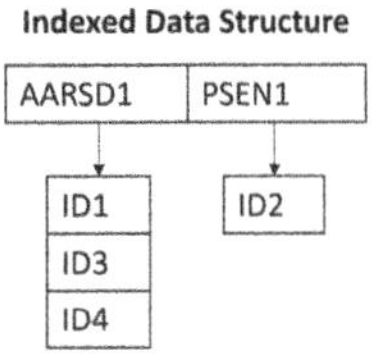

Fig. 10. Example of blocking keys and how records are indexed and distributed in blocks.

Block 1		
Record 1	Record 2	Similarity
ID1	ID3	4
ID1	ID4	1
ID3	ID4	1

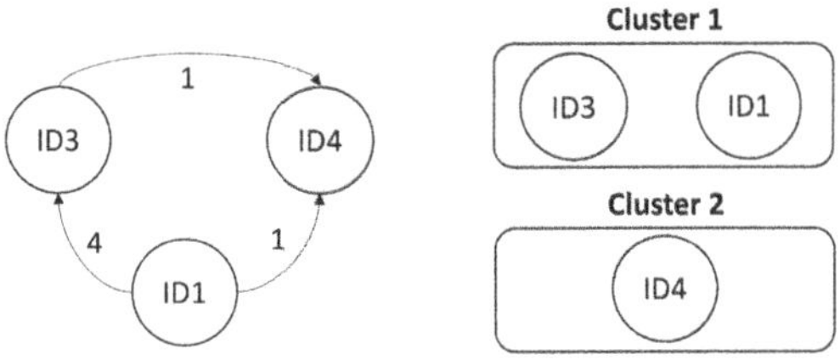

Fig. 11. Example of similarity graph and clusters for entities in block 1.

At the end of this task, the dataset is ready to be sent to the user as a result of the performed query.

4.4 Rewriting System

The user interacts with the VDIS by executing queries over the global schema and getting the corresponding answers. To achieve this functionality, user queries must be translated into a set of specific queries over the sources, considering the particular data schema of each source. Therefore, the system must provide the required rewriting algorithms. Examples of rewriting algorithms can be found in [13,14].

Table 2. Example of a dataset with conflict records.

	Name	Age	UserID	online_time	Department	Source
1	Alice	23	A764	1:02:33	D1	Source_1
2	Charles	32	S432	4:20:02	D2	Source_2
3	Charles	32	S432	4:20:02	D3	Source_3
4	Alice	23	A764	2:02:46	D1	Source_4

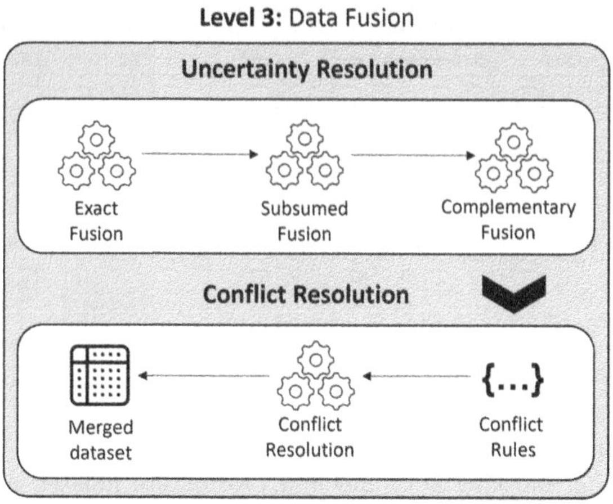

Fig. 12. Data Fusion pipeline.

The Rewriting System uses the wrappers created during the Wrapper Generation stage to extract the required data from the sources. After the Incremental Record Linkage and the Data Fusion levels, the data is returned to the user as the answer to the query.

5 Conclusion and Future Work

Genomic data management implies collecting, integrating, and querying many data sources with different data schemas, quality, and formats. To ease the addition of new sources, improve query flexibility, and reduce the maintainability efforts, a virtual data integration approach can be helpful. To obtain the integrated dataset, the platform can benefit from using an incremental record linkage approach based on clustering techniques to reduce the number of entities to be compared when measuring similarity. Finally, to improve the conflict-solving process, the platform must consider data fusion techniques that evaluate the accuracy, the freshness, and the dependence of the sources. To ensure its usefulness and applicability, the entire process must be as automatic as possible.

Based on the current state of the art, this work describes a theoretical proposal for creating an Automated Virtual Data Integration system (AVDIS). The proposed AVDIS aims to solve or mitigate the main problems derived from data integration in data-intensive domains such as Genomics. Compared to current related works, the AVDIS improves the schema integration by allowing the possibility of considering domain conceptual models as global schemas to ensure semantic integrity and data standardization. From an automation perspective, this approach enhances performance by eliminating the need for schema merging and post-processing, as the global schema is inherently consistent and aligned with domain knowledge. Regarding record linkage, defining the blocking rules is typically a manual process that requires domain expertise. Some approaches incorporate machine learning or heuristic methods to automate or assist in defining optimal blocking strategies. Including a conceptual model improves the efficiency of the automation process because it already represents domain knowledge in a machine-readable format that can be used to automatically select the blocking keys. The conceptual model also represents the possibility of an attribute of having null values simplifying the uncertainty resolution step in Data Fusion.

For future work, we plan to develop an ETL process based on the proposed theoretical approach. The development will be organized into three distinct phases:

- Analysis: During the analysis phase, a comprehensive evaluation of technological requirements will be conducted to ensure scalability and performance, leveraging distributed processing, parallel computing, and architectures such as Spark and cloud—based solutions. At the same time, available algorithms-particularly AI algorithms—will be examined to identify the most accurate and efficient options.
- Architecture implementation: After the technological requirements and algorithms are clear, the technological architecture will be implemented and tested.
- Evaluation: The evaluation of the ETL process will involve measuring various quality metrics with a focus on performance, accuracy, and efficiency.

By reducing manual effort and integrating existing independent solutions, we aim to enhance the overall performance of the data integration process.

References

1. Abiteboul, S., Manolescu, I., Rigaux, P., Rousset, M.C., Senellart, P.: Data Integration, pp. 193–225. Cambridge University Press (2011)
2. Bernasconi, A., García S., A., Ceri, S., Pastor, O.: A comprehensive approach for the conceptual modeling of genomic data. In: Ralyté, J., Chakravarthy, S., Mohania, M.K., Jeusfeld, M.A., Karlapalem, K. (eds.) ER 2022. LNCS, vol. 13607, pp. 194–208. Springer, Cham (2022). https://doi.org/10.1007/978-3-031-17995-2_14
3. Bernstein, P.A., Madhavan, J., Rahm, E.: Generic schema matching, ten years later. Proc. VLDB Endow. **4**(11), 695–701 (2011). https://doi.org/10.14778/3402707.3402710 https://doi.org/10.14778/3402707.3402710
4. Bleiholder, J., Naumann, F.: Data fusion. ACM Comput. Surv. **41**(1), 41 (2009). https://doi.org/10.1145/1456650.1456651
5. Cheatham, M., Pesquita, C.: Semantic data integration. In: Handbook of big data technologies, pp. 263–305 (2017). https://doi.org/10.1007/978-3-319-49340-4_8
6. Christen, P.: A survey of indexing techniques for scalable record linkage and deduplication. IEEE Trans. Knowl. Data Eng. **24**(9), 1537–1555 (2011). https://doi.org/10.1109/TKDE.2011.127
7. Cohen, S., Sagiv, Y.: An incremental algorithm for computing ranked full disjunctions. J. Comput. Syst. Sci. **73**(4), 648–668 (2007). https://doi.org/10.1016/j.jcss.2006.10.015
8. Dong, X.L., Naumann, F.: Data fusion-resolving data conflicts for integration (2009). https://lunadong.com/publication/fusion_vldbTutorial.pdf. Accessed 3 Aug 2023

9. Dong, X.L., Srivastava, D.: Big data integration. In: 2013 IEEE 29th International Conference on Data Engineering (ICDE) (2013). https://doi.org/10.1109/ICDE.2013.6544914
10. Flores, J., et al.: Incremental schema integration for data wrangling via knowledge graphs. Semant. Web 1–38 (2023). https://doi.org/10.3233/SW-233347
11. García S, A., Guizzardi, G., Pastor, O., Storey, V.C., Bernasconi, A.: An ontological characterization of a conceptual model of the human genome. In: International Conference on Advanced Information Systems Engineering, pp. 27–35. Springer (2022)
12. Gruenheid, A., Dong, X.L., Srivastava, D.: Incremental record linkage. Proc. VLDB Endow. **7**(9), 697–708 (2014). https://doi.org/10.14778/2732939.2732943
13. Hai, R., Quix, C., Zhou, C.: Query rewriting for heterogeneous data lakes. In: Benczúr, A., Thalheim, B., Horváth, T. (eds.) ADBIS 2018. LNCS, vol. 11019, pp. 35–49. Springer, Cham (2018). https://doi.org/10.1007/978-3-319-98398-1_3
14. Imprialou, M., Stoilos, G., Cuenca Grau, B.: Benchmarking ontology-based query rewriting systems. In: Proceedings of the AAAI Conference on Artificial Intelligence, vol. 26, no. 1, pp. 779–785 (2021). https://doi.org/10.1609/aaai.v26i1.8215
15. Katsis, Y., Papakonstantinou, Y.: View-based data integration. In: Liu, L., Özsu, M.T. (eds.) Encyclopedia of Database Systems, pp. 3332–3339. Springer, Boston (2009). https://doi.org/10.1007/978-0-387-39940-9_1072
16. Lenzerini, M.: Data integration: a theoretical perspective. In: Proceedings of the Twenty-First ACM SIGMOD-SIGACT-SIGART Symposium on Principles of Database Systems, PODS 2002, pp. 233–246. Association for Computing Machinery, New York (2002). https://doi.org/10.1145/543613.543644
17. Leon, A., Pastor, O.: Towards a shared, conceptual model-based understanding of proteins and their interactions. IEEE Access **9**, 73608–73623 (2021)
18. León, A., Reyes, J., Burriel, V., Valverde, F.: Data quality problems when integrating genomic information. In: Link, S., Trujillo, J.C. (eds.) ER 2016. LNCS, vol. 9975, pp. 173–182. Springer, Cham (2016). https://doi.org/10.1007/978-3-319-47717-6_15
19. Nadal, S., Abelló, A., Romero, O., Vansummeren, S., Vassiliadis, P.: Graph-driven federated data management. IEEE Trans. Knowl. Data Eng. **35**(1), 509–520 (2021). https://doi.org/10.1109/TKDE.2021.3077044
20. Osman, I., Yahia, S.B., Diallo, G.: Ontology integration: approaches and challenging issues. Inf. Fusion **71**, 38–63 (2021). https://doi.org/10.1016/j.inffus.2021.01.007
21. O'Hare, K., Jurek-Loughrey, A., Campos, C.D.: A review of unsupervised and semi-supervised blocking methods for record linkage. In: Linking and Mining Heterogeneous and Multi-view Data, pp. 79–105 (2019). https://doi.org/10.1007/978-3-030-01872-6_4
22. Peffers, K., Tuunanen, T., Rothenberger, M.A., Chatterjee, S.: A design science research methodology for information systems research. J. Manag. Inf. Syst. **24**(3), 45–77 (2007). https://doi.org/10.2753/MIS0742-1222240302
23. Saeedi, A., Nentwig, M., Peukert, E., Rahm, E.: Scalable matching and clustering of entities with famer. Complex Syst. Inform. Model. Q. **16**(95), 61–83 (2018). https://doi.org/10.7250/csimq.2018-16.04
24. Steorts, R.C., Ventura, S.L., Sadinle, M., Fienberg, S.E.: A comparison of blocking methods for record linkage. In: Domingo-Ferrer, J. (ed.) PSD 2014. LNCS, vol. 8744, pp. 253–268. Springer, Cham (2014). https://doi.org/10.1007/978-3-319-11257-2_20

Human Factors in Information Systems

Developing a Design Features Taxonomy of Human-Computer Interaction in Social Media that Affect User Engagement and Addictive Behaviors

Maria Fernanda Granda(✉), María-Belén Sarmiento, Ana-Gabriela Nuñez, Ricardo Maldonado, and Otto Parra(✉)

Department of Computer Science, University of Cuenca, Cuenca, Ecuador
{fernanda.granda,mbelen.sarmiento,anag.nunez,ricardo.maldonado, otto.parra}@ucuenca.edu.ec

Abstract. **[Context and motivation]** The rapid growth of social media has transformed how users, especially young people, engage with digital platforms and applications, raising concerns about their addictive potential. **[Question/problem]** However, there remains a limited understanding of what design features of Human-Computer Interaction in social media platforms most strongly encourage addictive behaviors and which user interface design elements are used to support engagement on these platforms. **[Principal ideas/results]** To close this gap, this study aims to develop a comprehensive design features taxonomy of Human-Computer Interaction within social media that influences addictive user behaviors and increases the time and frequency of usage. In developing our taxonomy, we drew from existing literature, incorporating insights from experienced Human-Computer Interaction, psychology, and software engineering practitioners to validate our taxonomy. By identifying specific design features that encourage prolonged engagement, this taxonomy provides a foundation for guiding design decisions that must be made to develop non-addictive social media. **[Contribution]** This taxonomy is a valuable tool for designing, regulating, and using social media more responsibly, balancing user engagement, user experience, and digital well-being while addressing the risks of compulsive behaviors and technology addiction.

Keywords: HCI Design Taxonomy · Social Media · Addictive User Behaviors

1 Introduction

In today's digital age, information and communication technologies (ICTs) have fundamentally changed how people interact, communicate, and participate in various activities. The prevalence of digital devices like smartphones and computers has led to increased use of digital platforms, resulting in significant changes in daily routines and social interactions [1]. This situation causes the issue of digital addiction, which is gaining more attention as studies show a considerable increase in the time people spend on digital

J. Grabis et al. (Eds.): RCIS 2025, LNBIP 547, pp. 313–330, 2025.
https://doi.org/10.1007/978-3-031-92474-3_19

platforms, mainly social media[1] [2]. Concerns have been raised about the potential for these technologies to encourage addictive behaviors, especially among young people who are the most frequent users of these digital platforms and applications.

The global issue of digital addiction, characterized by excessive use of digital applications such as social media, online games, and other internet-based platforms, poses a significant challenge to mental health and well-being [3]. Around 3.8 billion people worldwide use the internet, spending an average of 2.5 h daily on social media [1]. Social media includes numerous applications and online channels such as Facebook, YouTube, Instagram, X, WhatsApp, TikTok, etc. These applications allow users to spend time interacting, creating, sharing, and consuming content [4].

To support responsible design through Requirements Engineering, it's essential to identify and integrate design features that promote healthy engagement while minimizing the risk of addictive behaviors [5]. Addressing a general research question (RQ0) like "How do HCI design features in social media platforms reinforce user retention mechanisms that drive engagement and addictive behaviors?" can guide the creation of social media prioritizing user well-being. From this general research question (RQ0), we defined two research questions that guided the study:

RQ1: Which user retention mechanisms in social media platforms have the strongest impact on engagement and addictive behaviors?
RQ2: What HCI design features reinforce and support these user retention mechanisms?

These questions directed the literature review and study selection criteria, shaping the taxonomy's focus on features influencing user engagement and potentially addictive behaviors.

The main objective of this study is to explore the design features of HCI within software designs that may contribute to developing addictive behaviors in users. Specifically, we focus on identifying and categorizing these design features into a comprehensive taxonomy. This taxonomy will provide a structured framework for analyzing the HCI features in software designs that may inadvertently lead to digital addiction. By highlighting these design features, the study aims to guide the development of non-addictive information systems (NAIS) [6] and contribute to ethical guidelines in software design, ensuring a balanced and healthy interaction with digital technologies.

The contribution of this paper is twofold: (i) to provide a design features taxonomy of HCI in social media that affect user engagement and addictive behaviors, and (ii) to what they are doing to empower users to take control over them and reduce compulsive interaction so that these apps are attractive but also respect the mental health and autonomy of users.

The paper is structured as follows: In Sect. 2, the background and related work are briefly described. In Sect. 3, the research methodology is described. In Sect. 4, the process to define the taxonomy is summarized. In Sect. 5, some recommendations for non-addictive social media are proposed. Finally, Sect. 6 summarizes our study's conclusions, limitations, and future work.

[1] In this work the terms "social media platform", and "social media" are considered similar.

2 Background and Related Work

In this section, we will first briefly define and outline topics related to addictive software design.

2.1 Definitions

HCI is the academic topic within Computer Science under which addictive software design falls [7]. Interaction design involves organizing and utilizing available resources to define, shape, and determine all user-focused qualities—structural, functional, ethical, and aesthetic—of a digital artefact intended for one or more users [8]. Our proposal is focused on identifying the design features of HCI.

Research indicates that certain design features of HCI in software can encourage user engagement, sometimes leading to addictive behaviors. These design features include continuous notifications, infinite scrolling, and personalized recommendations, which aim to maximize user interaction but may also promote compulsive usage patterns [6]. Furthermore, the psychological mechanisms driving these behaviors, such as the need for social validation through "likes" and the desire for gaming achievements, add complexity to the issue [9].

The rise of digital addiction highlights the importance of user-centered design in software development. NAIS frameworks promote ethical design by encouraging balanced usage, transparency in usage patterns, and restrictions to prevent excessive use [10]. As researchers, software designers, policymakers, and individuals interested in digital addiction, our proposal in this process is vital.

2.2 Related Work

The second part of this section contains work related to studies reporting taxonomies proposed in games. Although our work is related to social media, these papers are closest to our research.

Flayelle et al. [11] propose a theory-driven general taxonomy of the design features of online applications that might promote uncontrolled and problematic online behaviors. The authors critically examine the available research on the relationships between technology design features and the loss of control and harms experienced by those who engage in online video gaming, online gambling, cyber sexual activities, online shopping, social networking, and on-demand TV streaming.

To enhance psychological understanding of problematic video gaming and to inform future research directions, King et al. [12] propose a new taxonomy of video game features, which includes (a) social features, (b) manipulation and control features, (c) narrative and identity features, (d) reward and punishment features, and (e) presentation features.

Windleharth and Ja Lee [13] present two taxonomies developed from studying 65 mobile games—a taxonomy of the types of transactions between game players and companies that transfer or create value for the gamer owner and a taxonomy of methods companies use to drive engagement and retention with mobile games. The paper also presents an example of applying these taxonomies to map the value transfer in the game.

These contributions highlight the critical relationship between design features in digital environments and user engagement, emphasizing the need for ongoing research in this area to understand better and mitigate problematic online behaviors.

To the best of the authors' knowledge, there is no detailed taxonomy of design features of HCI in social media that characterize user engagement and drive addictive behaviors.

3 Research Methodology

The present study aims to define a taxonomy of design features of HCI in social media using a three-phase approach. First, in the Data Collection Phase, we collected data from a literature review and explanatory expert interviews. Second, to address the RQs of this study adequately, we developed the taxonomy employing the methodology proposed by Nickerson [14] and Kundisch [15] called the "Extended Taxonomy Design Process" (ETDP), which adheres to the design science research paradigm and consists of six activities. Third, the taxonomy was evaluated by identifying the interaction design features of HCI on the most used social networks such as Facebook, Instagram, WhatsApp, Twitter-X, YouTube, and TikTok.

In this section, we describe each phase and the steps of this methodology applied to define our taxonomy.

3.1 Phase 1: Data Collection

This phase consists of two steps: (i) identify relevant studies on HCI in addictive software design strategies and (ii) conduct a literature review on HCI + addictive design + social media. Therefore, we began by identifying relevant studies on HCI in the context of addictive software design strategies. This was followed by a focused literature review that examined HCI, addictive design, and social media.

To establish a research-based foundation, we considered prior studies on HCI and addictive behaviors [1, 6, 16], ensuring that taxonomy effectively addresses user engagement and the potential for addictive behaviors.

We did a comprehensive literature review to identify studies relevant to the research questions on social media addiction and HCI features influencing user engagement. To achieve this, we employed the following search query: ((addict* AND (feature* OR characteristic* OR factor* OR trait*)) AND ("social media" OR "social network*" OR "social platform*") AND (hci OR "Human-Computer Interaction" OR "user interface")). This search was designed to capture studies examining addiction-related features and user interaction patterns within social media environments. We searched databases such as Scopus, Taylor and Francis, and IEEE, identifying an initial pool of 314 studies. The studies followed a multi-stage filtering process. First, duplicated entries were removed, leaving 313 unique studies. Then, three reviewers screened titles and abstracts to assess their eligibility, which narrowed the selection to 97 studies.

Finally, a thorough full-text review led to the selection of 45 studies that closely aligned with the focus of the taxonomy on engagement and potential addictive behaviors. Table 1 below summarizes each stage in the literature review and the study selection process.

Table 1. Summary of Literature Review Results Across Databases

Source	Potential Studies	Removing duplicates	Scanning title and abstract	Selected Studies
Scopus	25	25	14	9
Taylor and Francis	287	287	83	36
IEEE	2	1	1	0
Total	**314**	**313**	**97**	**45**

From the analysis of 45 selected studies, we identified key HCI features influencing user engagement and addictive behaviors on social media platforms.

3.2 Development of the Taxonomy

To develop the taxonomy, five activities of the extended taxonomy design process (ETDP) provided by Kundisch et al. [15] were used, which enhances the taxonomy development method of Nickerson et al. [14]. The ETPD can be placed within the design science research paradigm, outlining relevant steps in this paradigm's six activities.

Activity 1: Identify the Problem and Motivate
In this first activity, the problem identified is the lack of a taxonomy to define the design features of HCI in social media, reducing the effect of the addictive property that many social media platforms could include. So, the taxonomy stakeholders could include UX designers, software developers, policymakers, mental health professionals, researchers, and users.

The motivation for defining the taxonomy is related to balancing the addictive impact of social media platforms when the designers define the user interface elements to include in the social media platforms.

Activity 2: Define the Objectives of a Solution
It was an iterative process along with the activity 3. Items identified in the literature were used to develop initial ideas of meta-characteristics and characteristics. Various grouping methods were explored to name items, characteristics, and meta-characteristics, leading to a three-iteration process.

The first iteration analyzed a list of item descriptions (e.g., recommendations, ads, reminders, labels) from the literature. Items were initially grouped based on graphical interface elements (e.g., links, buttons, scroll bars) and hardware devices (e.g., touchscreen, keyboard, mouse), with meta-characteristics classified as input or output elements. However, this approach was not orthogonal, so some elements appeared in multiple categories.

In the second iteration, items were classified based on their interaction elements (e.g., likes, comments) and their role, for example, in gratification, attention capture, and usability enhancement. Emotional and monetary gratification elements were grouped under the Gratification meta-characteristic, and new items were added or adjusted as needed. This allowed for better mapping of the different items found in the literature. Some meta-characteristics and characteristics resulted from the literature review (i.e., Reward from Reward and Punishment Features [12], Reinforcement and Reward [13]), Time from Temporarily Available Information [11], Attention Capture from Appeal Factors [13]) and the others were chosen by the authors as a result of several group sessions.

In the third iteration, the names of some meta-characteristics were adjusted; for example, Gratification came before Reward, and it was felt to represent better the descriptions of the elements identified in the literature that can be grouped under this meta-characteristic.

A similar process was performed for the rest of the meta-characteristics. The taxonomy development process is complete when all meta-characteristics and characteristics are unique, each characteristic is represented by at least one item, and no further changes are made during the final iteration. Non-exclusive characteristics across meta-characteristics are allowed to enhance flexibility. Additionally, taxonomy must meet subjective standards of being concise, robust, comprehensive, extendable, and explanatory [17].

Activity 3: Design and Development

Using the results from the literature review, a conceptual-to-empirical approach was employed, with iterative cycles refining the taxonomy until the predefined criteria were met. Each meta-characteristic (see Table 2) is a structured guideline to pinpoint essential characteristics and elements. For example, Notifications and Alerts fall under the Immediate Reward criterion, while Infinite Scrolling is categorized under User Engagement Design. These items are the design features of HCI in social media that most influence the promotion of addictive behaviors. They serve as a starting point for identifying features that encourage or discourage addictive behaviors from helping protect users from addictive behaviors and offer strategies for managing these effects.

This approach provides a detailed view of the various features within each criterion, enhancing our taxonomy's ability to reflect the diversity of HCI features across the studies. Table 2 summarizes our initial taxonomy approach's meta-characteristics, characteristics, items, and examples. For a complete overview of the selected studies, refer to Appendices A and B, which list all studies from S001 to S045 alphabetically, along with the number of studies referencing each feature and their prevalence (N = 45).

Activity 4: Demonstration

This was applied to each social media platform considered in this research to ensure that the taxonomy's characteristics are mutually exclusive and collectively exhaustive. By using some iterations, the ending conditions were met.

Activity 5: Evaluation

To evaluate the proposed taxonomy, we designed an expert-based evaluation. Eight experts from Computing, Design, Psychology, and Education participated voluntarily. This group included professor-researchers and professionals with experience in HCI, habit formation, interface design, and taxonomy development.

Firstly, we designed a demographic questionnaire using Google Forms and distributed it by e-mail. Each expert answered the questionnaire to provide his/her information regarding their experience in the area (details can be seen in Appendix E on file named "Expert_Evaluation_EN.pdf"). Then, a Likert scale survey was conducted to identify the most suitable experts regarding their experience in HCI, social networks, addiction, and taxonomies. Since the goal was to determine the relevance of the experts, a simple scale was sufficient. Secondly, we designed a questionnaire containing three sections on meta-characteristics, characteristics, and items, all included in the taxonomy (results can be seen in Appendix F on file named "validation of the taxonomy by experts.xlsx"). The expert validation was carried out through content validation [18], which consisted of an instrument that included the proposed taxonomy's meta-characteristics, characteristics, and items. For each item, the user was asked to select clarity, relevance, and coherence levels according to a table containing the meaning of each element of the proposed taxonomy. In addition, the sufficiency had to be evaluated at the item, characteristic, and meta-characteristic level to determine whether the items were sufficient to measure the group to which they belonged in the proposed taxonomy. The format employed was adapted from the model of the expert judgment template proposed by Escobar-Pérez and Cuervo-Martínez [19] (see Appendix D). In addition, the experts could add observations and suggest additional features for the taxonomy.

Expert feedback was used to refine the taxonomy through a dual approach, empirical-to-conceptual and conceptual-to-empirical, guided by the ETDP framework. Experts suggested improving some item descriptions (e.g., notifications and alerts), adjusting characteristic names (e.g., intuitive to attractive presentation), and meta-characteristics (e.g., Time for Time management). Also, items under the Persuasion and Manipulation characteristic, such as Hidden Charges, have been added to promote transparency, and under Permanent Connectivity, such as Cloud Sync, to support device interoperability.

Essential adjustments were made to enhance the taxonomy's practical application and align it with real-world user interaction patterns. Also, Quick Context Recovery was added to improve usability by allowing users to resume their activities smoothly after interruptions within the Usage Frequency characteristic. Items like Color Palette and Smooth Navigation were included in the Intuitive design characteristic to create a cohesive user experience (UX). After validation, some items added through expert input mentioned in this section were integrated into their respective characteristics and aligned with relevant meta-characteristics. These additions, marked with an asterisk (*), can be seen in Table 2 and Appendix C as part of our final taxonomy. The authors reviewed the refinements until they reached a consensus, ensuring they met the study's objectives, were mutually exclusive and collectively exhaustive, and effectively captured social media applications' functional and ethical aspects.

Table 2. Summary Meta-characteristics (MC), Characteristics, Items, and Examples

RQ1		RQ2	
MC	Characteristic	Item	Examples of user interface elements
Reward	Immediate Reward	Notifications and alerts	Facebook: Notification of a new post from a friend and birthday reminders, messages, push alerts, and suggestions
		Instant rewards	Reactions, likes
		Auditory or visual responses	WhatsApp-red dot with a number to notify that there are new messages, sounds when reacting, and animations (GIFs)
	Variable Reward	Random or unpredictable reward systems	Likes (similar/surprise reactions), mentions and tags, viral content, and suggestions from friends or followers
		Periodic reinforcements	Timing elements (e.g. hourglass, counter, etc.) Influencers, artists, and celebrities provide discount codes
	Interactive Reward	Point and achievement systems	Streaks, offers. Facebook "Super Fan" badges, Twitter follower levels, Snapchat "streaks"
		Measurable progress	Progress bars or statistics that visualize progress toward a goal. Instagram: engagement stats (i.e. reach, impressions, interactions)
Time Management	Usage Frequency	Quick context recovery *	WhatsApp: featured messages, Instagram: save posts
		Phase unlocks	Badge systems that users earn by accumulating time or certain actions (e.g. featured fan)
		Time spent metrics	Timer
	Exclusivity and Lack of Time	Time-limited content	Messages, alerts, reminders, stories, or short statuses that disappear after 24 h, seasonal offers
		Exclusive access	Features are available only to premium users, and content viewing is restricted to certain users, such as Instagram best friends
	Real-Time	Real-time updates	Instagram and TikTok: comments on live broadcasts
		Live content	Video and live interactions

(continued)

Table 2. (*continued*)

RQ1		RQ2	
MC	Characteristic	Item	Examples of user interface elements
Attention Capture	User Engagement Design	Infinite scrolling	Scroll and auto play videos
		Fresh content cycles	Suggestion to refresh posts (e.g., pull to refresh)
		Recommendation navigation	Recommendations on streaming platforms or social media based on user preferences (e.g., Instagram shows suggestions for users)
		Sticky content	Reading recommendations: horoscopes, tests, interesting news, games, suggestive videos
	Persuasion and Manipulation	Dark patterns	Locked or late access item, unlockable with points (e.g., money)
		Hidden charges or in-app purchases *	Timers, shopping cart, content unlock announcement
		FOMO	Notifications, alarms (e.g., TikTok: "Everyone is watching", WhatsApp: One-time viewing content)
	Customization	Content algorithms	Publications content according to the tastes and preferences of users (e.g.,"People you may know", "For you" in WhatsApp)
		Customizable avatars/profiles	Profile photos, avatar, appearance
		privacy options *	Audience settings, online status control, video and comment restrictions
Emotional Dependency	Social Interaction	Content reactions	Likes, dislikes, comments, and sharing posts
		Social comparison	Leader boards, ranking lists, friends list, followers list;
		Real-time messaging	Real-time messaging and chat
		Stalker	TikTok: profile views, X(Twitter): Publications, subscriptions, featured, multimedia, followers when viewing a profile
	Emotional Engagement	Storytelling narratives	Control to publish chronological photos, location, etc., of the user (e.g., featured stories on Instagram)

(*continued*)

Table 2. (*continued*)

RQ1		RQ2	
MC	Characteristic	Item	Examples of user interface elements
		Self-presentation	Apps that constantly encourage users to share what they are doing, photos, or their location (e.g., Facebook: information on education, profession, work, marital status, messages like "What are you thinking?")
Usability	Permanent Connectivity	Cross-platform access	Responsive graphical interface elements (display size, navigation size, textual content, visual content, brand identifiers, action buttons)
		Cloud sync *	Option to enable/disable synchronization
	Intuitive	Consistency	Fixed key locations on buttons, menus, and layouts
		Simplicity in design	There are fewer buttons, fewer popups, and simple icons
		Color palette *	Buttons with familiar or striking colors
		Clear language	Button labels, messages, descriptions (e.g., "Add to your story", "Search" in Instagram)
		Smooth navigation *	Navigation bars, drop-down menus

3.3 Phase 3: Evaluation

The proposed taxonomy was evaluated by applying it to social networks like Facebook, Instagram, WhatsApp, TikTok, YouTube, and Twitter-X. Systematically analyzing each platform made it possible to map their features to specific categories within the taxonomy. However, not all characteristics are equally applicable to every platform, as shown by the evaluation conducted (details can be seen in Appendix F on file named "social_media_taxonomy_examples.xlsx"). Certain platforms lack some items from the taxonomy, or the importance of these items may differ significantly depending on the platform. Despite this, the taxonomy retains its utility, as most elements are consistently found across platforms. Although some characteristics are not included in each social media, moreover, since Meta Platforms, Inc. Owns Facebook, Instagram, and WhatsApp, there is evidence of gradual feature integration among these platforms, such as adding message reactions and status updates to WhatsApp.

4 Taxonomy of Design Features of HCI in Social Media: Results and Discussion

This section provides detailed explanations of the proposed taxonomy of the design features of HCI in social media platforms. Additionally, it includes answers to each of the research questions proposed in Sect. 1.

4.1 Meta-characteristic 1: Reward

It is a positive reinforcement or outcome that follows a behavior, increasing the likelihood of that behavior being repeated.

Immediate rewards (e.g., notifications, likes) trigger curiosity and dopamine-driven motivation, encouraging repeated app use (S021, S045). Variable rewards (unpredictable reinforcement) create surprise, keeping users engaged through anticipation (S021). Periodic rewards (e.g., spaced notifications) sustain interaction via timed incentives (S014). Interactive rewards (gamified points, achievements, progress tracking) foster accomplishment through measurable metrics (e.g., follower counts, badges), enhancing motivation (S021).

Discussion: This meta-characteristic of social media platforms plays a significant role in fostering addictive behaviors among users. The platforms employ various reward mechanisms to boost user engagement and encourage addiction. Many platforms also integrate other types of reward points, achievements, and measurable progress metrics to cultivate a sense of accomplishment and encourage more frequent user interaction.

4.2 Meta-characteristic 2: Time Management

The temporal structuring of content is key in driving continuous engagement and reinforcing the platform's relevance in users' lives.

Temporal structuring drives engagement through content pacing and relevance. Usage frequency measures login/interaction rates, while quick context recovery helps users refocus post-distraction (e.g., minimizing notifications via Focus Assist). Scarcity leverages phase unlocks (e.g., time-limited bonuses) and time spend metrics to optimize retention (S023). Exclusivity fuels urgency via restricted access (e.g., VIP features) and short-form content (15–60-s videos) for rapid attention capture (S021, S016). Real-time functionality (live updates, streaming) fosters immediacy, enhancing addictive appeal (S012, S025).

Discussion: The organization of time-related features is crucial for driving user engagement. Platforms monitor how frequently users log in and interact with content, optimizing their offerings based on this data to enhance engagement. This immediacy contributes to the addictive nature of social media by keeping users engaged with ongoing content.

4.3 Meta-characteristic 3: Attention Capture

It refers to software features designed to draw or hold a user's focus, often automatically or without conscious effort.

User engagement design captures and maintains users' attention, encouraging them to interact meaningfully and repeatedly with a product or platform. Infinite scrolling (S030): Automatically loads fresh content without pagination, creating an "endless" stream to prolong the interaction. Fresh content cycles: Systematically introduce new material to sustain interest and encourage repeat visits. Recommendation algorithms (S016): Personalize content via behavioral data, fueling addiction by aligning with user preferences. Sticky content: Horoscopes, quizzes, or games build loyalty through low-effort, high-engagement interactions.

Persuasion and manipulation can have significant social and psychological effects on social media because of the platform's broad reach and the use of personal data to create highly targeted content. Dark patterns: Interfaces designed to trick users into unintended actions (e.g., trick questions, "roach motel" traps) that exploit cognitive biases and serve platform interests. Hidden charges: Financial transactions are obscured during app use, often via unclear prompts or default settings. FOMO: Anxiety about missing updates drives compulsive connectivity, fueling addiction by targeting emotional/social needs (S021).

Customization gives users greater control over their interactions and engagement, making their social media experience more relevant and enjoyable. Content algorithms (S021): Tailor feeds via user behavior/preferences, enhancing engagement and retention. Profile customization (S011): Avatars/updates boost self-esteem, potentially driving excessive use and dependency. Privacy options: Added by experts to balance personalization with data control.

Discussion: Features that capture and maintain user attention are essential in promoting addiction. Platforms systematically introduce new content to sustain user interest and prolong interaction time, which is vital for encouraging users to return. Additionally, algorithms that customize content based on user behavior provide an endless supply of engaging material, further fueling addiction.

4.4 Meta-characteristic 4: Emotional Dependency

It refers to how certain design features are crafted to create a psychological reliance on the platform for emotional support, validation, or fulfilment. These features are often strategically designed to encourage frequent engagement by appealing to users' need for connection, recognition, and self-esteem.

Content reactions (likes, comments, shares): Users gauge social validation via others' responses (S011, S020, S024). Social comparison: Algorithms amplify life/achievement comparisons, driving engagement (S021). Real-time messaging: Typing indicators, reading receipts and status updates simulate face-to-face immediacy (S011, S030). Stalking: Monitoring others' activity (e.g., Facebook's second-most common use) fuels passive engagement (S011).

Emotional engagement occurs when users feel emotionally invested in their interactions on the platform, such as feeling happy, inspired, validated, or motivated by the content or people they follow. In cultures where video consumption dominates, creators prioritize engaging storytelling narratives and co-creation initiatives (S015). In a self-presentation, the users utilize videos to present themselves to the public while sharing content. This gratification activates the brain's reward system, reinforcing continued social media use and potentially leading to social media addiction (S027).

Discussion: Emotional dependency in social media addiction refers to users' psychological reliance on platforms for emotional support, validation, and fulfilment. Social media design features encourage constant engagement by appealing to users' connection, recognition, and self-esteem needs. A key factor is social comparison, driven by algorithms that promote content leading users to compare themselves to others, causing feelings of inadequacy or superiority. Additionally, real-time communication features enhance the sense of closeness and personal connection, deepening emotional dependency.

4.5 Meta-characteristic 5: Ease of Use/Usability

Design features that promote positive user experience encourage engagement and reduce frustration by making actions and information readily accessible.

Permanent connectivity refers to features that remain constantly connected to social media, enabling continuous access to information, social interactions, and online services. Cross-platform access: Seamless use across devices (e.g., mobile/desktop) via responsive design (S026, S028, S029). Cloud sync: Automatically updates data across devices, validated by experts for real-time access and management.

Attractive presentation characteristics ensure a smooth and enjoyable experience, encouraging more frequent use and deeper engagement by minimizing confusion or effort. Consistency: Fixed button/menu layouts reduce confusion (S018, S001). Simplicity: Minimal visual elements (e.g., fewer buttons, clear icons) minimize distractions (S012). Clarity: Concise language and labels ensure confident interaction (S018). Visual hierarchy: Expert-validated color palettes and smooth navigation guide attention and ease transitions.

Discussion: The discussed features enhance user engagement and reduce frustration. Permanent connectivity, cloud synchronization, and responsive design ensure seamless device use. An intuitive interface, appropriate color palette, and smooth navigation improve usability and encourage prolonged use. These usability features also support psychological mechanisms that contribute to addictive behaviors.

Final Discussion: It is important to recognize that almost any interaction feature in social media can be linked to addictive behaviors, depending on how it is designed and implemented. Items like notifications, infinite scrolling, variable rewards, and social validation mechanisms (e.g. likes, shares, comments) can enhance user experience and engagement and reinforce compulsive usage patterns.

This highlights the delicate balance between user experience, engagement, and addiction. While designers aim to create intuitive and engaging interfaces, excessive or manipulative use of these features can lead to problematic behaviors, digital overconsumption, and negative mental health outcomes.

Therefore, the proposed taxonomy will help developers and UX designers balance user interaction and well-being by identifying features that empower users to control themselves and reduce compulsive interaction so that these apps are attractive and respect users' mental health and autonomy.

5 Some Recommendations for Non-addictive Social Media

We apply two strategies to obtain recommendations for designing non-addictive social media: (i) A literature mapping to get recommendations from authors for the design of non-addictive user interfaces (e.g., [6]). (ii) Consult users and experts regarding which elements of the taxonomy should be avoided in the design of non-addictive interfaces.

The results obtained suggest that the recommendations for non-addictive social media can involve:

Prioritize Transparency. Providing users with information about addictive current risks and how to avoid them. This helps users understand the system (i.e. algorithms) and make informed decisions about their engagement.

Continuous Monitoring. Keep track of usage time (e.g., *time spend metrics*) and suppress some interaction features that trigger addictive behaviors. For example, timers or reminders can help users monitor their time on the platform.

Design for Well-Being. Introducing intentional interruptions to hinder flow is an important element in the development of addictive tendencies. For example, (i) encouraging scheduled breaks and offering options to limit daily usage can help create healthier habits, and (ii) a feature that gradually dims the screen or reduces color intensity after a set amount of time could subtly remind users to take a break.

Customization Options. Adjusting or adapting HCI features to avoid remaining constantly connected to social media. For example, (i) allowing users to customize their notification settings can help mitigate attention capture, and (ii) instead of constant alerts, platforms could offer digestible summaries at specific times, which promotes more intentional engagement with content rather than reactive browsing.

Some examples are included below (see Table 3) to evaluate whether these design recommendations have been integrated into the platforms used in the study.

Table 3. Evaluation of design recommendations of social media platforms used in the study

Strategy	Facebook	Instagram	X	TikTok	WhatsApp	YouTube
Recommendation 2: Continuous Monitoring						
Set time limits	Users can set time to monitor and restrict usage	Users can set usage limits	It does not enforce time-based usage restrictions. It implements technical usage limits to ensure the stability of the platform	Users can set usage limits	It does not enforce time-based usage restrictions. It implements technical usage limits to prevent spam and maintain service quality	It does not enforce native time limits or usage restrictions for users but offers voluntary tools to promote healthier viewing habits
Recommendation 3: Design for Well-Being						
Use restricted mode	It offers user-controlled privacy tools to limit content visibility	Reduce engagement pressure by disabling likes on posts	Hide likes and retweets to minimize compulsive interactions	Its algorithm and platform dynamics influence how engagement is measured and valued	Reduce distractions by muting groups	Limits access filtering content (violence, mature, etc.)
Limit Engagement Metrics	Disable auto-play to reduce visual stimuli	Reduce engagement hidden likes and comments	Hide likes and retweets to minimize compulsive interactions	It does not impose direct limits on engagement metrics. It has restrictions to infinite scrolling	Hide your online presence to avoid engagement pressure	Limit content exposure to avoid excessive engagement
Recommendation 4: Customization Options						
Manage notifications	Disable or limit push notifications	Temporarily disable notifications	Manage settings	Automatically limit notifications after 10 pm for younger users	Customize or disable notifications to prevent constant interruptions	It provides user-controlled notification management through customizable settings
Feed preferences	Prioritize or unfollow content to avoid compulsive scrolling	Adjust algorithmic content by muting certain keywords	Avoid algorithmic feeds by creating personalized lists of accounts	Users can adjust recommendations	It provides tools to customize conversations and groups. It does not support customizable preferences to help users	It supports limited feed customization through user-driven preferences. In this case, native feed customization is reactive, not proactive (to help users limit usage)

6 Conclusions, Limitations, and Future Work

This paper gives a deeper understanding of the design features of HCI on social media that most effectively promote addictive behaviors and the user interface design elements used to drive engagement. To address this, we present a design features taxonomy of HCI in social media that influence user engagement and addictive tendencies.

Analyzing 45 research papers and conducting expert surveys, we identified five meta-characteristics—reward, time, attention capture, emotional dependency, and ease of use/usability—encompassing 13 features and 37 specific items. Key features identified, such as notifications, infinite scrolling, personalized algorithms, and real-time interactions, highlight specific design practices that increase user engagement but may also lead to compulsive usage.

Our study highlights the need for further research to analyses the interaction features of individual social media in greater detail. It underscores the importance of extending research to other categories of addictive applications, such as online video games, gambling, and other Internet-based services. Furthermore, our research establishes a foundation for comparing the characteristics of social media with other platforms, such as online games. Another contribution of this work is offering a basis for formulating design guidelines to develop social media that balances user engagement with ethical considerations like reducing addictive tendencies.

Despite its contributions, this study has certain limitations that point to areas for future research. The reviewed literature reflects our interpretation of the state of the art but represents only a portion of the rapidly evolving social media and HCI design landscape. Also, it does not differentiate between social media platforms; the platforms analyzed as examples of features promoting addiction were Facebook, X (formerly Twitter), Instagram, TikTok, YouTube, and WhatsApp, selected for their widespread use. However, due to this lack of specificity, not all characteristics may apply equally to every platform, as evidenced by the evaluation. Some platforms do not encompass all items in the taxonomy, or the relevance of these items may vary significantly depending on the platform. Nevertheless, the taxonomy remains valuable as most elements are present across platforms. We followed the methodology proposed by Kundisch D. et al. [15] for taxonomy development and validated its applicability through expert surveys, deeming it theoretically robust and relevant to current digital platforms.

Future research could enhance and expand the proposed taxonomy by utilizing qualitative and quantitative methods for more in-depth case analyses. Additionally, we plan to propose/extend a software quality model to evaluate addictiveness in social media platforms. Finally, this taxonomy and design criteria can be the basis for extending research to other categories of addictive applications, such as online video games, gambling, etc.

Acknowledgments. This work is part of the research project "Evaluation framework for identifying and quantifying the level of potential addiction to software based on its human-computer interaction design features - EvAddictSoft" which is supported by Vicerrectorado de Investigación e Innovación de la Universidad de Cuenca - Ecuador.

Appendix

They can be accessed at https://acortar.link/rwnBUs.

References

1. Allcott, H., Gentzkow, M., Song, L.: Digital addiction. Am. Econ. Rev. **112**(7), 2424–63 (2022). https://doi.org/10.1257/aer.20210867
2. Mujica, A., Crowell, C., Villano, M., Uddin, K.: Addiction by design: Some dimensions and challenges of excessive social media use. Med. Res. Arch. **10**(2), 29 (2022)
3. Raj, G., Kumar Sharma, A., Arora, Y.: Analyzing the effect of digital technology on mental health. In: Strategies for E-commerce data security, p. 695. IGI Global Scientific Publishing, India (2024)
4. Yue, H., Yue, X., Zhang, X., Liu, B., Bao, H.: Exploring the relationship between social exclusion and social media addiction: the mediating roles of anger and impulsivity. Acta Physiol (Oxf.) **238**, 103980 (2023). https://doi.org/10.1016/j.actpsy.2023.103980
5. Zannat, N., Mahmud, M.: Mindful UX design in industry 4.0: mitigating addiction and enhancing user ell-being in social media and AI environments. J. Inf. Syst. Digit. Technol. **5**(2), 24 (2023)
6. Kloker, S.: Non-addictive information systems. Inf. Syst. Front. **22**(3), 549–562 (Jun.2020). https://doi.org/10.1007/s10796-020-10011-w
7. Neyman, C.: A survey of addictive software design, vol. 1, no. 1, p. 12 (2017)
8. Löwgren, J., Stolterman, E.: Thoughtful Interaction Design. A Design Perspective on Information Technology. The MIT Press, Cambridge (2004). https://www.droit-technologie.org/wp-content/uploads/2017/04/Jonas_Lo%CC%88wgren_Erik_Stolterman_Thoughtful_InterBookZZ.org_.pdf
9. Cai, H., Ding, X., Zhang, S.: Application of human-computer interaction technology in rehabilitation treatment of mental and nervous system diseases. In: SHS Web Conference, vol. 165, p. 02007 (2023). https://doi.org/10.1051/shsconf/202316502007
10. Montes, B., Luis, J., Riera, F.E.S.: Trabajo de Titulación para optar al título de Ingeniero en Electrónica y Telecomunicaciones
11. Flayelle, M., Brevers, D., King, D.L., Maurage, P., Perales, J.C., Billieux, J.: A taxonomy of technology design features that promote potentially addictive online behaviours. Nat. Rev. Psychol. **2**(3), 136–150 (2023). https://doi.org/10.1038/s44159-023-00153-4
12. King, D., Delfabbro, P., Griffiths, M.: Video game structural characteristics: a new psychological taxonomy. Int. J. Ment. Health Addict. **8**(1), 90–106 (2010). https://doi.org/10.1007/s11469-009-9206-4
13. Windleharth, T., Lee, J.H.: Taxonomies for transactions and user engagement in mobile games (2020)
14. Nickerson, R.C., Varshney, U., Muntermann, J.: A method for taxonomy development and its application in information systems. Eur. J. Inf. Syst. **22**(3), 336–359 (2013). https://doi.org/10.1057/ejis.2012.26
15. Kundisch, D., et al.: An update for taxonomy designers: methodological guidance from information systems research. Bus. Inf. Syst. Eng. **64**(4), 421–439 (2022). https://doi.org/10.1007/s12599-021-00723-x
16. WHO. Health Topics. Addictive behaviours. https://www.who.int/health-topics/addictive-behaviour#tab=tab_1
17. Jacob, A., Teuteberg, F.: Towards a taxonomy for design options of social networking technologies in sustainable business models. Sustainability **13**(1), 81 (2020). https://doi.org/10.3390/su13010081

18. Almanasreh, E., Moles, R., Chen, T.F.: Evaluation of methods used for estimating content validity. Res. Social Adm. Pharm. **15**(2), 214–221 (2019). https://doi.org/10.1016/j.sapharm.2018.03.066
19. Escobar-Pérez, J., Cuervo-Martínez, A.: Validez de contenido y juicio de expertos: una aproximación a su utilización (2008)

User Correction of Misinformation on Social Media: Perceived and Actual Social Norms

Selin Gurgun[1], Emily Arden-Close[1], Keith Phalp[1], and Raian Ali[2](✉)

[1] Faculty of Science and Technology, Bournemouth University, Poole, UK
[2] College of Science and Engineering, Hamad Bin Khalifa University, Doha, Qatar
raali2@hbku.edu.qa

Abstract. Despite the effectiveness of the User Correction approach in combating misinformation, people on social media often avoid correcting others due to fears of consequences and doubts about its usefulness. Users' perceptions and misperceptions regarding challenging misinformation—a form of User Correction—remain underexplored. This study, based on data from 250 UK-based social media users, examined misperceptions about correcting misinformation and their impact on the likelihood of taking action. The findings reveal that individuals hold misperceptions about the negative consequences of challenging others and the acceptability of such behavior. These consequences include perceived negative effects on relationships, harm to the person being challenged, and the belief that challenging misinformation is ineffective. Participants believed that challenging others could damage relationships more than being challenged themselves, and they perceived challenging others as more futile than being challenged. Those who viewed correcting misinformation as more socially acceptable for others were more likely to engage in such actions. Injunctive norms and perceived negative impacts on the person being challenged also influenced the likelihood of participating in User Correction. This study highlights the importance of addressing users' attitudes and misperceptions to develop effective user-centric countermeasures on social media. The findings suggest that a social norms approach to behavior change may be a promising strategy to encourage corrective actions.

Keywords: Misinformation · User Correction · Challenging Misinformation · Misperception · Pluralistic Ignorance

1 Introduction

Although evidence suggests that the User Correction approach, which involves people actively debunking, flagging and correcting misinformation is one of the most effective ways to combat its dissemination [1], people are reluctant to take action when they encounter misinformation on social media [2–5]. This may be due to Pluralistic Ignorance (PI), which is when the majority erroneously believe that others have different thoughts or behaviours compared to themselves [6]. While there are many approaches to tackle misinformation such as machine learning [7, 8], natural language processing [9, 10] and algorithmic approaches [11], limited research focuses on users' perspectives and human

J. Grabis et al. (Eds.): RCIS 2025, LNBIP 547, pp. 331–346, 2025.
https://doi.org/10.1007/978-3-031-92474-3_20

factors to combat misinformation. To encourage users to challenge misinformation, it is crucial to explore their attitudes, perceptions and misperceptions around challenging misinformation.

Social norms, commonly referred to as informal guidelines directing social behaviours [12] are divided into two categories: a) descriptive norms (what is common) and b) injunctive norms (what is approved or disapproved) [13]. While social norms help people to guide their own behaviour as they are tied to perceived social pressures [14], they are also used to predict how others will behave. In other words, social norms serve as a framework through which people understand and judge the behaviours of those around them. As in other social spaces, norms in Social Network Sites (SNS) are formed through discourse, constructed by individuals, given meaning within similar groups, and learned through observing other users' actions [15, 16]. Users within SNS are expected to adhere to implicit norms such as refraining from clicking "like" on a death-related post news [16] or withholding negative feedback related to appearance [17]; otherwise, they may face sanctions [18, 19]. There are two main reasons for this conformity - normative influence and interpersonal influence [12]. People conform because they implicitly recognise the group's norms and conformity is often interpersonally rewarding, as groups tend to consist of like-minded individuals. Therefore, nonconformity, such as challenging misinformation, can be seen unfavourable for two reasons: desire to conform to social norms (important standards for behaviour) and negative interpersonal consequences. Understanding user perceptions regarding challenging behaviour on social media will help determine whether users' decision to challenge others who post misinformation and do user correction is based on misperceptions.

In this study, we hypothesised that users might hold misperceptions towards challenging misinformation which could negatively influence their likelihood to challenge. When deciding whether to challenge misinformation people may weigh the anticipated negative consequences of challenging misinformation and decide whether to challenge depending on the benefits outweigh the drawbacks [20]. We categorised the anticipation of negative consequences into three categories: relationship consequences, negative impact on the person being challenged, and futility of the act of correction. We also examined how acceptable challenging misinformation is perceived by participants.

By limiting exposure to dissenting opinions through algorithms, SNS may lead individuals to become more entrenched in their existing beliefs and less receptive to alternative viewpoints, resulting in the creation of "echo chambers" [21]. Consequently, in an environment where questioning is not actively encouraged, constructive criticism is rarely received and correcting someone might be perceived as violating an important interpersonal norm. Indeed, studies showed that instead of confronting, users prefer to use indirect and invisible communication such as blocking, unfollowing and unfriending [22]. This can promote a culture of politeness and conformity where individuals are reluctant to voice dissenting viewpoints or challenge others, even when it is necessary for healthy discussion. To address this issue, we aim to explore users' views and perceptions of others, while also investigating potential misperceptions.

It is important to acknowledge that confronting might lead to tension or conflict, resulting in negative effects on relationships [23]. These "relationship consequences" refer to the users' views and perceptions about the potential negative impact on the relationships. Furthermore, we seek to understand the potential harm that people might experience when they are challenged. Some people may feel offended or embarrassed when their viewpoints are contradicted [5, 24], which we categorise as "negative impact on the person being challenged". In addition, we aim to investigate users' beliefs about the effectiveness of challenging misinformation to gain insights into whether individuals think their attempts to correct misinformation are likely to be successful or view such efforts as futile (hopeless/ineffective), resulting in no significant change [25]. Lastly, we examine the acceptability of challenging behaviour, often referred to as "injunctive norms" [26]. We aim to understand users' attitudes regarding the acceptability of challenging misinformation and their beliefs about whether such behaviour is acceptable to others.

The primary aim of this paper is not to explore reasons for not challenging misinformation [27]. It focuses on users' perceptions and potential misperceptions concerning correcting misinformation and whether these misperceptions impact likelihood to challenge, in a UK-based sample. By exploring users' attitudinal and behavioural misperceptions to explain their influence on challenging misinformation, we integrate and organize a previously disjointed and dispersed set of studies from diverse fields. While relationship consequences have been studied in face-to-face settings in China [28–30], and both individualist (UK) and collectivist (Arab) cultures misperceive the negative consequences of challenging misinformation [31], no study has assessed whether this influences likelihood to challenge.

Based on the above discussion, we pose the following research questions as depicted in Fig. 1:

RQ1: To what extent do individuals' own attitudes, their perception of others and misperception of the following influence likelihood to challenge misinformation?

RQ1.a: relationship consequences, ***RQ1.b:*** negative impact, ***RQ1.c:*** futility.

RQ1.d: injunctive norms.

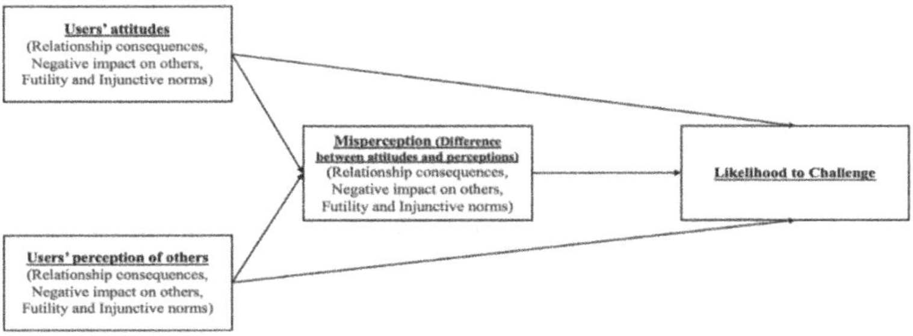

Fig. 1. Diagram illustrating the research questions

2 Methodology

We provided definitions of key concepts including "misinformation" and "challenging" to ensure participants had a clear understanding. We defined the term "acquaintance" as connections in the territory between strangers and intimates on Facebook such as former co-workers, neighbours, or someone you know from a Facebook group. This clarification was important, as research suggests that individuals may be less likely to challenge misinformation shared by acquaintances than that shared by strangers, given the potential impact on their social relationships.

Facebook was selected due to its extensive global usage (approximately 2.91 billion monthly active users) [32] and its prevalence across all age groups [33]. Moreover, 67% of internet users mentioned that they encountered fake news on Facebook [34].

The questionnaire assessed both participants' own attitudes towards being challenged when posting misinformation, which we refer to as "self-report" and their perceptions of others' attitudes towards being challenged when posting misinformation, which we refer to as "perceived".

2.1 Recruitment and Data Collection

The questionnaire for this study was designed using Qualtrics™ (https://www.qualtrics.com), a web-based survey platform. Participants were recruited through Prolific™ (www.prolific.co), an established platform for online recruitment for research studies, to participate in an online survey exploring attitudes to and perceptions of challenging misinformation on social media. Inclusion criteria included being aged 18+ years, fluent in English, UK-based, using Facebook with an authentic identity, and having encountered misinformation on Facebook. Eligible participants received £4 for their time.

Participants who met the inclusion criteria were given the link to the questionnaire after reading the participant information sheet and giving consent and informed of their right to withdraw from the survey at any stage. The survey included three attention checks to ensure the integrity of responses. Participants who did not pass at least two checks were excluded from the analysis. The study received approval from Bournemouth University Research Ethics Committee, UK. The final dataset comprised 250 participants (age 18–77; 143 female, 104 male, 3 non-binary).

2.2 Measures

Demographic Measures. Participants provided their gender, age and educational level.

Fake News Presented. The content presented featured an article about an asteroid potentially hitting Earth, which had been previously circulated on CNN's iReport news hub in 2014[1]. We made it clear that the news provided was false as the study objective was not to assess ability to detect misinformation but to gain insights regarding challenging false content and participation in user correction. We avoided political or social issues, which could influence participants' responses [35]. To enhance the scenario's authenticity and avoid participant perception of anonymity, we deliberately disclosed the name and photo of the Facebook user sharing the content. We selected the gender-neutral name Alex to minimise gender confounding. We instructed participants to answer as if they were participating in social media and user correction using their real identity, to control for the online disinhibition effect.

[1] https://shorturl.at/ZJqiX.

Likelihood to Challenge. Likelihood to challenge was measured using a single self-report question *"Think about a specific time when you saw misinformation on your Facebook feed, shared by an acquaintance in your Facebook network. How likely were you to challenge the misinformation in a way that others can see?"* on a 7-point scale from "extremely unlikely" to "extremely likely" [5, 34]. In later analysis, we dichotomized the scores as unlikely to challenge, scored 1–3 on the likelihood question (n = 147), and at least neutral, scored 4–7 on the likelihood of challenging question (n = 103) as in previous research utilizing a similar approach [5].

Relationship Consequences. We adapted the items from the scale of Perceived Relationship Costs [29, 30]. Sample items include "It would offend me (them)" and "The relationship between us will deteriorate". Internal reliability was high for both actual, i.e., how the participants think regarding being challenged ($\alpha = 0.92$) and perceived, i.e., how they believe others think ($\alpha = 0.92$).

Negative Impact on the Person Being Challenged. Based on previous research [4, 36, 37], actual negative impact and perceived negative impact were assessed with two items ($\alpha = 0.82$ and $\alpha = 0.77$ respectively) on a 7-point Likert scale from Strongly Disagree to Strongly Agree. The items are "*I (they) would feel embarrassed*' and '*I (they) would feel that I (they) will be viewed as untrustworthy by other social contacts*".

Futility of the Act of Challenging. Actual and perceived futility were assessed with two items ($\alpha = 0.68$ and $\alpha = 0.66$ respectively) on a 7-point Likert scale from "strongly disagree" to "strongly agree", adapted from Milliken et al.'s study on employee silence [37]. Items include "*Challenging would change my (their) mind that the information I (they) shared is true*" and "*Challenging would make me (them) delete the post*".

Injunctive Norms. Two items assessed perceptions of others' attitudes regarding the acceptability of challenging misinformation: "How do you find challenging others on Facebook when they share misinformation?" and *"How would a typical person from your Facebook network find challenging others when they share misinformation?"* on a 7-point scale from (1) very unacceptable to (7) very acceptable.

2.3 Data Analysis

Paired-samples t-tests evaluated within-participant discrepancy and compared personal reactions towards being challenged and perceived reactions of others when they are challenged. Chi-square tests assessed differences between nominal variables, and Spearman's correlation associations between continuous variables. Misperception score was computed for each participant by subtracting the self-report mean from the participants' perception (them score). A positive misperception score was interpreted as an overestimation, while a negative score an underestimation of the attitudes. Further details on the method are presented elsewhere [38]. Ordinal logistic regression evaluated the impact of the predictor variables on the ordinal dependent variable. Analysis was conducted in SPSS, Version 28. The study design and data associated with this paper are available on the Open Science Framework (OSF) at https://osf.io/uny7g/.

3 Results

3.1 Descriptive Statistics

Overall, 250 participants completed the survey, of which 57.2% (143) identified as female, 41.6% (104) male, and 1.2% (3) non-binary. Regarding age, 17.8% (44) were aged 18–24 years, 37.2% (92) 25–34, 26.7% (66) 35–44, and 18.2% (45) over 45. Most (62.8%, 157) had at least a university degree, 22.8% (57) a college degree, and 14.4% (36) had completed secondary education.

3.2 Own Attitudes and Perceptions of Others' Attitudes vs. Likelihood to Challenge

A Spearman's rank-order correlation analysis investigated whether likelihood to challenge is related to self-reported attitude and perceived attitude of others. Likelihood to challenge was negatively correlated with self-reported relationship consequences $r_s(248) = -.214$, $p < .01$, self-reported negative impact $r_s(248) = -.185$, $p < .01$ and perceived negative impact $r_s(248) = -.147$ $p < .05$. There was also a statistically significant, positive correlation between self-reported injunctive norms and likelihood to challenge $r_s(248) = .347$, $p < .01$ and perceived injunctive norms and likelihood to challenge $r_s(248) = .167, p < .01$. Respondents who report fewer negative consequences were more willing to challenge, and individuals who think that challenging is socially acceptable are also more willing to challenge misinformation (See Table 1).

Table 1. Correlations between likelihood to challenge and self-reported attitudes and perceptions of others

	Self-reported relationship consequences	Self-reported futility	Self-reported negative impact	Self-reported injunctive norms	Perceived relationship consequences	Perceived futility	Perceived negative impact	Perceived injunctive norms
Likelihood to challenge	$-.214^{**}$	0.010	$-.185^{**}$	$.347^{**}$	−0.120	0.034	$-.147^{*}$	$.167^{**}$

* p < .05, ** p < .01;

After examining the correlation between the variables, two ordinal logistic regression analyses were performed to further explore their relationship. In the first model we analysed whether people's self-reported attitudes impact likelihood to challenge misinformation controlling for age. The assumption of proportional odds was not met, as assessed by a full likelihood ratio test comparing the fit of the proportional odds model to a model with varying location parameters, $\chi2(25) = 54.081$, $p < .001$. Therefore, an examination of the assumption of proportional odds was undertaken by running separate binomial logistic regressions on cumulative dichotomous dependent variables. This examination showed that for most variables the assumption of proportional odds appears tenable. The deviance goodness-of-fit test indicated that the model was a good fit to the observed data, $\chi2(1489) = 866.955$, $p = 1$, but most cells were sparse with zero frequencies in 85.7% of cells. However, the final model statistically significantly predicted the dependent variable over and above the intercept-only model, $\chi2(5) = 53.819$, $p < .001$. In terms of the individual predictor variables, injunctive norms and age had a significant positive effect on likelihood to challenge (odds ratio of 1.570 (95% CI, 1.315 to 1.873) $p < .001$ and odds ratio of 1.036 (95% CI, 1.015 to 1.058) $p < .001$ respectively). Overall, the results suggest that self-reported injunctive norms and age were significant predictors of likelihood to challenge misinformation (See Table 2).

In the second model we analysed whether people's perceptions of others impacts their likelihood to challenge misinformation, controlling for age. The assumption of proportional odds was not met, as assessed by a full likelihood ratio test comparing the fit of the proportional odds model to a model with varying location parameters, $\chi2(25) = 42.759$, $p < .05$. Therefore, an examination of the assumption of proportional odds was undertaken by running separate binomial logistic regressions on cumulative dichotomous dependent variables. This examination showed that for most of the variables the assumption of proportional odds appears tenable. The deviance goodness-of-fit test indicated that the model was a good fit to the observed data, $\chi^2(1483) = 891.180$, $p = 1$, but most cells were sparse with zero frequencies in 85.7% of cells. However, the final model statistically significantly predicted the dependent variable over and above the intercept-only model, $\chi^2(5) = 26.821$, $p < .001$. Regarding the independent variables, "age", "perceived injunctive norms" and "perceived negative impact on others" significantly predicted likelihood to challenge. An increase in age was associated with an increase in the odds of likelihood to challenge with an odds ratio of 1.038 (95% CI, 1.017 to 1.060), $p < .001$ and an increase in perceived injunctive norms was associated with an increase in the odds of likelihood to challenge with an odds ratio of 1.255 (95% CI, 1.055 to 1.493), $p < .05$. A decrease in perceived negative impact on others was associated with an increase in the odds of likelihood to challenge, with an odds ratio of 0.749 (95% CI, 0.591 to 0.949), $p < .05$. Perceived futility is marginally significant with the odds ratio of 1.228 (95% CI, 0.998 to 1.510), $p = .052$. Overall, the results suggest likelihood to challenge misinformation is influenced by age, perceived injunctive norms, and perceived negative impact on others (See Table 2).

Table 2. Ordinal logistic regression results for predicting the likelihood to challenge

Model	Predictors	B	SE	Odds Ratio	95.0% CI	
					Lower Bound	Upper Bound
	Age	0.036	0.011	1.036**	1.015	1.058
1 Self-reported	Relationship consequences	−0.130	0.098	0.878	0.723	1.065
	Futility	−0.037	0.089	0.963	0.824	1.171
	Negative impact on oneself	−0.152	0.098	0.850	0.808	1.149

(*continued*)

Table 2. (*continued*)

Model	Predictors	B	SE	Odds Ratio	95.0% CI	
					Lower Bound	Upper Bound
	Injunctive norms	0.451	0.901	1.570**	1.315	1.873
	Age	0.038	0.011	1.038**	1.017	1.060
2 Perceived	Relationship consequences	0.034	0.115	1.035	0.825	1.297
	Futility	0.205	0.106	1.228	0.998	1.510
	Negative impact on others	−0.289	0.121	0.749*	0.591	0.949
	Injunctive norms	0.227	0.089	1.255*	1.055	1.493

* p < .05, ** p < .001

3.3 Overestimating Others' Negative Attitudes (Misperception) vs. Likelihood to Challenge

A Spearman's rank-order correlation analysis was conducted to investigate whether likelihood to challenge is related to the amount of discrepancy between self-reported attitude and perceived attitude of others. As seen in Table 3, the analysis showed that there was a statistically significant, negative correlation between the misperception of negative impact and likelihood to challenge $r_s(248) = -.147, p < .05$ and. The greater the participants perceive the difference between self and others regarding negative impact, the lower the likelihood to challenge. There was also a statistically significant, positive correlation between misperception of injunctive norms and likelihood to challenge $r_s(248) = .167, p < .01$. The more the respondents perceived a discrepancy between self and others regarding the acceptability of the behaviour, the greater their intention to challenge misinformation.

Table 3. Correlations between likelihood to challenge and misperceptions

	Misperception of relationship consequences	Misperception of futility	Misperception of negative impact	Misperception of injunctive norms
Likelihood to challenge	−0.120	0.034	−.147*	.167**

* p < .05, ** p < .01.

As the correlation provided us with the relationship between the differences (misperception) and likelihood to challenge, we further explored the differences by forming a hypothesis of overestimation. Misperception can include both overestimation and underestimation. According to our hypothesis, individuals' likelihood to challenge may be influenced by their tendency to overestimate negative reactions of others or to perceive that others find challenging misinformation acceptable. Therefore, prior to regression analysis, we operationalised misperception data as overestimation. To do this, we computed a "misperception score" for each participant by subtracting perception of the group attitude ("them" score) from the actual mean of the group's responses to the attitude questions (average of all "me" scores). Norm misperception scores ranged from − 3.95 to 3.54. Positive norm misperception scores (actual attitude of participants < perceived attitude of others) indicated a tendency to overestimate others' negative attitude when challenged whereas negative misperception scores (actual attitude of participants > perceived attitude of others) or zero score (actual attitude of participants = perceived attitude of others) were classified as having no overestimation. This approach enabled grouping of participants based on whether they overestimated or not.

A cumulative odds ordinal logistic regression with proportional odds was run to determine the effect of overestimating relationship costs, overestimating futility, overestimating negative impact, and overestimating injunctive norms controlling for age on the likelihood to challenge, while controlling for age (See Table 4). The assumption of proportional odds was not met, as assessed by a full likelihood ratio test comparing the fit of the proportional odds model to a model with varying location parameters, $\chi2(25) = 119.451$, $p < .001$. Therefore, an examination of the assumption of proportional odds has undertaken by running separate binomial logistic regressions on cumulative dichotomous dependent variables. This examination showed that for most of the variables the assumption of proportional odds appears tenable. The deviance goodness-of-fit test indicated that the model was a good fit to the observed data, $\chi^2(1081) = 750.234, p = 1$, but most cells were sparse with zero frequencies in 81.9% of cells. However, the final model statistically significantly predicted the dependent variable over and above the intercept-only model, $\chi^2(5) = 16.808$, $p < .01$. Regarding the independent variables, "age" and overestimating "injunctive norms" significantly predicted likelihood to challenge (odds ratio of 1.034 (95% CI, 1.013 to 1.056) $p < .01$ and odds ratio of 0.613 (95% CI, 0.390 to 0.963), $p < .05$) respectively.

Table 4. Ordinal logistic regression results for predicting the likelihood to challenge with predictors including age, overestimating negative consequences and injunctive norms

Predictors	B	SE	Odds Ratio	95.0% CI	
				Lower Bound	Upper Bound
Age	0.034	0.010	1.034*	1.013	1.056
Overestimating Relationship Consequences	−0.193	0.274	1.213	0.709	2.076
Overestimating Futility	0.165	0.261	0.848	0.508	1.415
Overestimating Negative Impact on Themselves	−0.258	0.240	1.295	0.809	2.072
Overestimating Injunctive Norms	0.489	0.230	0.613*	0.390	0.963

Findings related to the factors influencing the likelihood to challenge misinformation are summarized in Table 5, highlighting how various attitudes, perceptions, and misperceptions impact individuals' likelihood to challenge misinformation.

Table 5. Key relationships and findings for each factor related to likelihood to challenge misinformation.

Factor	Impact on Likelihood to Challenge Misinformation	Key Findings
Self-Reported Relationship Consequences	Negative Impact	People are less likely to challenge misinformation if they believe it will harm their relationships
Self-Reported Negative Impact on Themselves	Negative Impact	People are less likely to challenge misinformation if they think it will have a negative impact on themselves
Self-Reported Injunctive Norms	Positive Impact	People are more likely to challenge misinformation if they believe it is socially acceptable to do so
Perceived Negative Impact (on others)	Negative Impact	People are less likely to challenge misinformation if they perceive it will negatively impact others
Perceived Injunctive Norms	Positive Impact	People are more likely to challenge misinformation if they believe others find it socially acceptable to do so
Misperception: Negative Impact	Negative Impact	Overestimating the harm of challenging misinformation decreases the likelihood of challenging it
Misperception: Injunctive Norms	Positive Impact (Paradoxical)	Overestimating others' disapproval paradoxically increases the likelihood of challenging misinformation
Self-reported Predictors	Positive Impact	Belief in injunctive norms (social acceptability) is the strongest predictor; age also positively predicts likelihood to challenge misinformation

(continued)

Table 5. *(continued)*

Factor	Impact on Likelihood to Challenge Misinformation	Key Findings
Perceived Predictors	Positive Impact	Perceived injunctive norms and lower perceived negative impact on others significantly increase likelihood to challenge
Overestimation of Others' Disapproval	Positive Impact	People who overestimate others' disapproval of challenging are more likely to challenge misinformation

4 Discussion

We aimed to determine the extent to which people's attitudes, perceptions of others' attitudes and misperceptions of relationship consequences, negative impact, futility and injunctive norms influence likelihood of challenging misinformation. This is an essential step towards the development of socio-technical countermeasures that help mitigate barriers to challenging misinformation and facilitate users' ability to critically analyse and express their opinions about misinformation. Identifying users' barriers, needs and motivations to challenging misinformation is a crucial starting point for a user-centred approach to design communication systems to ensure that technology meets the users' needs [39]. Such design approaches encounter numerous challenges, including understanding the users' interaction with technology across contexts, gathering and analysing user requirements, and translating those requirements into practical designs [40]. To overcome these challenges, we examined users' assumptions and decision-making processes when interacting with others on social media. This enhanced understanding of users' behaviours and preferences regarding confronting in the digital sphere.

One key finding is that there is indeed pluralistic ignorance regarding negative consequences (relationship consequences and futility) and injunctive norms in terms of challenging misinformation on social media, in line with prior research findings that individuals overestimate others' negative reactions [29]. Our results show people both overestimate the damage corrections will cause to their relationships and believe that correcting others is more futile than it is. This pluralistic ignorance may act as a barrier to challenging misinformation and in turn lead to the propagation of misinformation.

Likelihood to challenge misinformation is influenced by perceptions of the potential negative impact on others (e.g., causing embarrassment), as well as perceptions of whether challenging the misinformation is socially acceptable. Results suggests that people avoid challenging misinformation due to altruistic concerns, such as prioritising others' well-being and happiness over their own personal gain or benefits [41]. Reasons for these altruistic intentions might be to help others [41], enhance social status [42] or avoid punishment [43]. If someone anticipates that challenging someone may cause harm or distress to the other person, they may choose to avoid conflict to protect the other person from negative consequences. In this way, the decision not to challenge misinformation is driven by a concern for the other person's well-being, rather than fear of negative consequences for themselves. One potential explanation for prioritising others' image over challenging misinformation is that individuals might be unsure or poorly informed about the topic and thus feel it is more important to protect others' reputations rather than potentially cause embarrassment or harm by challenging misinformation. People's likelihood to challenge is also influenced by their opinions about how acceptable challenging is in the online environment, in line with research that suggests people are more inclined to adopt a behaviour if they perceive it to be commonly accepted by others [12].

Likelihood to challenge was associated with the discrepancy between the self and others' beliefs regarding negative impact and injunctive norms. Our results suggest that people's misperceptions about social norms predicted likelihood to challenge. Specifically, when people think challenging misinformation is more acceptable for others, they are more likely to challenge misinformation themselves, in line with findings that obligation is a cognitive reference point for evaluating the appropriateness of behaviour [44]. As individuals internalise their sense of obligation, they develop a greater sense of responsibility for correcting misinformation, ultimately contributing to a normative culture of correcting misinformation.

These results highlight the importance of promoting positive norms around speaking out against misinformation to encourage more people to challenge it. Effectively promoting norms is primarily reliant on how one's behaviour is perceived by other members of the pertinent group, in a way that leads them to alter their normative expectations. People might experience discomfort when challenging misinformation, as they perceive it to be against the injunctive norm, in line with the argument that people experience dissonance when they disagree with each other. In one study, participants who held extreme opinions on certain issues were asked to discuss those opinions with people who had opposing views. Before the discussion, participants reported negative emotions (e.g., discomfort, tension), indicating cognitive dissonance [45].

Cognitive dissonance arising from nonconformity can lead to a decrease in misinformation being challenged. This issue could be addressed by decreasing stigma around conflict. While stigma is not necessarily attached to challenging others, there are social norms around correcting others' mistakes. Social norms around conflict avoidance can make it difficult for people to speak out against vaccine misinformation due to concerns about risking relationships with friends and family [3]. However, conflicts can be resolved by engaging in conversations that aim to understand the issue better and finding a common ground, known as "deliberate discourse" [46], and can lead to growth and positive change. To that end, fostering a culture that prioritises openness and critical thinking could encourage people to challenge misinformation.

Although there are various interventions for dealing with misinformation online, including corrections and debunking, transparency measures and visibility reduction [47], limited research has focused on user-centric social media designs that encourage users to challenge misinformation [48]. Promoting critical thinking, constructive argumentation, and deliberate discourse by providing features on social media can facilitate the process of challenging misinformation. Taking a user-centric approach to change their attitudes and behaviours can be more effective as it considers their perspectives. In fact, giving users who will ultimately benefit from the design a central role in the development of ideas, concepts and designs can improve the chances of meeting their needs and expectations [49]. In that sense, co-design where the end users participate directly and proactively in the design development process has much to offer [49].

This study has some limitations. It was conducted with 250 UK-based social media users and may not represent attitudes and behaviours of other cultures. Open communication and confrontation are perceived as more socially acceptable in Western societies [50]. Additionally, the reliance on self-reports may introduce recall error [51] or social desirability bias [52], highlighting the need for observational and experimental designs to assess challenging misinformation. While this research provides a preliminary understanding of perceptions and norms surrounding challenging misinformation, future studies should explore cross-cultural differences, and platform-specific influences to develop a more comprehensive understanding.

5 Conclusion

This paper assessed the extent to which individuals' attitudes, perceptions of others' attitudes and misperception of relationship consequences, negative impact, futility and injunctive norms influence likelihood to challenge misinformation on social media. It also confirms previous findings that social media users anticipate more negative consequences from challenging misinformation than is the case. People are more likely to challenge misinformation not only if they believe it is socially acceptable to do so, but also if they believe others think it is socially acceptable to do so. Our study is novel in showing that the misperception of these consequences plays a significant role in discouraging users from acting. It also highlights how addressing these misperceptions can create opportunities to shift user behavior towards more proactive engagement with misinformation. These findings highlight the importance of developing positive norms regarding speaking out against misinformation. Designing social media in a way that facilitates the challenging of misinformation, supported by user-centric designs, is needed to change attitudes and behaviours around correcting misinformation.

Acknowledgements. This publication was supported by NPRP 14 Cluster grant # NPRP 14C-0916–210015 from the Qatar National Research Fund (a member of Qatar Foundation). The findings reflect the work and are solely the authors' responsibility.

Supplementary Material and Data Availability. The study design and data associated with this paper are available on the Open Science Framework (OSF) at the following link: https://osf.io/uny7g/.

References

1. Walter, N., Murphy, S.: How to unring the bell: a meta-analytic approach to correction of misinformation. Commun. Monogr. **85**(3), 423–441 (2018)

2. Bode, L., Vraga, E.K.: Correction experiences on social media during COVID-19. Soc. Media + Soc. **7**(2) (2021)
3. Chadwick, A., Vaccari, C.: News sharing on UK social media: misinformation, disinformation, and correction. Online Civic Culture Centre, Loughborough University, Loughborough (2019)
4. Tandoc, E.C., Lim, D., Ling, R.: Diffusion of disinformation: how social media users respond to fake news and why. Journalism **21**(3), 381–398 (2020)
5. Tully, M., Bode, L., Vraga, E.K.: Mobilizing users: does exposure to misinformation and its correction affect users' responses to a health misinformation post? Soc. Media + Soc. **6**(4) (2020)
6. Miller, D.T., Monin, B., Prentice, D.A.: Pluralistic ignorance and inconsistency between private attitudes and public behaviors. In: Attitudes, Behavior, and Social Context, pp. 95–114. Psychology Press (1999)
7. Ahmed, A.A.A., Aljabouh, A., Donepudi, P.K., Choi, M.S.: Detecting fake news using machine learning: a systematic literature review. Psychol. Educ. **58**(1), 1932–1939 (2021)
8. Ibrishimova, M.D., Li, K.F.: A machine learning approach to fake news detection using knowledge verification and natural language processing. In: Advances in Intelligent Networking and Collaborative Systems, pp. 223–234 (2020)
9. de Oliveira, N.R., Pisa, P.S., Lopez, M.A., de Medeiros, D.S.V., Mattos, D.M.F.: Identifying fake news on social networks based on natural language processing: trends and challenges. Information **12**(1), 38 (2021)
10. Nakov, P., et al.: Overview of the clef–2021 checkthat! Lab on detecting check-worthy claims, previously fact-checked claims, and fake news. In: Advances in Information Retrieval (ECIR), pp. 639–649 (2021)
11. Figueira, Á., Oliveira, L.: The current state of fake news: challenges and opportunities. Procedia Comput. Sci. **121**, 817–825 (2017)
12. Cialdini, R.B., Trost, M.R.: Social influence: social norms, conformity and compliance. In: Gilbert, D.T., Fiske, S.T., Lindzey, G. (eds.) The Handbook of Social Psychology, vols. 1–2, 4th edn., pp. 151–192. McGraw-Hill (1998)
13. Dempsey, R.C., McAlaney, J., Bewick, B.M.: A critical appraisal of the social norms approach as an interventional strategy for health-related behavior and attitude change. Front. Psychol. **9**, 2180 (2018)
14. Ajzen, I.: The theory of planned behavior. Organ. Behav. Hum. Decis. Process. **50**(2), 179–211 (1991)
15. McLaughlin, C., Vitak, J.: Norm evolution and violation on Facebook. New Media Soc. **14**(2), 299–315 (2012)
16. Wagner, A.J.M.: Do not click "like" when somebody has died: the role of norms for mourning practices in social media. Soc. Media+ Soc. **4**(1) (2018)
17. Paddock, D.L., Bell, B.T.: "It's better saying I look fat instead of saying you look fat": a qualitative study of UK adolescents' understanding of appearance-related interactions on social media. J. Adolesc. Res. **39**(2), 243–71 (2021)
18. Neubaum, G., Krämer, N.C.: What do we fear? expected sanctions for expressing minority opinions in offline and online communication. Commun. Res. **45**(2), 139–164 (2018)
19. Rashidi, Y., Kapadia, A., Nippert-Eng, C., Su, N.M.: It's easier than causing confrontation": sanctioning strategies to maintain social norms and privacy on social media. Proc. ACM Hum.-Comput. Interact. **4**, 1–25 (2020)
20. Sunstein, C.R.: Cognition and cost-benefit analysis. J. Leg. Stud. **29**(S2), 1059–1103 (2000)
21. Cinelli, M., Morales, G.D.F., Galeazzi, A., Quattrociocchi, W., Starnini, M.: The echo chamber effect on social media. Proc. Natl. Acad. Sci. **118**(9), e2023301118 (2021)
22. Wu, T.-Y., Xu, X., Atkin, D.: The alternatives to being silent: exploring opinion expression avoidance strategies for discussing politics on Facebook. Internet Res. **30**(6), 1709–29 (2020)

23. Sleeper, M., Balebako, R., Das, S., McConahy, A.L., Wiese, J., Cranor, L.F.: The post that wasn't: exploring self-censorship on facebook. In: Proceedings of the 2013 Conference on Computer Supported Cooperative Work, pp. 793–802 (2013)
24. Steen-Johnsen, K., Enjolras, B.: The fear of offending: social norms and freedom of expression. Society **53**, 352–362 (2016)
25. Van Houtte, M., Stevens, P.A.J.: Sense of futility: the missing link between track position and self-reported school misconduct. Youth Soc. **40**(2), 245–264 (2008)
26. Lapinski, M.K., Rimal, R.N.: An explication of social norms. Commun. Theory **15**(2), 127–147 (2005)
27. Gurgun, S., Cemiloglu, D., Arden-Close, E., Phalp, K., Nakov, P., Ali, R.: Why do we not stand up to misinformation? Factors influencing the likelihood of challenging misinformation on social media and the role of demographics. Technol. Soc. **76**, 102444 (2024)
28. Wei, X., Zhang, Z.-X., Chen, X.-P.: I will speak up if my voice is socially desirable: a moderated mediating process of promotive versus prohibitive voice. J. Appl. Psychol. **100**(5), 1641 (2015)
29. Zhang, Z.-X., Wei, X.: Superficial harmony and conflict avoidance resulting from negative anticipation in the workplace. Manag. Organ. Rev. **13**(4), 795–820 (2017)
30. Zhang, Y., Zhang, Z.-X., Wang, M.: Harmony, illusory relationship costs, and conflict resolution in Chinese contexts. In: Leung, A.K.Y., Chiu, C.-Y., Hong, Y.-Y. (eds.) Cultural Processes: A Social Psychological Perspective, pp. 188–209. Cambridge University Press, Cambridge (2011)
31. Gurgun, S., Norman, M., Arden-Close, E., Phalp, K., Ali, R.: How would i be perceived if i challenge individuals sharing misinformation? Exploring misperceptions in the UK and Arab samples and the potential for the social norms approach. In: Baghaei, N., Ali, R., Win, K., Oyibo, K. (eds.) Persuasive Technology, vol. 14636, pp. 133–150. Springer, Cham (2024)
32. Statista. Distribution of Facebook users worldwide as of January 2022, by age and gender (2022). https://www.statista.com/statistics/376128/facebook-global-user-age-distribution. Accessed March 2022
33. Statista. Number of Facebook users in the United Kingdom (UK) from 2018 to 2028(2024). https://www.statista.com/statistics/553538/predicted-number-of-facebook-users-in-the-united-kingdom-uk. Accessed January 2025
34. Cohen, E., et al.: To correct or not to correct? Social identity threats increase willingness to denounce fake news through presumed media influence and hostile media perceptions. Commun. Res. Rep. **37**(5), 263–75 (2020)
35. Geiger, N., Swim, J.: Climate of silence: pluralistic ignorance as a barrier to climate change discussion. J. Environ. Psychol. **47**, 79–90 (2016)
36. Altay, S., Hacquin, A.-S., Mercier, H.: Why do so few people share fake news? It hurts their reputation. New Media Soc. **24**(6), 1303–24 (2022)
37. Milliken, F.J., Morrison, E.W., Hewlin, P.F.: An exploratory study of employee silence: issues that employees don't communicate upward and why. J. Manage. Stud. **40**(6), 1453–1476 (2003)
38. Sandstrom, M., Makover, H., Bartini, M.: Social context of bullying: do misperceptions of group norms influence children's responses to witnessed episodes? Soc. Influ. **8**(2–3), 196–215 (2013)
39. Muller, M., Shami, N.S., Millen, D.R., Feinberg, J.: We are all lurkers: consuming behaviors among authors and readers in an enterprise file-sharing service. In: Proceedings of the ACM (2010)
40. Stephanidis, C., Antona, M., Ntoa, S.: Human factors in ambient intelligence environments. In: Gavriel Salvendy, W.K. (ed.) Handbook of Human Factors and Ergonomics, pp. 1354–1373. Wiley Online Library (2012)

41. Batson, C.D., Van Lange, P.A.M., Ahmad, N., Lishner, D.A., Hogg, M.A., Cooper, J.: Altruism and helping behavior. In: The Sage Handbook of Social Psychology, pp. 241–258. Sage (2007)
42. Kafashan, S., Sparks, A., Griskevicius, V., Barclay, P.: Prosocial behavior and social status. In: Cheng, J.T., Tracy, J.L., Anderson, C. (eds.) The psychology of social status, pp. 139–158. Springer (2014)
43. Hoffman, M.L.: Empathy and prosocial behavior. Handb. Emot. **3**, 440–455 (2008)
44. Heywood, J.L.: The cognitive and emotional components of behavior norms in outdoor recreation. Leisure Sci. **24**(3–4), 271–281 (2002)
45. Matz, D.C., Wood, W.: Cognitive dissonance in groups: the consequences of disagreement. J. Pers. Soc. Psychol. **88**(1), 22 (2005)
46. Johnson, D.W., Johnson, R.T., Smith, K.A.: Constructive controversy: the educative power of intellectual conflict. Change Mag. High. Learn. **32**(1), 28–37 (2000)
47. Hartwig, K., Doell, F., Reuter, C.: The landscape of user-centered misinformation interventions-a systematic literature review. arXiv preprint arXiv:2301.06517 (2023)
48. Gurgun, S., Arden-Close, E., McAlaney, J., Phalp, K., Ali, R.: Can we re-design social media to persuade people to challenge misinformation? An exploratory study. In: Persuasive Technology. Lecture Notes in Computer Science, vol. 13832, pp. 123–141. Springer, Cham (2023)
49. Sanders, E.B.N., Stappers, P.J.: Co-creation and the new landscapes of design. CoDesign **4**(1), 5–18 (2008)
50. Friedman, R., Chi, S.-C., Liu, L.A.: An expectancy model of Chinese-American differences in conflict-avoiding. J. Int. Bus. Stud. **37**, 76–91 (2006)
51. Sudman, S., Bradburn, N.M., Schwarz, N.: Thinking About Answers: The Application of Cognitive Processes to Survey Methodology. Jossey-Bass (1996)
52. Grimm, P.: Social desirability bias. Wiley Int. Encycl. Mark. (2010)

Do Social Media Simultaneously Contribute to Well-Being and Use Disorder? Empirical Evidence and Design Challenges

Tourjana Islam Supti[1], Ala Yankouskaya[2], Sameha Alshakhsi[1], Areej Babiker[1], Dena Al-Thani[1], and Raian Ali[1(✉)]

[1] College of Science and Engineering, Hamad Bin Khalifa University, Doha, Qatar
raali2@hbku.edu.qa

[2] Department of Psychology, Bournemouth University, Poole, UK

Abstract. Social media has played a significant role in enhancing social well-being (SWB) across its five dimensions: integration, acceptance, contribution, actualization, and sense of coherence. However, research has shown that certain usage patterns and relationships with social media can lead to what is termed Social Media Disorder (SMD), a non-clinical term describing a compulsive, excessive, and obsessive relationship with social media that causes harm to individuals, their social circles, and society as a whole. This study examined how participants from the UK and the Arab Gulf Cooperation Council (GCC) region perceived the contribution of social media to the five dimensions of SWB and its role in fostering SMD. A cross-sectional survey of 563 participants (255 from the UK, 308 from the GCC) revealed differences between the two populations in their perceptions of social media as both an enabler of SWB and a facilitator of tendencies toward SMD. Additionally, the findings showed a clear intersection between these dual roles, hence posing major design challenges. This research calls for a shift in social media design to enhance SWB without triggering or facilitating SMD. It emphasizes the complexity of achieving this balance and advocates for novel human-computer interaction features and modalities to address the issue.

Keywords: Social well-being · Social media · Social media disorder · Human-centered design

1 Introduction

Since the rise of social media, how we make friends, communicate, and keep in touch has changed drastically. Many social media platforms, such as Facebook (FB), Instagram, and Twitter, have become integral to our daily lives, allowing users to share and stay informed like never before. The rapid advancement of technology has connected the world, making interactions more convenient but often superficial [1]. Beyond personal connections, social media has transformed global communication and fostered community building while also presenting challenges such as misinformation, social isolation,

J. Grabis et al. (Eds.): RCIS 2025, LNBIP 547, pp. 347–361, 2025.
https://doi.org/10.1007/978-3-031-92474-3_21

and cultural tensions [2]. Besides its positive impact on life, it has also introduced challenges that can undermine individuals' well-being and mental health issues, such as anxiety, depression, loneliness, FoMO (Fear of Missing Out) and cyberbullying [3–5]. Excessive reliance on social media for social connection has been linked to feelings of isolation, dissatisfaction, and mental health issues. In particular, there have been growing concerns about the compulsive or problematic use of social media, commonly known as 'social media addiction' or 'social media disorder' [6–8].

Given these challenges, it is important to consider the broader concept of well-being, often described from an individual's perspective, commonly referred to as 'subjective well-being' [9]. However, it is also helpful to think of well-being as something shaped by the social environment beyond the mental or emotional state of the individuals separately [9]. Social well-being (SWB) refers to an individual's perception of their connectedness, contribution, and sense of belonging within their social environment [10]. It encompasses dimensions such as integration, coherence, contribution, actualization and acceptance [10]. SWB is not limited to physical interactions; it can also be fostered online. Online environments may play an important role in shaping an individual's sense of connectedness, belonging, and contribution to society.

Social media platforms have been widely recognized for enhancing well-being through fostering social connections and facilitating access to information and resources. Research indicates that engaging with supportive online communities can bolster self-esteem, reduce loneliness, and improve psychological health [11]. For example, platforms like Facebook and Instagram enable users to maintain meaningful relationships, particularly when face-to-face interactions are limited, such as during the COVID-19 pandemic [12]. The Uses and Gratifications Theory (U&G) [13] further posits that people actively seek out media to satisfy various needs, such as social interaction, self-presentation, and entertainment. This theory has been widely applied to understand how users engage with social networking sites and online communities. For instance, a study on social networking sites highlighted that users derive significant satisfaction from fulfilling their needs for social interaction, self-presentation, and entertainment [14]. Another research found that the gratifications obtained from social media use, such as information seeking, incentive, and brand likeability, strongly affect customer participation in social media brand communities [15].

Moreover, social media provides opportunities for advocacy and self-expression, empowering individuals to share personal narratives and amplify marginalized voices, thus contributing. This participatory aspect fosters a sense of belonging and purpose, which are critical components of well-being as defined by frameworks such as the PERMA model (Positive Emotion, Engagement, Relationships, Meaning, and Accomplishment) [16]. Additionally, social media serves as a valuable educational tool, offering resources on mental health, career development, and other life-enhancing topics. Studies by Primack et al. and Meshi et al. have shown that social media use decreases social isolation [17, 18]. Furthermore, it helps cultivate a feeling of connectedness with others who share similar interests [19], further reducing social isolation. Studies state that social media is essential in shaping one's presence, identity, and reputation, facilitating social interaction, forming and maintaining relationships, sharing ideas, and significantly correlating with social support [20, 21]. Notably, behaviors like FoMO, rooted in social media, are not solely trivial; they often relate to meaningful pursuits such as job opportunities and scarce chances [22].

Despite these benefits, an increasing body of evidence highlights the negative side of social media use. Social Media Disorder (SMD), often considered a form of behavioral addiction, is characterized by compulsive engagement, difficulty disengaging, loss of control, neglect of offline responsibilities, and negative consequences on daily life [8]. The SMD Scale is a validated measurement tool based on conceptualizing SMD as a condition characterized by compulsive, excessive, and problematic use of social media [8]. It highlights that SMD is similar to Internet Gaming Disorder (IGD) and can be measured using similar symptoms and diagnostic criteria [23]. Van den Eijnden's study defines SMD through nine criteria (preoccupation, tolerance, withdrawal, etc.), with a diagnosis requiring five or more within 12 months, highlighting its multifaceted impact on emotional, social, and functional well-being [8].

Features such as push notifications, endless content feeds, and gamified interactions are designed to capture and sustain user attention, inadvertently fostering dependency [24]. This problematic usage has been linked to adverse psychological outcomes, including heightened levels of anxiety, depression, and FoMO [25, 26]. For instance, frequent exposure to idealized portrayals of others on platforms like Instagram can lead to negative self-comparisons, reducing self-esteem and overall life satisfaction. Additionally, excessive time spent on social media often displaces activities critical to well-being, such as physical exercise, sleep, and in-person social interactions [27, 28]. These behaviors align with the criteria for SMD, such as preoccupation, withdrawal, and displacement. However, at the same time, such behaviour can be justified by the need to harness chances and seek timely information which is needed for better well-being [22].

The dual impact of social media as both a tool for enhancing well-being and a source of psychological distress represents a central paradox of contemporary digital life [29]. Social media platforms, such as Facebook, exemplify this duality by fostering SWB through features that enable connection, belonging, and social validation, while simultaneously contributing to risks such as SMD. Active engagement, such as posting and interacting with others, is more likely to yield positive outcomes, while passive consumption, like endless scrolling, often exacerbates feelings of inadequacy and dissatisfaction [30]. The pursuit of SWB, driven by the desire for social acceptance and belonging, can paradoxically lead to compulsive use when individuals rely excessively on digital interactions for validation [31]. Features like curated personas, push notifications, and endless content feeds further reinforce dependency, displacing offline activities and deepening the psychological toll [32].

To address the potential negative consequences of excessive screen time, there are platforms and tools specifically designed to help individuals reduce their digital consumption and promote healthier habits [33]. Popular apps like *Freedom*, *YourHour*, *DigitalDetox*, *Digital Wellbeing*, *AntiSocial*, *Beeminder*, and *Calm* provide users with tools to track, limit, and set goals for their device usage [33, 34]. These platforms offer app usage monitoring, time limits, and reminders to take breaks, encouraging users to be more mindful of their screen time. Additionally, *Do Not Disturb* modes and *Focus Mode* on various devices help reduce distractions by silencing notifications during designated times [35]. These tools empower individuals to regain control over their digital habits, fostering a healthier balance between online and offline activities. However, empirical evidence of their success remains unclear, and they are often viewed as regulatory tools rather than effective instruments for fundamentally changing attitudes and behaviors in a way that enhances well-being while reducing problematic use [36].

While social media platforms have become ubiquitous globally, their impact may vary significantly across cultural contexts. In particular, the UK and Arab cultures, with their distinct social, historical, and technological backgrounds, may experience social media use differently [37], leading to varying outcomes regarding Social Media Contribution to Social Well-Being (SM-SWB) and SMD. In the UK, social media is widely embraced, with its use being integral to maintaining social connections, staying informed, and participating in public discourse. The cultural emphasis on individualism in the UK may lead users to prioritize personal expression, identity construction, and social validation online [38]. In contrast, Arab cultures are often characterized by stronger collectivist values, which emphasize family ties, community, and social cohesion [39]. These cultural norms may shape social media use differently, such as maintaining family bonds and staying connected with community networks. This study explores these cultural differences in SM-SWB and SMD. Differences shall entail the need for culturally sensitive design and socio-technical interventions. This study aims to address the following research questions:

RQ1: How do Arab and UK populations differ in their perceptions of SM-SWB across its five dimensions and its role in SMD across its nine symptoms?

RQ2: Is there an association between perceptions of social media's contribution to SWB across its five dimensions and SMD, including its nine symptoms?

RQ3: What are the design challenges in building social media platforms that facilitate SWB while controlling SMD?

2 Methodology

2.1 Participants and Procedure

Participants for this study were recruited from the United Kingdom (UK) and the Arab Gulf Cooperation Council (GCC) through TGM Research (https://tgmresearch.com/), a company specializing in data collection for research purposes. These two cultural contexts were selected for their contrasting moral principles and societal values, offering a rich foundation for comparative analysis [40]. The GCC countries were selected for the Arab sample due to their close relationships, geographic proximity, and shared cultural values, norms, and political frameworks [41]. These countries are also comparable because of their relatively stable economies and political systems. Moreover, they have made significant progress in digital transformation, fostering technologically advanced societies that shape both business operations and consumer behaviors [42, 43]. This study is part of a larger research project. Additional information about the overall project can be found at: https://osf.io/n47xg/. The survey was developed and administered using SurveyMonkey (www.surveymonkey.com), a platform for creating and managing questionnaires. To ensure the clarity and accuracy of the survey questions, the research team engaged in an iterative development process. Initially drafted in English, the survey was translated into Arabic by two team members following the recommended back-translation method [44]. A pilot test was subsequently conducted with a small group of participants to identify and resolve any ambiguity or unclear language.

Eligibility criteria for participation required individuals to be over 18 years old, born and currently residing in the UK (England, Scotland, Wales, or Northern Ireland) or the GCC (Saudi Arabia, Qatar, Bahrain, Kuwait, Oman, or the UAE), and to self-identify with the cultural norms and practices of either the UK or the Arab GCC. All participants provided informed consent and were given the option to withdraw from the survey at any time. To maintain high data quality, the survey included embedded attention checks, and participants who failed more than two checks or completed the survey too quickly were excluded from the analysis. Participants who completed the survey received compensation for their time. The Institutional Review Board (IRB) of the institution affiliated with the last author granted the study's ethical approval.

Participants identifying as non-binary gender were excluded from the analysis in both datasets due to the small sample size. Additionally, individuals aged 60 and older were excluded from the UK sample as we could not recruit participants from that age group in the Arab population. Only about 7% of the population in many Arab countries falls within this age group [45]. Homogeneity of age is essential for a meaningful cross-cultural comparison. Consequently, the age range for analysis was limited to 18–60 years. The final dataset comprised 563 participants, 255 from the UK and 308 from the Arab GCC.

2.2 Measures

Demographic Measures. Participants were asked to provide their age and gender. Age was recorded as a continuous variable in years, while gender was reported using an open-text field that has been manually checked and coded by the authors. Both fields were mandatory for completion.

Social Media Contribution to Social Well-Being (SM-SWB). The survey questions used in this study to evaluate SM-SWB were adapted from the SWB Scale developed by Keyes [10]. This scale encompasses five dimensions of SWB: social integration (feeling as a part of the community), social contribution (a feeling that one has a contribution to make to society), social coherence (understanding the social world as predictable and comprehensible), social actualization (positive comfort level with society), and social acceptance (accepting others as they are). The original scale includes 15 items, with three questions dedicated to each dimension of SWB. In the adjusted version used in the current study, the scale was contextualized to the online realm by adjusting terms related to social structures (e.g., replacing 'community' with 'online community' and similar changes) to reflect the social media context. Participants were also given an introductory statement "Regarding your social media experience, indicate the extent to which you agree with the following statements:". Participants responded to these items using a seven-point Likert scale (1 = strongly disagree to 7 = strongly agree). Scores for each dimension were calculated by summing the responses to the three associated questions. Reliability for each of the five dimensions was assessed using Cronbach's alpha, with values of 0.81 for the UK sample and 0.79 for the Arab sample, indicating good reliability across both samples.

Social Media Disorder (SMD). The SMD scale [8] consists of nine items such as preoccupation, tolerance, withdrawal, persistence, displacement, problems, deception, escape, and conflict. The first item is defined as "…regularly found that you can't think of anything else but the moment that you will be able to use social media again?". Participants responded to each item on a 5-point Likert scale, ranging from 1 (Never) to 5 (Always). Total SMD scores were calculated by summing participants' responses, with higher scores reflecting greater levels of SMD. The scale has shown good internal consistency, with Cronbach's alpha ranging from 0.76 to 0.82 [46]. In this study, Cronbach's alpha was 0.89 for the UK sample and 0.81 for the Arab sample.

All the questionnaires (in both Arabic and English) used in this paper can be accessed through the Open Science Framework (OSF) provided in the supplementary material of this manuscript.

2.3 Data Analysis

All analyses were performed for UK and Arab participant samples using *JASP* software version 0.19.0.0 (https://jasp-stats.org/). Skewness and kurtosis for all variables in both samples were within the range of $+2$ to -2 except for SMD Deception and SMD Conflict in the UK sample, indicating that most of the data followed an approximately normal distribution [47]. The selection of statistical tests was supported by the Central Limit Theorem (CLT) [48], which states that, with sample sizes greater than 30, the sample means will approximate a normal distribution, making parametric tests suitable. Prior performing inferential statistics, we checked the assumption of equality of variances (assessed using Levene's test). Levene's test was significant ($p > 0.05$) suggesting a violation of equal variance assumption for all factors in SMD and SWB. Therefore, to assesses mean differences of these factors between UK and Arab samples, we employed Welch's t-test, which can account for the unequal variances and imbalanced sample sizes. We performed additional analysis to quantify the evidence for these differences using Bayesian independent t-test with Bayes Factor robustness check [49]. For each comparison, we report BF_{10} indicating evidence for the alternative hypothesis.

Pearson's correlations were used to examine the relationships between the SM-SWB and SMD symptoms. While the SMD symptoms data are ordinal, the large sample size permits parametric tests. Typically designed for interval data, parametric tests are robust and can be appropriately applied to ordinal data, mainly when the sample size is large [50]. We performed additional analysis to quantify the evidence for this correlation using a Bayesian correlation robustness check.

All the data and analysis files associated with this paper can be accessed through the Open Science Framework (OSF) link in the supplementary material.

3 Results

3.1 Descriptive Statistics

The demographic characteristics of the participants in the study are presented in Table 1 for both the UK and Arab samples.

Table 1. Participant Demographic

	Variables	Participants (N = 563)	
		UK (N = 255)	Arab (N = 308)
Gender (%)	Male	109 (42.75%)	139 (45.13%)
	Female	146 (57.25%)	169 (54.87%)
Age	M (SD)	38.39 (10.99)	34.02(7.69)
	Range	18–59	18–57
Education (%)	Basic Education	121 (47.45%)	58 (18.83%)
	Higher Education	134 (52.55%)	250 (81.17%)
Employment (%)	Student	9 (3.53%)	9 (2.92%)
	Employed/Self-Employed	200 (78.43%)	248 (80.52)
	Unemployed	46 (18.04%)	51 (16.56%)

3.2 Comparing Mean Scores for SMD and SM-SWB Between the Samples

A Welch's approximation t-test was employed to examine the mean scores SWB and SMD scores between samples. Additionally, evidence for the alternative hypothesis was assessed using the Bayes Factor (BF_{10}) with default priors. This analysis reveals notable differences between the UK and Arab samples in SM-SWB and SMD (Table 2). In terms of SM-SWB, Arab participants generally report higher levels across most dimensions, particularly in areas related to Integration, Contribution, Actualization, and Coherence, suggesting that social media contributes significantly to their sense of personal growth, clarity, and social connection. These findings indicate that social media may be a more effective platform for fostering connection, participation, and a sense of belonging within the Arab cultural context than the UK.

Conversely, Arab participants also exhibit significantly higher levels of SMD across all dimensions, including Preoccupation, Tolerance, and Withdrawal. This suggests a heightened vulnerability to problematic social media use, characterized by excessive engagement and difficulty disengaging from social media activities. The consistent differences between the two groups highlight potential cultural influences on the role of social media, where the Arab sample appears to experience both enhanced well-being and a greater risk of adverse effects.

It has to be noted that the differences between these samples were not due to extreme values in both samples. To analyse the distribution of responses and identify dominant categories, Pareto charts were plotted for each variable in both samples. The cumulative frequency lines showed that no individual category accounted for more than 10% of all responses in the SWB factors in either sample. In SMD, none of the categories contributed more than 50%, and this pattern was consistent across the samples. This analysis suggests a relatively even distribution of responses across categories, without one dominating disproportionately. All the Pareto chats can be found on the OSF link provided in the supplementary material.

Table 2. Independent t-test results of variables between the UK and Arab samples

Variables	UK	Arab	t-test (*W, Cohen's d, p)	**BF_{10}
	$M(SD)_{min/max}$	$M(SD)_{min/max}$		
SM-SWB Integration	$12.18\ (4.51)_{3/21}$	$13.16\ (4.15)_{3/21}$	-2.64, $d = -0.22$, $p = .01$	$BF_{10} = 2.89$
SM-SWB Acceptance	$12.27\ (3.19)_{3/21}$	$12.39\ (2.99)_{3/20}$	-0.46, $d = -0.04$, $p = .65$	$BF_{10} = 0.10$
SM-SWB Contribution	$12.68\ (4.33)_{3/21}$	$13.86\ (3.82)_{3/21}$	-3.42, $d = -0.29$, $p < .001$	$BF_{10} = 30.76$
SM-SWB Actualization	$12.22\ (3.92)_{3/21}$	$14.53\ (3.52)_{3/21}$	-7.32, $d = -0.63$, $p < .001$	$BF_{10} > 100$
SM-SWB Coherence	$13.07\ (3.28)_{3/20}$	$14.51\ (3.08)_{5/21}$	-5.36,$d = -0.46$,$p < .001$	$BF_{10} > 100$
SMD-Preoccupation	1.99 (1.07)1/5	$2.83\ (1.03)_{1/5}$	-9.41, $d = -0.79$, $p < .001$	$BF_{10} > 100$
SMD-Tolerance	$1.74\ (0.87)_{1/5}$	$2.75\ (1.12)_{1/5}$	-12.08, $d = -1.01$, $p < .001$	$BF_{10} > 100$
SMD-Withdrawal	$1.78\ (0.94)_{1/5}$	$2.82\ (1.17)_{1/5}$	-11.68,$d = -0.98$,$p < .001$	$BF_{10} > 100$
SMD-Persistence	$2.33(1.18)_{1/5}$	$3.03(1.18)_{1/5}$	-7.07,$d = -0.59$,$p < .001$	$BF_{10} > 100$
SMD-Displacement	$2.04(1.15)_{1/5}$	$2.71(1.22)_{1/5}$	-6.67, $d = -0.56$, $p < .001$	$BF_{10} > 100$
SMD-Problem	$1.55(0.82)_{1/5}$	$2.07(1.05)_{1/5}$	-6.54, $d = -0.55$,$p < .001$	$BF_{10} > 100$
SMD-Deception	$1.35(0.79)_{1/5}$	$1.97(1.23)_{1/5}$	-7.31, $d = -0.61$,$p < .001$	$BF_{10} > 100$
SMD-Escape	$2.50(1.21)_{1/5}$	$3.21(1.19)_{1/5}$	-7.02, $d = -0.59$, $p < .001$	$BF_{10} > 100$
SMD-Conflict	$1.35(0.80)_{1/5}$	$1.82(1.05)_{1/5}$	-5.99, $d = -0.50$, $p < .001$	$BF_{10} > 100$

Note. Small effect size ($d = 0.2$), Medium effect size ($d = 0.5$), Large effect size ($d = 0.8$). * Negative values for Welch's test indicate a smaller mean for the UK sample than the Arab sample. ** According to Lee & Wagenmakers [51] Classification scheme, $BF_{10} > 100$ provides extreme evidence for the alternative hypothesis (H_1), 30–100 – very strong evidence, 10–30 – strong evidence, 3–10 moderate evidence, 1–3 and 1/3-1– anecdotal evidence, 1 – no evidence

3.3 Correlation Analysis

The correlation analysis reveals distinct patterns between SM-SWB factors and SMD symptoms in the UK and Arab samples, highlighting cultural differences in their interactions (Table 3 and Table 4). Additionally, evidence for the alternative hypothesis was assessed using the Bayes Factor (BF_{10}).In the UK sample, most SMD symptoms positively correlate with SWB factors, particularly Integration and Actualization. This suggests that individuals who feel socially integrated or derive personal growth through social media may also experience greater vulnerability to problematic behaviors such as Preoccupation, Withdrawal, and Tolerance. Coherence, representing a sense of clarity and purpose, has minimal associations with SMD symptoms, indicating its limited connection to problematic use in this context. In contrast, the Arab sample exhibits weaker and often negative correlations, with factors like Contribution and Coherence showing protective effects against symptoms such as Displacement, Deception, and Problem behaviors. While some positive correlations are observed, particularly between Withdrawal and SM-SWB factors like Integration and Actualization, they are less pronounced than in the UK sample. These findings underscore cultural variations in how social media fosters well-being and its potential risks, emphasizing the need for culturally tailored strategies to mitigate the negatives while enhancing the positives.

Table 3. Pearson's correlation between social media contribution to SWB factors (columns) and SMD symptoms (rows) in the UK sample.

	$Integration_{BF10}$	$Acceptance_{BF10}$	$Contribution_{BF10}$	$Actualization_{BF10}$	$Coherence_{BF10}$
$Preoccupation_{BF10}$	$0.37^{***}{}_{BF10>100}$	$0.16^{**}{}_{BF10=2.19}$	$0.15^{*}{}_{BF10=1.45}$	$0.29^{***}{}_{BF10>100}$	$0.01_{BF10=0.08}$
$Tolerance_{BF10}$	$0.36^{***}{}_{BF10>100}$	$0.13^{*}{}_{BF10=0.73}$	$0.21^{***}{}_{BF10=23.19}$	$0.24^{***}{}_{BF10>100}$	$0.14^{*}{}_{BF10=0.92}$
$Withdrawal_{BF10}$	$0.32^{***}{}_{BF10>100}$	$0.13^{*}{}_{BF10=0.61}$	$0.19^{**}{}_{BF10=6.34}$	$0.29^{***}{}_{BF10>100}$	$0.14^{*}{}_{BF10=0.91}$
$Persistence_{BF10}$	$0.21^{***}{}_{BF10=17.34}$	$-0.04_{BF10=0.09}$	$0.07_{BF10=0.13}$	$0.10_{BF10=0.28}$	$0.00_{BF10=0.08}$
$Displacement_{BF10}$	$0.25^{***}{}_{BF10>100}$	$0.09_{BF10=0.25}$	$0.01_{BF10=0.08}$	$0.15^{*}{}_{BF10=1.24}$	$0.06_{BF10=0.12}$
$Problem_{BF10}$	$0.23^{***}{}_{BF10=92.03}$	$0.02_{BF10=0.08}$	$0.08_{BF10=0.17}$	$0.07_{BF10=0.15}$	$-0.01_{BF10=0.08}$
$Deception_{BF10}$	$0.28^{***}{}_{BF10>100}$	$0.09_{BF10=0.21}$	$0.05_{BF10=0.11}$	$0.19^{**}{}_{BF10=8.23}$	$0.04_{BF10=0.09}$
$Escape_{BF10}$	$0.33^{***}{}_{BF10>100}$	$0.17^{**}{}_{BF10=2.48}$	$0.11_{BF10=0.38}$	$0.19^{**}{}_{BF10=8.92}$	$0.09_{BF10=0.22}$
$Conflict_{BF10}$	$0.23^{***}{}_{BF10>100}$	$0.07_{BF10=0.15}$	$-0.03_{BF10=0.09}$	$0.14^{*}{}_{BF10=1.00}$	$0.03_{BF10=0.09}$

Note. Significance levels: * $p < .05$, ** $p < .01$,*** $p < .001$. BF_{10}: According to Lee & Wagenmakers [51] Classification scheme, $BF_{10} > 100$ provides extreme evidence for the alternative hypothesis (H_1), 30–100 – very strong evidence, 10–30 – strong evidence, 3–10 moderate evidence, 1–3 and 1/3-1– anecdotal evidence, 1 – no evidence

For a more nuanced analysis, we conducted an additional analysis controlling age and gender, which is detailed in the OSF link provided in the supplementary materials. Most correlations remained consistent, but some differences emerged. In the UK sample, correlations for Tolerance and Acceptance weakened, while those for Persistence and Actualization showed minor declines or lost significance, indicating partial demographic influence. In the Arab sample, correlations for Persistence, Displacement, and Problem with outcomes like Contribution and Coherence weakened or lost significance, while negative associations, such as Displacement and Coherence, became more pronounced. These findings highlight the demographic sensitivity of these relationships across cultural contexts.

Table 4. Pearson's correlation between social media contribution to SWB factors (columns) and SMD symptoms (rows) in the Arab sample.

	Integration$_{BF10}$	Acceptance$_{BF10}$	Contribution$_{BF10}$	Actualization$_{BF10}$	Coherence$_{BF10}$
Preoccupation$_{BF10}$	0.10$_{BF10=0.38}$	0.07$_{BF10=0.16}$	0.06$_{BF10=0.12}$	0.07$_{BF10=0.15}$	−0.05$_{BF10=0.11}$
Tolerance$_{BF10}$	0.09$_{BF10=0.24}$	0.05$_{BF10=0.09}$	0.06$_{BF10=0.11}$	0.12*$_{BF10=0.73}$	0.02$_{BF10=0.08}$
Withdrawal$_{BF10}$	0.18**$_{BF10=10.92}$	0.15*$_{BF10=1.92}$	0.14*$_{BF10=1.32}$	0.18**$_{BF10=11.76}$	−0.03$_{BF10=0.08}$
Persistence$_{BF10}$	−0.05$_{BF10=0.11}$	−0.03$_{BF10=0.08}$	−0.15*$_{BF10=1.79}$	−0.09$_{BF10=0.26}$	−0.09$_{BF10=0.29}$
Displacement$_{BF10}$	−0.11$_{BF10=0.45}$	−0.04$_{BF10=0.09}$	−0.17**$_{BF10=5.86}$	−0.08$_{BF10=0.20}$	−0.17**$_{BF10=6.62}$
Problem$_{BF10}$	−0.10$_{BF10=0.34}$	−0.18**$_{BF10=8.18}$	0.10$_{BF10=0.35}$	−0.11$_{BF10=0.42}$	−0.13*$_{BF10=0.96}$
Deception$_{BF10}$	−0.05$_{BF10=0.10}$	−0.07$_{BF10=0.16}$	−0.14*$_{BF10=1.51}$	−0.09$_{BF10=0.24}$	−0.10$_{BF10=0.33}$
Escape$_{BF10}$	−0.02$_{BF10=0.08}$	−0.06$_{BF10=0.13}$	−0.12*$_{BF10=0.70}$	0.01$_{BF10=0.07}$	−0.04$_{BF10=0.09}$
Conflict$_{BF10}$	−0.05$_{BF10=0.10}$	−0.09$_{BF10=0.27}$	−0.08$_{BF10=0.17}$	−0.11*$_{BF10=0.49}$	−0.11$_{BF10=0.47}$

Note. Significance levels: * $p < .05$, ** $p < .01$,*** $p < .001$. BF_{10}: According to Lee & Wagenmakers [51] Classification scheme, $BF_{10} > 100$ provides extreme evidence for the alternative hypothesis (H_1), 30–100 – very strong evidence, 10–30 – strong evidence, 3–10 moderate evidence, 1–3 and 1/3-1– anecdotal evidence, 1 – no evidence

4 Discussion

The results of this study reveal significant cultural differences in various dimensions of SM-SWB. In the Arab sample, higher levels of integration, contribution, actualization, and coherence were reported compared to the UK sample. This suggests that in collectivist cultures, such as those in Arab contexts, social media influences social cohesion by transforming the structure of networks and individual perceptions while maintaining community and family ties that enhance engagement and cooperation [52]. These findings align with prior research emphasizing the role of social media in fostering community and belonging within collectivist societies [37]. In contrast, the UK sample, reflecting a more individualistic cultural orientation, reported comparatively lower SM-SWB scores.

The significant cultural differences in SMD symptoms further underscore the role of societal norms and values in shaping social media experiences. Arab participants exhibited higher levels of Preoccupation, tolerance, and withdrawal, suggesting heightened susceptibility to compulsive behaviors than the UK sample. This may reflect the duality of social media's role in collectivist cultures, enhancing connectedness; it may also exacerbate dependency due to the cultural emphasis on maintaining close-knit relationships [29]. Such findings align with the work of van den Eijnden et al. [8], who identified similar types of behavioral patterns associated with SMD.

The correlation analyses provide deeper insights into the relationship between SM-SWB and SMD. In the UK sample, positive correlations between SWB factors such as integration and actualization and almost all SMD symptoms suggest that individuals who derive personal growth and social connection from social media may simultaneously experience heightened vulnerability to Preoccupation, tolerance, withdrawal, displacement, deception, escape, and conflict. This duality aligns with the uses and gratifications theory [15], emphasizing the pursuit of social and psychological benefits through active engagement, and the dual-process model [29], which highlights that benefits are contingent on moderated use. Conversely, the Arab sample exhibited weaker and often negative correlations, with SM-SWB factors such as contribution and appearing to serve as protective factors against SMD symptoms. In other words, individuals who experience these positive dimensions of social media use are less likely to show signs of problematic behaviors, like excessive perception, displacement, problem, deception, and escape. These findings suggest people who feel that their social media use serves a purpose (e.g., contributing to a cause or fostering clear, goal-oriented connections) may be more resilient to the negative psychological impacts of social media (e.g., feelings of inadequacy, anxiety, or dependency). This aligns with research on Kuwaiti and US students, where US students exhibit higher trust and engagement with social media, reflecting personal autonomy, while Kuwaiti students, influenced by collectivist values, show more caution and lower trust due to conformity to group norms and awareness of risks like privacy invasion and cyberbullying [53]. This divergence highlights the importance of culturally tailored approaches to addressing the risks associated with social media use, emphasizing the need for strategies that consider the cultural context and values of different societies.

The findings underscore the need for social media platforms to redesign their features to promote SM-SWB while mitigating SMD risks. Features like push notifications, endless content feeds, and gamified interactions enhance engagement but can foster dependency and compulsive behaviors. Deactivating these features may negatively impact User Experience (UX) and SWB. While strategies to promote digital well-being, such as mindful usage, ethical algorithm design, digital literacy tools, and discouraging passive consumption, show promise, they come with trade-offs. Existing research on e-health interventions for digital addiction aligns with the view that these approaches may inadvertently lead to negative consequences like lower self-esteem and alternative addictive behaviors [54]. Features like time caps, usage dashboards, and "Take a Break" reminders are designed to help users regulate their social media habits [55], but they often fail to address deeper behavioral tendencies. Users can easily dismiss notifications or ignore usage reports, rendering these tools less effective. Ethical algorithm design, which prioritizes meaningful content from close friends or educational material, aims to reduce Preoccupation with addictive content [56]. However, these algorithms may still inadvertently promote prolonged engagement through subtle design techniques that keep users hooked. Digital literacy tools, such as in-app guides, gamified tutorials, and quizzes, aim to educate users about healthy social media habits and the risks of SMD [57], but a lack of user interest or oversaturation of gamified features often limits their impact.

Moreover, efforts to discourage passive consumption, like scroll-ending prompts, autoplay limits, and focus modes, attempt to nudge users toward taking breaks [58]. Yet, these interventions often fall short of the addictive design of infinite scrolls and autoplay. Even features like grayscale mode and muting notifications, which aim to limit overstimulation and curb overuse, may have limited effectiveness due to user reluctance to opt in, potentially reducing the intermittent rewards and reinforcement users typically receive from social media [13]. This creates a loss of gratification, akin to the experience of gambling addicts who feel like 'dead bodies' walking when they quit, as no stimuli or source of gratification can replace what they found in games. Sensation seeking, along with intermittent rewards, reinforcement, and even the variable ratio reinforcement schedule seen in gaming, gambling, and social media, plays a key role in keeping users engaged through unpredictability and reward cycles. This mechanism makes it challenging to design platforms that balance well-being without facilitating SMD.

While this study provides valuable insights, it is not without limitations. The cross-sectional design limits causal inferences, and the reliance on self-report measures may introduce common method bias. Additionally, due to the absence of a validated scale to measure SM-SWB, we adapted the SWB scale proposed in [59] by including a brief introductory text indicating that the questions relate to life on social media. However, some questions may not fully align with the context of social media, and additional questions may need to be introduced to better reflect the nature of communication and activities on these platforms. Lastly, our sampling process could have included individuals who refrain from using social media, as well as those with limited internet access due to financial or technological constraints, lifestyle factors, or job requirements. Future research could employ longitudinal designs to explore the temporal dynamics of SWB and SMD, e.g., during events such as festive seasons, crises, or political occurrences like elections. We also need experimental studies to assess the impact of the presence and the absence of social media on both well-being and SMD [60]. Moreover, we need to expand the cultural scope to include diverse regions.

5 Conclusion

This study highlights the complex interplay between social media's capacity to enhance SWB and its potential to foster disorder. The findings underscore significant cultural variations, with Arab participants reporting higher levels of both SWB and SMD compared to their UK counterparts. These results emphasize the need for balanced platform designs and culturally sensitive interventions that address the dual impact of social media. By fostering mindful usage and promoting meaningful engagement, social media platforms can better align with the well-being goals of diverse user populations. Future efforts should focus on longitudinal and experimental research to further explore these dynamics and refine design strategies for healthier online experiences.

Supplementary Material: The questionnaires and data for this paper are available on the Open Science Framework (OSF) at the following link: https://osf.io/8kghc.

Acknowledgement. This publication was supported by NPRP 14 Cluster grant # NPRP 14C-0916-210015 from the Qatar National Research Fund (a member of Qatar Foundation). The findings herein reflect the work and are solely the responsibility of the authors.

References

1. Forgeard, V.: How social media impact our life (explained) - brilliantio. https://brilliantio.com/how-social-media-impact-our-life/. Accessed 16 Dec 2024

2. Omar, A.S., Ondimu, K.O.: (PDF) The impact of social media on society: a systematic literature review. https://www.researchgate.net/publication/381800701_The_Impact_of_Social_Media_on_Society_A_Systematic_Literature_Review, https://doi.org/10.9790/1813-130696106. Accessed 16 Dec 2024
3. Kross, E., Verduyn, P., Sheppes, G., Costello, C.K., Jonides, J., Ybarra, O.: Social media and well-being: pitfalls, progress, and next steps. Trends Cogn. Sci. **25**, 55–66 (2021). https://doi.org/10.1016/j.tics.2020.10.005
4. Sadagheyani, H.E., Tatari, F.: Investigating the role of social media on mental health. Ment. Health Soc. Incl. **25**, 41–51 (2020). https://doi.org/10.1108/MHSI-06-2020-0039
5. Alutaybi, A., Arden-Close, E., McAlaney, J., Stefanidis, A., Phalp, K., Ali, R.: How can social networks design trigger fear of missing out? In: 2019 IEEE International Conference on Systems, Man and Cybernetics (SMC), pp. 3758–3765. IEEE (2019)
6. Andreassen, C.S., Torsheim, T., Brunborg, G.S., Pallesen, S.: (PDF) development of a facebook addiction scale. ResearchGate (2024). https://doi.org/10.2466/02.09.18.PR0.110.2.501-517
7. Kuss, D.J., Griffiths, M.D.: Online social networking and addiction - a review of the psychological literature. Int. J. Environ. Res. Public Health **8**(9), 3528–3552 (2011)
8. van den Eijnden, R.J.J.M., Lemmens, J.S., Valkenburg, P.M.: The social media disorder scale. Comput. Hum. Behav. **61**, 478–487 (2016). https://doi.org/10.1016/j.chb.2016.03.038
9. Teghe, D., Rendell, K.: Social wellbeing: a literature review (2005). https://doi.org/10.13140/RG.2.2.28891.26406
10. Keyes, C.L.M.: Social well-being. Soc. Psychol. Q. **61**, 121–140 (1998)
11. Administrator: Online Support Groups and Mental Health – iResearchNet. https://psychology.iresearchnet.com/health-psychology-research/mental-health-stigma/. Accessed 19 Jan 2025
12. Choi, M., Choung, H.: Mediated communication matters during the COVID-19 pandemic: the use of interpersonal and masspersonal media and psychological well-being. J. Soc. Pers. Relatsh. **38**, 2397–2418 (2021). https://doi.org/10.1177/02654075211029378
13. McQuail, D.: McQuail's Mass Communication Theory. SAGE, Los Angeles (2010)
14. Gallion, A.: Applying the uses and gratifications theory to social networking sites: a review of related literature (2010)
15. Kamboj, S.: Applying uses and gratifications theory to understand customer participation in social media brand communities: Perspective of media technology. Asia Pac. J. Mark. Logist. **32**, 205–231 (2019). https://doi.org/10.1108/APJML-11-2017-0289
16. Seligman, M.: PERMA and the building blocks of well-being. J. Posit. Psychol. **13**, 333–335 (2018). https://doi.org/10.1080/17439760.2018.1437466
17. Primack, B.A., et al.: Social media use and perceived social isolation among young adults in the U.S. Am. J. Prev. Med. **53**, 1–8 (2017). https://doi.org/10.1016/j.amepre.2017.01.010
18. Meshi, D., Cotten, S.R., Bender, A.R.: Problematic Social Media Use and Perceived Social Isolation in Older Adults: A Cross-Sectional Study | Gerontology | Karger Publishers. https://karger.com/ger/article-abstract/66/2/160/148340/Problematic-Social-Media-Use-and-Perceived-Social. Accessed 19 Jan 2025
19. Twenge, J.M., Campbell, W.K.: Media use is linked to lower psychological well-being: evidence from three datasets. Psychiatr. Q. **90**, 311–331 (2019). https://doi.org/10.1007/s11126-019-09630-7
20. Chen, H.-T., Li, X.: The contribution of mobile social media to social capital and psychological well-being: examining the role of communicative use, friending and self-disclosure. Comput. Hum. Behav. **75**, 958–965 (2017). https://doi.org/10.1016/j.chb.2017.06.011
21. Holliman, A.J., et al.: Adaptability and social support: examining links with psychological wellbeing among UK students and non-students. Front. Psychol. **12** (2021). https://doi.org/10.3389/fpsyg.2021.636520

22. (PDF) Fear of Missing Out (FoMO) as Really Lived: Five Classifications and one Ecology. In: ResearchGate. https://doi.org/10.1109/BESC48373.2019.8963027
23. APA: Diagnostic and statistical manual of mental disorders: DSM-5. American psychiatric association, Washington (2013)
24. Potential risks of content, features, and functions: The science of how social media affects youth. https://www.apa.org/topics/social-media-internet/youth-social-media-2024. Accessed 19 Jan 2025
25. The Impact of Social Media on Teen Mental Health | Psychology Today. https://www.psychologytoday.com/us/blog/promoting-empathy-with-your-teen/202411/the-impact-of-social-media-on-teen-mental-health. Accessed 19 Jan 2025
26. Alshakhsi, S., Babiker, A., Montag, C., Ali, R.: On the association between personality, fear of missing out (FoMO) and problematic social media use tendencies in European and Arabian samples. Acta Psychol. (Amst.). **240**, 104026 (2023). https://doi.org/10.1016/j.actpsy.2023.104026
27. Zishan Khan, M.: The dangers of excessive social media usage. https://www.psychiatrictimes.com/view/the-dangers-of-excessive-social-media-usage. Accessed 19 Jan 2025
28. How social media can negatively affect your child. https://health.clevelandclinic.org/dangers-of-social-media-for-youth. Accessed 19 Jan 2025
29. Kumar, D.D.: The dual impact of social media on mental health: risks, benefits, and mediating factors (2024). https://papers.ssrn.com/abstract=4977644, https://doi.org/10.2139/ssrn.4977644
30. Godard, R., Holtzman, S.: Are active and passive social media use related to mental health, wellbeing, and social support outcomes? A meta-analysis of 141 studies. J. Comput.-Mediat. Commun. **29**, zmad055 (2024). https://doi.org/10.1093/jcmc/zmad055
31. Lervik-Olsen, L., Andreassen, T.W., Fennis, B.M.: When enough is not enough: behavioral and motivational paths to compulsive social media consumption. Eur. J. Mark. **58**, 519–547 (2023). https://doi.org/10.1108/EJM-12-2022-0898
32. The Psychological Cost of Curating Your Online Image | Mental Health and Social Media. https://www.ourmental.health/screen-time-sanity/curating-happiness-the-hidden-mental-health-toll-of-social-media. Accessed 19 Jan 2025
33. UMATechnology: The 9 Best Websites for Practicing Digital Minimalism. https://umatechnology.org/the-9-best-websites-for-practicing-digital-minimalism/. Accessed 19 Jan 2025
34. 10 Best Apps to Break Social Media Addiction – TechPP. https://techpp.com/2023/07/04/best-apps-to-break-social-media-addiction/. Accessed 19 Jan 2025
35. What Is Focus on iPhone and iPad, and How Is It Better Than Do Not Disturb?. https://www.howtogeek.com/733833/what-is-focus-on-iphone-and-ipad-and-how-is-it-better-than-do-not-disturb/. Accessed 19 Jan 2025
36. Digital wellbeing tools through users lens. Technol. Soc. **67**, 101778 (2021). https://doi.org/10.1016/j.techsoc.2021.101778
37. Hofstede, G.: Culture's Consequences: Comparing Values, Behaviors, Institutions and Organizations Across Nations. SAGE Publications (2001)
38. Williams, D.: Understanding the 6 Dimensions of UK Culture. https://toppandigital.com/translation-blog/understanding-6-dimensions-uk-culture/. Accessed 19 Jan 2025
39. Penn, L.: Collectivist Cultures and Societies in the Arab World. https://www.arabamerica.com/collectivist-cultures-and-societies-in-the-arab-world/. Accessed 17 Dec 2024
40. Hofstede, G.: Culture's Consequences: International Differences in Work-Related Values. SAGE (1984)
41. Jhon, C.: History and development of the gulf cooperation council: a brief overview – ProQuest. https://www.proquest.com/openview/cea0a336eb5510891175cc71bf25c640/1?cbl=36744&pq-origsite=gscholar. Accessed 25 Jan 2025

42. Telecom Review: From Deserts to Digital Dominance: Tech Transformation in the GCC. https://www.telecomreview.com/articles/reports-and-coverage/7409-from-deserts-to-digital-dominance-tech-transformation-in-the-gcc/https%3A%2F%2Fwww.telecomreview.com%2Farticles%2Freports-and-coverage%2F7409-from-deserts-to-digital-dominance-tech-transformation-in-the-gcc%2F. Accessed 25 Jan 2025
43. Smith, S.: Digital Transformation in the GCC – EGIC. https://egic.info/information-sheet/digital-transformation-in-the-gcc/. Accessed 25 Jan 2025
44. Brislin, R.W.: Back-Translation for Cross-Cultural Research - Richard W. Brislin (1970). https://journals.sagepub.com/doi/10.1177/135910457000100301. Accessed 23 No 2024
45. Benamer, H.T.S.: The Arab world. In: Benamer, H.T.S. (ed.) Neurological Disorders in the Arab World, pp. 3–12. Springer, Cham (2014). https://doi.org/10.1007/978-3-319-07257-9_1
46. (PDF) The Social Media Disorder Scale: Validity and psychometric properties. ResearchGate (2024). https://doi.org/10.1016/j.chb.2016.03.038
47. Curran, P.J., West, S.G., Finch, J.F.: The robustness of test statistics to nonnormality and specification error in confirmatory factor analysis (1996)
48. Kwak, S.G., Kim, J.H.: Central limit theorem: the cornerstone of modern statistics. Korean J. Anesthesiol. **70**, 144 (2017). https://doi.org/10.4097/kjae.2017.70.2.144
49. (PDF) Model-Averaged Bayesian t-Tests. https://www.researchgate.net/publication/363629648_Model-Averaged_Bayesian_t-Tests. Accessed 24 Jan 2025
50. Norman, G.: LIkert scales, levels of measurement and the "laws" of statistics. Adv. Health Sci. Educ. Theory Pract. **15**, 625–32 (2010). https://doi.org/10.1007/s10459-010-9222-y
51. Lee, M.D., Wagenmakers, E.-J.: Bayesian Cognitive Modeling: A Practical Course. Cambridge University Press, New York (2013). https://doi.org/10.1017/CBO9781139087759
52. Hendrix, J.: Reviewing the Evidence on Social Media and Social Cohesion | TechPolicy.Press. https://techpolicy.press/reviewing-the-evidence-on-social-media-and-social-cohesion. Accessed 21 Jan 2025
53. Alsaleh, D.A., Elliott, M.T., Fu, F.Q., Thakur, R.: Cross-cultural differences in the adoption of social media. J. Res. Interact. Mark. **13**, 119–140 (2019). https://doi.org/10.1108/JRIM-10-2017-0092
54. Alrobai, A., McAlaney, J., Phalp, K., Ali, R.: Exploring the risk factors of interactive E-health interventions for digital addiction. Int. J. Sociotechnol. Knowl. Dev. **8**, 1–15 (2016). https://doi.org/10.4018/IJSKD.2016040101
55. Hunt, M.G., Marx, R., Lipson, C., Young, J.: No more FOMO: limiting social media decreases loneliness and depression. J. Soc. Clin. Psychol. **37**, 751–768 (2018). https://doi.org/10.1521/jscp.2018.37.10.751
56. Pariser, E.: The Filter Bubble: What the Internet is Hiding from You. Viking/Penguin Press, London/New York (2011)
57. Theophilou, E., Lobo-Quintero, R., Hernández-Leo, D., Sánchez-Reina, R., Ognibene, D.: Embedding educational narrative scripts in a social media environment. IEEE Trans. Learn. Technol. **17**, 1780–1793 (2024). https://doi.org/10.1109/TLT.2024.3409063
58. Ole, R., et al.: The Loop and Reasons to break it: investigating infinite scrolling behaviour in social media applications and reasons to stop. Proc. ACM Hum.-Comput. Interact. (2023). https://doi.org/10.1145/3604275
59. Keyes, C.L.M.: Social well-being scale (2013). https://doi.org/10.1037/t13598-000
60. Media use and attitudes. https://www.ofcom.org.uk/media-use-and-attitudes/. Accessed 25 Jan 2025

Business Process Engineering and Management

Multi-perspective Next Event Prediction in PPM via Heterogeneous Graph Neural Networks

Sebastiano Dissegna[1(✉)], Chiara Di Francescomarino[1], and Massimiliano Ronzani[2]

[1] Department of Computer Science Engineering, University of Trento, Trento, Italy
{sebastiano.dissegna,c.difrancescomarino}@unitn.it
[2] Fondazione Bruno Kessler, Trento, Italy
mronzani@fbk.eu

Abstract. Predictive Process Monitoring (PPM) is a subfield of Process Mining that mainly focuses on using standard machine learning (ML) and deep learning (DL) methods to learn from historical data and predict the future of an ongoing process execution based on its early stages. While different ML and DL methods have been extensively explored in the PPM literature, a new type of technique has emerged in recent years. These techniques involve Graph Neural Networks (GNNs), an innovative type of neural network that only a few approaches in PPM have applied so far. GNN models, and even more Heterogeneous GNN (HGNN), offer the advantage of working with a more natural representation of complex sequences such as execution traces, as well as dynamic systems such as process models, thus allowing for a more expressive and semantically rich encoding. This work presents SEPHIGRAPH, an approach that utilizes an HGNN model to process a graph encoding of the traces to tackle the multi-perspective next event prediction task, which involves predicting the full set of attributes of the next event to be performed. We evaluate its performance on multiple real-world datasets and compare it against other state-of-the-art approaches. The results show that SEPHIGRAPH is able to outperform existing approaches both in the next activity prediction task, as well as in the multi-perspective next event task.

Keywords: Heterogeneous Graph Neural Networks · Predictive Process Monitoring · Next Event Prediction

1 Introduction

Process Mining is an interdisciplinary field that combines data science and process management, enabling organizations to analyze and improve their operational processes using data extracted from event logs. Within this field, Predictive Process Monitoring (PPM) has gained attention for its ability to forecast future events and outcomes in ongoing process executions. PPM mainly utilizes

J. Grabis et al. (Eds.): RCIS 2025, LNBIP 547, pp. 365–382, 2025.
https://doi.org/10.1007/978-3-031-92474-3_22

ML and DL techniques to make predictions about various aspects of a process execution, including its outcome, cycle time, and next event(s). These predictions are learned from patterns found in historical data, providing significant value in decision-making and resource management.

Traditional PPM approaches typically rely on standard ML and DL models that represent process traces as sequences. However, these methods can be limited when applied to more complex processes, as they often fail to capture the intricate relationships between process components. Recent advancements in neural network architectures, particularly Graph Neural Networks (GNNs), offer promising alternatives for PPM by addressing some of these limitations. GNNs are designed to process data represented in graph structures, which makes them well-suited for modeling the complex dependencies and relationships that can be found in process traces. This advantage is further evident with Heterogeneous GNNs (HGNNs), which can encode diverse types of nodes and relationships, offering a deeper insight into the complexities of process dynamics. For example, representing trace prefixes as graphs can reveal relationships among attributes within the trace that may be difficult or impossible to capture using traditional sequence-based encodings. These hidden relationships among attributes are important for all types of predictions, but even more for more complex types of predictions such as the prediction of the next activity with all its attributes.

In this work we introduce SEPHIGRAPH, which aims at tackling the problem of next event prediction, that is predicting not only the next activity but also the full set of attributes of the next event, something that to the best of our knowledge has not been explored before – except for a few works. Predicting the next event, that is not only the next activity, but also all the attributes of the event, could be extremely useful from a business perspective. We can think of management scenarios, in which predicting not only the next activity but also its duration and the used resource(s), would allow for a more detailed analysis and for an improved planning of the work. Similarly, we can think of strategical scenarios, in which predicting not only the next activity of the customer but also their requests, the complexity or other information about their requests, would help the management of the process. To this aim, we propose a new encoding that translates trace prefixes into a heterogeneous graph representation, which is then processed by an HGNN to provide multi-perspective next event predictions. This type of encoding allows SEPHIGRAPH to take full advantage of the graph representation rich semantic and to outperform stat-of-the-art approaches on different real-world datasets both in the next activity prediction task, as well as in the multi-perspective next event task.

The paper is structured as follows. In Sect. 2 the background theory on PPM and GNNs is introduced. Section 3 contains an analysis of the current state-of-the-art approaches. In Sect. 4 heterogeneous graph encoding of the prefixes is detailed, along with the HGNN model employed. Section 5 and Sect. 6 discuss the evaluation carried out and the obtained results, respectively. Finally, Sect. 7 concludes the paper and discusses future direction.

2 Background

In this section we discuss the background theory on PPM and GNN.

Table 1. Example of a trace belonging to the BPI 2012A dataset

org:resource	activity	time:timestamp	case:reg_date	CaseID	amount
112	a_submitted	2011-09-30 22:38:44.546	2011-10-01 00:38:44.546	173688	20000
112	a_partlySubmitted	2011-09-30 22:38:44.88	2011-10-01 00:38:44.546	173688	20000
112	a_preaccepted	2011-09-30 22:39:37.906	2011-10-01 00:38:44.546	173688	20000
10862	a_accepted	2011-10-01 09:42:43.308	2011-10-01 00:38:44.546	173688	20000
10862	a_finalized	2011-10-01 09:45:09.243	2011-10-01 00:38:44.546	173688	20000

2.1 Predictive Process Monitoring (PPM)

Process Mining is a discipline positioned between Process Science and Data Science [26]. It leverages the event data collected in information systems to tackle compliance and performance problems. The focus of Process Mining is on operational processes, which are processes that require a repeated execution of activities to deliver products or services. These processes can be represented in event logs, which are a description of the events in the process of interest.

Definition 1 (Event Log). *An event log $\mathcal{E}$ is a collection of events that correspond to the selected process. Each event $e \in \mathcal{E}$ has a series of attributes $a \in \mathcal{AT}$, where $\mathcal{AT}$ is the set of all possible attributes. The three main attributes are as follows:*

- *$Case_{ID}$ (e^{id}): this identifies the unique case a specific event belongs to;*
- *activity ($e^{act} \in \mathcal{A}$): this is the activity (in the set of activities $\mathcal{A}$) performed in that event;*
- *timestamp (e^{ts}): this indicates the time in which the event occurred.*

Along with the three aforementioned main attributes, each event may contain other additional (categorical or numerical) attributes. Other examples of attributes belonging to an event include the resource that has carried out the activity, e.g., the person responsible for it; the cost of executing a certain event during the process; or information about how much time has elapsed since the start of the process or since the previous event occurred.

An event log is a collection of traces that are identified via the $Case_{ID}$ attribute, which is unique to that case.

Definition 2 (Trace). *A trace $\sigma \models \langle e_1, .., e_{|\sigma|} \rangle$ is a non-empty sequence of events $\sigma \in \mathcal{E}^* \backslash \{\emptyset\}$, such that $\forall e_i, e_j \in \sigma, 1 \leq i \leq j \leq |\sigma| : e_i^{id} = e_j^{id} \wedge e_i^{ts} \leq e_j^{ts}$, i.e., they all have the same $Case_{ID}$ and are ordered according to the timestamps.*

We distinguish between *trace attributes* (static attributes), whose value is constant throughout the whole trace and *event attributes* (dynamic attributes), whose value can change among different events of a trace. In Table 1, we report an example of a trace contained within the $BPI2012A$ [8] event log. This particular event log contains information about the loan application process, from submission to approval or rejection, of a Dutch financial institution. In the example, for instance, $Case_{ID}$ is a trace attribute and *org:resource* is an event attribute.

PPM is a subfield of Process Mining, which aims at forecasting the future of a process execution [5]. PPM approaches are usually split into two stages. In the first stage, we have a *learning phase*, in which an ML or a DL model is built on the event log of interest. In the second stage, we have a *runtime phase*, where the learned model is queried to obtain a prediction on the future of an ongoing case. Historical traces contained in the event log correspond to complete process executions. To train the predictive model we need incomplete trace executions, or trace prefixes that are defined as in the following.

Definition 3 (Trace Prefix). *Given a trace* $\sigma \models \langle e_1, .., e_{|\sigma|} \rangle$*, the trace prefix of length* k*, where* $k \leq |\sigma|$*, is the sub-trace* $tp(\sigma, k) \models \langle e_1, .., e_k \rangle$*.*

ML and DL models cannot directly process prefixes in their raw form; they must first be encoded. Selecting an appropriate encoding technique is therefore a crucial design decision. There are various methods for encoding a prefix. For instance, the *simple-index encoding* approach treats the prefix as a basic symbolic sequence, taking into account the position of the activity within the trace and disregarding data and data flow information [13]. Another option is the *complex-index encoding*, where the data associated with events in a sequence are divided into static and dynamic information, thus including the evolution of the event attribute values in the prefix [13].

2.2 Graph Neural Networks

Compared to other DL architectures Graph Neural Networks are specifically built to work with data that are represented as graphs. We define a graph as in the following.

Definition 4 (Graph). *A graph* G *is defined via a tuple* (V, E)*, where* V *is the set of nodes (vertices), and* E *is the set of edges (arcs). With* $v_i \in V$ *we denote a node and with* $e_{ij} \in E$ *we denote the edge that goes from node* v_i *to node* v_j*.*

We can represent the connectivity of a graph with a corresponding adjacency matrix A. An adjacency matrix is a matrix of boolean values that has dimension $n \times n$, where n is the number of nodes. It follows that, if $A_{ij} = True$ then there exists an edge $e_{ij} \in E$ connecting node v_i and node v_j, $A_{ij} = False$ otherwise. In conjunction with the adjacency matrix, we need two more matrices to represent the graph features. The first one $X^V \in \mathbb{R}^{n \times d}$ contains the node features, with $x_v \in \mathbb{R}^d$, where d is the feature dimensionality of each node. The second one, which is optional in many architectures, is the edge attribute matrix X^E, where

$X^E \in \mathbb{R}^{m \times c}$, with x^e_{ij} representing an edge feature vector, m the number of edges, and c the dimensionality of each edge vector. The neighborhood of a node v is defined as $N(v) = \{u \in V | e_{vu} \in E\}$, which is the set of nodes such that there exists an edge from v to them.

GNNs can be used to address three different types of problems. The first one regards *node level tasks*, that is node classification or node regression. With a GNN we can indeed obtain high-level node representations, which can be fed to a Multi-Layer Perceptron (MLP) to predict a node class in an end-to-end pipeline. The second type deals with *edge level tasks*, namely edge classification or link prediction, with the latter predicting the presence or not of an edge between two nodes. The last type concerns *graph classification tasks*, which summarize the information contained in the entire graph in input and classify it according to the specific class. Moreoever, the graphs provided as input to a GNN can be of two types, *homogeneous* or *heterogeneous*. In the former case, all nodes in the graph are of the same type, and it follows that their node feature vectors contain the same amount and type of information, i.e., they have the same dimensionality. In the latter case, nodes and edges can be of different types.

Convolutional GNN (CGNN) are one of the most used types of GNNs. By leveraging the developments that Convolutional Neural Networks obtained in the Computer Vision field, these approaches extend the convolution operation to graphs. Similarly to what happens in a convolutional step in image processing, where a pixel representation is updated according to the value of the other pixels inside the filter's window, in a CGNN, the node representation is updated according to the value of the nodes inside its neighborhood. An advantage of CGNNs compared to the 2D convolutional operation lies in the fact that there is no fixed size of the node's neighborhood as in the pixel case, thus removing the need to employ zero padding by adding some fake nodes, which is what happens when working with images, where pixels with values set to zero are needed for the operation to work correctly.

From CGNNs, two different branches of approaches are being developed, namely, *Spectral-based approaches* and *Spatial-based approaches*. We are interested in the Spatial-based family of approaches.

Spatial Based Convolutional Graph Neural Network. This family of architectures defines the graph convolutional operation based on a node's local neighborhood. In general, a node's representation is updated by considering both its representation from the previous layer and the representations of its neighboring nodes. When applied to the problem of graph classification, graph pooling layers can be employed with this family of architectures. These layers are responsible for reducing the graph dimensionality by aggregating groups of nodes into a single one. This helps make the end-to-end training process more efficient.

The Message Passing Neural Network MPNN is a general spatial-based model upon which more refined architectures can be built to address specific problems. With this model, we perform a K-step message-passing operation to propagate the information across the entire graph, where K is the number of layers. During the message passing phase, hidden states h_v^k at each node v in the graph are updated as follows

$$h_v^k = UP_k\left(h_v^{k-1}, \sum_{u \in N(v)} AGRR_k\left(h_v^{k-1}, h_u^{k-1}, x_{vu}^e\right)\right). \tag{1}$$

By looking at Eq. 1 we can notice two functions. The first one, $AGGR$, is responsible for aggregating the information obtained by taking as input the representation at the previous layer of the node we are updating, along with the representation of each neighbor and the edge feature vector that connects them. As stated before the feature vector on the edge is optional. The second function in (1) is an update function, UP, which is applied to the sum of the message information from the neighborhood and to the previous node representation. Both $AGRR$ and UP have learnable parameters.

Finally, if we are performing graph classification, we apply a $READOUT$ function, which also has learnable parameters and summarizes the information contained in the entire graph

$$h_G = READOUT\left(h_v^K \mid v \in V\right). \tag{2}$$

Graph Attention Network (GAT) [27]. All the aforementioned methods assume that the contribution of neighboring nodes is either identical or predetermined. In the GAT architecture, an attention mechanism, described in Eq. 3, is introduced to make the weights between two connected nodes learnable:

$$h_v^k = \zeta\left(\sum_{u \in N(v) \cup v} \alpha_{vu}^k W^k h_u^{k-1}\right) \tag{3}$$

with ζ a non-linear function. The attention coefficient α_{vu}^k, which represents the importance of node u to node v at layer k, is learned via an end-to-end neural network architecture. This mechanism assigns higher weights to more important nodes, as described by the formula:

$$\alpha_{vu}^k = SOFTMAX(LeakyReLU(a^T[W^k h_v^{k-1} || W^k h_u^{k-1}])) \tag{4}$$

where a and W are learnable parameters, and $||$ denotes the concatenation operator.

3 Related Work

In this section we report a general overview of the works about PPM and we then focus on the two categories of works closer to the current one, that is those dealing with multi-perspective next event predictions and the ones leveraging GNNs.

PPM literature can be classified into three main groups [5] depending on the objective of the prediction.

- *Outcome-oriented.* The first group addresses the problem of outcome-oriented PPM [25], which aims to classify an ongoing process execution, or prefix, based on a set of possible categorical outcomes. Due to the relatively simple nature of this type of predictions, traditional ML methods are often preferred over DL techniques. Decision trees (DTs) and Random Forests (RFs) are the ML models most widely used for this task.
- *Time related.* A second group deals with time-related predictions [28]. The focus is mainly on two attributes. The first one is the next timestamp, representing the time in which the next activity will be performed, while the other is the remaining cycle time, which is the amount of time until the trace will be completed.
- *Event related.* The third group is interested in predicting the next event to be performed – and in particular the next activity [3,4,6,14,19,24] with some of them extending the prediction also to the full suffix [3,14,24].
  Many DL models are employed in the next activity prediction task. Typical DL models are, for instance, Convolutional Neural Networks [6,16], LSTM models [3,14,24], and more recently GNNs [4,19].

3.1 Multi-perspective Next-Event Prediction

One important group of works in the third category that is relevant for this paper is the one that predicts not only the activity to be performed or the timestamp of the next event, but also next event attributes in an end-to-end manner. Only few works tackle the prediction of more than one attribute of the next event. For instance, [24] predicts the next activity and the next timestamp. The work reported in [3] aims at predicting the activity, the role performing that activity, the next timestamp and, the remaining time both for the next event and for the full trace completion (suffix), employing a Long Short Term Memory (LSTM) model. [1] extends the work in [3], by introducing a sampling approach aimed at reducing repetitions of activities in the predicted case suffixes, which improves the overall performance of the previous approach for all the predicted attributes. An LSTM is also used in [14] to predict the activity and the entire set of attributes, both for the next step and the full suffix.

3.2 Graph Neural Networks Approaches

Few recent approaches in the literature leverage GNNs for PPM [4,12,19,20,29]. Most of them use GNNs with Object-Centric Event-Logs (OCELs) [12,20]. The work in [12] compares two techniques based on gradient boosting against two based on GNNs and evaluates them on more than 20 different Key Performance Indicator (KPI) definitions. In [20], an HGNN is employed for predicting the remaining cycle time.

Most of the approaches that work on single CaseID event logs employ a homogeneous GNN: [4,19,29] focus on predicting the next activity, while [11] focuses on predicting the remaining cycle time. A rather recent work [17] differently from the aforementioned ones, employs a heterogeneous model of a GNN tackling the problem of next activity prediction while also providing an explanation on the output of the network. In their approach, named *PROPHET*, the authors build a heterogeneous graph, which includes four types of nodes containing information about the prefix events: for the activity, resource, timestamp, and trace attributes. In their representation all trace attributes are inserted into a single node. As for the edges, each action node is connected with the subsequent action node, according to the order of the prefix, and this holds true also for the time and resource nodes. Moreover, the trace attribute node is connected to all the others in the graph. From the comparison they carried out, their method results the best performing one on the task of next activity prediction, so we decided to consider it as the baseline to evaluate our approach.

4 Graph-Based Encoding Solution

In this section we present our graph-based solution SEPHIGRAPH. Specifically, we first discuss the heterogeneous graph encoding of the traces extracted from the event logs, and then we introduce the architecture of the HGNN model employed.

4.1 Encoding

To work with an HGNN we must first define a procedure that takes as input a trace prefix and returns a heterogeneous graph. Let C denote the set of categorical and numerical attributes of the trace. In the encoding we propose, we have a node type for each attribute in the trace. For instance, the encoding of the trace in the example in Table 1, will contain five different types of nodes: two categorical and three numerical. To convert a categorical feature into a numerical vector we rely on the one-hot encoding, which transforms a categorical feature into a vector of length equal to the number of classes of that specific category, where only one value is set to one and all the others are set to zero. With this, a node corresponding to a categorical value will contain the one-hot vector of the corresponding class. For numerical attributes, we apply the logarithm function to bring all the features to a similar scale, which helps the model converge when predicting their values. It follows that we have a node for each attribute at each event. For trace attributes, which are constant throughout the trace, we have a single node.

Moreover, in the proposed encoding, we connect nodes of the same type following the sequential order established in the trace, according to the timestamp. Moreover, each activity node is connected to all the nodes of the other attributes belonging to the same event. Finally, a further edge is used to connect the activity node to the trace attribute nodes. An example of the encoding graph for a trace with one trace attribute and timestamp and resource as event attributes is depicted in Fig. 1.

Besides the node features, we can also compute some edge attributes. In our case, we add edge attributes only on edges between nodes of the same type. For categorical nodes, we add a boolean flag, which is set to True if the target node belongs to a different class than the source node. Concerning nodes representing numerical attributes, we use the difference between the target and the source node feature vectors. Following this procedure we obtain, for each trace, a set of graphs, one for each prefix length.

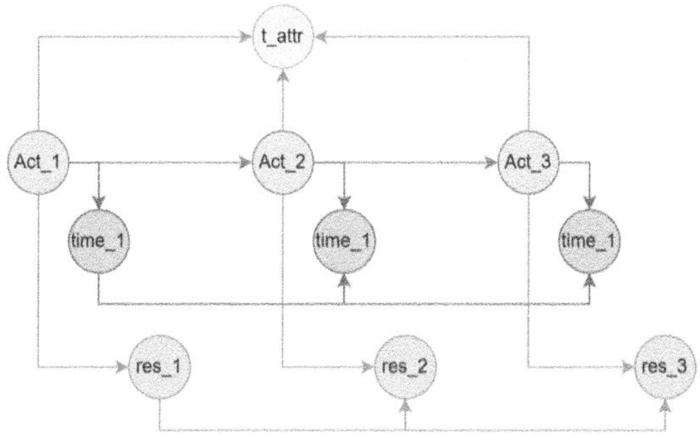

Fig. 1. Simple example of a heterogeneous graph encoding of a trace.

4.2 Heterogeneous Graph Model

When working with GNNs the most important operation is the convolutional one. As stated in Sect. 2 we can have different variants of the aggregation and message update function. Compared to a homogeneous GNN, in the heterogeneous case, we have edges that connect nodes of different types. This requires us to define a convolutional operator for each single tuple of node types connected in the graph, each operator with its own set of parameters for each convolutional layer. Considering Fig. 1 as an example, we would need a total of 6 different operators:

- $Activity_i \rightarrow Activity_{i+1}$
- $Activity_i \rightarrow TraceAttribute_i$
- $Activity_i \rightarrow Time_i$
- $Activity_i \rightarrow Resource_i$
- $Time_i \rightarrow Time_{i+1}$
- $Resource_i \rightarrow Resource_{i+1}$

In our case we use the *GATv2* operator [2] to compute the convolution operation, which is a more robust operator compared to the *GAT* one, discussed in Sect. 2. This was chosen since it learns the importance of each edge in the graph when updating a node representation. The update function is detailed in (5), while the computation of the attention coefficients is presented in (6).

$$h_i^{'} = \alpha_{ii}\Theta_s h_i + \sum_{j \in \mathcal{N}(i)} \alpha_{ij}\Theta_t h_j \tag{5}$$

$$\alpha_{ij} = \frac{\exp(a^T LeakyReLU(\Theta_s h_i + \Theta_t h_j + \Theta_e e_{ij}))}{\sum_{k \in \mathcal{N}(i) \cup i} \exp(a^T LeakyReLU(\Theta_s h_i + \Theta_t h_j + \Theta_e e_{ik})))} \tag{6}$$

The parameters are as follows:

- h_i is the representation of node i at the current layer
- α_{ij} is the attention coefficient between node i and node j
- Θ is a set of learnable parameters, for s (source node), t (target node) and e (edge)

Like many other convolutional operators, the information on the edges is optional. In this case, only edges between nodes of the same type have a set of attributes, as discussed in the Sect. 4.1. We also enable the multi-head attention functionality, choosing to average the attention outputs, instead of concatenating them.

After an arbitrary number of convolutional layers, we need to extract the predictions for the attributes in $\mathcal{C}$ we are interested in. The standard approach when performing graph classification is to summarize the entire graph into a single vector, via a pooling function. However, since we are working with a heterogeneous graph, we opted to perform a global mean pooling for each node type. It follows that we obtain one vector summarizing all the information for a specific type of node. This allows us to predict a certain attribute, for example, the next activity to be performed, using only the information contained within the activity nodes of the graph. This holds also for the timestamp, the resource, and all other attributes. This is achieved by feeding each feature vector to its linear layer.

Finally, we need to define the loss function to train our model. Since we are dealing with a classification problem for categorical attributes and a regression problem for real value attributes, we need to combine them, via a sum. The total loss is as follows

$$LOSS_{TOTAL} = \sum_{k \in \mathcal{C}_c} CrossEntropy(x_k, y_k) + \sum_{k \in \mathcal{C}_T} L1(x_k, y_k), \tag{7}$$

where $L1$ is the Mean Absolute Error Loss, x_k and y_k are respectively the predicted class or value, and the ground truth for attribute k.

5 Evaluation Setting and Implementation

In this section we present the research questions we are going to address in the evaluation of our approach, the datasets we employed, and how the evaluation was carried out.

5.1 Research Questions

In this work, we are interested in investigating the performance of our method SEPHIGRAPH by answering the following research questions.

RQ1 How does SEPHIGRAPH compare to state-of-the-art approaches for next activity prediction?
RQ2 How does SEPHIGRAPH compare to state-of-the-art approaches for multi-perspective next event prediction?

To address the first research question, we compare our approach against PROPHET, the most recent work predicting the next activity and leveraging GNNs [17], which has been shown to be the best-performing approach in the state of the art for next activity prediction. Along with this, we also include in the comparison TACO, the best homogeneous GNN approach [19], according to a benchmark we carried out in a previous work [7]. Regarding the second research question, two works perform multi-perspective next-event prediction: [1] and [15].

Table 2. Dataset information

	HELPDESK	BPI-2012-A	BPI-2012-WC	BPI-2013-O	BPI-2017-O	BPI-2020-R	SP2020
trace #	3804	13087	9658	819	42995	6886	23906
event #	13710	60849	72413	2351	193849	36796	178078
activity #	9	10	6	5	8	19	13
avg. case length	4	5	7	3	5	5	7
avg. dur. (days)	8.8	8.1	11.4	58.7	19.1	12	20.2
attribute #	2	4	4	9	9	7	5

5.2 Datasets

The evaluation is carried out on seven real-life event logs reported in the following, which are also used in [17]. They are all available on the 4TU Centre for Research[1], except for SP2020 that is available on Zenodo[2].

- **HELPDESK** [18]: an event log concerning the ticketing management process of the Help desk of an Italian software company.
- **BPI-2012-A** [8]: an event log containing the sub-processes "Application" of a loan application process.
- **BPI-2012-WC** [8]: an event log containing the sub-processes"Work" of a loan application process.
- **BPI-2013-O** [22]: an event log containing traces of the problem management system of Volvo IT located in Belgium.

[1] https://data.4tu.nl/portal.
[2] https://zenodo.org/records/3928487.

- **BPI-2017-O** [10]: an event log pertaining to a loan application process of a Dutch financial institute.
- **BPI-2020-R** [9]: an event log containing traces related to Requests for Payment on two years of travel expense claims.
- **SP-2020** [23]: an event log from a service process of a home appliances vendor for repairing faulty devices.

The main characteristics of the datasets are reported in Table 2.

5.3 Environment and Procedure

SEPHIGRAPH has been developed on Python 3.9, using PyTorch 2.1 and PyTorch Geometric with CUDA 12.7. The hyperparameter optimization has been carried out using the Adaptive Experimentation platform.[3] The source code is available on a public GitHub repository.[4] All experiments have been carried out on a laptop equipped with an Intel (R) Core (TM) i7-13700H, 32 GB of RAM, and a 4070 Nvidia GPU.

The hyperparameter optimization, whose objective is to minimize the total loss as detailed in (7), has been conducted on the following set of parameters: *Number of layers* in the range $[2, 5]$, *Learning rate* in the range $[10^{-4}, 10^{-1}]$ on a logarithmic scale, *Batch size* with values $\{256, 512, 1024, 2048\}$, *Number of attention heads* with values $\{1, 2, 3, 4\}$, *Aggregation type* with values {sum, mean, max}, and *Random seed* set to 123.
The total number of optimization trials has been set to 20. The size of the node hidden representation has been set to 128 since it provided a boost in execution speed during the training phase of the network, and it has therefore been kept constant throughout all the experiments. The maximum number of epochs has been set to 20, with an early stopping when the loss did not decrease for 4 consecutive epochs.

The next activity prediction task was carried out on all seven datasets, taking into account every category of activity, while the multi-perspective next event prediction task only on five of them, excluding the HELPDESK [18] and SP-2020 [23] datasets since they do not include the resource attribute in any of their events. Following the work done in [17], we split the dataset in (80%/20%) for train and test set according to the timestamp of the traces, assigning the most recent ones for the test set. The measure employed for answering **RQ1**, that is for evaluating the accuracy of the activity prediction is the Macro average F1 (Macro F1), a multiclass classification metric typically used in the case of imbalanced domains. The Macro F1 ($\frac{1}{k}\sum_{i=1}^{k} F1Score_i$) averages the F1 score of each activity class observed in the event log, where k is the total number of classes.

Differently from [17], which performs the hyperparameter optimization to maximize this metric, we aim at minimizing the total loss for all the attributes at the same time rather than only optimizing the activity attribute.

[3] https://ax.dev/.
[4] https://github.com/sebdisdv/SEPHIGRAPH.

In **RQ2**, we focus on predicting three attributes of the next event: the activity, the timestamp, and the role of the resource extracted from the event log via the algorithm described by Song and van der Aalst [21]. For evaluating the predictions related to the categorical variables, that is activity and role (extracted from the resource), we use the accuracy while for the timestamp we use the Mean Absolute Error (MAE). Moreover, to evaluate the capability of SEPHIGRAPH to predict the resource attribute in the next event with respect to state-of-the-art works, which only predict roles, we also compute the corresponding role by mapping the predicted resource to its role using the algorithm proposed in [21], in addition to computing the accuracy of predicting the resource attribute. All described metrics are computed on the test set.

6 Evaluation Results

In this section we discuss the results obtained from the comparison of SEPHIGRAPH with the considered state-of-the-art approaches for both next activity and multi-perspective next event prediction described in Sect. 5.

Table 3. RQ1 results: Macro F1 on the Activity attribute

	Helpdesk	BPI-2012-A	BPI-2012-WC	BPI-2013-O	BPI-2017-O	BPI-2020-R	SP2020
PROPHET [17]	26.3	69	70.5	40.6	72.1	50.5	60.3
TACO [19]	25.8	67.2	68.7	39.8	50.5	51.2	59.6
SEPHIGRAPH(loss)	56.66	66.05	73.06	43.15	75.91	88.04	65.58
SEPHIGRAPH(MacroF1)	**87.86**	**70.03**	**76.08**	**62.03**	**88.92**	**88.04**	**69.38**

6.1 Answering Research Question 1

In Table 3 we report the results of the Macro F1 score associated with the next activity prediction obtained by the three approaches on the seven real-world datasets. For SEPHIGRAPH, we report two different Macro F1 measures. The first one (*loss*) is obtained by leveraging the configuration that achieved the lower total loss during the hyperparameters optimization; the second one is obtained with the configuration that during the optimization achieved the highest Macro F1 measure on the next activity prediction, but not the lowest total loss. We decided to include the second one, since the other two approaches, which exclusively predict the next activity and not all the other attributes, set as objective of their training the maximization of the Macro F1 measure. The best approach for each dataset is highlighted in bold, while the second best is underlined.

From the evaluation, we can see that SEPHIGRAPH can beat the other two approaches most of the times only considering the configuration that minimizes the total loss. The BPI-2012-A dataset is the only dataset in which SEPHIGRAPH loses, even if by a small margin, compared to the other two approaches. This is mainly due to the hyperparameter optimization based on the loss. On the SP2020 dataset we observe an increase of 5.28 points, on BPI-2017-O an increase of 3.81, on BPI-2012-WC an increase of 2.56, and on BPI-2013-O an increase of 2.55. The other two datasets (HELPDESK and BPI-2020-R) are the ones in which SEPHIGRAPH obtains the highest performance increase w.r.t. the other two approaches. On the HELPDESK dataset there is an increase of 30.36 points, and on the BPI-2020-R an increase of 37.54 points. Based on these results, we can positively answer **RQ1**. SEPHIGRAPH beats the best homogeneous GNN approach TACO [19], and the best approach in the literature PROPHET [17] on most datasets by a margin, except for one dataset, where the results are within 3 points of difference. While, if we consider the best configuration for the Macro F1 metric, our approach is capable of beating the other methods across all datasets, by a larger margin compared to the aforementioned results.

6.2 Answering Research Question 2

In Table 4 we report the results on the five datasets that contain the *resource* attribute. For each of them, we extracted the role, which is the group a resource belongs to, using the algorithm reported in [21]. We compare SEPHIGRAPH against the state-of-the-art LSTM model in [1], which predicts the next activity along with the timestamp and the role of the resource. In detail, the table reports for both methods the activity and role (*Role direct (acc.)*) accuracy, as well as the MAE of the timestamp in seconds, which are the same measures employed in [1]. Moreover, for SEPHIGRAPH, it also reports some performance metrics related to the Resource attribute, that is (i) the accuracy of the role obtained by mapping the predicted resource to the corresponding role through the algorithm in [21] (*RoleMap*) and (ii) the resource accuracy (*acc.*), as described in Sect. 5.

Table 4. RQ2 Results on the next event prediction: activity and role accuracy, timestamp MAE and indirect and direct resource accuracy

Dataset	Activity (acc.)		Role direct (acc.)		Timestamp (MAE)		Resource	
	[1]	SEPHIGRAPH	[1]	SEPHIGRAPH	[1]	SEPHIGRAPH	RoleMap	acc.
BPI-2012-A	54.42	**68.75**	78.37	**80.68**	4.19	**0.06**	33.15	52.78
BPI-2012-W	68.96	**81.22**	**100**	85.24	6.63	**0.05**	82.82	11.33
BPI-2013-O	**41.03**	34.35	88.45	**98.77**	6.39	**0.07**	98.77	9.2
BPI-2017-O	85.71	**92.72**	85.33	**90.77**	5.59	**0.01**	77.06	33.78
BPI-2020-R	77.68	**87.39**	88.62	**99.81**	8.7	**0.15**	99.87	99.87

We can see that SEPHIGRAPH is able to obtain a higher accuracy on all datasets, except for the BPI-2013-O, with an average increase of 7.32 points. For the role (*Role direct (acc.)*), we obtain a higher accuracy for all datasets except for the BPI-2012-W, with an average increase of 2.9 points in accuracy overall. Regarding the timestamp, SEPHIGRAPH achieves a lower MAE on all datasets. Focusing on the prediction of the resource attribute, we can observe that when measuring the accuracy of the mapped role (RoleMap) and comparing it with the direct prediction of the role, although the comparison is unfair, the accuracy is lower only for the BPI-2012-A and the BPI-2017-O, while in the other cases we have competitive results. The obtained low results are mainly due to the high number of resource classes of these datasets. For example in the BPI-2012-A dataset, we have 61 classes of resources and only 3 roles. When looking at the overall resource accuracy, instead, the worst result is related to the BPI-2013-O dataset, in which we have an accuracy of 9.2. This is due to the fact that we have 240 classes of resources in this log, and most of them appear only once in the entire log. Based on the results, we can also positively answer to **RQ2**, since SEPHIGRAPH is able to achieve higher accuracy for both categorical attributes, as well as a much lower MAE for the timestamp, across all datasets.

SEPHIGRAPH is also able to predict all the other attributes in addition to the aforementioned three, which we do not report here due to space constraints, although they are available on the source code repository.[5] The performance we achieve on the other attributes is also overall quite good ranging from 0.92 to 1 for categorical variables (excluding the outlier attribute *RfpNumber* of the dataset BPI20R, which has 6322 class values) and between 0.2 to 0.02 in terms of logarithmic MAE for numerical attributes.

6.3 Training Times of GNN Approaches

One of the issues of state-of-the-art GNN-based methods for PPM is their long training times [7]. For instance TACO [19] on the BPI-2012-A dataset, which is of medium size, would require more than 30 min for a single epoch. However, SEPHIGRAPH, by leveraging more recent tools available in the PyTorch Geometric library compared to those used in previous GGN techniques, is able to complete each epoch in approximately 12 s. These tools indeed enable the execution of the forward pass through the HGNN model by grouping graphs into batches. This eliminates the temporal execution disadvantage of GGN methods compared to standard ML and DL techniques in the PPM literature.

6.4 Limitations

We can identify some areas of the work in which there is room for improvement. Firstly, for each hyperparameter configuration, we only analyzed the performance of a single model, instead of averaging multiple models trained using the same configuration. Moreover, the number of configurations tested along with the number of epochs for each training phase could be increased to achieve better performance at the cost of longer training times. Lastly, as shown in Table 4, the accuracy of the next resource prediction leaves room for improvement. One solution for this limitation might be to use a weighted cross-entropy loss, to give higher importance to classes whose support is low. This could be done also for all other categorical attributes.

[5] https://github.com/sebdisdv/SEPHIGRAPH/tree/main/results.

7 Conclusion

In this work, we propose a new approach to tackle the multi-perspective next-event prediction problem in PPM and compare it to state-of-the-art approaches for both the next activity and the next event prediction. From the results, we can conclude that SEPHIGRAPH can outperform the state-of-the-art approaches that focus on predicting the next activity, while also achieving higher performance on the multi-perspective next event prediction. Furthermore, compared to other GNN approaches for PPM, the training time is extremely reduced.

Future directions for this work could include extending the model to predict the full remaining suffix of the traces along with their complete data payload. Additionally, another possible extension is the integration of background knowledge. By leveraging the expressivity of GNN models, this could guide the model during training or even at inference time.

Acknowledgement. This work has been partially funded by the PNRR project FAIR - Future AI Research (PE00000013), under the NRRP MUR program funded by the NextGenerationEU, the HORIZON 2020 HumanE-AI project (Grant 952026), and the PRIN project PINPOINT Prot. 2020FNEB27, CUP H23C22000280006 and H45E2100021000.

References

1. Ali, M.A., Dumas, M., Milani, F.: Enhancing the accuracy of predictors of activity sequences of business processes. In: Research Challenges in Information Science, pp. 149–165. Springer (2024)
2. Brody, S., Alon, U., Yahav, E.: How attentive are graph attention networks? (2022). https://arxiv.org/abs/2105.14491
3. Camargo, M., Dumas, M., González-Rojas, O.: Learning accurate LSTM models of business processes. In: Hildebrandt, T., van Dongen, B.F., Röglinger, M., Mendling, J. (eds.) BPM 2019. LNCS, vol. 11675, pp. 286–302. Springer, Cham (2019). https://doi.org/10.1007/978-3-030-26619-6_19
4. Chiorrini, A., Diamantini, C., Genga, L., Potena, D.: Multi-perspective enriched instance graphs for next activity prediction through graph neural network. J. Intell. Inf. Syst. **61**, 5–25 (2023). https://doi.org/10.1007/s10844-023-00777-1
5. Di Francescomarino, C., Ghidini, C.: Predictive process monitoring. In: van der Aalst, W.M.P., Carmona, J. (eds.) Process Mining Handbook. LNBIP, vol. 448, pp. 320–346. Springer, Cham (2022). https://doi.org/10.1007/978-3-031-08848-3_10
6. Di Mauro, N., Appice, A., Basile, T.: Activity prediction of business process instances with inception CNN models. In: Alviano, M., Greco, G., Scarcello, F. (eds.) AI*IA 2019. LNCS (LNAI), vol. 11946, pp. 348–361. Springer, Cham (2019). https://doi.org/10.1007/978-3-030-35166-3_25
7. Dissegna, S., Di Francescomarino, C.: Graph neural networks for ppm: review and benchmark for next activity predictions. In: Gdowska, K., Gómez-López, M.T., Rehse, J.R. (eds.) Business Process Management Workshops, pp. 31–43. Springer, Cham (2025)

8. van Dongen, B.: BPI challenge 2012 (2012). https://doi.org/10.4121/uuid:3926db30-f712-4394-aebc-75976070e91f
9. van Dongen, B.: BPI challenge 2020: Request for payment (2020).https://doi.org/10.4121/uuid:895b26fb-6f25-46eb-9e48-0dca26fcd030
10. van Dongen, B.: BPI challenge 2017 - offer log (2021). https://doi.org/10.4121/12705737.v2
11. Duong, L.T., Travé-Massuyès, L., Subias, A., Merle, C.: Remaining cycle time prediction with graph neural networks for predictive process monitoring. In: International Conference on Machine Learning Technologies, ICMLT 2023, pp. 95–101 (2023). https://doi.org/10.1145/3589883.3589897
12. Galanti, R., de Leoni, M.: Predictive analytics forÂ object-centric processes: do graph neural networks really help? In: Business Process Management Workshops, pp. 521–533. Springer (2024)
13. Leontjeva, A., Conforti, R., Di Francescomarino, C., Dumas, M., Maggi, F.M.: Complex symbolic sequence encodings for predictive monitoring of business processes. In: Motahari-Nezhad, H.R., Recker, J., Weidlich, M. (eds.) BPM 2015. LNCS, vol. 9253, pp. 297–313. Springer, Cham (2015). https://doi.org/10.1007/978-3-319-23063-4_21
14. Lin, L., Wen, L., Wang, J.: MM-Pred: a deep predictive model for multi-attribute event sequence. In: SIAM International Conference on Data Mining, pp. 118–126. SIAM (2019). https://doi.org/10.1137/1.9781611975673.14
15. Lin, L., Wen, L., Wang, J.: MM-Pred: a deep predictive model for multi-attribute event sequence, pp. 118–126. Society for Industrial and Applied Mathematics (2019). https://doi.org/10.1137/1.9781611975673.14
16. Pasquadibisceglie, V., Appice, A., Castellano, G., Malerba, D.: Using convolutional neural networks for predictive process analytics. In: International Conference on Process Mining, ICPM 2019, pp. 129–136. IEEE (2019). https://doi.org/10.1109/ICPM.2019.00028
17. Pasquadibisceglie, V., Scaringi, R., Appice, A., Castellano, G., Malerba, D.: PROPHET: explainable predictive process monitoring with heterogeneous graph neural networks. IEEE Trans. Serv. Comput. **17**(6), 4111–4124 (2024). https://doi.org/10.1109/TSC.2024.3463487
18. Polato, M.: Dataset belonging to the help desk log of an Italian company (2017). https://doi.org/10.4121/uuid:0c60edf1-6f83-4e75-9367-4c63b3e9d5bb
19. Rama-Maneiro, E., Vidal, J.C., Lama, M.: Embedding graph convolutional networks in recurrent neural networks for predictive monitoring. IEEE Trans. Knowl. Data Eng. **36**(1), 137–151 (2024). https://doi.org/10.1109/TKDE.2023.3286017
20. Smit, T.K., Reijers, H.A., Lu, X.: HOEG: a new approach for object-centric predictive process monitoring. In: Advanced Information Systems Engineering, pp. 231–247. Springer (2024)
21. Song, M., van der Aalst, W.M.: Towards comprehensive support for organizational mining. Decis. Support Syst. **46**(1), 300–317 (2008). https://doi.org/10.1016/j.dss.2008.07.002
22. Steeman, W.: BPI challenge 2013, open problems (2013). https://doi.org/10.4121/uuid:3537c19d-6c64-4b1d-815d-915ab0e479da
23. Stierle, M., Matzner, M.: Customer service - device repair process (SP2020) (2020). https://doi.org/10.5281/zenodo.3928487
24. Tax, N., Verenich, I., La Rosa, M., Dumas, M.: Predictive business process monitoring with LSTM neural networks. In: International Conference on Advanced Information Systems Engineering, pp. 477–492. Springer (2017)

25. Teinemaa, I., Dumas, M., La Rosa, M., Maggi, F.M.: Outcome-oriented predictive process monitoring: review and benchmark. ACM Trans. Knowl. Discov. Data (TKDD) **13**(2), 1–57 (2019)
26. Van Der Aalst, W.M.P., Carmona, J. (eds.): Process Mining Handbook, Lecture Notes in Business Information Processing, vol. 448. Springer, Cham (2022). https://doi.org/10.1007/978-3-031-08848-3
27. Velickovic, P., Cucurull, G., Casanova, A., Romero, A., Liò, P., Bengio, Y.: Graph attention networks. In: International Conference on Learning Representations. OpenReview.net (2018)
28. Verenich, I., Dumas, M., Rosa, M.L., Maggi, F.M., Teinemaa, I.: Survey and cross-benchmark comparison of remaining time prediction methods in business process monitoring. ACM Trans. Intell. Syst. Technol. **10**(4), 34:1–34:34 (2019). https://doi.org/10.1145/3331449
29. Weinzierl, S.: Exploring gated graph sequence neural networks for predicting next process activities. In: Marrella, A., Weber, B. (eds.) BPM 2021. LNBIP, vol. 436, pp. 30–42. Springer, Cham (2022). https://doi.org/10.1007/978-3-030-94343-1_3

OCPQ: Object-Centric Process Querying and Constraints

Aaron Küsters(✉) and Wil M. P. van der Aalst

Chair of Process and Data Science (PADS), RWTH Aachen University, Aachen, Germany
{kuesters,wvdaalst}@pads.rwth-aachen.de

Abstract. Process querying is used to extract information and insights from process execution data. Similarly, process constraints can be checked against input data, yielding information on which process instances violate them. Traditionally, such process mining techniques use case-centric event data as input. However, with the uptake of Object-Centric Process Mining (OCPM), existing querying and constraint checking techniques are no longer applicable. Object-Centric Event Data (OCED) removes the requirement to pick a single case notion (i.e., requiring that events belong to exactly one case) and can thus represent many real-life processes much more accurately. In this paper, we present a novel highly-expressive approach for object-centric process querying, called *OCPQ*. It supports a wide variety of applications, including OCED-based constraint checking and filtering. The visual representation of nested queries in OCPQ allows users to intuitively read and create queries and constraints. We implemented our approach using (1) a high-performance execution engine backend and (2) an easy-to-use editor frontend. Additionally, we evaluated our approach on a real-life dataset, showing the lack in expressiveness of prior work and runtime performance significantly better than the general querying solutions SQLite and Neo4j, as well as comparable to the performance-focused DuckDB.

Keywords: Object-Centric Process Mining · Querying · Constraints

1 Introduction

In organizations, process execution data contain valuable insights that are often not leveraged to their full extent. The domain of process querying is concerned with methods and techniques for extracting such insights from event data. For example, given data of an order management process, a simple query for interesting cases could be formulated in natural language as "Find all cases where `pay order` is executed more than once". Process querying of execution data also has a strong correspondence to process constraints: Identifying violations of a process constraint, e.g., "`pay order` should be executed exactly once per case", corresponds to querying its violations (i.e., a query with the negated constraint).

J. Grabis et al. (Eds.): RCIS 2025, LNBIP 547, pp. 383–400, 2025.
https://doi.org/10.1007/978-3-031-92474-3_23

To promote using this opportunity of gaining insights, it is important to allow stakeholders to query interesting scenarios in their processes themselves. For that reason, graphical notations for constraints or queries are often introduced, as a way to also allow stakeholders without programming experience to utilize them. Similarly, there are also graphical declarative process models, like DECLARE [6], in which visual models describe an underlying set of constraint rules.

More recently, process querying and constraint approaches based on *Object-Centric Event Data* (OCED) have been proposed [2–4]. OCED no longer assumes a single case notion in data, i.e., that events belong to exactly one defined case. Instead, OCED contains a set of objects and a set of events of specified types. OCED also allows for relationships between objects and events, as well as between objects and objects. As such, OCED can represent many real-life processes much more accurately than traditional, flat event logs. For example, in an order management process, the different objects interacting in the process could include `customers`, `orders`, `items`, `packages`, and `employees`. Therefore, classical case-centric approaches, which assign one object per event, are too limiting [1]. These advantages of OCED translate directly to process querying techniques based on OCED. In particular, a more accurate and interconnected representation of the underlying real-life process can be queried, instead of only a flat representation that could lead to inaccurate results or misleading conclusions.

In this paper, we present an object-centric nested querying approach with an accompanying full graphical tool implementation, focusing on high expressiveness, fast runtime performance, and easy usability. An overview of the approach, including inputs and outputs, is shown in Fig. 1. First, stakeholders design queries or constraints, optionally based on some regulation or specification document. The created queries can then be evaluated based on input OCED, yielding query results. The query results are visually shown in aggregation (e.g., with the total number of results or the percentage of violating instances) but can also be explored in the tool individually or exported (e.g., to a CSV or XLSX file).

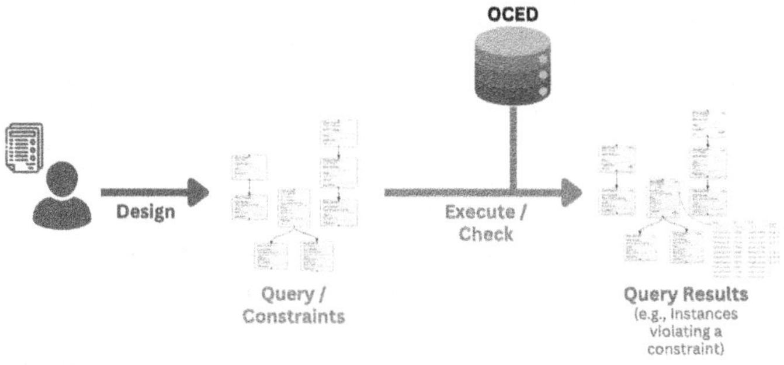

Fig. 1. Overview of the object-centric process querying and constraint approach.

Consider the following example of an object-centric constraint for an order management process: "If a third payment reminder for an order is sent to a customer, no (other) order by this customer should be confirmed afterwards".

What at first seems to be a rather simple rule actually involves multiple different object and event types, as well as multiple instances of the same object type (i.e., multiple order objects). In fact, to the best of our knowledge, none of the so far proposed graphical process querying or constraint techniques can express such a rule. Even in all-purpose querying languages, like SQL or Cypher, expressing such a rule is not only unapproachable for people without programming skills, but also largely impractical for larger data because of their poor performance for such types of queries.

We fill this gap by introducing *OCPQ*, an object-centric process querying approach, that leverages the full flexibility of the OCED data model, while focusing on easy, visual usability and a very fast runtime performance. We also present a full graphical tool implementation of the approach, which is publicly available at https://github.com/aarkue/ocpq. In contrast to related work, OCPQ can also express advanced queries and constraints spanning multiple object and event types while still allowing for easy visual modeling. For instance, the previously introduced example can be modeled visually as a nested query constraint in OCPQ, as shown in Fig. 2.

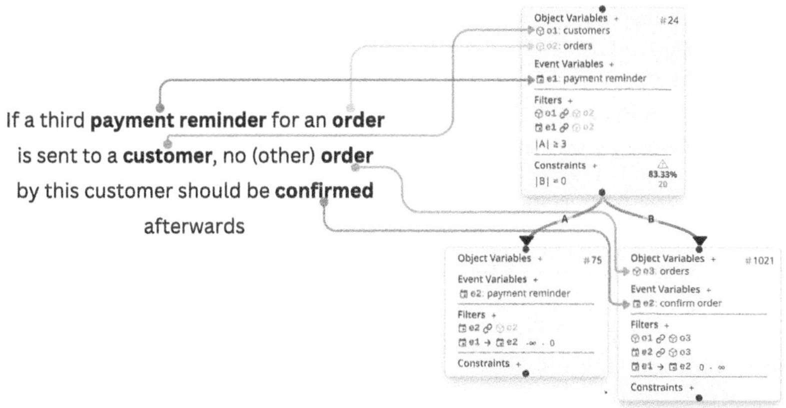

Fig. 2. The mapping of an example constraint in natural language to a visual nested query constraint in OCPQ. The included arrows indicate how objects or events mentioned in the textual description are modeled in the visual constraint.

In our tool implementation, nested queries are evaluated through a recursive, parallelizable algorithm, achieving good runtime performance and demonstrating the feasibility of our approach, even for larger, real-life datasets.

The remainder of this paper is structured as follows. In Sect. 2, we first discuss related work. Next, we introduce preliminaries in Sect. 3. Section 4 describes the main concepts of our approach. Our tool implementation is covered in Sect. 5, followed by an evaluation of the expressiveness and runtime in Sect. 6. Finally, we conclude this paper in Sect. 7.

2 Related Work

In this section, we present related work on process querying, process constraints and declarative process models.

Process querying research covers filtering and manipulation of process repositories, which can contain process models as well as process executions (i.e., event data), based on a (formal) query. In this paper, we focus on querying of event data without corresponding process models. In [5], the authors describe the *Process Instance Query Language* (PIQL), which allows querying the number of cases or events fulfilling specified criteria. The proprietary *Celonis Process Querying Language* (Celonis PQL) introduced in [12] is another domain-specific process querying language heavily inspired by SQL. Through the integration of Celonis PQL with the underlying data model schema, table joins do not need to be specified explicitly, as would be the case with `JOIN` statements in SQL. All the previously mentioned process querying research is primarily focused on traditional, flat event data. However, in [3], Esser et al. describe storing multi-dimensional (i.e., object-centric) event data in graph databases and using the universal graph querying language *Cypher* of the *Neo4j* database system to query entities or subgraphs.

Apart from querying, process constraints as well as declarative process models are also important related fields for this paper. Both of these fields are concerned with checking if input event data satisfies specified rules, and returning violating fragments of data if not. Most notably, *DECLARE* [6] was the first declarative approach for business process management. While internally, DECLARE uses Linear Temporal Logic (LTL), parametrized LTL templates are represented visually (e.g., as specific types of arrows) to ease usability. An initial constraint template language, also called DECLARE, is included by default, but custom template languages can be created and used as well.

In [10], Schöning et al. present an SQL-based approach for discovering declarative process constraints. By creating subqueries for returning activity combinations or computing support and confidence values, the discovery of constraints that satisfy given thresholds can be implemented as a simple SQL `SELECT` query. The authors report generally competitive performance metrics for discovering constraints. However, expanding the set of discoverable constraints, for example to ternary instead of only binary constructs, would lead to significant higher execution times. In [9], the authors extend this approach to also consider data attributes, resource, and time perspectives. Similarly, in [8], the authors present a declarative process mining framework based on SQL queries, allowing for discovery and checking of constraints. The authors additionally evaluated the time required for executing the resulting query sets across different database schemas.

There are also a few object-centric process constraint approaches. OCBC models, introduced in [2], combine behavioral constraints, inspired by DECLARE patterns, with object class data models imposing cardinality rules. As such, OCBC models can express that relationships between objects on the class-level should also manifest on the behavior-level (e.g., only items belonging to a purchase order should be associated with its pick item events). In [4], the authors introduce *Object-Centric Constraint Graphs* (OCCGs). These constraint graphs can capture interactions between objects and events, as well as control-flow between event types, based on a given object type. Additionally, performance metrics regarding events can be included.

3 Preliminaries

As preliminaries, we first define some basic mathematical concepts.

For a set X, the *powerset* of X is $\mathcal{P}(X) = \{Y \mid Y \subseteq X\}$. We also use partial functions: Given two sets A and B, a partial function $f\colon A \nrightarrow B$ maps *some* of the elements of A to values in B. For elements x that are not mapped to a value (i.e., $x \notin \text{dom}(f)$), we write $f(x) = \bot$. If two partial functions have disjoint domains (i.e., $\text{dom}(f) \cap \text{dom}(g) = \emptyset$), we write $f \cup g$ for their symmetric union. Moreover, we use subset notation (i.e., $f \subseteq g$) if it holds that $\forall_{x \in \text{dom}(f)}\ g(x) = f(x)$.

Next, we define the universes that form the basis of our formalization.

Definition 1. *Let $\mathcal{U}_\Sigma$ be the universe of strings. We use the following pairwise disjoint universes:*

- $\mathcal{U}_{ev} \subseteq \mathcal{U}_\Sigma$ *Universe of events* *(e.g., e_1)*
- $\mathcal{U}_{obj} \subseteq \mathcal{U}_\Sigma$ *Universe of objects* *(e.g., o_1)*
- $\mathcal{U}_{etype} \subseteq \mathcal{U}_\Sigma$ *Universe of event types (*activities*)* *(e.g.,* `confirm order`*)*
- $\mathcal{U}_{otype} \subseteq \mathcal{U}_\Sigma$ *Universe of object types* *(e.g.,* `orders`*)*
- $\mathcal{U}_{attr} \subseteq \mathcal{U}_\Sigma$ *Universe of attribute names* *(e.g.,* `time`*)*
- $\mathcal{U}_{qual} \subseteq \mathcal{U}_\Sigma$ *Universe of relationship qualifiers* *(e.g.,* `places`*)*
- $\mathcal{U}_{\text{obVar}} \subseteq \mathcal{U}_\Sigma$ *Universe of object variable names* *(e.g.,* `o1`*)*
- $\mathcal{U}_{\text{evVar}} \subseteq \mathcal{U}_\Sigma$ *Universe of event variable names* *(e.g.,* `e1`*)*
- $\mathcal{U}_{\text{setName}} \subseteq \mathcal{U}_\Sigma$ *Universe of set variable names* *(e.g.,* `A`*)*

We write $\mathbb{T}$ for all possible timestamps and durations and $\mathcal{U}_{val}$ for the universe of all attribute values (with, for instance, $\mathbb{T} \subseteq \mathcal{U}_{val}$ and $\mathcal{U}_\Sigma \subseteq \mathcal{U}_{val}$).

Next, we introduce Object-Centric Event Data (OCED) formally. Our definition is inspired by the OCEL 2.0 specification[1], but the presented approach is not limited to any particular OCED model. At the very core, OCED contains a set of objects and a set of events, each of which have additional attributes. For objects, these attributes can also change over time. Certain types of attributes (e.g., for the object type or relationships between objects) are mandatory.

Definition 2. *Object-Centric Event Data (OCED) can be described as a tuple $L = (E, O, eaval, oaval)$ of the following components:*

- ***Events*** *$E \subseteq \mathcal{U}_{ev}$ as the set of events.*
- ***Objects*** *$O \subseteq \mathcal{U}_{obj}$ as the set of objects.*
- ***Event Attributes*** *$eaval : E \to (\mathcal{U}_{attr} \nrightarrow \mathcal{U}_{val})$, which provides attribute values for events. For convenience, we write $eaval_e = eaval(e)$ for an $e \in E$ as a shorthand. The following properties have to hold for eaval:*

[1] https://www.ocel-standard.org/.

- $\forall_{e \in E}$ $eaval_e(\texttt{activity}) \in \mathcal{U}_{etype}$: *each event has exactly one* event type.
- $\forall_{e \in E}$ $eaval_e(\texttt{objects}) \subseteq \mathcal{U}_{qual} \times O \wedge eaval_e(\texttt{objects}) \neq \emptyset$: *each event has at least one qualified reference to an object.*
- $\forall_{e \in E}$ $eaval_e(\texttt{time}) \in \mathbb{T}$: *each event has a timestamp.*

– ***Object Attributes*** *oaval* : $O \rightarrow (\mathcal{U}_{attr} \times \mathbb{T} \not\rightarrow \mathcal{U}_{val})$, *which provides the attribute values of an object* $o \in O$ *at a concrete timestamp. For convenience, we write* $oaval_o^t(attr) = oaval(o)(attr, t)$ *for a given* $o \in O, t \in \mathbb{T}$ *and* $attr \in \mathcal{U}_{attr}$ *as a shorthand. The following properties have to hold for oaval:*
- *For the* time-stable *attributes* $a \in \{\texttt{objects}, \texttt{type}\} \subseteq \mathcal{U}_{attr}$, *the assigned value must not change over time. In particular, it should hold that* $\forall_{o \in O} \forall_{t \in \mathbb{T}} \forall_{t' \in \mathbb{T}}$ $oaval_o^t(a) = oaval_o^{t'}(a)$. *We also write* $oaval_o(a) = oaval_o^t(a, t)$ *with an arbitrary timestamp* $t \in \mathbb{T}$ *for these attributes.*
- $\forall_{o \in O}$ $oaval_o(\texttt{type}) \in \mathcal{U}_{otype}$: *every object has exactly one* object type.
- $\forall_{o \in O}$ $oaval_o(\texttt{objects}) \subseteq \mathcal{U}_{qual} \times O$: *an object can, optionally, contain qualified references to (other) objects.*

As an example OCED, consider $L = (E, O, eaval, oaval)$ with a set of objects $O = \{o_1, o_2, o_3, o_4\}$ and a set of events $E = \{e_1, e_2, e_3, e_4, e_5, e_6\}$. The attribute values of all objects and events, assigned by *eaval* and *oaval*, are shown in Fig. 3. For *oaval*, time-stable object attributes are marked with an $*$ in the timestamp column. Apart from the mandatory attributes, the following two custom attributes are present in L: The customer o_1 has an attribute `city`, indicating the city of residence of the customer. After providing the city initially in 2016, the attribute was updated in 2018, as the customer moved their residence. Moreover, the `payment reminder` event e_5 has an attribute `fee`, indicating the additional fine incurred by the late payment (e.g., 15€).

Object	Attribute	Timestamp	Value
o_1	type	*	customers
o_1	objects	*	{(places, o_2)}
o_1	city	2016-01-06T14:15	Bonn
o_1	city	2018-09-03T10:32	Aachen
o_2	type	*	orders
o_2	objects	*	{(contains, o_3), (contains, o_4)}
o_3	type	*	items
o_4	type	*	items

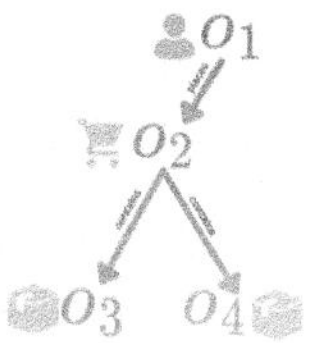

(a) *oaval* with a visualization of the O2O relationships.

Event	Attribute	Value
e_1	activity	place order
e_1	objects	{(customer, o_1), (order, o_2), (item, o_3), (item, o_4)}
e_2	activity	pack item
e_2	objects	{(item, o_3)}
e_3	activity	pack item
e_3	objects	{(item, o_4)}
e_4	activity	ship items
e_4	objects	{(ships, o_3), (ships, o_4)}
e_5	activity	payment reminder
e_5	objects	{(recipient, o_1), (order, o_2)}
e_5	fee	15
e_6	activity	pay order
e_6	objects	{(order, o_2)}

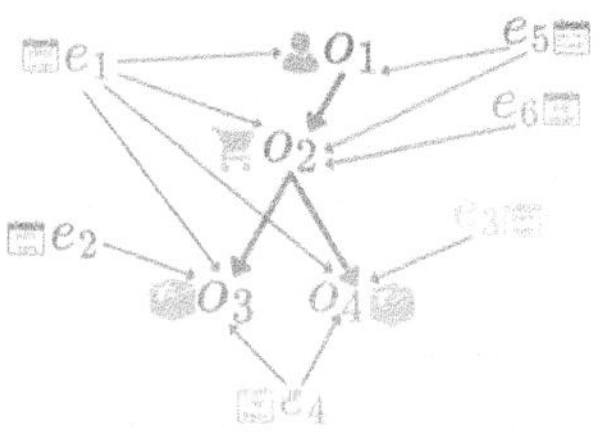

(b) *eaval* with a visualization of the O2O and E2O relationships.

Fig. 3. Example *oaval* and *eaval* tables with corresponding relationship visualizations.

The graph in Fig. 3(a) shows all objects in O as nodes. Edges between nodes indicate the object-to-object (O2O) relationships, that are formally expressed through the `objects` attributes. In the graph of Fig. 3(b), events and event-to-object (E2O) relationships are included additionally. For readability, the qualifiers for the O2O and E2O relationships are omitted.

For an OCED $L = (E, O, eaval, oaval)$, we introduce these shorthands:

- $E_L = E$ and $O_L = O$ for the set of objects or events of L, respectively.
- Function $type_L \in (O \cup E) \rightarrow (\mathcal{U}_{otype} \cup \mathcal{U}_{etype})$, which assigns *object or event types* to objects or events, defined as:

$$type_L(x) = \begin{cases} oaval_x(\texttt{type}), & \text{if } x \in O \\ eaval_x(\texttt{activity}), & \text{if } x \in E \end{cases}$$

- Function $time_L \in E \rightarrow \mathbb{T}$ with $time_L(e) = eaval_e(\texttt{time})$, which maps an event to its timestamp.
- Given an optional qualifier $q \in \mathcal{U}_{qual} \cup \{*\}$, we define the function $obj_L^q \colon (E \cup O) \rightarrow \mathcal{P}(O)$, which assigns an event or object to its set of object references:

$$obj_L^q(x) = \begin{cases} \{o \mid (q', o) \in eaval_x(\texttt{objects}) \wedge (q = * \vee q = q')\}, & \text{if } x \in E \\ \{o \mid (q', o) \in oaval_x(\texttt{objects}) \wedge (q = * \vee q = q')\}, & \text{if } x \in O \end{cases}$$

For simplicity we also write $obj_L = obj_L^*$ for all object references without considering their qualifiers.

4 Object-Centric Querying and Constraints

In this section, we detail our approach to object-centric querying and constraints. First, we introduce the concept of *variable bindings*. They make up the output of our querying approach. A variable binding is a collection of concrete objects and events of an OCED that are referred to using variable names. Through these names, events or objects can be differentiated even if they are of the same type.

Definition 3. *Let L be an OCED. The set of variable bindings $\mathbb{B}_L$ under L is:*

$$\mathbb{B}_L = \{b_1 \cup b_2 \mid b_1 \in (\mathcal{U}_{\text{evVar}} \not\rightarrow E_L) \wedge b_2 \in (\mathcal{U}_{\text{obVar}} \not\rightarrow O_L)\}$$

Consider an example OCED L with $o_1, o_2, o_3 \in O_L$ and $e_1, e_2, e_3 \in E_L$, and let $\texttt{o1}, \texttt{o2}, \texttt{o3} \in \mathcal{U}_{\text{obVar}}$ and $\texttt{e1}, \texttt{e2}, \texttt{e3} \in \mathcal{U}_{\text{evVar}}$ be object and event variable names. Then the following example bindings are part of $\mathbb{B}_L$: $b_1 = \{\}$, $b_2 = \{\texttt{o1} \mapsto o_1\}$, $b_3 = \{\texttt{o2} \mapsto o_1\}$, and $b_4 = \{\texttt{e1} \mapsto e_1, \texttt{e2} \mapsto e_3, \texttt{o1} \mapsto o_1, \texttt{o3} \mapsto o_3\}$.

In the context of an OCED, we next introduce the concept of child and parent bindings. A child binding contains all the event and object variables of the parent, and also maps them to the same values as the parent.

Definition 4. *Let L be an OCED. We define the parent-child relation $\sqsubseteq_L$ as follows: For two bindings $p, c \in \mathbb{B}_L$, $p \sqsubseteq_L c \Leftrightarrow \forall_{x \in \text{dom}(p)}\ p(x) = c(x)$. When $p \sqsubseteq_L c$, we call c a* child binding *of p and p a* parent binding *of c. Clearly, $\sqsubseteq_L$ is a partial order (i.e., reflexive, antisymmetric and transitive).*

For every OCED L, the empty binding $\{\}$ is the smallest element in $\mathbb{B}_L$ regarding $\sqsubseteq_L$. Considering the previous example bindings, it holds that $b_2 \sqsubseteq_L b_4$ and $b_3 \not\sqsubseteq_L b_4$. The $\sqsubseteq_L$ relation is useful for describing the output of *nested* queries.

Next, we introduce *binding predicates.* They allow specifying which bindings a query should return, similar to a `WHERE` clause in SQL.

Definition 5. *Let L be an OCED. Given L, a* binding predicate *describes a set of bindings that satisfy this predicate. However, different predicates can be distinguished even if they induce the same set of satisfied bindings. We write $\mathbb{P}_L$ for the set of all binding predicates. If a binding $b \in \mathbb{B}_L$ satisfies a predicate $s \in \mathbb{P}_L$, we write $b \models s$. Additionally, we use the same notation for a set of predicates $S \subseteq \mathbb{P}_L$: $b \models S \Leftrightarrow \forall_{s \in S}\ b \models s$.*

As an example binding predicate for an OCED L, consider $s \in \mathbb{P}_L$ with $s \models b \Leftrightarrow b(\texttt{o1}) \in O_L \wedge b(\texttt{e1}) \in E_L \wedge b(\texttt{o1}) \in obj_L(b(\texttt{e1}))$ for all $b \in \mathbb{B}_L$. The binding predicate s encompasses all variable bindings where the variables `o1` and `e1` are bound to objects or events of L, respectively, such that the values are in an event-to-object relationship.

One can specify (pairwise disjoint) *collections of binding predicates* for different purposes. Initially, we define the collection $\mathbf{BASIC_L} \subseteq \mathbb{P}_L$ in the context of L. Note that our approach is easily extendable by adding more predicate types and collections. However, for brevity, we do not define further predicates (e.g., based on general data attributes, like the `price` of an `orders` object) here.

$\mathbf{BASIC_L}$ is made up of three predicate types:

- **Event-to-Object** For an event variable $v \in \mathcal{U}_{\text{evVar}}$, an object variable $v' \in \mathcal{U}_{\text{obVar}}$ and an optional relationship qualifier $q \in \mathcal{U}_{qual} \cup \{*\}$: $\text{E2O}(v, v', q) \in \mathbf{BASIC_L}$, with for any $b \in \mathbb{B}_L$:
$$b \models \text{E2O}(v, v', q) \;\Leftrightarrow\; b(v) \in E_L \wedge b(v') \in O_L \wedge b(v') \in obj_L^q(b(v))$$
- **Object-to-Object** For two object variables $v, v' \in \mathcal{U}_{\text{obVar}}$, and an optional qualifier $q \in \mathcal{U}_{qual} \cup \{*\}$: $\text{O2O}(v, v', q) \in \mathbf{BASIC_L}$, with for any $b \in \mathbb{B}_L$:
$$b \models \text{O2O}(v, v', q) \;\Leftrightarrow\; b(v) \in O_L \wedge b(v') \in O_L \wedge b(v') \in obj_L^q(b(v))$$
- **Time Between Events** For two event variables $v, v' \in \mathcal{U}_{\text{evVar}}$ and a duration interval $t_{min}, t_{max} \in \mathbb{T}$: $\text{TBE}(v, v', t_{min}, t_{max}) \in \mathbf{BASIC_L}$, with for any $b \in \mathbb{B}_L$:
$$b \models \text{TBE}(v, v', t_{min}, t_{max}) \;\Leftrightarrow\; b(v) \in E_L \wedge b(v') \in E_L \wedge t_{min} \leq time_L(b(v')) - time_L(b(v)) \leq t_{max}$$

Consider the predicate $s_1 = \text{O2O}(\texttt{o1}, \texttt{o2}, *) \in \mathbf{BASIC_L}$ and the bindings $b_5 = \{\texttt{o1} \mapsto o_1\}$, and $b_6 = \{\texttt{o1} \mapsto o_1, \texttt{o2} \mapsto o_2, \texttt{e1} \mapsto e_1\}$. As b_5 does not assign `o2`, it holds that $b_5 \not\models s_1$. Assuming an object-to-object relation between o_1 and o_2 exists in L, $b_6 \models s_1$ would hold.

Next, we introduce the concept of *binding boxes*, which correspond to simple queries, yielding sets of variable bindings as output.

Definition 6. *Let L be an OCED. A binding box $\mathfrak{b}_L = (\text{Var}, \text{Pred})$ over L is a tuple consisting of:*

- $\text{Var} \in \{ev \cup ob \mid ev \in \mathcal{U}_{\text{evVar}} \nrightarrow \mathcal{P}(\mathcal{U}_{etype}) \land ob \in \mathcal{U}_{\text{obVar}} \nrightarrow \mathcal{P}(\mathcal{U}_{otype})\}$, *a partial function which specifies to values of which event or object types selected variables should be bound.*
- $\text{Pred} \subseteq \mathbb{P}_L$, *a set of binding predicates.*

Intuitively, $\mathfrak{b}_L$ *binds the variable names* dom(Var) *to all combination of values (i.e., events or objects of L) such that they are of the specified types and the predicate set S holds. For convenience, we sometimes write* $\text{Var}(\mathfrak{b}_L) = \text{Var}$ *and* $\text{Pred}(\mathfrak{b}_L) = \text{Pred}$. *Additionally, we write* $\mathfrak{BOX}_L$ *for the set of all binding boxes under L. We define when a binding $b \in \mathbb{B}_L$ satisfies the binding box, written as $b \models \mathfrak{b}_L$, as follows:*

$$b \models \mathfrak{b}_L \iff b \models \text{Pred} \land \text{dom}(b) = \text{dom}(\text{Var}) \\ \land \forall_{v \in \text{dom}(\text{Var})} \big(b(v) \in E_L \cup O_L \land type_L(b(v)) \in \text{Var}(v)\big)$$

Next, we present a simple example binding box involving one event variable and one object variable. For this and further examples, consider an OCED $L = (E, O, eaval, oaval)$ of an order management process that is not fully specified here for brevity. Consider the simple binding box $\mathfrak{a}_L = (\text{Var}, \text{Pred})$, with:

- $\text{Var} = \{\texttt{e1} \mapsto \{\texttt{place order}, \texttt{confirm order}\}, \texttt{o1} \mapsto \{\texttt{orders}\}\}$
- $\text{Pred} = \{\text{O2E}(\texttt{e1}, \texttt{o2}, \texttt{order})\}$

For convenience, we will use a visual notation schema for further examples: Given a binding box $\mathfrak{a}_L = (\text{Var}, \text{Pred})$, the elements of Var are split into multiple lines on the top. On the bottom (i.e., in the predicate; below the line), the filter predicates Pred are listed using their representation (e.g., E2O).

$\mathfrak{a}_L$

$\texttt{o1} : \text{OBJECT}(\texttt{orders})$

$\texttt{e1} : \text{EVENT}(\texttt{place order}, \texttt{confirm order})$

$\text{E2O}(\texttt{e1}, \texttt{o1}, \texttt{order})$

Given this example binding box $\mathfrak{a}_L$, we can also construct its output set. Assume that L contains only the objects o_1, o_2, o_3 of type `orders`, where o_1 and o_2 are associated with a `place order` event (i.e., with e_1 and e_2, respectively). Additionally, assume that o_1 is the only object that is also associated with a `confirm order` event e_3. Then we can construct the output bindings of $\mathfrak{a}_L$, $\{b \in \mathbb{B}_L \mid b \models \mathfrak{a}_L\}$, as: $out_L(\mathfrak{a}_L) = \big\{\{\texttt{o1} \mapsto o_1, \texttt{e1} \mapsto e_1\}, \{\texttt{o1} \mapsto o_2, \texttt{e1} \mapsto e_2\}, \{\texttt{o1} \mapsto o_1, \texttt{e1} \mapsto e_3\}\big\}$.

To facilitate nested queries, we define a relation $\preceq_L$ between binding boxes over an OCED L. It encompasses the concept of *refined* binding boxes, where new object or event variables can be introduced, and the filter is at least as strict as before.

Definition 7. *Let L be an OCED. Let $\mathfrak{a}_L, \mathfrak{b}_L \in \mathfrak{BOX}_L$. We say $\mathfrak{a}_L \preceq_L \mathfrak{b}_L$ holds if:* $\mathrm{Var}(\mathfrak{a}_L) \subseteq \mathrm{Var}(\mathfrak{b}_L)$ *and* $\mathrm{Pred}(\mathfrak{a}_L) \subseteq \mathrm{Pred}(\mathfrak{b}_L)$.

For instance, with $\mathfrak{a}_L$ from before and $\mathfrak{b}_L = (\{\texttt{o1} \rightarrow \{\texttt{orders}\}\}, \emptyset)$, it holds that $\mathfrak{b}_L \preceq_L \mathfrak{a}_L$. The concept of refined binding boxes enables nested queries, where the first binding box only queries a subset of the overall involved objects or events (e.g., only an `orders` object but no events).

Next, we define the restriction of binding boxes on only a subset of considered predicates. They enable ignoring certain predicate types for the $\preceq_L$ relation.

Definition 8. *Let L be an OCED, let $\mathfrak{a} = (\mathrm{Var}, \mathrm{Pred}) \in \mathfrak{BOX}_L$ be a binding box over L and let $X \subseteq \mathbb{P}_L$ be a set of binding predicates. The filter-restriction of $\mathfrak{a}$ to X, denoted as $\mathfrak{a}|_X$, is the binding box $\mathfrak{a}|_X = (\mathrm{Var}, \mathrm{Pred} \cap X)$ over L.*

These concepts allow defining a tree structure of binding boxes, where children are refined versions of their parents when considering only basic predicates. These *query trees* are the core of our approach and enable declarative, nested querying of objects and events.

Definition 9. *Let L be an OCED. A query tree is a tuple $T = (V, F, r, l, box)$:*

- *V is a finite set of nodes.*
- *$r \in V$ is the designated root node (i.e., the only node with no parent).*
- *$F \subseteq V \times V$ is a set of edges between nodes, such that in the directed graph (V, F) there is exactly one path from the root r to a for all $a \in V$.*
- *$l \colon F \rightarrow \mathcal{U}_{\mathrm{setName}}$ is an injective function, assigning unique names to edges.*
- *$box \colon V \rightarrow \mathfrak{BOX}_L$ is a function which maps each node in V to a binding box over L, such that for all edges $(a, b) \in F$ with $box(a) = \mathfrak{a}$ and $box(b) = \mathfrak{b}$, it holds that $\mathfrak{a}|_{\mathbf{BASIC_L}} \preceq_L \mathfrak{b}|_{\mathbf{BASIC_L}}$.*

Next, in Fig. 4, we show an example query tree only using predicates from $\mathbf{BASIC_L}$. Afterwards, we introduce more complex queries and constraint examples which motivate the specified restriction of the *box* function.

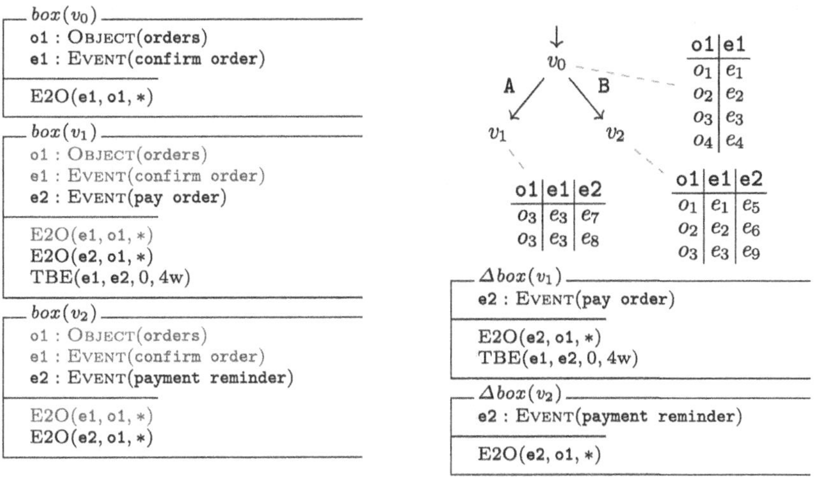

Fig. 4. A query tree with three nodes. On the left, the binding box of each node is shown. On the top right, the tree structure is visualized with example output tables. The boxes on the bottom right, marked with Δ, only show additions to their parents. (Color figure online)

In Fig. 4, the query tree $T_1 = (V, F, r, l, box)_L$ is shown, with $V = \{v_0, v_1, v_2\}$, $r = v_0$, $F = \{(v_0, v_1), (v_0, v_2)\}$, $l((v_0, v_1)) = \texttt{A}$, and $l((v_0, v_2)) = \texttt{B}$. The top right of Fig. 4 shows the graph of T_1 with exemplary output binding tables, while *box* is presented on the left. Naturally, the child binding boxes contain many duplicates (shown in gray). To ease readability, we omit variables and predicates that are already present in the binding box of the parent node in future examples. With these omissions (marked using Δ), $box(v_1)$ and $box(v_2)$ can be presented more compactly, as shown on the bottom right.

On the top right of Fig. 4, we show exemplary output sets of $box(v_0)$, $box(v_1)$, and $box(v_2)$ as tables next to the corresponding nodes. The rows for v_0 are colored in four different colors. For v_1 and v_2, each output row is colored based on $\sqsubseteq_L$, indicating from which parent binding in the output set of v_0 the row is derived. The first two output binding rows for v_0 (in cyan and magenta) have exactly one child binding in the output set of v_1 and none in the output set of v_2. For the third output binding row of v_0 (in green), one child binding in v_1 exists, and there are also two child bindings in the output set of v_2. The last output row of v_0 (in orange) has no child binding in the output sets of v_1 or v_2.

In a nested query, oftentimes, the result of the inner query is used in the outer query in an aggregated way. For instance, in a query for all customers with more than 100 orders, the outer query (all customers) uses the result count of the inner query (all orders by the customer) as its filter. To express such queries, we introduce a new set of binding predicates, **CHILD SET**$_u^T$, in the context of a query tree $T = (V, F, r, l, box)$ and one of its nodes $u \in V$. For every child node $v \in V$ with $(u, v) \in F$ and $l((u, v)) = A$, predicates of the form CBS(A, n_{min}, n_{max}) with $n_{min}, n_{max} \in \mathbb{N}_0$ are available in **CHILD SET**$_u^T$. They are fulfilled for a binding $b \in \mathbb{B}_L$, when the set of child bindings of b in v, $S = \{x \in \mathbb{B}_L \mid x \models box(v) \wedge b \sqsubseteq_L x\}$, is in the specified size range (i.e., $n_{min} \leq |S| \leq n_{max}$). Note, that these predicates can be recursive, as the predicates of a child node are already considered when evaluating the predicates of the parent node. In particular, they are also only well-defined for the binding box of a specified node, which is why they are not considered for the $\preceq_L$ relation in the tree, which only considers the predicates in $\mathbf{BASIC_L}$.

Figure 5 shows such an extended version of the previous tree example.

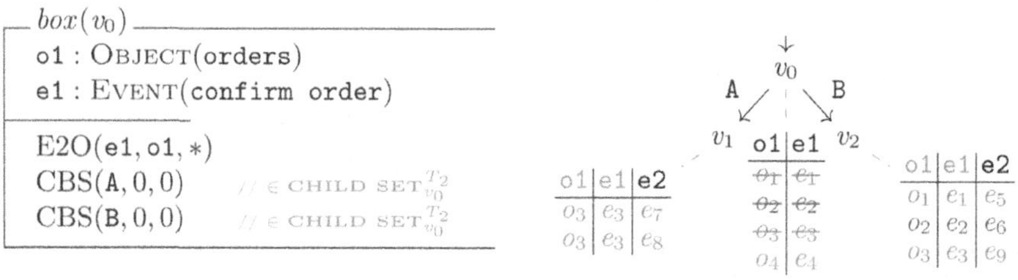

Fig. 5. An extension of the query tree from Fig. 4 with child filter predicates (CBS).

Figure 5 shows a new query tree $T_2 = (V, F, r, l, box)$, with $V = \{v_0, v_1, v_2\}$, $r = v_0$, and $F = \{(v_0, v_1), (v_0, v_2)\}$. While v_1 and v_2 are the same as in Fig. 4, the root node box (v_0) now contains two additional predicates of type **CHILD SET**$_{v_0}^T$. T_2 queries placed orders that *were not paid fast* (i.e., within 4 weeks after confirmation) and for which also *no payment reminder was sent.* Again, we annotate example output tables for each node in the tree on the right of Fig. 5. If bindings are removed only by a child set predicate of a binding box but fulfill the basic predicates of it (i.e., $\mathbf{BASIC_L}$), we indicate this by including this binding with a strikethrough. Generally, children of the node might still contain child bindings for crossed out parent binding rows (e.g., all rows for v_1 and v_2 in Fig. 5); however, in practice, they are often not of particular interest and thus sometimes ignored or omitted from the result tables.

Next, we want to outline how the presented querying approach can be extended. In general, the output tables of the query tree nodes can be augmented and filtered freely. As augmentation, labels for each output row can be computed and added as columns (e.g., the total order volume of a customer). In the following, we will describe how *constraints* can be implemented. At its core, constraints are a special case of such general labels, defining for each output row if it should be considered *satisfied* or *violated.* As this classification is binary, a set of predicates for each node can be used for specifying the violation criteria. For a tree node $v \in V$, we write $constr(v) \subseteq \mathbb{P}_L$ for its set of constraint predicates. Fully defining these additions formally is outside the scope of this paper. However, in the following, we present a short example as a demonstration: Consider the query tree constraint $C = ((V, F, r, l, box), constr)$ shown in Fig. 6, with $V = \{v_0, v_1\}$, $r = v_0$ and $F = \{(v_0, v_1)\}$. The graph (V, F) with the edge labels l is shown on the right, while box with $constr$ is presented on the left.

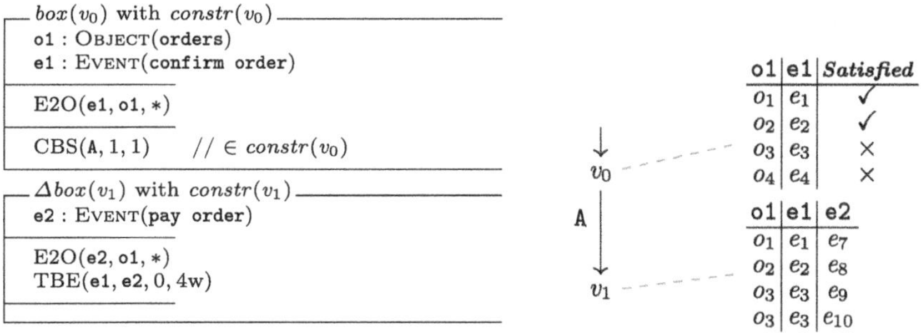

Fig. 6. Example constraint specifying that every confirmed order *should* be paid within 4 weeks after the confirmation exactly once. For each $v \in V$, the set $constr(v)$ is shown in the corresponding binding box below an extra line. In the example output tables shown on the right, this constraint is satisfied for all binding rows of $box(v_0)$ except the last two rows. The violation status of an output binding of $box(v_0)$ is annotated to the corresponding output row as either ✓ or ×.

5 Implementation

We implemented the querying and constraint checking approach introduced in Sect. 4 as the full-stack application *OCPQ* consisting of two parts: (1) A high-performance execution engine backend implemented in Rust and (2) an interactive user-friendly query editor frontend written in Typescript. The tool is publicly available at https://github.com/aarkue/OCPQ, and there are installers as well as further resources and guides available at https://ocpq.aarkue.eu.

Describing the implementation, and all its features, in detail is outside the scope of this paper. However, we briefly mention some key aspects:

Importing & Pre-processing: All file types of the OCEL 2.0 specification can be imported. The OCEDs are then pre-processed to enable fast query execution.
Representation of Bindings: Variable bindings are represented in a memory-efficient way, encoding the variable name and its value only through an 8-byte integer each. This enables faster execution times and usage for larger datasets.
Parallelized Recursive Query Execution Algorithm: Queries are executed by a recursive algorithm that allows for full parallelization between bindings.
HPC Deployment: To leverage this parallelization, the tool allows users to easily deploy query execution on a High-Performance Computing (HPC) cluster.
Intelligent Binding Order: New variables are bound in an efficient order, applying predicate filters as early as possible to remove unwanted bindings.

A screenshot of the tool is shown in Fig. 7. The frontend editor visualizes query trees similar to the notation style that is used throughout this paper.

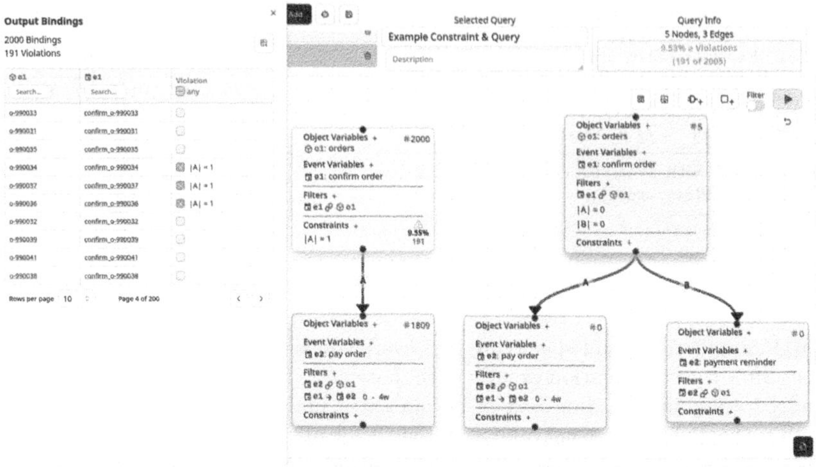

Fig. 7. Screenshot of the implemented OCPQ tool. The two shown trees correspond to the examples in this paper (i.e., Fig. 6 on the left, and Fig. 5 on the right). The root node of the left tree is colored according to the percentage of violations (9.55%) and its output binding results are shown in the table on the left.

6 Evaluation

For evaluation, we created seven queries and constraints for a real-life OCED dataset of a loan application process. The OCED was derived from the BPI Challenge 2017 log [11], and has more than 1,200,000 events and 100,000 objects.

The seven example queries (Q1 – Q7) are designed to be of real-life relevancy and cover a large variety of concepts, ranging from simple to more complex. In the following, we briefly introduce the used queries, first categorizing their concept or type and then describing the concrete query in natural language:

- **Q1** A simple constraint on the number of events that should be associated with an object: "Every application should be submitted exactly once".
- **Q2** A basic eventually-follows constraint (under a specified object type): "Every Offer should be returned at least once after creation".
- **Q3** A constraint on the number of objects of a type to be associated with events: "Each O_Returned event should involve exactly one Offer".
- **Q4** An eventually-follows constraint spanning across events of two related objects: "After an Application was accepted, there should be at least one associated Offer accepted afterwards".
- **Q5** A constraint enforcing that an object is associated with an event if it was involved with another event: "The Resource that accepts an Application should also create all Offers for that Application".
- **Q6** A query for the maximal duration between two events, both associated with an object: "What is the maximum delay between an Offer being created and accepted?".
- **Q7** A query for multiple object and event instances of the same type that are linked through another object: "Get all combinations of two Offers that are associated with the same Application and the corresponding Offer creation events".

As qualitative evaluation, to investigate expressiveness, we analyzed for each query whether it (or an equivalent constraint) can be modeled in DECLARE[2] [6],

[2] As DECLARE is based on traditional, flat event data, we assumed a reasonable flat representation of the OCED on only one object type for evaluating its expressiveness.

OCCG [4], or OCBC [2]. As the implementations of these approaches are primarily research prototypes, and are either very slow or completely unusable for larger datasets, we did not measure their execution times. Instead, as quantitative evaluation, we compared the query execution duration of OCPQ to the general querying platforms Neo4j (used in [3]), SQLite (one of the official OCEL 2.0 formats), and the performance-focused DuckDB [7], to demonstrate the runtime performance and scalability of OCPQ.

Figure 8 shows how Q4 has been formulated in OCPQ, SQL (for SQLite and DuckDB), and Cypher (for Neo4j), as an example. While evaluating the usability of OCPQ in detail is outside the scope of this paper, this example still demonstrates the complexity of implementing simple business queries in general querying languages, like SQL or Cypher.

For SQLite, the OCEL 2.0 database was completely loaded into memory, to be a more accurate comparison to OCPQ, and it was ensured that appropriate table indices were added. For Neo4j, the database dump[3] from [3], was imported

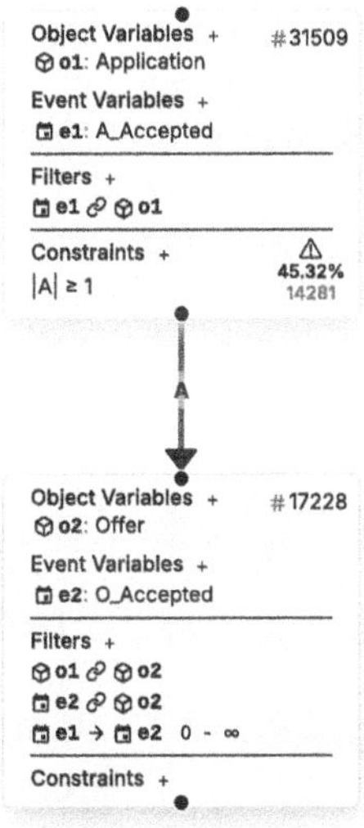

(a) Q4 in OCPQ

```sql
WITH
    AcceptedOffers AS (
        SELECT DISTINCT
            O2O.ocel_source_id,
            E2.ocel_id,
            E2.ocel_time
        FROM
            object_object AS O2O
            INNER JOIN object_Offer AS A2 ON O2O.ocel_target_id = A2.ocel_id
            INNER JOIN event_object AS E2O2 ON E2O2.ocel_object_id = A2.ocel_id
            INNER JOIN event_O_Accepted AS E2 ON E2O2.ocel_event_id = E2.ocel_id
    )
SELECT
    A1.ocel_id,
    E1.ocel_id,
    (
        SELECT
            COUNT(*)
        FROM
            AcceptedOffers AS AO
        WHERE
            AO.ocel_source_id = A1.ocel_id
            AND AO.ocel_source_id = A1.ocel_id
            AND E1.ocel_time <= AO.ocel_time
    ) >= 1 AS satisfied
FROM
    object_Application AS A1
    INNER JOIN event_object AS E2O ON E2O.ocel_object_id = A1.ocel_id
    INNER JOIN event_A_Accepted AS E1 ON E2O.ocel_event_id = E1.ocel_id;
```

(b) Q4 in SQL

```cypher
MATCH (o1:Entity { EntityType: "Application" }) <-[:CORR]- (e1:Event {Activity:
↪ "A_Accepted"})
OPTIONAL MATCH (e2:Event { Activity: "O_Accepted" }) -[:CORR]-> (o2:Entity {EntityType:
↪ "Offer"}) -[:REL]-> (o1)
WHERE e1.timestamp <= e2.timestamp
WITH o1, COUNT(e2) AS e2_c, COUNT(o2) AS o2_c
WHERE e2_c < 1 OR o2_c < 1
RETURN COUNT(o1) AS violationCount
```

(c) Q4 in Cypher (violation count only)

Fig. 8. Constraint Q4 formulated in OCPQ, SQL, and Cypher.

[3] See https://data.4tu.nl/datasets/5c9717a0-4c22-4b78-a3ad-d2234208bfd7/1.

in a compatible version of Neo4j (3.5.35) and additionally Neo4j was configured to allow extensive memory usage. To measure accurate execution times for Neo4j, we used query formulations that report either the total count of results (for queries) or violations (for constraints). This was done to exclude misleading execution times for when the result rows are being streamed instead of fully computed. Moreover, these numbers are also always calculated in OCPQ.

We used the full current implementation for formulating the example queries using OCPQ, also including features that were only briefly mentioned in this paper. All queries except Q6 are modeled fully visually and without programming in OCPQ. For Q6, a scripting feature of the tool is used to calculate the maximum duration of subquery results as a label. More details on the evaluation, including the dataset, raw execution times, and formulations of the queries across all applicable approaches are available at https://github.com/aarkue/ocpq-eval.

The combined results for both evaluations are shown in Fig. 9. The results show two things very clearly: First, there is a large gap in expressiveness, resulting in many relevant queries and constraints that are not representable in previously proposed visual approaches, like DECLARE, OCCG, or OCBC. In particular, the queries Q5 – Q7 cannot be modeled in any of them. Second, also general-purpose querying solutions, like SQLite or Neo4j, are not well suited for important types of OCED queries and are significantly slower than OCPQ for all tested queries. While DuckDB performs similar to OCPQ in terms of execution times, SQL formulations of OCED queries quickly grow complex and are difficult to write, read, and interpret. Moreover, a simple SQL translation also does not yield any subquery results, unlike OCPQ where results are available for each subquery.

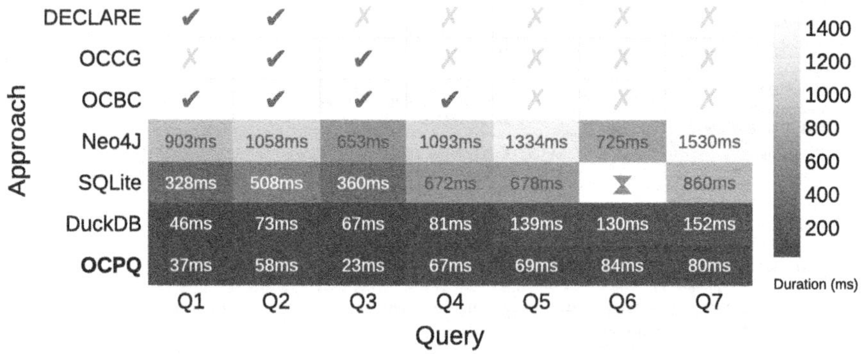

Fig. 9. Combined qualitative and quantitative evaluation of OCPQ. For the first three approaches, we only specify if the query/constraint can be expressed or not. For Neo4j, SQLite, DuckDB, and OCPQ, all queries are expressible, and we report their mean execution time across ten runs on the BPIC2017 OCED. Evaluating Q6 in SQLite took longer than four minutes and is thus omitted.

Of course, there are some limitations to our evaluation. As we created the queries Q1 – Q7 ourselves, they might exhibit certain biases. However, they still serve as typical examples with real-world relevancy and cover a wide range of constructs. Also, while we spent special attention to crafting well-performing queries for both SQLite and Neo4j, even more efficient formulations might be possible. Finally, we performed our evaluation only on one real-life OCED dataset.

7 Conclusion

In this paper, we proposed an object-centric querying and constraint approach, *OCPQ*, based on variable bindings of objects and events. Through the use of bindings, it can express more advanced queries and constraints than previous graphical approaches. We implemented our approach as a full-stack solution, supporting efficient execution of queries and constraints, as well as an interactive editor for creating them. As evaluation, we constructed several example queries and constraints and compared OCPQ to other approaches. The evaluation demonstrated the limitations of previous work in terms of expressiveness, and showed that OCPQ also significantly outperforms several general querying solutions, namely SQLite and Neo4j, in terms of runtime.

In future work, we plan to conduct a detailed performance analysis on more datasets. Additionally, the expressiveness of the proposed approach has to be studied systematically and in more depth. Furthermore, we see a lot of potential for extensions. Our concept of object-centric querying is very universal, allowing applications for *process constraints*, *OCED filtering*, *general annotations*, as well as for generating *situation tables* as input for machine learning techniques.

Acknowledgments. The authors gratefully acknowledge the German Federal Ministry of Education and Research (BMBF) and the state government of North Rhine-Westphalia for supporting this work as part of the NHR funding.

Disclosure of Interests. The authors have no competing interests to declare that are relevant to the content of this paper.

References

1. van der Aalst, W.M.P.: Object-centric process mining: unraveling the fabric of real processes. Mathematics **11**(12) (2023)
2. van der Aalst, W.M.P., Artale, A., Montali, M., Tritini, S.: Object-centric behavioral constraints: integrating data and declarative process modelling. In: Proceedings of the 30th International Workshop on Description Logics, Montpellier, France, 18–21 July 2017. CEUR Workshop Proceedings, vol. 1879 (2017)
3. Esser, S., Fahland, D.: Multi-dimensional event data in graph databases. J. Data Semant. **10**(1–2), 109–141 (2021)
4. Park, G., van der Aalst, W.M.P.: Monitoring constraints in business processes using object-centric constraint graphs. In: Process Mining Workshops - ICPM 2022 International Workshops, Bozen-Bolzano, Italy, 23–28 October 2022, Revised Selected Papers. Lecture Notes in Business Information Processing, vol. 468, pp. 479–492. Springer (2022)

5. Pérez-Álvarez, J.M., Díaz, A.C., Parody, L., Quintero, A.M.R., Gómez-López, M.T.: Process instance query language and the process querying framework. In: Process Querying Methods, pp. 85–111. Springer (2022)
6. Pesic, M., Schonenberg, H., van der Aalst, W.M.P.: DECLARE: full support for loosely-structured processes. In: 11th IEEE International Enterprise Distributed Object Computing Conference (EDOC 2007), 15–19 October 2007, Annapolis, Maryland, USA, pp. 287–300. IEEE Computer Society (2007)
7. Raasveldt, M., Muehleisen, H.: DuckDB. https://github.com/duckdb/duckdb
8. Riva, F., Benvenuti, D., Maggi, F.M., Marrella, A., Montali, M.: An SQL-based declarative process mining framework for analyzing process data stored in relational databases. In: Business Process Management Forum - BPM 2023 Forum, Utrecht, The Netherlands, 11–15 September 2023, Proceedings. Lecture Notes in Business Information Processing, vol. 490, pp. 214–231. Springer (2023)
9. Schönig, S., Di Ciccio, C., Maggi, F.M., Mendling, J.: Discovery of multi-perspective declarative process models. In: Service-Oriented Computing - 14th International Conference, ICSOC 2016, Banff, AB, Canada, 10–13 October 2016. Lecture Notes in Computer Science, vol. 9936, pp. 87–103. Springer (2016)
10. Schönig, S., Rogge-Solti, A., Cabanillas, C., Jablonski, S., Mendling, J.: Efficient and Customisable Declarative Process Mining with SQL. In: In: Advanced Information Systems Engineering - 28th International Conference, CAiSE 2016, Ljubljana, Slovenia, 13–17 June 2016. Lecture Notes in Computer Science, vol. 9694, pp. 290–305. Springer (2016)
11. van Dongen, B.: BPI challenge 2017 (2017). https://data.4tu.nl/articles/_/12696884/1
12. Vogelgesang, T., Ambrosy, J., Becher, D., Seilbeck, R., Geyer-Klingeberg, J., Klenk, M.: Celonis PQL: a query language for process mining. In: Process Querying Methods, pp. 377–408. Springer, Cham (2022)

Investigating Tailored Retraining for Online Process Predictions Using Log Features

Suhwan Lee(✉), Xixi Lu, and Hajo A. Reijers

Utrecht University, Utrecht, The Netherlands
{s.lee,x.lu,h.a.reijers}@uu.nl

Abstract. Predictive process monitoring leverages predictive models to forecast outcomes of ongoing processes. Such predictive models must be trained and often *retrained* to sustain high performance and adapt to the dynamic environments in which modern processes operate. While much research has focused on detecting drifts in process behavior, limited attention has been given to leveraging drift detection for tailored retraining. Consequently, this leaves a gap in understanding the impact of using log features for retraining the predictive models. This paper addresses this gap by examining two key log features for triggering tailored retraining: *label distribution* and *variant coverage*. Using 23 variations of ten event logs, we evaluate the impact of retraining methods based on these features against three baselines. Our findings show retraining using specific log features yields modest but consistent improvement in performance. These insights contribute to the development of more resilient predictive models, highlighting the potential of tailored retraining methods to mitigate performance degradation in dynamic process environments.

Keywords: Process mining · Event stream · Predictive process monitoring · Model retraining

1 Introduction

Modern business processes operate in dynamic environments, requiring adaptive techniques to forecast their future states effectively [24]. Predictive process monitoring (PPM) addresses this need by using machine learning models to anticipate the outcomes of ongoing processes. These models must be trained and often *retrained* to maintain high performance and adapt to the evolving conditions. Studies have shown that when deploying such predictive models in *online* settings, the performance of predictive models tends to fluctuate significantly [9,23], which confirms the need to retrain these models.

While most existing research on PPM assumes offline training—where models are trained once and then deployed statically—such approaches often fail to maintain performance over time in dynamic environments. Frequent retraining,

J. Grabis et al. (Eds.): RCIS 2025, LNBIP 547, pp. 401–417, 2025.
https://doi.org/10.1007/978-3-031-92474-3_24

on the other hand, introduces the risk of overfitting to the most recent data or resulting in unnecessary energy consumption and cost overheads. Thus, selecting suitable retraining strategies is important to ensure the predictive model remains effective. These strategies typically address two key questions: *when* to retrain and *which data* to use for retraining. In this paper, we focus on the question of *when* to retrain.

Regarding *when* to retrain, two main strategies are commonly identified: (1) periodic retraining (e.g., incremental algorithms) and (2) retraining triggered by concept drift. For the latter, numerous methods exist to detect concept drifts [5, 16]. However, to date, only one study explicitly uses drift detection as a retraining trigger [8], using an existing detection method that focuses solely on control-flow [2]. To the best of our knowledge, no studies have explored using other types of drift, such as drift in data attributes, or examined their effectiveness as a retraining trigger.

In the context of process predictions, in addition to control-flow, other characteristic of an event log—such as the distribution of the outcome labels-also influence predictive performance. We refer to such measurable properties of an event log as *log features.* Examples of log features include the number of distinct variants, the proportion of outcome labels, or the distribution of certain data attributes. Monitoring these features reveals the current behavior and content of the log. It also helps to identify the specific factors contributing to performance drops and variability in model performance. By identifying these factors, it becomes possible to explore more targeted model updates, reducing the risk of unnecessary retraining while maintaining model accuracy. This observation highlights an opportunity: moving beyond control-flow drift detection to leverage log features as retraining triggers.

Building on this motivation, we investigate the effect of using log feature drifts for retraining in this paper. We focus on two log features, i.e., *variant coverage* and *label distribution*, and monitor their fluctuations. The predictive model is automatically retrained when fluctuations are detected based on predefined conditions, such as when the label distribution exceeds a certain threshold.

To assess the effectiveness of these retraining conditions, we evaluated two retraining methods across ten event logs with various target labels in a total 23 logs. The performance was compared against three baselines: retraining once and two concept-drift detection methods, using four predictive models. The results show that the proposed retraining methods (*variant-based retraining* and *label-based retraining*) consistently yield modest accuracy improvements, with *label-based retraining* improving performance in 19 out of 23 logs and *variant-based retraining* in 16 out of 23 logs. The findings offer insights into the development of tailored retraining conditions to mitigate performance drops, enhancing the resilience of these models in real-world applications.

The remainder of the paper is organized as follows. Related work is discussed in the next section. Section 3 provides the preliminaries, followed by a detailed description of the proposed method in Sect. 4. The setup of the empirical analysis is reported in Sect. 5. The experimental results are provided in Sect. 6. Finally, conclusions are presented in Sect. 7.

2 Related Work

Predictive process monitoring has received substantial interest in the process mining community. Few studies have delved into the performance fluctuation of online predictive models [9,23]. These studies reveal that the performance of such models can decrease substantially in dynamic environments, highlighting the need for retraining to maintain high performance. In this section, we discuss existing retraining strategies for such an online setting. The approaches are categorized into two dimensions based on when the decision to retrain is triggered: (1) periodic updates vs. (2) concept drift detection. Table 1 provides an overview of existing approaches based on their retraining conditions and prediction goals.

Several studies use a *periodic retraining strategy*, where predictive models are updated at fixed intervals regardless of changes in the data. Rizzi et al. [21] introduced three strategies for updating predictive models, focusing on periodic rediscovery or incremental approach. They proposed two retraining strategies: one that uses consecutive batches of recent training data and another that uses incrementally accumulated training data over time. Márquez-Chamorro et al. [14] conducted a comparative study on various retraining strategies by incorporating fixed update intervals and concept drift detection. They proposed six strategies for process outcome prediction, including cumulative and sampling approaches. To cope with the potential concept drift, some methods use a periodically updated incremental model [13,19]. Maisenbacher and Weidlich [13] employed incremental classifiers to build predictive models. These models are designed to adapt to the changes in event streams by updating the model when new data becomes available. They analyze data selection points by comparing the average accuracy of the entire dataset. Similarly, Pauwels and Calders [19] developed a deep neural network with an incremental learning approach for next activity prediction in event streams. Their strategy enables fast model updates with prediction performance comparable to Long Short-Term Memory (LSTM)-based deep learning models. While these works aim to improve predictive performance and adapt to the potential changes in event streams, they struggle to identify the specific changes occurring in the event streams or determine the moment for model retraining.

Table 1. Overview of Related Works on Retraining Conditions and Prediction Goals

Literature	When to retrain			Prediction goal			Publication Year
	Periodically	Concept Drift Detected	Log Feature	Next Activity	Process Outcome	Throughput Time	
Rizzi et al. [21]	v				v		2022
Márquez-Chamorro et al. [14]	v	v			v		2022
Maisenbacher and Weidlic [13]	v				v		2017
Pauwels and Calder [19]	v			v			2021
Pasquadibisceglie et al. [17]		v		v			2023
Baier et al. [1]		v				v	2020
Kosciuszek and Hassani [8]		v		v			2024
This paper		*	v		v		

* Indicates the retraining strategy has been included in the evaluation but was not proposed in this paper.

Regarding *concept drift detection*, the model retraining is triggered by the identification of concept drift in event streams. Baier et al. [1] investigated optimal data selection points to retrain a predictive model for throughput time prediction in the presence of concept drift. Their study offered insights by presenting the improved accuracy of the retrained model when the concept drift is detected. Pasquadibisceglie et al. [17] developed an LSTM-based model for next activity prediction, where retraining is triggered upon detecting concept drift. Their algorithm adjusts by fine-tuning based on drift detection to cope with the deviated process variants. Similarly, Kosciuszek and Hassni [8] introduced a prediction framework for the next activity prediction to detect the concept drift and update the model training parameters. Their framework enables dynamic model updates by adjusting training data size, training intervals, retraining frequency, and model parameters. These adjustments are based on the frequency and proximity of detected drifts. However, their works remain limited in control-flow drift, focusing primarily on the process changes rather than data-level variations, which may also impact model performance.

In summary, despite various research on retraining strategies for online predictive process monitoring, there remains a need to investigate retraining triggers based on data-level variations using log features. In the case of (1) periodic retraining, the focus has primarily centered on strategies to enhance prediction performance, with less emphasis on determining the optimal timing for retraining. While (2) concept drift detection can signal control-flow changes and adjust the model to it, no existing study has explored log features, such as changes in trace variant distributions and outcome label balance, which can significantly impact predictive performance. To address this gap, we propose tailored retraining strategies based on *variant coverage* and *label distribution*, which aim to account for both control-flow and data-level changes to improve the predictive accuracy in dynamic event streams.

3 Preliminaries

In this section, we define the preliminaries, including the notations for the event log, event stream, and the key components of pre-processing to build a predictive model.

Basic Notations. Let $\mathbb{N}_n^+$ represent the first n positive natural numbers, and let S be a target set. A sequence s is a function $s : \mathbb{N}_n^+ \rightarrow S$, mapping integer indexes to elements of S.

Given a set of activity labels A, the domain of timestamps $\mathbb{N}^+$, and a set of I attribute domains D_i, we define the set of event attributes as $E = A \times \mathbb{N}^+ \times [D_1 \times \ldots \times D_i \times \ldots \times D_I]$. A trace σ is a sequence of n events $\sigma : \mathbb{N}_n^+ \rightarrow E$. We denote with $\mathcal{E}$ the event universe, with $\mathcal{E} = E \times J$, where J is a set of possible case ids, and with $\mathcal{E}^*$ the universe of sequences of events. An event stream is an infinite sequence $\Psi : \mathbb{N}^+ \rightarrow \mathcal{E}$.

For simplicity, we write events as $e_{k,j}$, where k indicates their position in a trace and traces as $\sigma_j = \langle e_{1,j}, \ldots, e_{i,j}, \ldots, e_{N_j,j} \rangle$, where N_j is the number of events in the trace σ_j. The function $t : \mathcal{E} \to \mathbb{N}^+$ returns the timestamp of an event. The prefix function $pref : \mathcal{E}^* \times \mathbb{N}^+ \nrightarrow \mathcal{E}^*$ is a partial function that returns the first p events of a trace, i.e., $pref(\sigma_j, p) = \langle e_{1,j}, \ldots, e_{p,j} \rangle$, with $p \leqslant N_j$. For the evaluation, event streams are derived from the event logs where multiple events may share the same timestamp. In cases where events share the same timestamp, we assume that the ordering of events in the event log reflects their true sequence and use this ordering in the event stream to compute prefixes.

An *event log* L is a finite set of traces: $L = \{\sigma_1, \sigma_2, \ldots, \sigma_m\}$, where each trace σ_j consists of a sequence of events: $\sigma_j = \langle e_{1,j}, e_{2,j}, \ldots, e_{N_j,j} \rangle$. Each event in a trace corresponds to an activity execution, along with its associated attributes.

A *variant* in an event log is a unique sequence of activity labels representing a distinct process instance. Formally, the set of variants of L is defined as: $Var(L) = \{v_1, v_2, \ldots, v_r\}$, where each variant v_i is a distinct sequence of activities, i.e., $v_i = \langle a_1, a_2, \ldots, a_n \rangle$, with $a_1, a_2, \ldots, a_n \in A$.

Each trace σ in L is associated with a binary outcome label. Without loss of generality, we assume that the value of this label becomes known with the last event $e_{N_j,j}$ of a trace. Therefore, we define the labeling function as $y : E \nrightarrow \{0, 1\}$, which returns the label of a trace in correspondence with its last event.

Sliding Window. To manage model updates in a streaming context, we adopt a sliding window approach, in which the window size is denoted by l. A sliding window continuously maintains a subset of the most recent data, allowing the model to be retrained or re-evaluated based on recent observations [7]. In our setting, the sliding window contains the most recent l completed cases, corresponding to the latest l received labels.

At any given time point t, the sliding window W_t holds the l most recent completed cases. For clarity, we denote the sliding windows at different time points $t_1, t_2, \ldots, t_k$ as $W_{t_1}, W_{t_2}, \ldots, W_{t_k}$, respectively.

4 Method

As discussed, in online settings, the performance of a predictive model fluctuates over time. To address this challenge, we propose two log features, *variant coverage* and *label distribution*, and tailored retraining strategies based on the log features. In the following, we briefly explain the context, i.e., training a predictive model in online settings in Sect. 4.1. Afterwards, we explain the proposed two log features in Sect. 4.2. Then, we discuss how log features can be used to trigger a retraining in Sect. 4.3. Figure 1 shows an example of how label distribution is calculated and how it triggers model retraining.

4.1 Context-Predictive Model Training

To train a predictive model for an online setting, we transform the first k completed cases from a stream of events into training data. During the grace period,

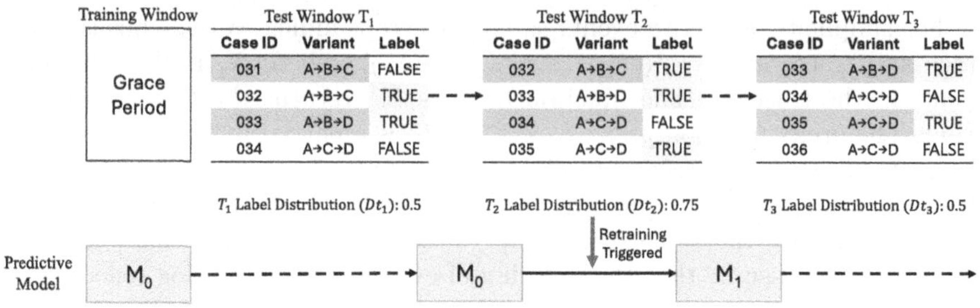

Fig. 1. Example of Label Distribution and Label-based Retraining

event data is preprocessed to construct a feature space for training an initial predictive model. For this feature space construction, we consider prefix-bucketing and index-based encoding, which is commonly adopted in the literature [10]. Afterward, a different predictive model is trained to predict the outcome by the prefix length, which is correspond to M_0 in Fig. 1.

Once the grace period ends, each newly completed case is allocated to a sliding window W_{t_i}. Given the test data evolves over time, the predictive models operate within the context of the continuous evaluation using an updated sliding window. For example, test window T_1 in Fig. 1 contains four cases with Case ID from 031 to 034. When new case, 035, is available, we have a new test window T_2. The predictive performance is computed to derive the accuracy of the model concerning the cases within the current sliding window.

4.2 Log Features

After defining the procedure for the predictive model training, we have retrieved the training data W_{train} and a sequence $W_{t_1}, W_{t_2}, \cdots, W_{t_n}$ of test data. Leveraging the test data updates, we use log features to continuously monitor event streams. Additional features can be explored for more comprehensive monitoring [24].

We propose **variant coverage** as the first log feature to track changes in the event stream. This feature measures the appearance of newly observed variants in the test data, which were not included during model training. The variant coverage is defined as the ratio of a set of variants in the test data that has been observed in the training data. For example, assuming 100 variants in the test data and 50 variants among them have been observed in the training data, then 0.5 is the variant coverage. Given a group of traces in the training data W_{train} and the test data W_{t_i}, the variant coverage at time t_i is defined as follows:

$$\mathcal{V}_{t_i} = \frac{|Var(W_{train}) \cap Var(W_{t_i})|}{|Var(W_{t_i})|}$$

When the ratio of variant coverage $\mathcal{V}_{t_i}$ is 1, all variants in the moving window (test data) W_{t_i} have already been seen in the training data.

The next log feature we propose is the **label distribution**, which measures the proportion of the specific traces, e.g., labeled *TRUE*, in the moving window (test data) W_{t_i}. This log feature is defined by the ratio of traces with a specific process outcome in the test data. Given traces in the test data W_{t_i}, the label distribution $\mathcal{D}$ for *TRUE* outcome at time t_i with respect to the test data W_{t_i} is defined as follows:

$$\mathcal{D}_{t_i} = \frac{|\{\sigma|\sigma \in W_{t_i} \wedge y(\sigma) = TRUE\}|}{|\{\sigma \in W_{t_i}\}|}$$

For example, let us consider three test windows, $T_1, T_2,$ and T_3, in event streams, as shown in Fig. 1. Test windows, T_1 and T_3 have two cases labeled as *TRUE* out of a total four cases, resulting in label distribution of 0.5. This indicates a equal distribution of process outcomes in the test data. Conversely, the label distribution of T_2 is 0.75, as it contains threes cases labeled as *TRUE*. It reflects a bias toward one of the binary classes in the test data.

Furthermore, it is important to note that our approach supports any other log feature. For example, one may use the *event level attributes coverage*, which measures the appearance of newly observed categorical event-level attributes in the test data.

4.3 Retraining Conditions

We introduce *retraining conditions* based on log features to decide the appropriate moments for updating the predictive model. The retraining triggers are evaluated individually using two log features: **variant coverage** and **label distribution**, each with predefined thresholds. These log features correspond to two distinct retraining strategies: *variant-based retraining*, which leverages variant coverage, and *label-based retraining*, which uses label distribution, respectively. Retraining is triggered when the feature falls below the respective threshold, indicating a significant deviation in the event streams. Notably, in this paper, these conditions are treated separately, ensuring that each log feature is assessed independently rather than being combined under a single decision rule.

For *variant-based retraining*, the model is retrained when the variant coverage feature falls below a predefined threshold, denoted as $\tau_{\mathcal{V}}$. At time point t_x, the model will be retrained if the following condition holds: $R_{\mathcal{V}} = (\mathcal{V}_{tx} < \tau_{\mathcal{V}})$

A drop in variant coverage indicates an increase in unseen process variants within the moving window of test data compared to the training data. This suggests significant changes in the underlying process variants that the model was trained on, necessitating model adaptation to maintain predictive performance.

Similarly, for *label-based retraining*, the model is retrained when the label distribution falls below or exceeds its predefined threshold, $\tau_{\mathcal{D}}$. At time point t_x, the condition for retraining based on label distribution is defined as: $R_{\mathcal{D}} = (\mathcal{D}_{tx} < \tau_{\mathcal{D},low}) \vee (\mathcal{D}_{tx} > \tau_{\mathcal{D},high})$

A reduction in label distribution indicates a shift in the balance of outcome labels within the moving window of test data. This may suggest a need for model adjustment to handle evolving data distributions.

As an example, assume the label distribution threshold is set to 0.7. In test window T_2, shown in Fig. 1, the label distribution is 0.75, which exceeds the given threshold. Since the condition is satisfied, the retraining of the predictive model is triggered, leading to a model update from M_0 to M_1.

5 Empirical Evaluation

In our empirical evaluation, we focus on assessing tailored retraining strategies aimed at improving model performance in online process outcome prediction. To achieve this, we compare the proposed *variant-based retraining* and *label-based retraining* strategies against baseline methods. We measure their impact on predictive accuracy, retraining frequency, and computation time across multiple event logs and classifiers. The code and datasets to reproduce these experiments are available at https://github.com/ghksdl6025/tailored_retraining.

5.1 Datasets

For our evaluation, we have considered ten publicly available event logs, of which eight are real-life event logs and two are synthetic datasets. The selected real-life event logs are widely used in predictive process monitoring for process outcome prediction [22], where additional details about the logs can also be found. To assess the effectiveness of our retraining strategies in handling concept drift, we include the two synthetic logs. These synthetic logs, IRO5K and OIR5K[1], simulate the application processes with concept drift intentionally introduced [11]. Although both datasets are derived from the same process, they differ in the type and occurrence of concept drift within the processes they represent. This distinction allows us to evaluate how well our strategies adapt to varying drift conditions.

Regarding the labeling of the process outcome, the process outcome can be formulated in different ways depending on the prediction goals and event logs. That means that the same event log can be used for different prediction tasks depending on how the process outcome is labeled. In this paper, we adopt the same outcome labeling function for the real-life event logs as in [22]. In total, we derived 23 different outcome prediction tasks based on the ten original event logs. Descriptive statistics for all event logs are provided in Table 2, which illustrates significant variability among them.

5.2 Setup

In our evaluation, we selected four classification algorithms as our predictive model to be trained and retrained: Hoeffding Adaptive Tree Classifier (HATC), eXtreme Gradient Boosting (XGB), Random Forest (RF), and Long Short-Term Memory (LSTM). HATC is an incremental model commonly used for streaming

[1] at: https://data.4tu.nl/articles/dataset/Business_Process_Drift/12712436.

Table 2. Descriptive statistics of event logs used in the evaluation

Dataset		# Cases	Early prefix	Middle prefix	Late prefix	Trunc length	# Events	# Activity	# True labels	# False labels
Real-life event log	BPIC2011 1	1143	10	19	27	36	150291	624	460	683
	BPIC2011 3	1143	9	16	24	40	150291	624	185	958
	BPIC2011 4	1143	11	21	30	31	150291	624	321	822
	BPIC2015 1	1199	11	21	30	40	52217	289	289	910
	BPIC2015 2	832	11	21	30	40	44354	304	163	669
	BPIC2015 3	1409	11	21	30	40	59681	277	288	1121
	BPIC2015 4	1053	11	21	30	40	47293	272	167	886
	BPIC2015 5	1156	11	21	30	40	59083	285	339	817
	Production	225	7	12	18	23	4543	55	122	103
	Sepsis 1	782	8	16	23	30	13902	16	111	671
	Sepsis 2	782	4	7	10	13	11706	16	107	675
	Sepsis 3	782	6	12	17	22	13245	16	671	111
	BPIC2012 1	4685	11	21	30	40	191378	37	2243	2442
	BPIC2012 2	4685	11	21	30	40	191378	37	1640	3045
	BPIC2012 3	4685	11	21	30	40	191378	37	802	3883
	BPIC2017 1	2000	6	11	15	20	79130	26	796	1204
	BPIC2017 2	2000	6	11	15	20	79130	26	920	1080
	BPIC2017 3	2000	6	11	15	20	79130	26	284	1716
	Road Traffic	2000	3	6	7	10	8770	11	808	1192
	Hospital billing1	2000	2	4	5	6	13423	18	183	1817
	Hospital billing2	2000	3	5	6	8	12813	17	116	1884
Synthetic event log	IRO5K	1000	5	8	12	15	10756	20	237	763
	OIR5K	1000	4	7	10	13	11042	20	243	757

classification, while XGB and RF are traditional machine learning algorithms. LSTM, on the other hand, is a deep learning model specifically designed for time-series and sequential data, offering a diverse range of predictive capabilities.

In streaming classification, it is a common approach to implement a grace period to allow classifiers to initialize properly [4]. In our experiments, we set a grace period of 200 labels, where these first 200 observed labels are dedicated solely to training the initial predictive model. Both in the grace period and when retraining, we applied the Synthetic Minority Oversampling Technique (SMOTE) [3] to ensure a balanced training set and prevent potential model training failures, in order to address the issue of the data imbalance.

For trace bucketing and encoding, we employed the index-based encoding method [10] and considered different prefix length buckets for each event log based on a truncation threshold. The truncation length was set at the point where 90% of cases in the minority class had been completed, ensuring balanced class representation. For computation reasons, we limited the maximum truncation length to 40. We then selected three prefix points -early, middle, and late- corresponding to the 1st, 2nd, and 3rd quartiles of this truncation length, respectively. This allows us to assess predictions at different stages of the process while managing computational costs. By deploying three prefix buckets across 23 logs and four models, we tested a total of 276 model configurations.

To trigger model retraining, the proposed variant coverage and label distribution strategies (detailed in Sect. 4.3) rely on the retraining conditions. For variant coverage, the predictive model is retrained when the coverage exceeds a predefined threshold of 0.6. For label distribution, retraining occurs when the label distribution falls below 0.1 or exceeds 0.9. These thresholds ensure the model adapts in response to changes in the event log data, particularly when significant deviations in variant coverage or label distribution are detected.

We used three baselines to evaluate the performance of our methods. The first baseline is a *train-once*, where models were trained immediately after the grace period and subsequently monitored without further updates. This allows us to observe how static models respond to evolving log features and process changes over time. The other two baselines focused on detecting concept drift to assess how process changes influenced predictive accuracy. The first, ProDrift [12], is an online concept drift detection method tailored for event streams, while PrefixCDD [6] monitors prefixes of event sequences to detect concept drift. Both baselines served as a comparison to evaluate the adaptability of our proposed techniques to process changes with stable predictive performance.

Regarding the performance evaluation, we considered a sliding test window with a size of 30 traces. For each test set, we compute the weighted F1-score of the model and the accuracy [20]. To simulate the streaming environment and implement the incremental classifier, we relied on the Python River package [15], which supports online machine learning. The classical machine learning classifiers, RF and XGB, were implemented using the Scikit-learn packages [20]. For the LSTM model, we used the PyTorch package [18].

6 Results and Discussion

For ease of discussion, the empirical results are structured around three evaluation objectives. First, we analyze the performance differences between the proposed retraining strategy and the baseline methods in predicting the process outcome (O1). Second, we evaluate the effect of trigger methods to train-once (O2). Finally, we assess the frequency of retraining and computation time by the retraining strategy (O3).

6.1 Accuracy Comparison Across Retraining Trigger Methods (O1)

We begin by analyzing the predictive performance of the proposed retraining strategies: *variant-based retraining* and *label-based retraining*. The performance is compared against baseline methods across event logs and classifiers.

Table 3 provides a comprehensive overview of the performance metrics across 23 logs under multiple retraining methods. The metrics include the baseline performance (*train-once*), performance of two proposed retraining methods (*variant-based retraining*, *label-based retraining*), two concept-drift based (*Prodrift* and *PrefixCDD*), and their differences with respect to *train-once* (Column *Diff*). Note that each value is the average accuracy of four models used for the three buckets. Standard deviations (SD) are also included to measure consistency and variability.

Overall, the results indicate that the *label based retraining* and *variant-based retraining* typically lead to modest improvement in accuracy (an average increase of 0.037 and 0.020, respectively, with respect to *train-once*), showing that these two retraining methods generally help models adapt to new conditions. For *label-based retraining*, an improvement is achieved for 19 out of 23 logs.

Table 3. Performance Metrics Across Event Logs: Accuracy, Standard Deviation, and Improvement Compared to Train-Once

Event logs	Train-Once	Variant-based	Diff	Label-based	Diff	ProDrift	Diff	PrefixCDD	Diff
bpic2011_1	0.589 (0.088)	0.654 (0.085)	**0.065**	0.636 (0.073)	**0.047**	0.589 (0.088)	0.000	0.586 (0.088)	−0.003
bpic2011_3	0.578 (0.233)	0.554 (0.199)	−0.024	0.660 (0.160)	**0.082**	0.578 (0.233)	0.000	0.589 (0.236)	**0.011**
bpic2011_4	0.518 (0.028)	0.536 (0.039)	**0.018**	0.546 (0.042)	**0.028**	0.518 (0.028)	0.000	0.516 (0.052)	−0.002
bpic2012_1	0.540 (0.033)	0.552 (0.053)	**0.012**	0.532 (0.035)	−0.008	0.539 (0.033)	−0.001	0.520 (0.036)	−0.020
bpic2012_2	0.624 (0.107)	0.570 (0.131)	−0.054	0.620 (0.101)	−0.004	0.624 (0.107)	0.000	0.569 (0.149)	−0.055
bpic2012_3	0.663 (0.233)	0.666 (0.207)	**0.003**	0.711 (0.188)	**0.048**	0.663 (0.233)	0.000	0.711 (0.205)	**0.048**
bpic2015_1	0.756 (0.290)	0.790 (0.212)	**0.034**	0.832 (0.180)	**0.076**	0.774 (0.248)	**0.018**	0.668 (0.350)	−0.088
bpic2015_2	0.683 (0.319)	0.756 (0.198)	**0.073**	0.779 (0.223)	**0.096**	0.683 (0.319)	0.000	0.740 (0.243)	**0.057**
bpic2015_3	0.789 (0.285)	0.820 (0.168)	**0.031**	0.821 (0.173)	**0.032**	0.789 (0.285)	0.000	0.803 (0.245)	**0.014**
bpic2015_4	0.805 (0.312)	0.817 (0.242)	**0.012**	0.839 (0.239)	**0.034**	0.813 (0.287)	**0.008**	0.815 (0.283)	**0.010**
bpic2015_5	0.659 (0.258)	0.732 (0.165)	**0.073**	0.731 (0.175)	**0.072**	0.683 (0.264)	**0.024**	0.663 (0.231)	**0.004**
bpic2017_1	0.549 (0.067)	0.536 (0.072)	−0.013	0.561 (0.067)	**0.012**	0.549 (0.067)	0.000	0.540 (0.061)	−0.009
Bpic2017_2	0.536 (0.032)	0.544 (0.019)	**0.008**	0.540 (0.019)	**0.004**	0.536 (0.032)	0.000	0.533 (0.033)	−0.003
Bpic2017_3	0.723 (0.228)	0.702 (0.261)	−0.021	0.738 (0.209)	**0.015**	0.723 (0.228)	0.000	0.724 (0.224)	**0.001**
Hospital_billing_1	0.736 (0.407)	0.833 (0.233)	**0.097**	0.879 (0.146)	**0.143**	0.736 (0.407)	0.000	0.736 (0.407)	0.000
Hospital_billing_2	0.823 (0.267)	0.810 (0.230)	−0.013	0.830 (0.213)	**0.007**	0.823 (0.267)	0.000	0.823 (0.267)	0.000
IRO5K	0.704 (0.109)	0.749 (0.142)	**0.045**	0.692 (0.138)	−0.012	0.705 (0.125)	**0.001**	0.704 (0.109)	0.000
OIR5K	0.611 (0.124)	0.614 (0.126)	**0.003**	0.633 (0.121)	**0.022**	0.628 (0.127)	**0.017**	0.611 (0.124)	0.000
Production_Data	0.748 (0.234)	0.789 (0.187)	**0.041**	0.723 (0.215)	−0.025	0.748 (0.234)	0.000	0.748 (0.234)	0.000
Sepsis_cases_1	0.742 (0.145)	0.769 (0.114)	**0.027**	0.759 (0.147)	**0.017**	0.742 (0.145)	0.000	0.742 (0.145)	0.000
Sepsis_cases_2	0.760 (0.214)	0.739 (0.247)	−0.021	0.799 (0.179)	**0.039**	0.760 (0.214)	0.000	0.760 (0.214)	0.000
Sepsis_cases_3	0.654 (0.234)	0.724 (0.134)	**0.070**	0.742 (0.150)	**0.088**	0.654 (0.234)	0.000	0.654 (0.234)	0.000
Traffic_fines_1	0.891 (0.168)	0.881 (0.118)	−0.010	0.928 (0.075)	**0.037**	0.876 (0.119)	−0.015	0.876 (0.119)	−0.015
Overall Avg.	0.682 (0.231)	0.702 (0.196)	**0.020**	0.719 (0.186)	**0.037**	0.684 (0.227)	**0.002**	0.680 (0.227)	−0.002

Standard deviations are show in parentheses. Positive difference (Diff) values are highlighted in bold.

Table 4. Performance Metrics Across Four Classifiers: Accuracy, Standard Deviation, and Improvement Compared to Train-Once

Classifiers	Train-Once	Variant-based	Diff	Label-based	Diff	ProDrift	Diff	PrefixCDD	Diff
Hatc	0.464 (0.236)	0.521 (0.168)	**0.057**	0.548 (0.160)	**0.083**	0.469 (0.231)	0.005	0.456 (0.215)	−0.009
Lstm	0.727 (0.210)	0.731 (0.187)	**0.004**	0.745 (0.156)	**0.018**	0.728 (0.205)	0.001	0.717 (0.208)	−0.010
Rf	0.781 (0.163)	0.790 (0.154)	**0.009**	0.806 (0.161)	**0.025**	0.782 (0.158)	0.001	0.778 (0.160)	−0.003
Xgb	0.755 (0.157)	0.764 (0.151)	**0.010**	0.777 (0.149)	**0.022**	0.757 (0.155)	**0.002**	0.768 (0.151)	**0.013**
Overall Avg.	0.682 (0.231)	0.702 (0.196)	**0.020**	0.719 (0.186)	**0.037**	0.684 (0.227)	**0.002**	0.680 (0.227)	−0.002

Standard deviations are show in parentheses. Positive difference (Diff) values are highlighted in bold.

On the other hand, retraining using *Prodrift* and *PrefixCDD* shows much fewer improvements, with an overall average increase of 0.002 and −0.002 across the logs, respectively, with respect to *train-once*.

Table 4 summarizes the performance metrics of four machine learning models (HATC, LSTM, RF, and XGB). The table presents the average accuracy across 23 logs (with three buckets) and compares the results across different retraining methods. Considering *label-based retraining*, we observe an improvement across all four models, with HATC the highest (an increase of 0.083) and RF the second highest (an increase of 0.025). The improvement achieved by *label-based retraining* for each model is consistently higher than other retraining methods, while *variant-based retraining* is ranked second.

Figure 2 presents the average accuracy of process outcome prediction for the RF classifier. Triangle markers have been added to represent the baseline *train-once* method to enhance clarity. The results reveal that *variant-based retraining* and *label-based retraining* show higher accuracy compared to the other retraining methods across most event logs.

The results reveal that (i) the two proposed retraining strategies achieve higher or comparable accuracy than baseline methods, and (ii) the traditional machine learning classifiers, RF and XGB, show better predictive performance compared to HATC and LSTM classifiers across most event logs. The tailored

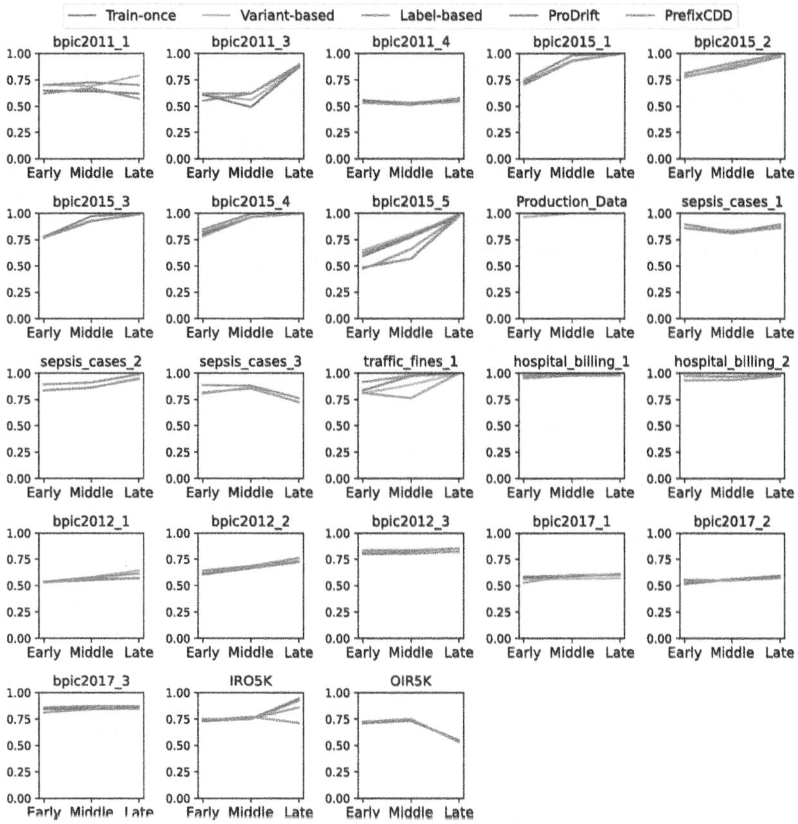

Fig. 2. Comparison of Accuracy Scores for Random Forest Across Event Logs by Trigger Method

retraining strategies, show higher or comparable average accuracy compared to train-once and concept drift detection methods. Specifically, the label distribution trigger demonstrates higher accuracy scores across all prefix lengths in datasets, e.g., bpic2011 4, bpic2015 2, and sepsis cases 2, when using the RF classifier.

The results also reveal variations in predictive performance across classifiers. Traditional machine learning classifiers, RF and XGB, perform better than HATC and LSTM in process outcome prediction across most event logs. The trend is observed under all retraining strategies. The potential explanation for the trend is the small training window size. RF and XGB adapt more efficiently to frequent retraining, while LSTM and HATC may face challenges, such as overfitting or slower adaptation, due to their complexity and bigger data size requirements.

6.2 Comparison Trigger Methods Against Train-once (O2)

We evaluate the predictive performance of each retraining trigger method, including tailored retraining and concept drift detection, by comparing them against the train-once baseline. Table 5 summarizes the comparison results, presenting the number of wins and descriptive statistics of average accuracy differences between the retraining triggers and the train-once method. For the descriptive statistics in the table, we calculate the median, mean, and standard deviation using only the cases where the trigger method achieved higher accuracy than train-once.

The results reveal that the proposed strategies improve the accuracy of the model more frequently than the concept drift detection methods. The label distribution trigger achieves the highest number of wins (158) across all event logs and classifiers, followed by variant coverage with 135 wins. In contrast, *Prodrift* and *PrefixCDD* achieve fewer wins, with 28 and 66, respectively.

Regarding the descriptive statistics of accuracy differences, the tailored retraining strategies show comparable improvements to the concept drift detection methods. When the trigger methods surpass the train-once baseline, all four trigger methods achieve approximately 10% of higher average accuracy improvement. However, the variability in accuracy gains differs across retraining meth-

Table 5. Performance Comparison of Trigger Methods Against Train-Once

	Variant-based Retraining	Label-based Retraining	ProDrift	PrefixCDD
Number of Wins	135	158	28	66
Median Difference*	0.031	0.048	0.057	0.056
Mean Difference*	0.091	0.103	0.103	0.091
Standard Deviation*	0.124	0.143	0.167	0.134

* Calculated only for cases where the trigger method outperforms the Train-Once

ods. The standard deviation for concept drift detection, *Prodrift* (0.167) and *PrefixCDD* (0.134), indicates bigger performance fluctuations when they exceed the train-once baseline. In contrast, the proposed strategies show relatively stable results, with lower standard deviations of 0.124 and 0.134 for variant coverage and label distribution, respectively.

6.3 Retraining Frequency and Computation Time (O3)

We assess the frequency of model retraining and the computation time required for each predictive algorithm under different retraining strategies.

We find that (i) the label distribution trigger requires more frequent retraining compared to the baseline methods, and (ii) the retraining time varies significantly depending on the predictive algorithm. Figure 3 shows the frequency of model updates across event logs for each retraining trigger method. The analysis reveals that the label distribution trigger results in the highest number of model retraining instances across nearly all event logs. A possible explanation is the non-stationary nature of the event logs, as the trigger is highly sensitive to changes detected within the sliding window for the test data. In contrast, the ProDrift method is triggered far less frequently, indicating that the method is less responsive to the changes in the event logs over time.

The frequency of model retraining varies across event logs. For example, BPIC 2012 and hospital billing event logs exhibit particularly high retraining counts compared to other datasets. This implies that logs with a high number of updates are more variable on the label distribution of the test sliding window, leading to more frequent trigger activations.

Regarding the computation time, the average retraining time differs significantly across the classifiers as shown in Fig. 4. Traditional machine learning models, RF and XGB, demonstrate notably shorter retraining times compared to the LSTM and HATC classifiers. Specifically, RF achieves the shortest average

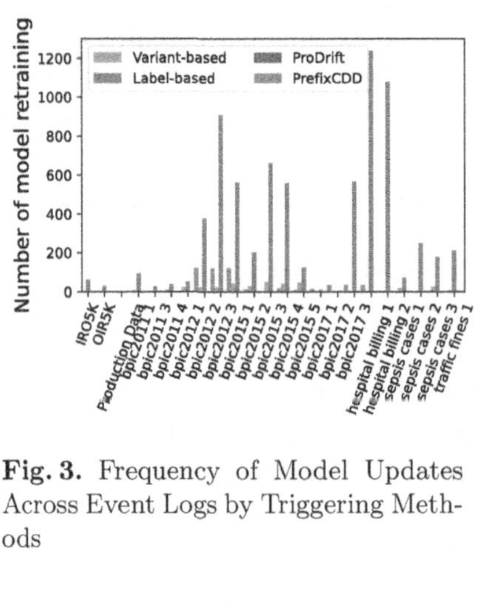

Fig. 3. Frequency of Model Updates Across Event Logs by Triggering Methods

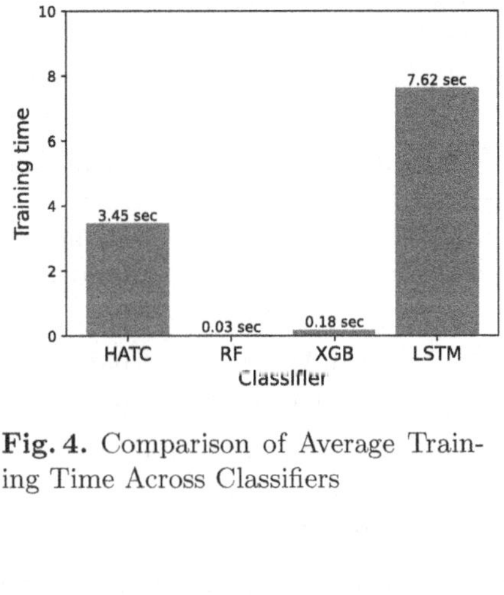

Fig. 4. Comparison of Average Training Time Across Classifiers

retraining time of 0.03 s, while LSTM requires approximately 7.62 s per retraining, making it 254 times slower than RF. The difference highlights the computational efficiency of traditional algorithms, which do not require iterative updates or complex mechanisms. For example, RF's decision tree structure allows rapid updates without iterative calculations. However, LSTM's sequential learning and HATC's concept drift adaptation lead to their higher computational costs.

7 Conclusion and Challenges

We have presented an approach for the tailored retraining strategy for process outcome prediction, designed to improve model performance in online settings. The proposed retraining strategies, *variant-based retraining* and *label-based retraining*, trigger model updates based on event log features, enabling the predictive model to adapt to changes in the data. Our empirical results show that both strategies achieve higher accuracy than baseline methods across various datasets and classifiers. Notably, the tailored retraining strategies yield modest yet consistent improvements in accuracy, with 19 out of 23 logs demonstrating enhanced performance.

Despite the performance gains achieved through tailored retraining, several challenges remain open for further investigation. We identified multiple challenges in online predictive process monitoring that require deeper exploration:

(***RC1***) *How can we identify the major cause of fluctuations in model performance?* Predictive performance can become volatile due to various factors, such as model overfitting, insufficient training data, or concept drift in the test data. Understanding the root cause of these fluctuations is important for developing effective retraining strategies and maintaining stable performance.

(***RC2***) *How can we balance the retraining frequency with computational efficiency, particularly in the continuous learning environment?* The approach proposed in this paper triggers the model retraining whenever the defined conditions are met. However, excessively frequent retraining can reduce the efficiency of online machine learning models.

(***RC3***) *How can we design a practical "when and how" strategy for updating the predictive model?* Even after determining the conditions for triggering model updates, there is room for improvement in customizing the training data used for retraining.

References

1. Baier, L., Reimold, J., Kühl, N.: Handling concept drift for predictions in business process mining. In: IEEE 22nd Conference on Business Informatics (CBI), vol. 1, pp. 76–83. IEEE (2020)

2. Bose, R., van der Aalst, W., Žliobaitė, I., Pechenizkiy, M.: Handling concept drift in process mining. In: Mouratidis, H., Rolland, C. (eds.) CAiSE 2011. LNCS, vol. 6741, pp. 391–405. Springer, Heidelberg (2011). https://doi.org/10.1007/978-3-642-21640-4_30
3. Chawla, N.V., Bowyer, K.W., Hall, L.O., Kegelmeyer, W.P.: Smote: synthetic minority over-sampling technique. J. Artif. Intell. Res. **16**, 321–357 (2002)
4. Domingos, P., Hulten, G.: Mining high-speed data streams. In: Proceedings of ACM SIGKDD International Conference on Knowledge Discovery and Data Mining (ACM SIGKDD), pp. 71–80 (2000)
5. Gallego-Fontenla, V., Vidal, J.C., Lama, M.: A conformance checking-based approach for sudden drift detection in business processes. IEEE Trans. Serv. Comput. (IEEE TSC) **16**(1), 13–26 (2021)
6. Huete, J., Qahtan, A.A., Hassani, M.: Prefixcdd: effective online concept drift detection over event streams using prefix trees. In: IEEE Computers, Software, and Applications Conference (COMPSAC), pp. 328–333. IEEE (2023)
7. Hulten, G., Spencer, L., Domingos, P.: Mining time-changing data streams. In: ACM SIGKDD, pp. 97–106 (2001)
8. Kosciuszek, T., Hassani, M.: Online next activity prediction under concept drifts. In: CAiSE, pp. 335–346. Springer, Cham (2024)
9. Lee, S., Comuzzi, M., Lu, X., Reijers, H.A.: Measuring the stability of process outcome predictions in online settings. In: 2023 5th International Conference on Process Mining (ICPM), pp. 105–112. IEEE (2023)
10. Leontjeva, A., Conforti, R., Di Francescomarino, C., Dumas, M., Maggi, F.M.: Complex symbolic sequence encodings for predictive monitoring of business processes. In: Business Process Management: 13th International Conference (BPM), pp. 297–313. Springer, Cham (2015)
11. Maaradji, A., Dumas, M., La Rosa, M., Ostovar, A.: Fast and accurate business process drift detection. In: BPM, pp. 406–422. Springer, Cham (2016)
12. Maaradji, A., Dumas, M., La Rosa, M., Ostovar, A.: Detecting sudden and gradual drifts in business processes from execution traces. IEEE Trans. Knowl. Data Eng. (IEEE TKDE) **29**(10), 2140–2154 (2017)
13. Maisenbacher, M., Weidlich, M.: Handling concept drift in predictive process monitoring. In: 2017 IEEE International Conference on Services Computing (SCC), vol. 17, pp. 1–8 (2017)
14. Márquez-Chamorro, A.E., Nepomuceno-Chamorro, I.A., Resinas, M., Ruiz-Cortés, A.: Updating prediction models for predictive process monitoring. In: CAiSE, pp. 304–318. Springer, Cham (2022)
15. Montiel, J., et al.: River: machine learning for streaming data in python. J. Mach. Learn. Res. **22**(1), 4945–4952 (2021)
16. Pasquadibisceglie, V., Appice, A., Castellano, G., Fiorentino, N., Malerba, D.: Stardust: a novel process mining approach to discover evolving models from trace streams. IEEE TSC **16**(4), 2970–2984 (2022)
17. Pasquadibisceglie, V., Appice, A., Castellano, G., Malerba, D.: Darwin: an online deep learning approach to handle concept drifts in predictive process monitoring. Eng. Appl. Artif. Intell. **123**, 106461 (2023)
18. Paszke, A., et al.: Pytorch: an imperative style, high-performance deep learning library. In: Advances in Neural Information Processing Systems, vol. 32, pp. 8024–8035 (2019)
19. Pauwels, S., Calders, T.: Incremental predictive process monitoring: the next activity case. In: BPM, pp. 123–140. Springer, Cham (2021)

20. Pedregosa, F., et al.: Scikit-learn: machine learning in Python. J. Mach. Learn. Res. **12**, 2825–2830 (2011)
21. Rizzi, W., Di Francescomarino, C., Ghidini, C., Maggi, F.M.: How do i update my model? On the resilience of predictive process monitoring models to change. Knowl. Inf. Syst. **64**(5), 1385–1416 (2022)
22. Teinemaa, I., Dumas, M., La Rosa, M., Maggi, F.M.: Outcome-oriented predictive process monitoring: review and benchmark. ACM Trans. Knowl. Discov. Data (TKDD) **13**(2), 1–57 (2019)
23. Teinemaa, I., Dumas, M., Leontjeva, A., Maggi, F.M.: Temporal stability in predictive process monitoring. Data Min. Knowl. Disc. **32**(5), 1306–1338 (2018). https://doi.org/10.1007/s10618-018-0575-9
24. Wuyts, B., Weytjens, H., vanden Broucke, S., De Weerdt, J.: Dylopro: profiling the dynamics of event logs. In: BPM, pp. 146–162. Springer, Cham (2023)

Navigating the Challenges of Process Mining Use Case Definition: A Qualitative Study

Kerstin Haug[1](✉), Pol Schumacher[2], Holger Wittges[1], and Stefanie Rinderle-Ma[1]

[1] Technical University of Munich, Arcisstr. 21, 80333 Munich, Germany
{kerstin.haug,holger.wittges,stefanie.rinderle-ma}@tum.de
[2] Celonis Labs GmbH, Theresienstr. 6, 80333 Munich, Germany
p.schumacher@celonis.com

Abstract. Defining accurate use cases is a critical yet underexplored phase of a process mining project. This study investigates the challenges project teams face in identifying, selecting, and specifying process mining use cases. Through semi-structured interviews with 13 experienced process mining consultants from various industries, the research identifies key barriers to effective use case definition.

The study highlights that key challenges are primarily related to stakeholder involvement and collaboration. These challenges include a heavy reliance on process experts, overcoming stakeholder resistance, convincing stakeholders of the tangible value of process mining, addressing differing perspectives on the same process, and dealing with variations in domain-specific terminology. Additionally, the findings emphasize the need for practical tools, such as a comprehensive knowledge bases, to support use case definition. We derive four main directions for future research: first developing frameworks for stakeholder onboarding and enablement, second exploring collaboration dynamics within process mining teams, third offering practical tool support for better guidance in the use case definition phase and fourth providing a comprehensive, consolidated knowledge base of process mining use cases. These insights provide a foundation for paving the way for more impactful process mining initiatives.

Keywords: Process Mining · Process Mining Challenges · Process Mining Projects · Process Mining Use Cases · Use Case Definition · Use Case Identification · Stakeholder Commitment · Stakeholder Collaboration · Organizational Alignment

1 Introduction

Process mining (PM) is a data-driven technique that analyzes and visualizes business processes by extracting insights from event logs generated by enterprise systems like ERP or CRM tools. It provides an end-to-end view of how processes are executed in reality, identifying inefficiencies, bottlenecks, and deviations from the intended workflows. By enabling organizations to compare real-world processes with their designed models, PM helps improve process transparency, optimize process performance and process compliance, and support fact-based decision-making [15].

J. Grabis et al. (Eds.): RCIS 2025, LNBIP 547, pp. 418–434, 2025.
https://doi.org/10.1007/978-3-031-92474-3_25

Defining a precise and proper use case is critical for the success of a PM project because it ensures alignment between business objectives and technical efforts, maximizes value realization, and minimizes risks associated with unclear goals or scope [11]. For example, Deloitte's Global Process Mining Survey highlights that selecting the appropriate processes to initiate a PM project significantly enhances adoption rates and maximizes value realization [4].

Process mining consultants (PMCs) typically play a central role in defining process mining use cases (PMUCs). They act as intermediaries between diverse stakeholders in PM project teams, align business objectives with technical feasibility, identify suitable processes for analysis, structure project scopes effectively, and help in selecting the right methods for PM implementation [8]. However, defining a PMUC is not a straightforward task. The lack of experience in PM in many organizations makes it an abstract tool for many people because they do not understand its capabilities and how to match them with their problem space. Additionally, PM introduces unique characteristics that require a strong process-oriented perspective. However, employees often lack a fully process-oriented understanding as they see the process only from the lens of their distinct business functions [7]. Understanding the obstacles PMCs face in defining PMUCs with their project teams is essential to overcoming these challenges and enabling them to support organizations in establishing effective use cases.

While existing research, such as [7, 12] and [2] propose frameworks for particular aspects of PMUC selection and definition, it remains unclear whether these methodologies effectively support the practical approach of PMUC definition followed by project teams. Other studies, including [10] and [17], explore the broader managerial and organizational implications of PM through end-to-end project analyses but do not explicitly provide insights into the challenges project teams face when defining PMUCs. To bridge this gap, our research follows three main objectives: first, we want to create a foundational understanding of the approach that PM project teams follow when identifying and deriving PMUCs to assess if current frameworks adequately address this phase. Second, we examine which challenges are faced in this phase of a PM project. Third, based on these challenges, we want to identify opportunities for improved tool support. For these three objectives, we define the following three research questions (RQs):

- **RQ.1:** *What approach is followed to define the scope of a use case for process mining?*
- **RQ.2:** *What are the fundamental challenges in defining use cases for process mining?*
- **RQ.3:** *What practical tools would help to solve these challenges?*

We conduct an interview study with 13 PMCs and PM experts in similar roles to answer these research questions.

Our research provides other researchers and academics with insights into real-world PMUC definition approaches, highlighting key challenges of PMCs and gaps in existing methodology and tool support. Additionally, we outline a research agenda to address these gaps and challenges. These insights contribute to the development of improved methodologies and tools that enable PMCs to define more effective and impactful use cases.

The remainder is organized into the following sections: in Sect. 2, we explain the characteristics of PMUCs and present existing research related to our problem definition. In Sect. 3, we describe our data collection and data analysis methodology. In Sect. 4, we present the observed results, and a discussion follows in Sect. 5. We conclude our work in Sect. 6.

In the following sections, we will summarize the steps of identifying, selecting, and specifying a PMUC using the term *process mining use case definition.*

2 Background

In Sect. 2.1, we outline the characteristics and types of PMUCs, highlighting the need for dedicated research on frameworks, methodologies and tools. In Sect. 2.2 we position our work within the broader landscape of research on PMUCs and practical PM studies.

2.1 Process Mining Use Cases

When defining PMUCs, it is essential to consider various factors, including organizational objectives, process requirements, data availability, and the expectations of diverse stakeholders. These use cases are typically driven by specific process goals such as efficiency improvement, process compliance monitoring, or process automation. Given that process mining (PM) adopts a strongly process-oriented perspective, this distinction must be taken into account when defining PMUCs, particularly in comparison to traditional Business Intelligence (BI) use cases [7]. Since PM relies on event log data, considering the quality and granularity of this data plays a crucial role in determining feasibility and success of a PMUC [15]. Additionally, meaningful PMUC specifications depend heavily on collaboration between IT teams, business analysts, and process owners, as multiple stakeholders contribute to defining and refining the analysis [8].

PMUCs can be categorized into different types. Process discovery identifies actual process flows from event logs and allows the user to analyze a process' frequent behavior, such as routing probabilities or most frequent paths. Conformance checking evaluates deviations from predefined process models or rules against which the actual process flow is checked to ensure process compliance. Process Enhancement aims at gaining insights into a process' organizational perspective (e.g., resources involved in a case, resources per task or work handovers) and its overall performance (e.g., bottlenecks or delays) and the underlying factors contributing to inefficiencies or compliance violations [1]. More advanced applications include predictive and prescriptive PM, which analyze historical process data to forecast future process outcomes [15].

The diverse characteristics of PMUCs underscore the need for dedicated research efforts to establish a well-defined approach that ensures the development of high-quality use cases.

2.2 Related Work

As relevant to our work, we examined research on PMUCs as well as studies that draw practical implications on the initialization phase of PM projects, in which project requirements are defined.

We identified three approaches that explicitly contribute to identifying, selecting, and specifying PMUCs. The PMUC canvas [7] is a structured template designed to guide the development and specification of PMUCs. It extends common data canvases to address the characteristics of a PMUC. In [2], the authors follow the objective of designing a workshop method along with a data canvas to guide the identification and mapping of process knowledge and relevant data to define clear data requirements for a PMUC. The research of [12] defines a set of criteria for selecting and prioritizing business processes for PM adoption. While these three publications provide structured frameworks for specific aspects of PMUC definition, current research does not assess whether these frameworks align with the approach that PM project teams follow in practice and what challenges project teams face. Therefore, we aim to establish a broader understanding of how PMUCs are defined in real-world scenarios. In Sect. 4.1 we outline the approach described by our interviewees. In Sect. 5.1 we discuss whether existing PMUC methodologies effectively support this practical approach. Beyond these methodologies, we also aim to identify opportunities for tool support based on the PMUC definition approach and the challenges associated with it.

Additionally, we examined studies on managerial and organizational aspects in PM projects to identify barriers and drivers for PMUC definition. We found the studies of [10] and [6] examining practical challenges in PM projects. The research of [17] and [3] analyze challenges during PM projects and additionally offer mitigation measures and practical recommendations. In [18], the authors investigate the similarities and differences between PM project methodologies and their application in practice and present practical obstacles. Further relevant research deals with identifying factors of successful PM initiatives and enablers for PM adoption [9, 13, 14]. In [8], the authors investigate the capabilities and competencies required in a PM project by performing an interview study among interviewees having differing roles related to PM. Besides role-specific insights, they also draw general observations across all interviewees on challenges they face.

While these studies provide practical insights into PM projects at a holistic level, some of which may relate to PMUC definition, they do not offer explicit insights on defining PMUCs. Our goal is to generate more granular insights, focusing specifically on the requirements identification phase of a PM project. In Sect. 5.2 we discuss how the insights of the examined studies support our results.

3 Methodology

In Sect. 3.1 we describe the interview setup we followed to collect the data for our study. In Sect. 3.2, we explain our approach to analyze the interview transcripts.

3.1 Data Collection

We aim to foster an open-ended discussion to establish a foundational understanding of PMUC definition in practice before developing more structured hypotheses. Therefore, we decided on expert interviews as our data collection methodology. We performed semi-structured interviews with PM experts, which took place virtually from May 2024 to August 2024. For our research, we selected experts who (1) have worked in a PMC role or a similar role in their organization and (2) who have, as part of this role, experienced the requirements definition phase of a PM project.

In total, 13 participants from different industries and organizations were interviewed. Table 1 gives an overview of the participants and their respective backgrounds.

Table 1. Overview of interview participants with participant Id, a description of their current role, the organization they work in, and their experience with PM

Id	Current role/ Organization	PM experience
P1	PMC/Science and technology company	7 years (industry experience)
P2	Business transformation manager/Manufacturer of optics and reprography products	4 years (industry experience)
P3	Principal data scientist/ PM vendor	8 years (industry experience)
P4	Senior PMC/Consulting company	8 years (industry experience)
P5	PMC/PM vendor	3 years (industry experience)
P6	Business process management consultant/ Semiconductor manufacturer	6.5 years (1.5 years industry experience, 5 years academic experience)
P7	Head of data & analytics/Airport operator	4 years (4 years academic experience at 50% and 3 years industry experience at 50%)
P8	Senior PMC/Airline operator	5 years (industry experience)
P9	PhD Student/Research institute	6 years (3 years industry expertise, 3 years academic expertise)
P10	Manager PM consulting/ PM vendor	5 years (industry experience)
P11	Senior PMC/Insurance company	6 years (industry experience)
P12	Product manager/PM vendor	7 years (industry experience)
P13	PMC/PM vendor	1 year (industry experience)

We interviewed six experts working at companies that implement PM within their organization, five experts working at a PM software vendor, one expert working at a research group that focuses on PM research, and one expert working at a consultancy firm, which provides PM consultancy services to other firms. On average, the participants had 5,4 years of experience in the PM field. Three participants had experience in PM from an academic perspective.

One week before the interview, the interview guideline, including the interview questions, was sent to the participants. The interviews lasted over 55 min on average, with the shortest interview lasting 43 min and the longest lasting 1 h and 1 min.

The interviews started with establishing the primary motivation of the study. Then, the interviewee was asked to describe their current role and experience with PM. In the main part of the interview, the participants were asked a total of ten questions and sub-questions, which were structured along with our three research questions. All questions were asked openly without predefined dimensions or categories.

3.2 Data Analysis

The interviews were recorded and transcribed using Speech-to-Text software. Since the machine-translated transcripts revealed quality issues, they were subsequently revised manually.

We followed the method proposed by [5] to analyze the data systematically. We chose a deductive-inductive approach for the development of the coding scheme where we defined the initial code set based on the dimensions of the PMUC canvas [7], the PM data canvas [2], the use case dimensions described by [12] and the high-level steps for the project initialization phase of PM projects described in the PM2 project methodology [16].

The text segments of the transcripts were categorized into these predefined categories. In cases where none of the categories matched the text segment, a new category was defined. In the end, unused codes were removed from the code set, and low-frequent codes were summarized into higher-frequent codes. The results of the qualitative data analysis are described in the following Sect. 4.

4 Results

In the following, we present the results from the qualitative analysis of the interview transcripts. Sect. 4.1 outlines the approach to PMUC definition, addressing **RQ.1**. Sect. 4.2 discusses the challenges associated with PMUC definition, answering **RQ.2**. Finally, Sect. 4.3 explores tool support for PMUC definition, presenting the findings related to **RQ.3**.

4.1 Results of RQ.1: Approach to PMUC Definition

To establish a commonly understood baseline between the interviewee and the interviewer, we started the interviews by asking participants about the main steps to define a PMUC. From their answers, we derived the steps depicted in Fig. 1 for identifying, selecting, and specifying a PMUC.

The steps *of Identification, Process Definition,* and *Source Data Identification* additionally encompass detailed sub-groups of essential information gathered during these phases. This information is structured according to the dimensions outlined in the PMUC canvas [7]. The component *KPIs and PPIs* is not part of the PMUC but has been identified as an essential component by our interview participants.

The step *of Stakeholder Onboarding* may not strictly fall within the scope of use case definition, but interview participants emphasized it as an important preparation in the overall process of defining meaningful use cases. Stakeholder onboarding involves educating and preparing key stakeholders before embarking on the collaborative use case definition. This step is crucial for setting expectations and aligning everyone on the goals

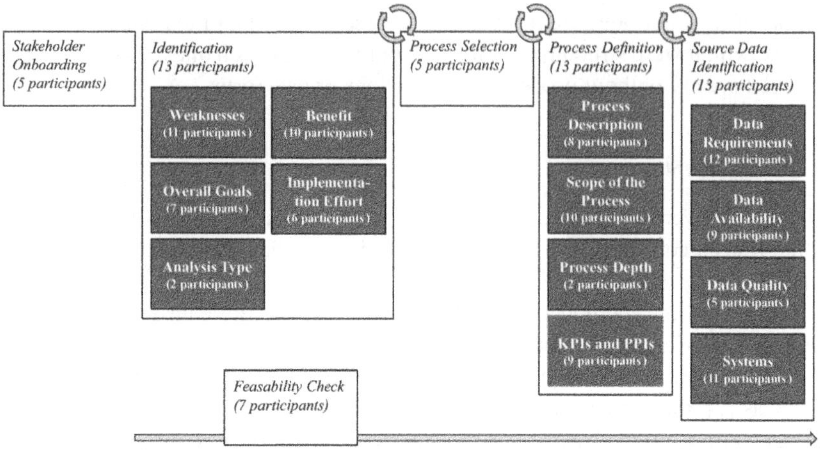

Fig. 1. Steps of defining a PMUC and gathered information (highlighted in blue). The data in parentheses indicates the number of participants who described this component.

and potential outcomes of the initiative. Participant **P8** pointed out, *"I often have the problem that I first need to explain the type of technology. [...] one of the most important points we always see—and the biggest challenge—is convincing experts, management, and stakeholders, and making it clear to them what process mining is and what they can gain from it"*. Explaining the technology, especially when it's something new to the organization, is essential to ensure stakeholders understand its capabilities, limitations, and potential impact. A thorough onboarding process fosters better understanding, ensuring that all stakeholders, whether in management, IT, or business units, are aligned and ready to actively contribute to defining use cases. This step also helps mitigate the challenge of ensuring that the necessary resources, such as process owners, are available and prepared to support the project effectively.

The *Identification* of relevant use cases involves an assessment of PM potential within an organization, focusing on already known weaknesses in their operating processes and defining clear business and process goals. This typically involves aligning the use case with ongoing strategic initiatives to ensure its relevance and impact. Process participants are engaged to describe their current situation, how the process operates, and identify pain points and challenges. These insights are then used to evaluate the improvement potential of a process across dimensions such as time, cost, quality, and flexibility. Finally, a business case is calculated by quantifying the benefit and value potential of the proposed process improvement and the use cases' implementation effort, ensuring that the investment in PM is justified with measurable benefits.

Process Selection entails prioritizing various use cases and choosing one for further analysis, guided by the criteria established in the preceding *Identification* step. A PMUC typically focuses on one specific process and aligns with one or multiple related improvement goals. Interviewee **P9** noted that *"sometimes opposing factors come into play. The company leadership might specifically want to improve a certain process that is not yet sufficiently digitized, making it difficult to analyze. However, they insist on* *focusing on this one process, even though it might not make sense for various reasons or is incredibly complex."*

Process Definition is a crucial step in PMUC definition to establish a shared understanding of how the current process operates. Process experts and process stakeholders are engaged to describe the process, including its key activities and structure. This involves scoping the process to identify and model relevant activities and defining the granularity level (Process Depth) at which they are needed. For the quality of the later analysis, *"it is crucial to understand the right level and define the milestones at the appropriate level. However, it is also important to have the necessary data for these steps at that specific level. This is another key challenge in properly scoping the entire process"* (**P12**). Key Performance Indicators (KPIs) and Process Performance Indicators (PPIs) are identified to measure the process's current state and provide a foundation for the analysis. These metrics help define measurable outcomes and assess the potential value of improvements.

Source Data Identification involves understanding the data sources and systems on which the process operates. This includes identifying the relevant IT systems, such as ERP, CRM, or others, and determining where the required data, such as tables, fields, and timestamps, resides. Once the systems are identified, the next task is to define the specific data requirements by outlining the key data points, their relationships, and their alignment with process objectives. This involves translating business definitions, such as activities or KPIs, into technical specifications that can be mapped to system fields. To ensure alignment, workshops are conducted with both business and technical stakeholders to refine these requirements, and ensure a shared understanding of how the data will support the use case. Identifying source data also requires assessing whether the data is available, accessible, and of the required quality.

Throughout these four phases, it is essential to continuously perform a PM *Feasibility Check*. Part of this step is to assess whether the problem space aligns with the capabilities of PM. Interview participant **P8** stated, it is important *"to create a mapping of the problems [to be addressed] because, in the end, this determines whether [actually process mining is needed] or if a simple KPI dashboard— which could be implemented with another solution—would be sufficient."* Ensuring PM feasability further involves checking that the targeted process meets the necessary characteristics for PM, and verifying the availability and quality of the data to support effective application.

It should be noted that these phases do not happen sequentially, but they may overlap, and steps are performed in a different order. PM experts described the whole approach as highly iterative and collaborative: it requires a lot of back-and-forth communication and iteration with stakeholders from different departments, such as process experts and IT specialists.

4.2 Results of RQ.2: Challenges of PMUC Definition

After establishing the steps in Sect. 4.1, we asked interview participants where they experience challenges. We identified eleven central challenges associated with defining a PMUC, that are summarized in Table 2.

Table 2. Identified challenges associated with defining a PMUC. No. Refers to the distinct number of participants who described this challenge. Participants could name multiple challenges.

Id	Challenge	No.
C1	Lack of an Integrated Skillset in Business Understanding and PM	7 participants
C2	PM Suitability	4 participants
C3	Poorly Documented Processes	4 participants
C4	High Dependency on Process Experts	8 participants
C5	Commitment of Key Stakeholders	8 participants
C6	Implicit Technical Knowledge	5 participants
C7	Different Perspectives on the Same Process	7 participants
C8	Project Stakeholders Speak Different Domain Languages	5 participants
C9	Determining Business Value	3 participants
C10	Aversion to PM	4 participants
C11	Data Availability and Quality	6 participants

Three distinct interview participants had to name each challenge at least to be considered significant—challenges that only one or two named were excluded from the following data analysis.

C1 - Lack of an Integrated Skillset in Business Understanding and PM: Experts who simultaneously understand the business process, PM capabilities, and know about the organization's systems are rare. Domain-specific knowledge is isolated in different roles. There is a lack of expertise in PM within the organization, and employees, especially business experts, do not understand the concepts and functionalities of PM. On the other hand, PMCs and PM analysts do not have the necessary business understanding. PMC **P1** stated that from their perspective, they *"lack the process understanding needed to truly offer the right solutions, and the business itself [...] understands the tool less at the beginning, making it harder to steer [the use case definition] in the right direction"*. This can lead to other issues, e.g., misusing PM for BI applications **(C2)** or overseeing potential use cases. In this context, participant **P10** stated that *"if [the business experts] really understood the technology [...] they would sometimes bring up completely different use cases [...] because they would actually consider it a possibility. If [the process mining consultant] didn't know it was a possibility—like that it could be a pain point—then one side says nothing, and the other says nothing"*.

C2 - PM Suitability: A common challenge to address in the use case definition phase is applying the tool to problems that match its capabilities. When describing their pain points, process experts do not assess if their problem is solvable by PM due to a lack of understanding of PM. If the process mining analyst or PMC does not truly understand the nature of the problem, this may result in applying the tool to problems where the capabilities of PM are not aligned with the requirements. Participant **P1** stated, "*Indeed, most of the use cases that come to [them] are not process mining use cases.*" For instance,

PMUCs are often confused with traditional BI or reporting use cases. This issue may arise because domain expertise is isolated in different roles (**C1**).

C3 - Poorly Documented Processes: Business processes within the organization are either not documented at all, inadequately documented, or the existing documentation is outdated. Interview participant **P4** said they dealt with big tech companies that *"hadn't defined processes at all."* Consequently, one of the initial steps during the use case identification is to model the process, where PMCs are heavily dependent on the implicit knowledge of process experts (**C4**).

C4 - High Dependency on Process Experts: Because processes are poorly documented (**C3**), much of the process knowledge within the organization is implicit, creating a heavy reliance on process experts who possess critical insights that may not be easily accessible or transferable. Interview participant **P10** stated that *"the less [they] know about the process, the more [they] have to ask the customer, which makes the definition [of a use case] much more difficult."* Hence, to define a PMUC, these process experts must be available in this critical phase (**C5**).

C5 - Commitment of Key Stakeholders: Key project stakeholders, whose input is crucial when defining the analysis target, are unavailable or only partially engaged due to capacity constraints. Particularly, operational people who hold a lot of process knowledge as they work in the process are *"always in conflict because they have urgent operational tasks, but [are needed] to do meaningful use case scoping"* (**P7**). Some stakeholders lack commitment because they don't yet recognize the value of PM. Interview participant **P6** stated, *"Without key facts and relying only on gut feeling, the relevant project members, such as a process owner, are not willing to commit, people whose expertise is actually needed for the project."* Some may support the project but are hesitant to acknowledge process weaknesses or commit to improvement targets out of fear that this information could be used against them.

C6 - Implicit Technical Knowledge: Knowledge about the information systems supporting the processes is deeply embedded in the minds of the IT personnel. They often lack a basic understanding of the process flows, which is needed when translating process operations into data requirements. Some IT system documentation may exist, but *"when it really [comes] down to figuring out which data fields [are needed] to use and what information is in which field, that's usually not documented. In those cases, it's often an SAP specialist who knows the system inside and out and just knows this information"* (**P4**).

C7 - Different Perspectives on the Same Process: Due to siloed thinking and a lack of end-to-end (E2E) process-oriented visibility, many stakeholders define the process solely from their perspective. In the first step of use case definition, when describing process weaknesses and defining process analysis goals, *"people tend to optimize their own area of the process, rather than focusing on the overall end-to-end outcome"* (**P9**). This issue refers to business stakeholders from different departments working on different parts of the process and external PMCs, who have *"a completely different perspective from how the customers view their processes"* (**P4**). When defining the overall analysis goal, this may lead *"[...] to focusing on the wrong areas"* (**P9**). This issue is additionally leveraged by project stakeholders using different business languages (**C8**).

C8 - Project Stakeholders Speak Different Domain Languages: In PM projects, stakeholders often come from different departments or expertise areas, leading to a mismatch in terminology and priorities. For instance, the typical project stakeholders involved when scoping a use case are process experts (process owners or operational process participants), a process mining analyst, and a technical expert who knows the systems supporting the process. All these stakeholders use different terminology to describe the same process. This can create communication barriers, challenging building a shared understanding of the underlying data and process. Not having a shared understanding of the process and its weaknesses impacts identifying a relevant use case and aligning its objectives.

C9 - Aversion to PM: Employees within the organization may resist PM due to fears of the transparency it brings. Defining a use case for PM often starts with identifying common process weaknesses. In this phase, employees *"may deliberately hide things that aren't going well because they know it's somehow related to the team, or similar reasons. Certain weaknesses might intentionally not be mentioned at all"* **(P9)**. This poses a challenge because it hinders the accurate definition of goals for a PMUC.

C10 - Determining Business Value: Interviewees stated that they struggle with estimating the business value of a PMUC simply based on a gut feeling without relying on data. This issue involves first understanding the actual weaknesses and cost drivers within a business process without relying on any data and second accurately estimating the impact of its improvements.

C11 - Data Availability and Quality: The transition between the use case definition and the data extraction phase was highly fluent for many interviewees. Defining a use case also requires defining technical requirements in terms of data requirements. Our interviewees hence mentioned identifying the relevant source data as a significant task when defining a use case for PM. In this context, participants stated that it is always a challenge to define a use case and remember that the required process data is available in the right quality. Some use cases might be highly valid for PM with promising business cases but cannot be implemented because the data is simply unavailable.

4.3 Results of RQ.3: Practical Tool Support in PMUC Definition

Building on the approach presented in Sect. 4.1 and the challenges derived in Sect. 4.2, the goal in the third part of the interview was to understand whether there are specific tasks that could be supported through practical tools such as software. To this end, PM experts were asked whether there were aspects of the challenges discussed where they would wish for additional assistance and what functional requirements they would envision for a software system. The answers to this question were very specific. Interviewees gave examples, including code generation, business interpretation of technical data, translating between different business terminology, and more. Since these tasks were mentioned too infrequently, we did not consider them significant. However, we identified two major themes: a *PMUC assistant* and a *knowledge database for PMUCs*, summarized in Table 3.

Table 3. Topics for better software support during PMUC definition. No. Refers to the number of participants who mentioned the topic. The same interviewee could mention multiple topics.

Topic	No.
T1 - PMUC assistant: An assistant who guides the user through the different steps of the use case definition to collect the required information	6 participants
T2 - Knowledge database for PMUCs: A knowledge database that documents typical PMUCs and their key information, such as process activities, common process inefficiencies, KPIs and PPIs, etc.	4 participants

A common theme mentioned by six participants is an assistant that guides the PM project team through the use case definition phase. Interviewees **P6** and **P12** described this process as a "*use case funnel*". In each step, the assistant should request the necessary information from the user and provide suggestions and examples for inspiration, e.g., common process weaknesses, standard activities in a specific process, or example calculations for a business case. Interviewee **P9** also mentioned that "*The system provides [...] feedback—what is still missing—and gives [...] some guidance on what [needs to be done] and when in order to structure the process*". Ultimately, the collected information should be standardized to easily compare different use cases. Interview participant **P1** also said, this assistant could "*function like a chatbot*" that poses questions and provides feedback e.g. "*suggestions [...] on how process mining could be used*".

A second topic four interview participants mentioned was a knowledge database for PMUCs. The interview participants stated that accessing a central repository, where typical PMUCs are documented, would significantly ease their work. For instance, the documentation of a PMUC should include a definition of the examined process, common process inefficiencies, and common KPIs and PPIs. Interviewee **P10** stated, "*The coolest thing would be to simply have a database, so to speak—a relatively simple database. [...] Which processes exist? [...] what are the typical KPIs and use cases defined within them? What are the typical cost drivers? [...]*".

5 Discussion, Outlook, and Limitations

In this section, we discuss our results in the context of state-of-the-art research for each research question, provide recommendations on how future research efforts should address our findings, and outline the limitations of our study.

5.1 PMUC Definition Support by Existing Methodologies

To evaluate whether the approach we derived to answer **RQ.1** is effectively supported in practice, we compare it with three established frameworks for defining PMUCs and discuss how these methodologies align with the steps described by our interview participants.

The PMUC canvas [7] structures the characteristics of a PMUC along five key dimensions: Initial Situation, Process, Target of Analysis, Data Management, and Involved People. These dimensions encompass all relevant information described by our interview participants in the phases of Identification, Selection, Process Requirements, and Data Requirements. However, the canvas does not explicitly include Key Performance Indicators (KPIs) and Process Performance Indicators (PPIs). The PMUC canvas incorporates additional considerations, such as Technological Acceptance, Technological Competence, Analysis Risks, Analysis Frequency, and Time Horizon. However, our interview participants mentioned these aspects too infrequently to be included in the final results. Furthermore, the PMUC canvas specifies the project team and stakeholders, but our interview participants did not highlight this aspect. At the time of defining the use case, they assumed that the team and stakeholders were already established. Notably, the PMUC canvas does not provide guidance on stakeholder onboarding, which emerged as a critical step in our findings.

The PM Data Canvas [2] primarily focuses on defining process and data requirements. For these two steps it offers a workshop method and a structured data canvas to detail aspects such as process activities, essential KPIs and PPIs, case ID definitions, and the mapping of process points to concrete data points within the source system. However, it does not address use case identification and selection or an overall feasibility assessment, nor does it provide any guidance on stakeholder onboarding.

The methodology of [12] on the other hand, primarily addresses use case identification and selection and evaluating the suitability of a process for applying PM. It specifies detailed criteria for business importance, challenges, employee skills, data quality, and optimization potential. Additionally, the methodology emphasizes the importance of organizational support in selecting the right process. However, while the established criteria provide a profound methodology for identifying and selecting use cases, it lacks guidance on how to secure organizational support, such as gaining management attention or employee buy-in.

In summary, the reviewed frameworks provide structured approaches to the phases of Identification, Process Selection, Feasability Check, Process Definition, and Source Data Identification in PMUC definition. However, none adequately addresses Stakeholder Onboarding, an aspect our interview participants highlighted as a crucial step before defining a PMUC. We explore this issue further in Sect. 5.2.

5.2 Key Challenges of PMUC Definition: Stakeholder Involvement, Education and Collaboration

As identification of source data, data availability, data access, and data quality issues have been consistently identified as major barriers in PM projects [6, 8, 10, 13, 14, 17], we initially hypothesized that defining precise data requirements would emerge as a primary challenge for **RQ2**. However, the insights shared by interview participants revealed a different perspective. They highlighted that the greater difficulty lies in identifying and defining the right use cases and achieving stakeholder alignment. They even stated that deriving data requirements becomes a procedural and diligent task once the use case goals and process requirements are established and aligned. While six participants mentioned data availability and quality **(C11)** as obstacles, they did not highlight these as the most critical challenges.

The challenges that were highlighted as particularly relevant are primarily managerial and organizational, often deeply interconnected. Different studies [8, 9] propose an effective team configuration as an essential success factor to mitigate such issues in PM projects. In addition to that, we argue that targeting stakeholder alignment and collaboration in such cross-functional teams is of high importance in this context, and future research should contribute to that. For instance, in challenge **C1**, our study participants described that even in cross-functional teams, results are suboptimal if stakeholders lack at least a basic understanding of knowledge outside of their domain of expertise.

In fact, the main challenges that we identified in defining effective PMUCs center around stakeholder involvement, education, and collaboration.

A significant dependency on process experts stems from the lack of explicit process knowledge within organizations, frequently linked to poor documentation quality **(C3)** [3, 17, 18]. Addressing this issue requires strong involvement from process experts **(C4)** and technical experts **(C6)**, but this effort can be hindered when key stakeholders are not fully committed to the project **(C5)**. A key finding from our research is that securing stakeholder commitment is one of the most significant challenges in defining PMUCs. This is consistently recognized as a significant barrier in PM projects [3, 9, 13, 14, 18], with existing studies emphasizing top management support as a critical success factor [3, 9, 13, 18]. In addition to that, our interview participants stressed the importance of proper stakeholder onboarding to improve commitment and alignment. Educating stakeholders on the potential and capabilities of PM could play a crucial role in overcoming these challenges. Despite recognizing these strategies, there is little guidance in the literature on how to effectively onboard stakeholders or secure top management support in PM projects. To address this gap, future research should prioritize developing actionable frameworks to educate stakeholders about the unique capabilities of PM. In this context, it is also crucial to highlight and articulate the distinct differences between PM and other BI applications, as these are often conflated **(C2)**. While the study of [17] also describes this issue, it does not elaborate on any mitigation strategy.

Future research should also provide practical recommendations on demonstrating the value of PM to build a persuasive case for executive buy-in. This would also help PM stakeholders in addressing the issue of determining the business value of PM, which has been described as a significant issue by our interview participants **(C9)** and other studies [3, 6, 10, 18]. In the context of stakeholder commitment, aversion to PM **(C10)** among key stakeholders further amplifies this lack of commitment, creating additional obstacles such as process experts being unwilling to share their domain knowledge. The studies of [6, 8, 10] describe this as an overall issue in PM projects, and [3] proposes to involve data protection stakeholders early on to build trust.

Collaboration among stakeholders with diverse expertise emerged as the second key area for improvement during the PMUC definition. Effective collaboration is often hindered by differing priorities **(C7)**, communication styles, and a lack of shared terminology **(C8)**. While existing research has been focused on exploring project members' expertise and capabilities [8], future research endeavors should explore the people involved in PM projects beyond their skills to include their goals, needs, and communication approaches. This could offer valuable insights for enhancing collaboration.

5.3 Improving Guidance in the PMUC Definition Phase

The insights from interview participants predominantly highlighted organizational and managerial challenges, and we could not identify specific tasks for direct tool support for **RQ.3** as we had initially anticipated.

Nevertheless, participants expressed a clear need for better guidance during the use case definition phase, suggesting the concept of a *PMUC assistant*. Addressing this need could involve extending practical tools, such as workshop templates and software tools, to provide clear guidance on navigating this phase. Additionally, integrating large language models (LLMs) as a chatbot interface could be a promising approach to explore. Such a system could guide users step-by-step through the use case definition process while also helping to structure and validate their inputs and give insightful recommendations.

Moreover, the absence of a consolidated, cross-industry knowledge base of PMUCs presents an opportunity for future research. Such a resource would help project teams draw on proven examples and better identify relevant use cases for their unique contexts.

5.4 Limitations

While our study provides valuable insights into the challenges of PMUC identification, it is essential to acknowledge its limitations.

First, our study is inherently limited by the focus on PMCs. Needs and challenges may differ from those of other PM stakeholders, such as business or technical experts, and may even conflict with them. Expanding the scope to incorporate the experiences and contributions of additional roles, such as IT specialists or process owners, could enrich our understanding and lead to more comprehensive strategies for addressing challenges in defining PMUCs.

Second, the interviews focused exclusively on experienced PMCs. While their expertise provided a deep understanding of the challenges in use case identification, the perspectives of less-experienced PMCs could offer an additional view of the challenges in use case identification.

Finally, our research was geographically restricted to the DACH (Germany, Austria, Switzerland) region. Organizational setups, communication styles, and cultural dynamics in this region may differ significantly from those in other parts of the world.

6 Conclusion

Our research explores how PM project teams approach the identification, selection, and definition of use cases for PM. We seek to uncover the challenges inherent in this process and highlight areas where PMCs perceive a need for enhanced tool support. These findings are derived from a qualitative data analysis collected through semi-structured interviews with 13 experienced PMCs. Our results indicate that the primary challenges defining PMUCs are predominantly managerial and organizational. These challenges center on securing the commitment of key stakeholders and fostering effective collaboration with them. Additionally, we identify a need for more guided assistance during the use case definition process and the availability of a reference base for use cases. We derive four key areas for future research: (1) exploring strategies to secure top management support and developing approaches for stakeholder enablement and education, (2) creating a better understanding of the different people involved in a PM project, their needs, and goals, to enhance collaboration, (3) developing structured methodologies and practical tool support to guide project teams through the use case definition phase and (4) providing a consolidated knowledge base of PMUCs.

Disclosure of Interests. The authors have no competing interests to declare that are relevant to the content of this article.

References

1. Ailenei, I., Rozinat, A., Eckert, A., van der Aalst, W.M.P.: Definition and validation of process mining use cases. In: Daniel, F., Barkaoui, K., Dustdar, S. (eds.) Business Process Management Workshops. Lecture Notes in Business Information Processing, vol. 99, pp. 75–86. Springer, Heidelberg (2012)
2. Brock, J., von Enzberg, S., Kühn, A., Dumitrescu, R.: Process mining data canvas: a method to identify data and process knowledge for data collection and preparation in process mining projects. Procedia CIRP **119**, 602–607 (2023). https://doi.org/10.1016/j.procir.2023.03.114
3. Eggert, M., Dyong, J.: Applying process mining in small and medium sized IT enterprises – challenges and guidelines. In: Di Ciccio, C., Dijkman, R., Del Río Ortega, A., Rinderle-Ma, S. (eds.) Business Process Management. Lecture Notes in Computer Science, vol. 13420, pp. 125–142. Springer, Cham (2022)
4. Galic, G., Wolf, M.: Delivering value with process analytics. Process Mining Adoption and Success Factors (2021). https://www.deloitte.com/de/de/services/financial-advisory/research/global-process-mining-survey.html. Accessed 18 Jan 2025
5. Gläser, J., Laudel, G.: Experteninterviews und qualitative Inhaltsanalyse als Instrumente rekonstruierender Untersuchungen, 4th edn. VS Verlag, Wiesbaden (2010)
6. Grisold, T., Mendling, J., Otto, M., vom Brocke, J.: Adoption, use and management of process mining in practice. BPMJ **27**(2), 369–387 (2021). https://doi.org/10.1108/BPMJ-03-2020-0112
7. Hardjosuwito, D., Braucks, F.L., Schröer, T., Schwanen, C., van der Aalst, W.M.: The Process Mining Use Case Canvas: A Framework for Developing and Specifying Use Cases. Publish-Ing, Hannover (2023)
8. Kipping, G., et al.: How to leverage process mining in organizations - towards process mining capabilities. In: Di Ciccio, C., Dijkman, R., Del Río Ortega, A., Rinderle-Ma, S. (eds.) Business Process Management. Lecture Notes in Computer Science, vol. 13420, pp. 40–46. Springer, Cham (2022)
9. Mamudu, A., Bandara, W., Wynn, M.T., Leemans, S.J.J.: A process mining success factors model. In: Di Ciccio, D., et al. (eds.) Business Process Management, vol. 13420, pp. 143–160 (2022)
10. Martin, N., et al.: Opportunities and challenges for process mining in organizations: results of a Delphi study. Bus. Inf. Syst. Eng. **63**(5), 511–527 (2021). https://doi.org/10.1007/s12599-021-00720-0
11. Reinkemeyer, L.: Process Mining in Action. Springer, Cham (2020)
12. Rott, J., Böhm, M.: Mine the right process - towards a method for selecting a suitable use case for process mining adoption. In: ECIS 2022 Research Papers, vol. 45 (2022)
13. Smit, K., Mens, J.: Process mining in the rail industry: a qualitative analysis of success factors and remaining challenges. In: BLED 2019 Proceedings, vol. 25 (2019)

14. Syed, R., Leemans, S.J.J., Eden, R., Buijs, J.A.C.M.: Process mining adoption. In: Fahland, D., Ghidini, C., Becker, J., Dumas, M. (eds.) Business Process Management Forum. Lecture Notes in Business Information Processing, vol. 392, pp. 229–245. Springer, Cham (2020)
15. van der Aalst, W.: Process Mining. Data Science in Action, 2nd edn. Springer, Heidelberg (2016)
16. van Eck, M.L., Lu, X., Leemans, S.J.J., van der Aalst, W.M.P.: PM2: a process mining project methodology. In: Zdravkovic, J., Kirikova, M., Johannesson, P. (eds.) Advanced Information Systems Engineering. Lecture Notes in Computer Science, vol. 9097, pp. 297–313. Springer, Cham (2015)
17. Zimmermann, L., Zerbato, F., Weber, B.: What makes life for process mining analysts difficult? A reflection of challenges. Softw. Syst. Model. (23), 1345–1373 (2024). https://doi.org/10.1007/s10270-023-01134-0
18. Zuidema-Tempel, E., Effing, R., van Hillegersberg, J.: Bridging the gap between process mining methodologies and process mining practices. In: Di Ciccio, C., Dijkman, R., Del Río Ortega, A., Rinderle-Ma, S. (eds.) Business Process Management Forum. Lecture Notes in Business Information Processing, vol. 458, pp. 70–86. Springer, Cham (2022)

Zero-Shot Approaches for the Extraction of Event Logs from Medical Notes

Allmin Susaiyah(✉) and Natalia Sidorova

Eindhoven University of Technology, Eindhoven, The Netherlands
{a.p.s.susaiyah,n.sidorova}@tue.nl

Abstract. In healthcare, documenting patient care activities through medical notes, such as those written by doctors and nurses, is a mandatory practice. These records contain valuable information that, when analyzed using data mining and process mining techniques, can improve the understanding of care processes, identify areas for improvement, address errors, and predict health-related care for the coming period. However, the textual nature of these documents poses a significant challenge, as it prevents the direct application of data and process mining techniques. To overcome this, the extraction of events—and subsequently, event logs—from medical notes is needed. We evaluate and compare two zero-shot approaches to event extraction: a sentence transformer–based approach and a large language model–based approach. The evaluation is carried out on a human-annotated dataset derived from an annotated subset of nurse notes in the MIMICS-III dataset. We highlight the strengths and weaknesses of each approach and discuss further steps necessary to effectively obtain event logs from medical notes.

Keywords: medical notes · unstructured data · information extraction · event extraction · process mining · event logs · sentence transformers · large language models

1 Introduction

A large amount of patient data, including administrative and clinical activities, is documented in electronic health records (EHR) in structured [8,27] and unstructured [11,17,20] formats. Structured data typically includes billing information, test results, prescriptions, and demographics, while unstructured data consist of medical records such as doctor and nurse notes, discharge summaries, radiology reports, and cardiology reports.

Nurse notes, like the one shown in Fig. 1a, are free texts recorded by nurses during their work shift, with invaluable information about patient status and progress [19]. They contain records about patient care activities and events, such as medication administration (e.g. "Lasix 20 mg" in Fig. 1a), responses to medication ("Urine output improved"), complaints ("not getting sleep"), activities of daily living ("OOB (= out of bed) to chair till 10 pm"), etc.

J. Grabis et al. (Eds.): RCIS 2025, LNBIP 547, pp. 435–451, 2025.
https://doi.org/10.1007/978-3-031-92474-3_26

#Patient Id: 3935; #Nurse Id: 14419; #Note Time: 2151-03-10 02:40:20

Heart failure (CHF), Diastolic, Acute
Assessment: **Urine output** low. Pt **drinking** lot of water and juice as she is **thirsty**.
Action: Lasix 20mg iv x 1
Response: **Urine output** improved
Plan: Watch I/O.
Alert and **oriented** x 3 ,**OOB to chair** till 10pm and then **to bed**. Rt LL **dressing changed**.
Slept on and off and at 2.30am
c/o not getting sleep as she is **coughing** a lot and **uncomfortable**. **morphine** 1mg iv given.
LS with ins/exp **wheez**, **Nebs PRN** and on **NC O2** 6lit/min

Negations | Time | Medical Abbreviations
Medical Terms | Other quantities | Common Abbreviations

(a) A nurse note from the MIMICS III dataset [17] with events shown in bold.

Pt. Id	Staff Id	Event Id	Event name	Event Type	Time	Further Info
3935	14419	e1	Urine Output	Observation	2151-03-10 02:40:20*	Quantity: Low
3935	14419	e2	Medication	Action	2151-03-10 02:40:20*	Quantity: 20mg, Drug: Lasix
3935	14419	e3	Urine Output	Observation	2151-03-10 02:40:20*	
3935	14419	e4	Sleep	Observation	2151-03-10 02:30:00*	Quality: disturbed

*de-identified note-time not event-time

(b) A fragment of the event log for the nurse note in Fig. 1a.

Fig. 1. (a) A nurse note and (b) an event log fragment.

Medical notes can be seen as an unstructured log of important events that occurred during patient care, making them an interesting source of data to analyze with process mining (PM) techniques [1,10]. These techniques enable the evaluation of the performance of the process, verification of compliance with care protocols, management of clinical risks, and prediction of subsequent activities, nurse workload, or patient outcomes. They cannot be applied directly to nurse notes, but need a structured event log, like the one shown in Fig. 1b, as input.

Event log extraction has been extensively used for structured health data [8, 27]. Extracting event logs from medical text is, however, a highly nontrivial task [13,14]. While it is easy to extract the hospital admission number of the patient for whom a note is written, the id of the staff member who wrote the note and the time when it was written, as they are stored in a structured way, it is difficult to detect events mentioned in the text. The automatic extraction of event logs from medical notes faces a number of challenges, partly caused by a purpose misalignment, as medical notes are not written with automatic event extraction in mind. Event descriptions lack consistency: the same event can be described in multiple ways. Medical notes can be inherently ambiguous and/or subjective due to the nature of natural language use. The frequent usage of common and medical abbreviations, as shown in Fig. 1a, further complicates the interpretation (e.g. "OOB" means out of bed and "iv" means intravenous in this note, but it can also mean "4" in different contexts).

The use of negation poses a separate challenge in extracting events from medical notes [28]. Negation can be expressed in various forms: directly using negative terms, such as "*not* getting sleep", or through negative quantifiers, like "barely slept". Moreover, phrases like "hardly any rest" imply a lack of sleep without explicitly mentioning it. Identifying these different types of negation is crucial in medical text processing, as they can significantly impact the interpretation of patient symptoms and conditions.

Various machine learning and NLP techniques have been utilized to transform unstructured text into structured event data [6,22,32]. Most of these techniques require large annotated datasets, while annotating is a costly and time-consuming process. Recent advances in NLP, such as zero-shot techniques, leverage pre-trained models that require no additional training and can be adopted to different tasks. These methods commonly use *transformers* (a type of deep learning architecture designed to process sequential data, such as text, efficiently) and, more recently, *large language models* (LLMs). Efforts to use generative AI to transcribe, compile and document medical interactions in a clear and concise manner are increasing, but general caveats of generative AI such as hallucinations and biases are not obsolete [5,18].

In this paper, we conduct an evaluation and comparison of a Sentence Transformer (ST)-based method and an LLM-based methods for event name extraction from medical notes. Events should belong to a set of event types predefined by our project partners, who are medical experts, like "pain", "sleep" or "eating". The evaluation is carried out on a dataset selected from [17] annotated by human annotators. Our research explores how key limitations of natural language impact the effectiveness of these two techniques in event detection and classification.

The remainder of the paper is organized as follows. In Sect. 2 we discuss relevant recent literature. In Sect. 3, the ST-based approach, the LLM approach, and the process of developing the evaluation dataset are presented. Section 4 covers the evaluation and comparison of both approaches. In Sect. 5 we discuss our observations, the practical limitations of both approaches, and describe directions for future research. Finally, Sect. 6 concludes the paper by summarizing key insights.

2 Related Works

The definition of an event varies between fields, including the fields of information extraction and process mining. Although these fields share similar concepts, they often use different terminologies to describe them.

Process mining [2,10] considers events in the context of business processes as initiations, progressions, or completions of activities. During the execution of a process, events are recorded in an event log with event records typically including a case ID, which indicates to which instance of a process (e.g. hospitalization, or patient) this event belongs, an event name, which describes the event (e.g. "eating"), a time stamp, agents and resources involved in the event execution (e.g. nurses or doctors), and possibly additional attributes that depend on the event type (e.g. medication name and dosage for the event "Medication"). The order of events is essential for deriving useful insights about the process. The extraction of events in process mining has typically been done on structured data [2,10], but recent works also look at unstructured text data [13,14].

In the field of *information extraction* (IE), an event is defined as the occurrence of an action or state transition, typically identified by verbs or gerunds [6,22,31,32]. In the nurse and doctor notes, an event is not necessarily described using specific parts of speech but by its contextual significance in the process (e.g. "Heart failure" or "Lasix" in Fig. 1a). The task of event extraction, also known as automatic content extraction [31], involves extracting such events from unstructured text and organizing them into a structured format. This structured representation typically includes several key components: the *event trigger*, which is a fragment of text that mentions the event, the *event name*, also known as the event type or trigger category [23], representing the class of the event, *attributes*, also known as *arguments* [6,23], which provide additional details about the event, and *attribute types*, or argument types, which denote the type of information conveyed by the attributes.

Event extraction has been extensively studied on benchmark datasets consisting of news articles, Wikipedia articles and other non-medical texts [31,32]. Unlike the sentences from the benchmark datasets used in the event extraction domain, the sentences in medical text are often informal, nongrammatical, fragmented, and full of jargon. Compare for example the sentence "The 1960 Concepción earthquakes were a succession of three destructive earthquakes that happened between 21 and 22 May 1960.", which comes from from the MAVEN dataset, popular in the IE community [32], with the sentence "LS with ins / exp wheez, Nebs PRN, and in NC o2 6lit / min." with multiple medical abbreviations ("LS" for lung sounds, "Nebs" for nebulizers, "PRN" for "as needed", and "NC o2" indicating nasal cannula oxygen) and several events mentioned in it.

Transformers are typically the opted approach for event extraction in the IE community. The survey presented in [23] discusses several methods of extracting events. A transformer-based approach [21] that uses BERT [7] and an attention-based RNN approach [33] showed the highest performance in terms of the F1 score on the ACE dataset [31], another popular dataset in IE.

Apart from the spoken-like language, which is frequently used in medical notes as discussed earlier, the medical domain adds additional complexity dimensions. Domain knowledge is crucial for understanding abbreviations, intended meaning and roles for each event. Context is important for correct interpretation of terms used: an abbreviation "LS" could have multiple meanings, including "Left Side", "Life Support" and "Lung Sounds", requiring the use of domain-aware and context-aware techniques. In [29], the authors investigate the extraction of a single type of medical event: adverse drug events from patient-written text. They explore various transformer architectures, including AutoEncoding, AutoRegressive, and Text-to-Text models, trained on annotated text data from patient written notes exhibiting varying levels of language formality. These notes are sourced from social networks such as Twitter (informal language) and AskAPatient (semi-formal language). The study shows that domain-pre-trained models improve the detection of adverse drug events. Furthermore, performance improves significantly when the language formality of the pre-training dataset aligns with that of the evaluation dataset. Although this work is closely related to our objective, it does not involve notes written by medical personnel, and the need for a large training dataset makes it unsuitable for our application.

Public datasets such as MIMIC-III [17] provide abundant medical notes but lack the event-level annotations necessary to develop and validate event extraction techniques. The closest alternative in this domain is the MedMentions dataset [24], which includes annotated sentences from abstracts of biomedical articles. However, these texts are grammatically structured and do not reflect the unique challenges presented by medical notes. This highlights the need for annotated datasets that contain actual medical notes. However, obtaining large-scale annotations for medical notes is very costly due to domain complexity, time required, and experts being overloaded with their regular work duties. In this context, zero-shot approaches, which do not require large amounts of annotated data for training, offer a promising solution. An additional advantage of zero-shot approaches is that the events types of interest can be determined during inference. This capability allows for customizing the extraction scope based on specific business goals for data analysis.

Recently, in [13,14], the authors addressed event extraction from nurse notes using dependency trees. They detect events with timestamps by extracting keywords through dependency rules and then use zero-shot ST-based semantic similarity to match these keywords with predefined event names. The reliance of the method on the presence of timestamps and proper grammar use complicates its applicability to nurse notes. In [9] sentences from nurse notes were compared with synonyms of predefined standard event names, using keyword match counts to assign the most relevant event name to the sentence. Semantic similarity using ST was also investigated. The BioLORD [26], a ST model fine-tuned using medical ontological concepts and descriptions, has recently shown state-of-the-art or near-state-of-the-art performance in the semantic representation of sentences from medical notes. However, the performance of the BioLORD model in the extraction of events from medical notes remains untested.

With the rise of LLMs, several zero-shot approaches using LLMs were used for event extraction. In [16], a three-stage approach using LLM was used to perform event extraction using an auto-chain of thought prompting strategy. In the results presented, this approach appears to outperform the transformer-based approach in the ACE dataset. In [6], it is demonstrated that while LLMs perform poorly in event detection compared to other transformer-based approaches, they serve effectively as data generators to tune transformer-based models. Although both methods are applied across various contexts and fields, their relevance to the issue of extracting events from medical notes remains unclear.

3 Materials and Methods

This study focuses on (1) the creation of an evaluation dataset by annotating a collection of nurse notes from the MIMIC-III dataset and (2) the evaluation and comparison of two approaches for extracting events from nurse notes. The first approach uses an ST for event name detection and a transformer model for attribute detection. The second approach uses an LLM for both event name detection and attribute detection.

3.1 Evaluation Dataset Creation

We limited the scope of the evaluation dataset to nurse notes from the MIMICS-III dataset, which includes 410,000 notes in English. The notes were acquired in a de-identified format and pertained to patients who were admitted to the critical care units at Beth Israel Deaconess Medical Center from 2001 to 2012. We replaced some common uniquely interpretable abbreviations in the notes with their expansions using a manually created lookup list, which contains different ways of writing (e.g. "Pt", "pt" and "PT" expand to "patient"). Then we extracted sentences from the complete set of nurse notes, which resulted in 8.2 million sentences. Subsequently, we calculated the number of words per sentence and used the sentence length in words as an inclusion criterion, taking only sentences between the 25th and 75th percentiles, which are 3 to 9 words long, resulting in 4.5 million sentences. From this, a random sample of 5000 sentences is selected for further annotation.

In consultation with medical experts, we selected five common event names of interest in the healthcare domain, namely "Eating", "Excretion", "Family", "Pain", and "Sleep". Based on the WordNet [4] synonym set, we created a dictionary with synonyms of event names. We selected all sentences that contain one or more synonyms of the event names. Additionally, a BioLORD ST is used to select sentences from the 5000 sentences that are semantically similar to the five predefined event names. This procedure resulted in the extraction of 478 sentences. Of the remaining 4522 sentences, another 478 were randomly selected to eliminate possible selection bias and include false negatives. Thus, a total of 956 sentences were obtained.

Each of the 956 sentences is subsequently annotated by one or two out of 12 volunteers who have a working knowledge of the English language. The volunteers were asked to assign an event name from the predefined list to each sentence mentioning an event, or a label "Other" if the event mentioned in the sentence was not on the list. Annotators were asked to indicate the correctness of a suggested event name by choosing one out of: 'Yes' (accurate), 'No' (inaccurate), and 'Not sure' (uncertain). In addition, the annotators had the option to provide details on any other event names found in the sentence. Most of the sentences were annotated by two annotators and were included in the evaluation dataset only when there was no disagreement between them on the event names. Finally, 395 sentences were selected out of the set of 956 sentences based on the following criteria: the annotators agreed on the label(s) and they did not choose 'Not sure' option when annotating sentences. 226 of the selected sentences were labeled by two annotators. Based on the annotation, 359 sentences (90.8%) describe one event, 35 sentences (8.9%) describe two events, and 1 sentence describes three events, yielding a total of 432 event instances. The distribution of event names across the dataset is as follows: "Eating": 75, "Excretion": 64, "Family": 77, "Pain": 71, "Sleep": 67, "Other": 78. The dataset is well-balanced, with each of the six classes containing between 64 and 78 sentences.

In addition, a data set of negated sentences is created to evaluate the negation detection capabilities of the methods considered in this study. Here, negation refers to linguistic constructs that reverse, nullify, or significantly contradict the meaning of a given proposition. Since manual rephrasing of sentences would take considerable time and effort, a Llama 3 [30] LLM is used to generate six negated sentences for each sentence from the evaluation dataset, three are generated using a prompt requesting the LLM to the sentence event-by-event, and the remaining three are negations of the sentence as a whole. Note that some of the input sentences contained negation. After that, illogical or unnatural sentences are discarded, and the one deemed most logical and non-trivial is selected for the inclusion in the dataset. For example, for the sentence "Sleeps well between cares", "Does not sleep well between cares" was selected. If all negated sentences were eliminated or the input sentence itself was irreparable, none of the negated sentences is selected. Additionally, the presence of negation in the input and the generated sentence is manually annotated. This resulted in 809 non-negated sentences out of which 384 had event names and 931 negated sentences out of which 346 had event names borrowed from the label assigned to the input sentences.

3.2 Transformer-Based Approaches

Traditional transformer models are fine-tuned to a predetermined task. For this reason, different models are to be used for event name detection and for attribute detection, unless the models are fine-tuned in a specific way, allowing to handle both tasks. Given the huge variety of attributes, the performance of such integrated end-to-end models is questionable. Therefore, we adopt two separate transformer-based models, one for event name detection and another one for attribute detection.

Sentence Transformer Approach for Event Name Detection. An ST converts a sentence or text fragment into vector embeddings, ensuring that text with similar semantics is placed closer together, while dissimilar text is placed farther apart within the vector space. To use an ST to detect the name of events in sentences [13,14], the following steps are taken. First, the ST model is used to generate vector embeddings for all predefined event names. Second, for each query sentence, the vector embeddings are obtained using the same model. Third, the similarity Sim_s is measured between the vector embedding of the sentence and the vector embedding of each of the event names. The event name that has

CONTEXT
You are trying to understand the sentences written by a nurse. Your task involves looking for specific event names within the text. Accurate detection of these event names is crucial for a clear understanding of the note.

TASK DESCRIPTION
1. Read the following sentence: "**{sentence}**"
2. Select the appropriate event name from the event name list based on its content. Event name list: **{event_name_list}**.
3. Evaluate the sentence carefully to determine the most appropriate event name that best matches the context and content of the sentence.

RESTRICTIONS
No matching event names - If none of **{event_name_list}** accurately describes the content of the sentence, classify it as "Other".

OUTPUT FORMAT
Format the output as {"event_name": "<event_name>"}, where <event_name> is the determined classification from the provided list or "Other" if no match is found.

Fig. 2. LLM prompt designed to extract event names from medical text, with bolded text serving as substitutes to be replaced with actual values.

the highest similarity is chosen if the similarity is greater than an empirically determined threshold τ. If none of the even names gives a similarity higher than τ, the sentence is assigned the event name "Other" to indicate that it does not have the predefined event names.

BERT-Based Attribute Detection—Negation. Since ST cannot detect negations, a BERT-based transformer model fine-tuned on medical text [3,7] is used to detect negations in sentences. This model checks each token with respect to the surrounding tokens and assigns a negation attribute to the token. If any token is attributed as negated, the whole sentence is considered negated.

3.3 LLM Based Approach

In the LLM based approach, the same model is used for both event name and attribute detection tasks due to its flexibility.

Event Name Detection. LLMs are used to assign event names to sentences using a prompting method following the guide presented in [15]. The template prompt contains four sections, shown in the example in Fig. 2. The Context section outlines the task of understanding sentences written by a nurse and emphasizes accuracy. The Task Description provides a step-by-step guide on how to evaluate each sentence and select the most appropriate event name from a predefined list. To address the event names that did not make into the predefined list, we introduce the "Restrictions" section instructing the classifier to assign "Other" when none of the predefined event names apply. Finally, the Output Format defines a structured response in JSON format, which is simpler to parse and analyze further. It is understood that there are various ways to create such prompts. However, we do not delve into these variations in this paper, as the focus is on the specific implementation described here.

Attribute Detection. In order to detect an attribute, the necessary prompt is added to the instruction and the output format. For detecting negation as an attribute, an additional point is inserted after the second point in the 'TASK DESCRIPTION' section as follows: "If any of the events are present but are negated, please still detect them, but turn on the negation flag". The 'OUTPUT FORMAT' is modified to:
{"event_name": "<event_name>", "is_negated":"<is_negated>"}.

4 Experiment and Results

In this section, we discuss the evaluation of the two approaches and compare the results obtained on the evaluation dataset.

4.1 Event Name Detection

For the task of detecting event names, the event names "Eating", "Excretion", "Family", "Pain", "Sleep" and "Other" are chosen. Both the ST and LLM models are instantiated with these predefined event names, and the sentences from the evaluation dataset are input to the two approaches.

The **transformer-based approach** is implemented using the BioLORD model employing the cosine similarity measuring the similarity between sentences and corresponding event names based on their vector representations. Table 1 shows the weighted average performance metrics for various choices of the threshold τ. As expected, precision increases with an increase of τ, since the model places greater emphasis on the similarity of the sentence-event name pairs. However, at higher τ values, many sentences are assigned the event name "Other", even when there were suitable matches in the set of event names. This results in a decrease in recall when $\tau > 0.2$. This trend is also observed in the F1 score, which increases with τ up to $\tau = 0.2$ and decreases when τ increases further. Therefore, $\tau = 0.2$ is chosen for the rest of the experiment.

Figure 3a shows a confusion matrix constructed to assess the performance of a classifier that assigns a single event label to each sentence, with annotations potentially including several labels per sentence. Each cell in the matrix contains the count of instances for which the predicted label was part of the set of annotation labels for a given sentence. The classes "Eating", "Family", and "Excretion" exhibit strong performance. However, "Pain" and "Other" show a high rate of

Table 1. The performance of ST-based approach against various threshold τ.

τ	Precision	Recall	F1 Score
0.1	0.82	0.82	0.79
0.2	**0.84**	**0.84**	**0.83**
0.3	0.86	0.81	0.82
0.4	0.86	0.72	0.74
0.5	0.86	0.59	0.59

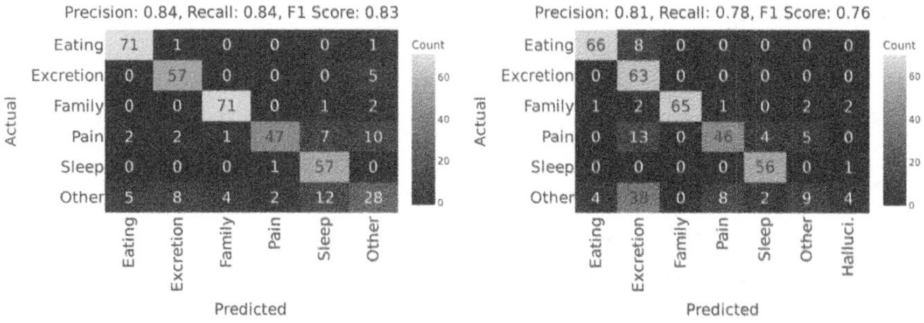

(a) ST (b) LLM

Fig. 3. Confusion matrices for the ST and LLM-based approaches.

false negatives, while the "Sleep" and "Other" classes are prone to a high number of false positives.

Negation turned out to be associated with false negatives. Examples of sentences annotated with the event "Pain" that were not classified as such include "NO COMPLAINTS OF PAIN OVERNIGHT", "No c/o pain.", "NO ABD PAIN." and "no pain meds required."

The **LLM-based approach** is prompted using the prompt templates shown in Fig. 2 for a sentence and five predefined event names on an 8 billion parameter Llama 3 LLM. All generations resulted in accurate and parsable JSON outputs of the format: $``event\_name'' : `` < event\_name >''$. Although most sentences resulted in either the predefined event names or "Other", there were instances in which the model generated hallucinated outputs. The hallucinated event names are "Birth", "Extraction", "Therapy", "Watching tv", and "Wean". On manual check, they al are appropriate event names for those sentences while not coming from the set of the predefined event names. Although hallucinations as such are generally considered a drawback of LLMs, our results show that they could be used as a feature in this context. After a validation by domain experts, hallucinated event names can be added to the predefined set iteratively, capturing more relevant events over time. Note that hallucinated event names are easy to detect in the LLM output using simple filtering.

The confusion matrix for the LLM approach, shown in Fig. 3b, was constructed similarly to the one for the ST approach. The hallucinated event names were replaced with "Hallucination" and are considered as "Other" for computing the performance metrics. The results show that the events "Eating", "Family" and "Sleep" were detected satisfactorily, while "Excretion" had a high rate of false positives, "Pain" had a high rate of false negatives and "Other" had a very high rate of false negatives. On further examination, some sentences labeled by the LLM with "Excretion" contained words like "tube" and "discharge", which can appear in the context of Excretion. Example of wrongly classified sentences include "Will remove feeding tube.", "Infant discharged to home."

The annotation of some sentences turned out to be incomplete. For example, the sentence "Feeding well and passing stool" was annotated with "Eating", and the sentence "Has feeling of urgency but denies pain on urination" was annotated with "Pain", missing out the label "Excretion", which was in this case correctly assigned by the LLM. Hallucinations produced by the model can be used as ideas for extending the set of the labels chosen for the classification task. For example, the LLM labeled the sentence "MALE BORN TO A 32-YEAR-OLD MOM, G4P2 NOW" with "Birth". All seven hallucinations in the evaluation experiment appear to be applicable to the sentences. Figure 3b shows the confusion matrix for the LLM approach after replacing hallucinated event names with "Others". The results show that the events "Eating", "Family", "Sleep" and "Pain" were detected satisfactorily, while "Excretion" had a high number of false positives.

General comparison of the two approaches is conducted by averaging performance metrics, with each class weighted according to its support. The weighted average precision, recall, and F1 score of the ST approach is observed to be higher than that of the LLM-based approach. The main cause is the better performance of the ST approach on sentences involving medical terminology.

Negation invariance test is performed to learn how robust the approaches are in detecting event names in the presence of negations in medical sentences. Figure 4 shows the confusion matrix and weighted average performance metrics for the two approaches using two groups of the negation dataset: 1) sentences that do not contain negation and 2) sentences that contain negation It is observed that, for the ST, there is a significant drop in recall in the presence of negation. However, the LLM seems to have not such a significant drop. In the presence of negation, the recall of the LLM approach is higher than the recall of the ST approach.

Unfamiliar concepts are often a challenge in zero-shot approaches. These are concepts that were not present in the pretraining data. To evaluate their impact on model performance, both zero-shot techniques were tasked with predicting the event name in sentences containing Llama 3-generated "painkiller names", which were manually verified to be nonexistent. These names, denoted as i, were formatted as: "The patient consumed i for good relief". The ST was accurate in 7 out of 10 cases, while the LLM produced only 2 out of 10 accurate classifications.

4.2 Attribute Detection

The **Negation** attribute is a binary attribute that indicates if a sentence has negations. The confusion matrix and binary performance metrics for the transformer and LLM approaches are shown in Table 2. They show that the LLM-based approach outperforms the transformer-based approach in terms of all metrics in the task of detecting negations in sentences. The biggest drawback of the transformer approach is the high rate of false negatives, which leads to a significantly lower recall and F1 score than the ones of the LLM.

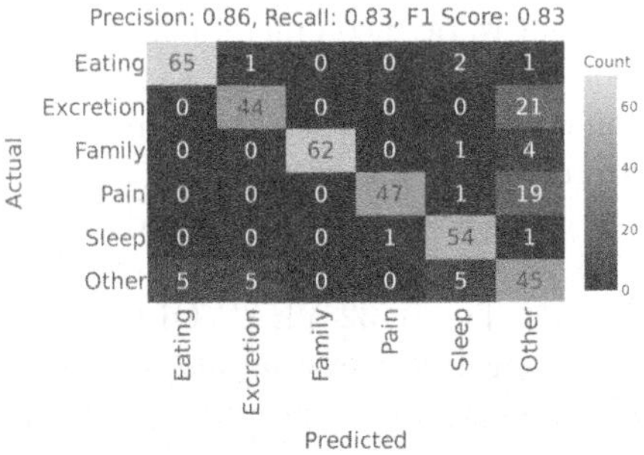

(a) Sentences w/o negation - ST

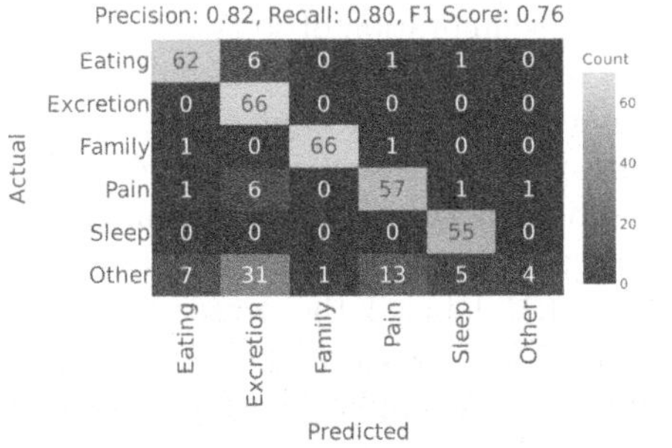

(b) Sentences w/o negation - LLM

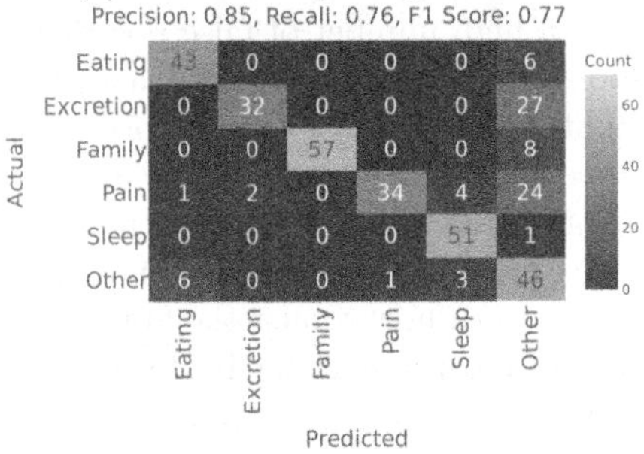

(c) Sentences w/ negation - ST

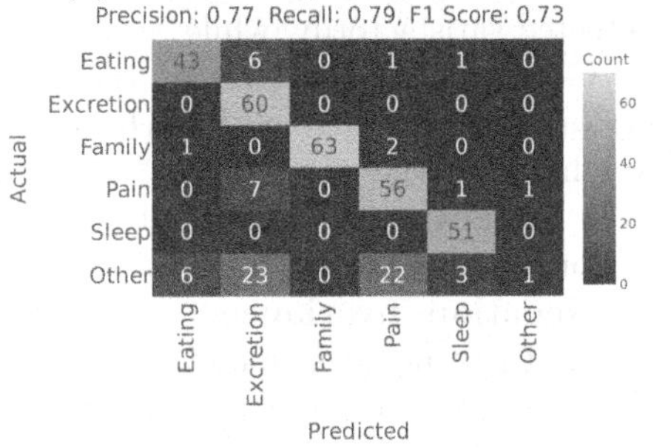

(d) Sentences w/ negation - LLM

Fig. 4. Confusion matrices and performance metrics comparing the ST-based approach with the LLM-based approach on the negation-evaluation dataset.

Event-related attributes are specific to event names. For the ST-approach, we used a separate model for each attribute to extract from sentences, while for the LLM-approach, the prompts were extended. Here, we evaluate the ability of the LLM to subclassify the sentences that belong to the event name "Pain" into subclasses following the SNOMED ontology [12] and identify an additional attribute: a body part. The SNOMED concept "pain management" (SNOMED identifier: 278414003) is chosen for this task. This concept has six children, namely, identification of cultural and value components related to pain management, patient-controlled analgesia management, pain and symptom management, pain control, pain rehabilitation, and pain relief. The prompt in Sect. 3.3 is modified to incorporate an additional instruction to detect body parts. An

Table 2. Confusion matrix and metrics for negation detection.

		Predicted	
		True	**False**
Act.	**True**	184	**162**
	False	25	359

(a) ST

		Predicted	
		True	**False**
Act.	**True**	320	26
	False	53	331

(b) LLM

Metric	**ST**	**LLM**
Precision	0.88	0.86
Recall	0.53	0.92
F1 Score	0.66	0.89

(c) Performance Metrics

evaluation dataset included sentences that have been labeled with "Pain" in the event name detection experiment. The sentences are manually labeled to the six subtypes, resulting in a dataset with 63 sentences.

In the evaluation, the LLM approach correctly subclassified 33 out of 63 sentences. Most of the misclassifications were due to missing contextual information. In practice, contextual information could be obtained using the whole nurse note instead of one sentence, or events extracted from other sentences. Furthermore, the LLM approach accurately identified the presence or absence of a body part in 91 out of 96 sentences. This demonstrates the convenience of the language model in attribute detection.

5 Discussion

The use of NLP techniques in extracting event logs from unstructured medical data, such as nurse notes, holds promise for improving process analytics in healthcare. However, several challenges and considerations need to be addressed to fully realize this potential. In this section, we discuss some key observations, alternative strategies, and limitations of our approach.

Evaluating Event Name Detection in the Medical Domain. The interpretation of intentions and observations in medical text can be highly temporal and subjective. Hence, labeling medical text with event names and attributes should ideally be performed by the same medical personnel who authored it. However, in this study, we use a de-identified public dataset of medical notes, making it impossible to involve the original authors in the annotation process. To address this limitation, we selected event names that are common and comprehensible even to nonmedical experts, while the dataset contains a significant amount of medical terminology and jargon.

We ensured that annotators had a good working knowledge of English, established inter-annotator agreement, and allowed them to reject incomprehensible or unsure sentences to minimize mislabeled event names from interpretation errors. Despite these precautions, the number of event names in our evaluation dataset remains significantly smaller than the diversity of events present in the medical field. Expanding research to larger and more representative datasets is essential for further progress. To create such datasets, we recommend collecting event names at the source. For instance, medical personnel could be provided with machine-generated event name suggestions (as demonstrated in this paper) immediately after entering a note. They could then assign or revise these event names based on their judgment. The resulting annotated data would be highly valuable for evaluating and fine-tuning event name detection models.

Event Log Extraction. This paper focuses on the extraction of event names and their associated attributes. Another important component of event log extraction that needs to be addressed is the chronological ordering of events. This task is particularly challenging because time is not always explicitly mentioned for individual events recorded in the notes. As a result, the ordering of events must be inferred through *temporal relationship extraction* [25], a distinct problem in IE. This task leverages temporal indicators such as "after", "before", as well as other temporal markers and additional contextual information, such as small process models and structured EHR entries with explicit timestamps. This complex and nontrivial task presents an interesting direction for future research.

Mentions of Event Combinations in a Sentence. Our current approach assumes that each sentence only mentions one event and that the representation of event names is flat. However, a sentence might describe multiple events. Mentions of separate events in one sentence, such as "Sleep" and "Eating" in the sentence "The patient slept a bit after lunch." can be addressed with rather straightforward extensions of our approach.

Sentences like "The patient experienced pain while eating, especially swallowing." present a more serious challenge. This sentence can be associated with the events "Pain" and "Eating" (both of which are considered in our study), but also with a more specific event "Odynophagia" (pain while swallowing). In such cases, assigning "Pain" and "Eating" as two independent events would lead to the loss of valuable information, potentially affecting the accuracy and completeness of the results. To ensure construct validity, ontology-like graph representations of event names can be employed, defining the connections between hyponyms and hypernyms.

Time Consideration of LLMs. A serious limitation of LLM-based methods is the computational cost and time required for inference. In our experiments, LLMs required an average of 4 s per sentence, which led to a total processing time of 8 h for a typical daily load for our partner hospital, consisting of 7,000 sentences. In contrast, ST-based methods were much faster, taking only 65 milliseconds per sentence, or 8 min in total for the same number of sentences. The substantial time difference makes such a straightforward use of LLMs impractical for large-scale real-time healthcare applications. This limitation could potentially be mitigated by optimizing the LLM inference process or incorporating faster, more efficient models such as distillation-based models. However, the trade-off between accuracy and speed remains a crucial consideration when choosing between LLMs and traditional methods.

6 Conclusion

In this paper, we explored the use of sentence transformer (ST)-based and large language model (LLM)-based approaches for extracting event names and attributes from medical notes. An evaluation dataset was created specifically for this purpose with volunteers who annotated MIMIC-III medical notes. We studied the performance of these two zero-shot approaches in the presence of negation and unfamiliar concepts in sentences.

Two publicly available models, BioLORD and Llama 3, were selected to represent ST and LLM, respectively. An embedding cosine similarity-based approach is used with the ST and a prompting-based approach is used with the LLM. Our results demonstrate that the ST approach over-performs the LLM-approach, particularly in detecting event names, but the ability of LLMs to perform well on extracting event names from sentences with negations and the negation attribute itself shows their potential as a supplement to traditional approaches. Additionally, leveraging LLM hallucinations to introduce new event names could further expand their utility in event log extraction. The computational cost and speed limitations of LLMs remain a barrier to their deployment in real-time applications. More research is needed to address these limitations and fully utilize their capabilities.

Finally, in this study, we limited the number of event types to five due to the tedious and time-consuming nature of annotating sentences with event names. This limitation poses a threat to the external validity of the results. To address these practical challenges in conducting case studies, it is important to facilitate dataset annotation at the source. Additionally, exploring graphical event representation and classification, subsentence analysis, and estimation of the event order using available structured data and process models can further improve accuracy and completeness. Based on our research, methods that combine ST and LLM approaches can open new perspectives on the application of data mining and process mining to medical texts.

Acknowledgement. This research was funded by the NWO project TACTICS (628.011.004) and Lunet in the Netherlands.

References

1. van der Aalst, W.: Business process management: a comprehensive survey. ISRN Softw. Eng. **2013**(1), 1–37 (2013)
2. van der Aalst, W.: Process Mining - Data Science in Action, 2nd edn. Springer, New York (2016)

3. van Aken, B., Trajanovska, I., Siu, A., Mayrdorfer, M., Budde, K., Loeser, A.: Assertion detection in clinical notes: Medical language models to the rescue? In: Proceedings of the Second Workshop on Natural Language Processing for Medical Conversations. Association for Computational Linguistics, Online (2021)
4. Bond, F., Foster, R.: Linking and extending an open multilingual wordnet. In: Proceedings of the 51st Annual Meeting of the Association for Computational Linguistics (Volume 1: Long Papers), pp. 1352–1362 (2013)
5. Bundy, H., et al.: Can the administrative loads of physicians be alleviated by AI-facilitated clinical documentation? J. Gen. Intern. Med. 1–6 (2024)
6. Chen, R., Qin, C., Jiang, W., Choi, D.: Is a large language model a good annotator for event extraction? In: Proceedings of the AAAI Conference on Artificial Intelligence, vol. 38, no. 16, pp. 17772–17780 (2024)
7. Devlin, J., Chang, M.W., Lee, K., Toutanova, K.: BERT: pre-training of deep bidirectional transformers for language understanding. In: Burstein, J., Doran, C., Solorio, T. (eds.) Proceedings of the 2019 Conference of the North American Chapter of the Association for Computational Linguistics: Human Language Technologies, pp. 4171–4186. Association for Computational Linguistics, Minneapolis, Minnesota (2019)
8. Diba, K., Batoulis, K., Weidlich, M., Weske, M.: Extraction, correlation, and abstraction of event data for process mining. Wiley Interdisc. Rev. Data Min. Knowl. Discovery **10**(3), e1346 (2020)
9. van Dijk, T.: Using Natural Language Processing for Event Classification in Dutch Medical Texts. Master's thesis, Eindhoven University of Technology (2024)
10. Dumas, M., Rosa, L.M., Mendling, J., Reijers, A.H.: Fundamentals of Business Process Management. Springer, Heidelberg (2018)
11. Elbattah, M., Arnaud, É., Gignon, M., Dequen, G.: The role of text analytics in healthcare: a review of recent developments and applications. Healthinf, 825–832 (2021)
12. Gaudet-Blavignac, C., Foufi, V., Bjelogrlic, M., Lovis, C.: Use of the systematized nomenclature of medicine clinical terms (SNOMED CT) for processing free text in health care: systematic scoping review. J. Med. Internet Res. **23(1)** (2021)
13. Geeganage, D.T.K., Wynn, M.T., ter Hofstede, A.H.: Text2el: exploiting unstructured text for event log enrichment. In: 2022 16th International Conference on Signal-Image Technology & Internet-Based Systems (SITIS), pp. 1–8. IEEE (2022)
14. Geeganage, D., Wynn, M.T., ter Hofstede, A.H.: Text2el+: Expert guided event log enrichment using unstructured text. ACM J. Data Inf. Qual. **16**(1), 1–28 (2024)
15. Giray, L.: Prompt engineering with chatgpt: a guide for academic writers. Ann. Biomed. Eng. **51**(12), 2629–2633 (2023)
16. Huang, F., et al.: A three-stage framework for event-event relation extraction with large language model. In: International Conference on Neural Information Processing, pp. 434–446 (2023)
17. Johnson, A., Pollard, T., Mark, R.: MIMIC-III clinical database (version 1.4). PhysioNet **10**(C2XW26), 2 (2016)
18. Koenecke, A., Choi, A.S.G., Mei, K.X., Schellmann, H., Sloane, M.: Careless whisper: speech-to-text hallucination harms. In: Proceedings of the 2024 ACM Conference on Fairness, Accountability, and Transparency, pp. 1672–1681. Association for Computing Machinery, New York (2024)
19. Korach, Z.T., et al.: Mining clinical phrases from nursing notes to discover risk factors of patient deterioration. Int. J. Med. Inform. **135**, 104053 (2020)

20. Kurniati, A.P., Hall, G., Hogg, D., Johnson, O.: Process mining in oncology using the MIMIC-III dataset. In: International Conference on Data and Information Science (ICoDIS) 2017, vol. 971, p. 012008. IOP Publishing (2018)
21. Li, F., et al.: Event extraction as multi-turn question answering. In: Findings of the Association for Computational Linguistics: EMNLP 2020, pp. 829–838 (2020)
22. Li, I., et al.: Neural natural language processing for unstructured data in electronic health records: a review. Comput. Sci. Rev. **46**, 100511 (2022)
23. Li, Q., Li, J., Sheng, J., Cui, S., Wu, J., Hei, Y., Peng, H., Guo, S., Wang, L., Beheshti, A., et al.: A survey on deep learning event extraction: approaches and applications. IEEE Trans. Neural Netw. Learn. Syst. **35**, 6301–6321 (2021)
24. Mohan, S., Li, D.: Medmentions: a large biomedical corpus annotated with UMLS concepts. In: Automated Knowledge Base Construction (AKBC) (2019)
25. Ning, Q., Feng, Z., Roth, D.: A structured learning approach to temporal relation extraction. In: Palmer, M., Hwa, R., Riedel, S. (eds.) Proceedings of the 2017 Conference on Empirical Methods in Natural Language Processing, pp. 1027–1037. Association for Computational Linguistics, Copenhagen (2017)
26. Remy, F., Demuynck, K., Demeester, T.: BioLORD-2023: semantic textual representations fusing large language models and clinical knowledge graph insights. J. Am. Med. Inform. Assoc. **31**(9), 1844–1855 (2024)
27. Remy, S., Pufahl, L., Sachs, J.P., Böttinger, E., Weske, M.: Event log generation in a health system: a case study. In: Fahland, D., Ghidini, C., Becker, J., Dumas, M. (eds.) BPM 2020. LNCS, vol. 12168, pp. 505–522. Springer, Cham (2020). https://doi.org/10.1007/978-3-030-58666-9_29
28. Rokach, L., Romano, R., Maimon, O.: Negation recognition in medical narrative reports. Inf. Retrieval **11**, 499–538 (2008)
29. Scaboro, S., Portelli, B., Chersoni, E., Santus, E., Serra, G.: Extensive evaluation of transformer-based architectures for adverse drug events extraction. Knowl.-Based Syst. **275**, 110675 (2023)
30. Touvron, H., et al.: Llama 2: open foundation and fine-tuned chat models. arXiv preprint arXiv:2307.09288 (2023)
31. Walker, C., Strassel, S., Medero, J., Maeda, K.: ACE 2005 multilingual training corpus (2006)
32. Wang, X., et al.: MAVEN: a massive general domain event detection dataset. In: Webber, B., Cohn, T., He, Y., Liu, Y. (eds.) Proceedings of the 2020 Conference on Empirical Methods in Natural Language Processing (EMNLP), pp. 1652–1671. Association for Computational Linguistics (2020)
33. Zhao, Y., Jin, X., Wang, Y., Cheng, X.: Document embedding enhanced event detection with hierarchical and supervised attention. In: Proceedings of the 56th Annual Meeting of the Association for Computational Linguistics, pp. 414–419 (2018)

Review of Design of Business Process Simulation Models

Samira Khraiwesh(✉) and Luise Pufahl

CIT, Technical University of Munich, Bildungscampus 2, 74076 Heilbronn, Germany
{samira.khraiwesh,luise.pufahl}@tum.de

Abstract. Business Process Simulation (BPS) serves as a critical tool for organizations to evaluate and optimize business process designs without incurring real-world risks or disruptions. It enables proactive assessment of the potential impacts of changes prior to implementation. In recent years, significant research has been conducted to enhance BPS across multiple perspectives; however, a comprehensive overview of the latest advancements and contributions in this domain remains limited. This systematic literature review examines 32 papers published between 2009 and 2025, focusing on advancements in building BPS models. It provides an overview of current research, covering key topics such as the main contributions of recent studies, the role of humans in BPS development, investigates how recent approaches support scenario generation, and provide key research directions and opportunities for advancing BPS model development. The findings indicate that recent research focuses on the design of BPS models from various perspectives, with automated techniques receiving the most attention, while limited efforts have been directed toward supporting scenario generation. These findings contribute valuable insights into existing methodologies, identify key research gaps, and provide guidance for advancing BPS model development.

Keywords: Business process simulation · Business process management · Literature review

1 Introduction

Organizations must continually improve and redesign their business processes to remain competitive in rapidly evolving environments [25]. As markets shift and new technologies emerge, businesses are required to stay agile, ensuring that their processes align with current demands [2]. *Business Process Simulation (BPS)* plays a crucial role in enabling organizations to model, analyze, and implement process changes with minimal risk [40]. BPS model consists of interconnected components that dynamically simulate task execution [23]. Scenario generation is the primary objective of BPS, as organizations seek to understand the consequences of applying changes to their processes by creating new designs, known

J. Grabis et al. (Eds.): RCIS 2025, LNBIP 547, pp. 452–469, 2025.
https://doi.org/10.1007/978-3-031-92474-3_27

as 'what-if' scenarios, and testing them through simulation prior to implementation. These scenarios are built starting from the system's current state called *as-is* model by adjusting the simulation model [1].

Despite the known benefits of BPS, it is still not widely applied in many organizations. The effectiveness of simulation results depends on the accuracy of the simulation model [21], which must closely reflect the actual dynamics of the processes. A key challenge for organizations is the lack of confidence in the reliability of BPS models. Traditionally, BPS models are manually discovered using expert input and techniques like interviews and observation [25], which is time-consuming [25] and may overlook uncommon scenarios [34]. Recently, significant research progress has been made in constructing BPS models from event logs and automated algorithms [8,12,14,26], supported by the increased availability of event logs [18]. This shift emphasizes data-driven methods within BPS, resulting in more accurate and easier-to-develop models. However, data quality issues in event logs still impact model reliability. Exploring how human expertise helps mitigate these issues is crucial.

To gain a deeper understanding of the current state of building BPS models, both individual studies and systematic reviews are essential. As noted by [33], the absence of a comprehensive literature review on BPS creates gaps in understanding the interrelation and progression of existing approaches. Rosenthal et al. [33] focus in their literature review on BPS on papers published between 1999 and 2016. While this review provided valuable insights into application purposes, simulation objectives, challenges in applying BPS, and the role of modeling languages, the field has advanced significantly since then. The increasing availability of event logs [18] has driven advancements in data-driven techniques and AI methodologies, such that also an increase in publications in the last 5 years can be observed[1]Further, these data-driven and AI techniques have innovated BPS model building where automation is highly involved. A certain increase in automation also raises the question of how human involvement should be designed to allow a particular trust in the simulation output. While previous studies on BPS [33] focused on BPS objectives and obstacles in applying BPS in general, this study addresses key aspects of innovating how a BPS model is built: (1) understanding how, also with recent technology, BPS model building is enhanced, (2) investigating the role of human involvement in the technology-driven approaches, and (3) analyzing how recent studies support scenario generation, one key aspect necessary to test different what-if scenarios. Additionally, our study identifies new opportunities for future research. By this, it aims to guide both researchers in identifying future directions and practitioners in applying these insights effectively.

To achieve the key aspects outlined earlier, this study addresses the following research questions.

- RQ1: What is the main focus of papers contributing to building BPS models?

[1] Based on published documents by year matching the search term "business process simulation" using ScienceDirect in 03/2025, we observed 23 between 2009–2012, 25 between 2013–2016, 25 between 2017–2020, and 52 between 2021–2025.

  - RQ1.1: Which BPS model components are emphasized in research?
  - RQ1.2: What are the recent advancements in methods for building BPS models?
- RQ2: What is the role of human involvement in developing BPS models?
- RQ3: How is scenario generation supported in the context of BPS?
- RQ4: What are the key research directions and opportunities for advancing BPS model development and addressing current gaps in the field?

The rest of the paper is structured as follows: Sect. 2 provides the background, and Sect. 3 describes the methodology. Section 4 presents our findings, followed by a discussion in Sect. 5, while Sect. 6 concludes the paper.

2 Background on Business Process Simulation Model Construction

Simulation is a descriptive method that allows decision-makers to assess the behavior of a model under various conditions [32]. When applied to business processes, It creates a virtual environment that simulates real-life business processes, allowing analysis of changes without real-world costs or effects. The BPS model is the core of the simulation study [1]. Consequently, the accuracy in building the BPS model significantly influences the results obtained from the simulation. For conducting a simulation study, literature has identified several phases that are typically executed iteratively: Following [38], and illustrated in Fig. 1, they are 1) modeling, 2) realizing, 3) verifying and validating and 4) experimenting and interpreting. In the following, we provide a brief overview of the separate phases.

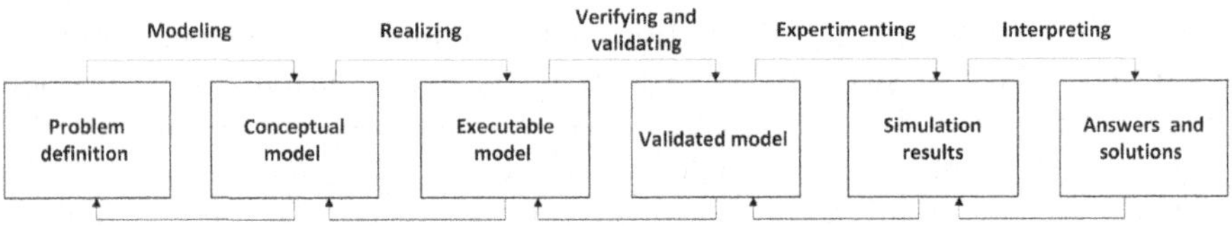

Fig. 1. Phases of a traditional simulation study [38]

Modeling Phase: The start point of this phase begins by identifying the objectives and scope of the simulation study. BPS can be used to compare the impact of two scenarios by answering *what-if* questions like: How would the cycle time change with a 5% increase in daily new cases? What if two essential resources became unavailable for an extended period? Defining the scope and objectives of the simulation study ensures that the necessary components are included. Once the scope and the objectives are defined, the next step in the modeling phase is to create a conceptual model.

A well-crafted conceptual model captures the critical aspects of a business process while avoiding unnecessary complexity, ensuring that it accurately represents essential elements. in this study, we aim to identify and analyze approaches that leverage automation, human involvement strategies, or other advanced techniques to enhance the building BPS model beyond traditional manual BPS modeling.

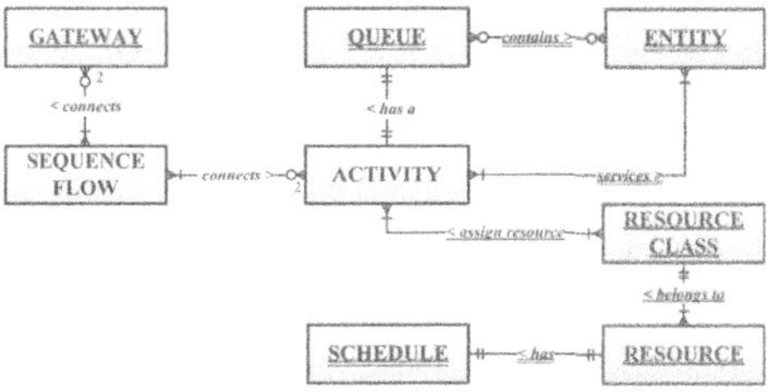

Fig. 2. General structure of a BPS model as defined by Martin et al. [24]

Many studies do not clearly explain the detailed internal structure of a BPS model [24]. The authors in [23] address this gap by presenting a general structure for BPS, shown in Fig 2. Martin et al. [23] highlights how process mining can support the construction of business process simulations. Process mining is a favored family of techniques to develop as-is BPS model because it can discover the previous described BPS components, such as the distribution of activity duration based on process data extracted from event logs in IT systems. *Event logs* are digital records generated by modern information systems, which systematically capture low-level details about the execution of various business processes within a company [16]. These logs document the sequence of activities, providing essential insights into how processes are carried out in real-time. Next to process mining, other techniques, such as machine learning are also used to improve the accuracy of the BPS model. Since we want to analyze techniques that are applied for discovering BPS models, it seems prudent to broaden our focus and take techniques beyond process mining into account. In terms of techniques, however, [13] observe that especially the human resources are reflected in a robotic kind of fashion. Tiredness effects, multi-tasking, preferences on working on certain tasks at specific times during the day are not taken into account. Thus, studies have made the effort to enhance the behaviour of BPS components.

Upon completing this phase, a conceptual model is created that contains classes of objects and the relationships between them, reflecting reality as accurately as possible [1].

Realization Phase: Transforming a conceptual model, constructed in the previous phase, into an executable model involves refining the theoretical framework of the system into a form that can be simulated and tested in practice. The choice of the simulation method, whether *Discrete Event Simulation (DES)*, *Agent-Based Simulation (ABS)*, or *System Dynamics (SD)*, depends on the nature of the business process being modeled [19]. DES is the traditional simulation method used to build a BPS model in the BPM domain. Also, DES is one of the most widely used approaches in operational research [37], a technique which focuses on simulating the events that occur over time within a business process.

A key advantage of DES is its straightforward modeling process, particularly when the organizational problem is well-defined. However, DES is not ideal for modeling human behavior, as its focus is primarily on analyzing processes within a system rather than the dynamics of human actions [36]. On the other hand, ABS uses agents to represent scenarios occurring within the system, with the behavior of the system understood by defining agent states and describing how they respond to actions within specific states. As to the third simulation method, SD, is best used to study business process behaviors over time, particularly for long-term stability analysis, feedback control structure evaluation, and optimizing performance in systems like supply chain management [3]. The choice between DES, ABS, and SD depends on the level of detail and the type of interactions within the system and the goal of the analysis.

Verifying and Validating Phase: In this phase, the executable model undergoes testing to identify and correct any errors [1]. Verification typically involves running small trial simulations step by step to ensure the model functions as expected. During validation, the simulation output is compared to real-world data, such as comparing simulated event logs with actual event logs. This phase may also involve human input to further validate the model's accuracy and ensure it aligns with real-world processes. Verification and validation are essential to ensure the model accurately represents the system.

Experimenting and Interpreting Phases: In these phases, experiments are conducted to ensure the reliability of the simulation results. These results are then analyzed to determine potential next steps. Feedback gathered during this phase can help improve the BPS model to better align with reality or address questions raised during the problem definition phase. This may open up new questions or lead to the proposal of new scenarios for comparison. It could also result in recommendations for improvements in the business process.

Scenario Generation: Analyzing the current state of a business process (*as-is* model), provides valuable insights for business owners and decision-makers, especially from a performance perspective. These insights are highly valuable, as they allow organizations to assess their processes and identify areas for improvement [31]. Based on the as-is model, new scenarios can be generated that reflect process improvements or changes of the process environment that are expected in the future, such as an increase in customers. Scenario generation can be achieved by modifying simulation parameters, such as resources and the number of incoming cases, or by adjusting the order of certain activities.

3 Research Methodology

To address the research questions mentioned above, we conducted a systematic review of the literature (SLR) using the eight-step method described in Okoli's methodology [27]: The main *purpose* of this SLR is to review the current state of development of the BPS model within the Business Process Management (BPM) and process mining domain, centering on four aspects: (1) recent focus and contributions in building BPS models, (2) the human role in BPS model development, (3) support for generating new scenarios, and (4) new research directions to advance BPS models and address gaps.

Table 1. Database, filters used, and the resulting number of studies in the primary search done in December 2024.

Database	Filters used	#
ACM DL	search term added to paper abstract filter	82
Science Direct	search term added to paper abstract filter, subject area = CS, publication title = information systems	22
Springer Link	search term added to paper abstract filter, Language: English, Subject: "Model building and simulation", Content type: Only 'Chapter' is not chosen.	287
IEEEExplore	search term added to paper abstract filter	20
Web of Science	search term added to paper abstract filter + web of science categories "Computer Science Information systems" and year starting from 2009	20
Total		**419**

For the *protocol and training* step, we identified recorded articles and reasons for inclusion or exclusion. The search was conducted using five databases: *Web of Science, ACM Digital Library, IEEEXplore, ScienceDirect, and SpringerLink.* For the search, we used *"business process"* AND *"simulation model"* to identify papers related to building the BPS model. Additional keywords like *"method"*, *"technique"* and *"approach"* were included to narrow our search in this way, we focus on papers that specifically address the development of BPS models by introducing new method or approach or using new technology. The keyword *"what-if"* is also included, enabling us to explore how recent approaches support generating new scenarios. The resulting search string: *"business process"* AND *"simulation model"* AND (*"method"* OR *"approach"* OR *"technique"* OR *"what-if"*), was applied across the five databases to the title and abstract, with filters shown in Table 1.

For the *practical screening*, we defined a set of inclusion and exclusion criteria as given in Table 2 which also provides the number of excluded papers in title-abstract screening. We aimed for studies that provide a contribution to the BPM and process mining field to enhance the BPS model or improve its building (IN1), such that we excluded the one mentioned under EX2. We focused on papers published from 2009 (EX1) onwards due to the increasing availability of event logs [18] and the pioneering work of Rozinat et al. [34], which is considered one of the earliest efforts to discover BPS models with process mining. We excluded studies providing domain-specific solutions (EX3) because focusing on solutions for any domain allows for consistent comparisons of methodologies and trends, emphasizing methodological advancements rather than context-specific adaptations. For the same reason, we excluded case studies and additionally reviews and surveys (EX4).

Applying the inclusion and exclusion criteria to the initial search by reading the titles and abstracts resulted in 35 papers. These results were further refined by *full-text screening* to ensure accuracy and relevance. It helps verify the content, assess quality, and eliminate studies that appear relevant initially but lack sufficient detail or alignment with the research focus. This step led to the exclusion of 15 papers after reading the full texts.

Table 2. Inclusion and exclusion criteria and excluded studies per criteria in the title-abstract screening

	Criteria	#
	Initial number of papers w/o duplicates	410
EX1	The study was published before 2009 and is not a conference or journal contribution	−120
EX2	The study is not related to BPS, has no contribution to building BPS models, or uses simulation merely as a tool are excluded	−196
EX3	The study is domain-specific as it narrows the simulation to a particular field or customizes the model to fit a specific context, limiting the findings' generalizability	−46
EX4	The study is a case study applying simulation in an application context or is a literature review or survey	−13
IN1	The study provides a contribution to enhancing or building BPS models within the BPM and process mining field	**35**

For the *quality appraisal*, discussions between the authors were conducted. The screening step was performed by the first author and verified through discussions with the second author. Additionally, a second author independently took random samples and double-checked the classification to ensure the accuracy and reliability of the results. Following these steps, we also conducted backward and forward searches. As a result, out of a total of 419 papers initially identified, 32 were ultimately included in our study (Fig. 3).

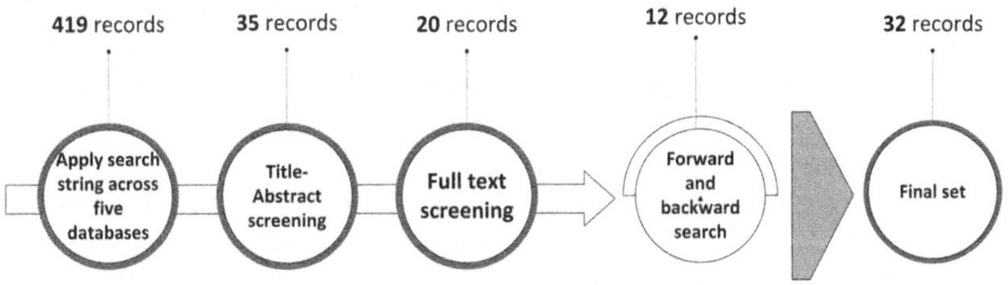

Fig. 3. Search process for relevant studies on building BPS models

For *data extraction*, we performed an inductive approach. Our inductive analysis began by extracting insights from each paper to address the reviews' purpose, allowing us to identify common patterns and themes. From there, we developed categories that naturally emerged from the content of the papers, ensuring a clear and accurate understanding of the research focus. To ensure reliability, the first author extracted data addressing each research question. Discussions with the second author led to the development and validation of inductive categories, reducing bias and ensuring consistency. To identify the main contributions of the studies related to the modeling of BPS models, we analyzed studies that provide an **enhancement** of existing BPS model components (see Martin et al. [23]), develop **new components**, or propose a **new BPS model type**, which we summarize as the **contributions regarding the BPS model**. We analyzed these works further to determine which components have received the most attention in recent research. Secondly, we have identified contributions regarding the **automation technique** of BPS model discovery in which we distinguish between **process mining**, combining **process mining and machine learning** and using **model transformation** techniques to build BPS model.

To explore the role of human involvement in constructing BPS models, we identified four distinct roles through inductive analysis. These categories provide a structured framework to understand how humans contribute at various stages of the modeling process. These include scenarios of **fully automated, pre-knowledge input, in-process adjustments, and post-simulation validation**. Fully automated refers to processes that predominantly rely on automated algorithms to operate, requiring minimal or no human interaction, pre-knowledge input involves human contributions before the process begins, such as providing domain knowledge or contextual data, and in-process adjustment captures human engagement during the process, where input is provided to refine or modify parameters and finally post-simulation validation reflects human involvement after the process, ensuring the validation.

Regarding generation of BPS scenarios, we classified the papers into three categories: **No Support For Scenario Generation** (papers that do not support scenario generation), **Potential for Scenario Generation** (papers that emphasize as-is model development and note the potential for scenario generation without elaborating on implementation), and **Detailed Scenario Generation** (papers that provide step-by-step guidance on generating new scenarios using their approaches).

4 Findings

In the following, we present our findings, shown in Fig. 4 and Fig. 5, based on the data extraction process described above. A full version of the concept matrix is available online[2]. The following subsections present the findings according to three main aspects defined previously.

[2] https://figshare.com/s/781346acac863d5f8fa1.

4.1 Focus and Contribution of Identified Studies

In this section, we focus on the key contributions of the papers regarding the design and modeling of BPS models. Most studies (21 in total) contribute to the design of a BPS model. As shown in Fig. 5 (a), among them, a significant portion (13 studies) enhance existing BPS model components, as defined by Martins et al. [23], six studies reinvent the BPS model in a way that differs from the traditional approach, and a smaller portion (two studies) develop new BPS model components. As depicted in Fig. 4, which shows the further analysis for papers are categorized under **Enhancement** category, each component is associated with at least one paper, represented as a circle. The authors in [20] proposed a BPS model that supports capability development of human resources, which is considered an enhancement to the resource component, while [9,10] introduced a method to enhance BPS models by addressing extraneous activity delays, which are delays that occur independently of resource contention or unavailability. The intersections between the circles indicate that at least one paper contributes to multiple components. For example, [14] introduces a method to address the multitasking behavior of resources. Addressing multitasking enhances how activities can be performed by resources. The activity and resource components are associated with the most portion of the papers (seven and six papers, respectively). Additionally, four papers focus on enhancing the availability schedule of resources, and three papers work on improving entity behavior by incorporating real-life entities, which helps build dynamic BPS models for short-term simulation. Five papers enhance more than one component (see circle intersections in Fig. 4). Furthermore, two papers are categorized under the **New Components** category, which comprises papers that introduce new components with regards to [23]. [16] introduce an environmental model that enhances business process simulations by integrating external factors such as economic indicators, weather conditions, or domain-specific variables, ensuring simulations reflect real-world dependencies. Furthermore, [22] introduced the data model as a new component, which includes global attributes that influence the entire process. The authors propose that global attributes act as an additional layer, ensuring consistency in managing data across activities, decision points, and resource allocations. They enable the model to incorporate data-aware behaviors, allowing simulations to dynamically adapt to changing data conditions. The third category is **New BPS model type**; papers in this category enhance BPS models by rethinking and rebuilding them from alternative viewpoints rather than relying on traditional approaches. They contribute to BPS modeling by focusing on the concept of *First-Citizenship*, which refers to identifying the primary component that should be prioritized as the starting point when building a BPS model, for example, [17] introduced a novel approach to BPS model that emphasizes a resource-first perspective. The authors in [17] worked to enhance the BPS model and address the limitations of traditional control-flow-first simulation models, which often fail to accurately capture the dynamics of real-world processes that involve distinct resource behaviors and decentralized decision-making. The **New BPS model type** category also includes papers that develop alternative simulation methods, moving beyond the traditional approach (particularly DES) to enhance BPS models. Two papers [30,31] used system dynamics to construct BPS models, while one paper [29] employed a hybrid approach that combines DES and SD methods. Research papers that primarily focus on the techniques used to build BPS models are assigned to **Techniques** category. 11 out of 32 papers are classified under this category. The results show that six of these 11 papers primarily focus on integrating machine learning with process mining. The authors in [7] introduced *DeepSimulator*, a method that combines *Data-Driven Simulation* and *Deep Learning* (DL) to learn accurate business process simulation models from event logs. Furthermore, The authors of [8] introduced a method for automating BPS model discovery from event logs. This approach improves accuracy by integrating machine learning to optimize hyper-parameters, reducing manual tuning and enabling efficient configuration exploration. Three papers explore alternative ways of applying process mining techniques to construct BPS models; for example, [34] explored the integration of multiple process mining methods to develop robust simulation models, leveraging insights from control-flow, data, performance, and organizational perspectives. Finally, two papers introduce model transformation methods, starting from BPMN models to create executable simulation models. [6] is one of the papers belonging to this subcategory that introduced a model-driven approach for automating the transformation of BPMN models into executable simulation code using the domain-specific language BPMN.

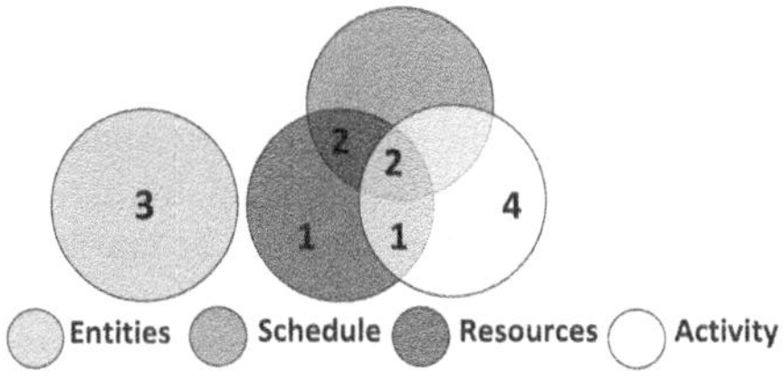

Fig. 4. Key Components in the Reviewed Papers. Each circle represents a component, with the number of contributing papers indicated. The numbers at the intersections reflect shared contributions.

4.2 The Role of the Human in Constructing BPS Models

This section examines the role of human involvement in the development of BPS models, specifically focusing on the BPM domain, where there is a notable tendency to prioritize automation. Figure 5 (c) provides an overview of the various roles of human involvement in building BPS models. Our analysis reveals that 15 papers introduced predominantly automated approaches that operate independently with minimal or no human input, reflecting the automation trend in the BPM domain. In contrast, ten papers utilized pre-knowledge or human input as initial data for simulations. [6] belongs to this category, where the authors introduced a method for building a BPS model starting from a BPMN model that is provided through documentation and human input. Furthermore, five papers include human involvement during the construction phase of BPS models; for example, [39] introduced a goal-oriented approach for dynamic BPS where humans guide activity selection and decision-making by defining and adjusting rules based on the business context. Lastly, three papers relied on human expertise for validation, where experts assessed the accuracy and reliability of the simulation results, often through iterative feedback loops.

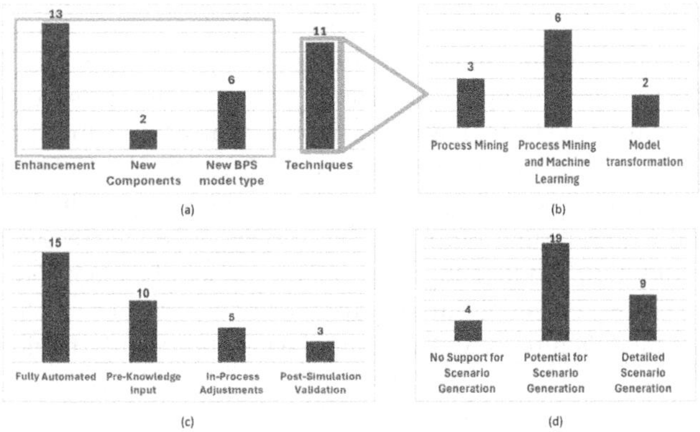

Fig. 5. Distribution of studies across BPS model development aspects: (a) Main focus and contributions, with the orange rectangle highlighting papers focused on design and modeling, and the green rectangle representing papers emphasizing techniques; (b) Breakdown of techniques from (a); (c) Role of human involvement in building BPS models; and (d) Levels of support for "what-if" scenario generation. (Color figure online)

4.3 Support for What-if Scenario Generation

Our analysis, shown in Fig. 5(d), indicates that most of the papers—19 out of 32 (i.e., more than half of the final set)—mention that their approach can support generating new scenarios. However, they do not provide detailed steps for generating these scenarios. Additionally, four papers do not clearly or explicitly indicate that their approach supports different designs. On the other hand, only nine papers (around a quarter of the total) focus extensively on how to generate new scenarios and provide detailed steps using their approaches. For example, [4] defines steps for generating new scenarios by modifying the control flow. The method involves specifying declarative control-flow changes, generating compliant event traces using deep learning, creating a stochastic process model, enriching it with simulation parameters, and analyzing performance metrics. While [30] proposes a step-by-step framework for generating new simulation scenarios by transforming event logs into aggregated SD-Logs and discovering relationships among process variables using statistical and machine learning techniques. These relationships are visualized as *causal-loop diagrams*, which are then converted into *stock-flow diagrams* for simulation and scenario-based analysis.

5 Discussion

Having analyzed the results obtained from our study, we now discuss these findings in light of the research questions outlined at the beginning of this work.

5.1 RQ1: What is the Main Focus of Papers Contributing to Building BPS Models?

Our literature review mainly focuses on BPS research in the BPM domain, and we observe three trends: *Firstly*, research recognizes that resource behavior, and especially the behavior of humans, is too robotic-like as reflected in the traditional BPS model. This has directed significant attention in the literature toward addressing the resource perspective and especially representing the behavior of human resources in a more realistic way. Similarly, activities are often represented in a rigid and oversimplified manner, failing to capture the dynamic and adaptive nature of human decision-making and interactions within processes. *Secondly*, with the increased availability of process data, research is promoting using data-driven techniques to create accurate and reliable BPS models. These techniques leverage event logs and other process-related data sets to enhance model precision, capture real-world variability, and improve the alignment of simulations with actual process behaviors. *Thirdly*, we observe that contributions promoting the usage of alternative BPS model paradigms, such as those focusing more on the resource perspective or employing alternative simulation methods like ABS, are gaining traction. These approaches are particularly helpful in scenarios where human behavior, interactions, or decentralized decision-making play a critical role, as they allow for a more granular and dynamic representation of processes than traditional methods.

5.2 RQ2: What is the the Role of Human Involvement in Developing BPS Models?

With the increasing availability of process data and automation capabilities, many approaches have primarily focused on exploring the possibilities of automation and pushing its boundaries. Fully automated BPS methods began emerging after 2020, driven by Camargo et al. [8] groundbreaking work, which sparked further research. Many studies have since built on his contributions, adding features like multitasking [14] and flexible calendars [21], leading to more adaptive and efficient simulations. However, the integration of human involvement in these processes has not been extensively researched as it requires expertise in organizational processes. The next phase of research should aim to bridge this gap by focusing on combining the strengths of both manual and automated approaches. Instead of treating them as separate or opposing methods, the goal should be to develop frameworks that seamlessly integrate human expertise with automation to achieve more effective and adaptable BPS. This finding also underscores the main motivation of this review: while BPS is a highly valuable tool for organizations, it is not widely used, even with significant advancements in model development and the increasing availability of event logs [18], which serve as a starting point for building BPS models. One of the main challenges lies in the need for tailored BPS models that reflect an organization's specific processes. These models often rely on real-world event logs, which frequently suffer from quality issues [15], such as missing or inconsistent data. This highlights the importance of incorporating human involvement in BPS model development to complement automated methods. The review also found that 16 papers incorporated some level of human involvement at different stages of BPS model development, depending on the specific requirements of the process. However, relying entirely on human involvement to build BPS models is often time-consuming and impractical, as highlighted by [11]. These findings align with the conclusion drawn by [5], which emphasizes the importance of achieving the optimal balance between human involvement and automated methods in data-driven processes. Human involvement does not mean replacing automation or relying solely on manual processes; rather, it involves integrating human expertise at critical points to address data quality issues and ensure the model reflects real-world processes accurately. This challenge could be considered an obstacle to implementing BPS in practice, emerging after 2020 from efforts toward automation. In comparison, the previous study identified tool complexity and the gap between users and experts as obstacles to BPS application.

5.3 RQ3: How is Scenario Generation Supported in the Context of BPS?

Our findings show that more than half of the reviewed papers provide potential support for scenario generation. This indicates that most of the papers focus primarily on building BPS model, which serves as the core and foundation of BPS [21]. While these papers mention that their approaches can be used to generate new scenarios, they do not provide detailed, step-by-step guidance on how to structure scenario generation. This highlights a gap in current research. This gap might be due to the complexity of predicting human behavior in business processes in future. Human behavior is dynamic and context-dependent, and forecasting it accurately is particularly challenging as circumstances evolve [28] also an observation mentioned in [33], where simulation complexity is identified as one of the obstacles to applying BPS in practice.

5.4 RQ4: What are the Key Research Directions and Opportunities for Advancing BPS Model Development and Addressing Current Gaps in the Field?

Having investigated studies on constructing BPS models within the BPM and process mining domains, we propose the following impulses (I) for researchers and practitioners in the field of BPS, as summarized in Table 3. (I1) While some work incorporates human involvement in simulation, human involvement is defined at specific points. However, achieving an optimal balance between automation and human involvement, with flexibility for intervention at different stages, is crucial. Inspired by the mode-switching concept for process discovery [35], this approach can be adapted to simulations. For example, during automated simulations, feedback may highlight an issue, allowing users to switch to manual mode, make adjustments, and return to automation. (I2) Although some papers provide step-by-step approaches for scenario generation, these efforts remain limited. Expanding on these initial contributions, future research should aim to develop comprehensive and structured frameworks for automated scenario generation, with a focus on ensuring rationality and addressing automation risks. (I3) Addressing the pervasive issue of poor data quality in event logs is critical. Researchers and practitioners should focus on developing data cleaning, augmentation, and validation tools supported by human oversight to ensure robust model inputs and outputs. (I4) While traditional simulation methods like DES dominate the field, existing work on alternative paradigms such as ABS and SD highlights the potential to further explore and expand their application. These approaches are particularly effective in capturing decentralized decision-making and human interactions.

Table 3. Research Opportunities for Advancing BPS Models

I	Research Opportunities
1	Investigate approaches to hybrid simulation models that integrate automation with human involvement, enabling adaptable intervention to address critical needs during simulations
2	Design structured frameworks for scenario generation that leverage automation while ensuring rational decision-making and mitigating risks
3	Expand the application of alternative simulation paradigms, including ABS and SD, to improve the modeling of decentralized processes and human-driven interactions
4	Address challenges in event log quality by developing robust methodologies for cleaning, enhancing, and validating data, with strategic human oversight as needed

This paper proposes a roadmap that outlines key research directions for advancing BPS model development; the key focus for future research in BPS is finding the optimal balance between human involvement and automation in BPS models. This balance is crucial for enhancing the accuracy and truthfulness of these models. While automation can efficiently handle repetitive tasks and process large datasets at scale, human involvement is essential for tasks requiring complex judgment and understanding of human behavior. Therefore, human involvement should be strategically integrated where it adds the most value. To move forward, the first step in this process should be to identify the specific requirements for human involvement—determining when and where human input is necessary. This includes understanding which stages of the process can be effectively automated and where human expertise is indispensable for decision-making. In the second step, future research can expand this framework to support the gap in scenario generation by integrating automated techniques, such as machine learning, with human decision-making processes. These techniques can serve as prediction tools, while human involvement ensures validation and adjustment of decisions during the process. By systematically studying these areas and building on these insights, BPS models can evolve to achieve a dynamic balance between human expertise and automation, enhancing their accuracy, flexibility, and applicability.

5.5 Threats to Validity

In a SLR, biases in study selection and data extraction can threaten the validity of the results. First, it was challenging to determine whether a paper should be included based solely on reading the abstract, particularly when assessing whether the contribution focused on building BPS. This uncertainty might have led to the inclusion of irrelevant or low-quality studies. To address this risk, we adopted a rigorous search strategy with a multi-step selection process. Initially, we included all papers that were undoubtedly relevant, as well as those with uncertain relevance. In a second round, we performed a full-text review to ensure the inclusion of papers that met the criteria. Moreover, the search terms may not have covered all related papers, potentially resulting in missed relevant studies. To mitigate this, we employed backward and forward snowballing techniques to identify additional studies that might not have been captured through the initial search. Another potential threat to the validity of this work is the subjective interpretation of extracted data, which may lead to incorrect classification. To mitigate this risk, we conducted multiple discussion and validation rounds among the authors to ensure consistency and reduce biases. Additionally, a second author independently took random samples and double-checked the classification to enhance the accuracy and reliability of the analysis.

6 Conclusion

To conclude, this review analyzed 32 papers published between 2009 and 2025, focusing on various aspects of building BPS models. Most papers concentrate on designing BPS models, particularly by enhancing BPS components to closely mimic reality. Resources and activities are the most studied components due to their critical importance in BPS models. Process mining remains the favored technique for building BPS models; however, it is increasingly enhanced by combining it with machine learning techniques. Human involvement is absent in many approaches, highlighting the need for an optimal balance between human input and automation to address data quality issues and improve model accuracy. Research on scenario generation remains limited, with many papers offering only potential support without detailed methodologies. This review is limited by its focus on the BPM domain and process mining, which may have excluded relevant interdisciplinary contributions. Additionally, search terms tailored to BPM and process mining may have overlooked alternative terminology, such as related concepts like discrete event simulation or process optimization, even though they might address aspects of simulating and analyzing business processes. Future work should aim to broaden the scope by incorporating insights from related fields,refining scenario frameworks, and identifying essential human input for optimal automation balance.

References

1. van der Aalst, W.M.: Business process simulation revisited. In: EOMAS (2010)
2. Amajuoyi, P., Benjamin, L.B., Adeus, K.B.: Agile methodologies: adapting product management to rapidly changing market conditions. GSC Adv. Res. Rev. **19**(2), 249–267 (2024)
3. An, L., Jeng, J.: On developing system dynamics model for business process simulation. In: Proceedings of the 37th Winter Simulation Conference, Orlando, FL, USA, 4–7 December 2005, pp. 2068–2077 (2005)
4. Barón-Espitia, D., Dumas, M., González-Rojas, O.: Automated generation of process simulation scenarios from declarative control-flow changes. PeerJ Comput. Sci. (2024)
5. Beerepoot, I., et al.: The biggest business process management problems to solve before we die. Comput. Ind. **146**, 103837 (2023)
6. Bocciarelli, P., D'Ambrogio, A., Paglia, E.: A language for enabling model-driven analysis of business processes. In: 2014 MODELSWARD (2014)
7. Camargo, M., Báron, D., Dumas, M., González-Rojas, O.: Learning business process simulation models: a hybrid process mining and deep learning approach. IS **117**, 102248 (2023)
8. Camargo, M., Dumas, M., González-Rojas, O.: Automated discovery of business process simulation models from event logs. Decis. Supp. Syst. (2020)
9. Chapela-Campa, D., Dumas, M.: Modeling extraneous activity delays in business process simulation. In: 2022 ICPM (2022)
10. Chapela-Campa, D., Dumas, M.: Enhancing business process simulation models with extraneous activity delays. IS **122**, 102346 (2024)
11. Dixit, P.M., Buijs, J.C., van der Aalst, W.M.: Prodigy: human-in-the-loop process discovery. In: RCIS 2018 (2018)
12. Dumas, M., Rosa, L.M., Mendling, J., Reijers, A.H.: Fundamentals of business process management (2018)
13. Estrada-Torres, B., Camargo, M., Dumas, M., García-Bañuelos, L., Mahdy, I., Yerokhin, M.: Discovering business process simulation models in the presence of multitasking and availability constraints. Data Knowl. Eng. **134**, 101897 (2021)

14. Estrada-Torres, B., Camargo, M., Dumas, M., Yerokhin, M.: Discovering business process simulation models in the presence of multitasking. In: RCIS, pp. 381–397 (2020)
15. Goel, K., Leemans, S.J., Martin, N., Wynn, M.T.: Quality-informed process mining: a case for standardised data quality annotations. ACM Trans. Knowl. Disc. Data (TKDD) **16**(5), 1–47 (2022)
16. Khodyrev, I., Popova, S.: Discrete modeling and simulation of business processes using event logs. Procedia Comput. Sci. **29**, 322–331 (2014)
17. Kirchdorfer, L., Blümel, R., Kampik, T., Van der Aa, H., Stuckenschmidt, H.: Agentsimulator: an agent-based approach for data-driven business process simulation. In: 2024 ICPM, pp. 97–104 (2024)
18. Landauer, M., Onder, S., Skopik, F., Wurzenberger, M.: Deep learning for anomaly detection in log data: a survey. ML Appl. (2023)
19. Law, A.M.: Discrete-event and agent-based simulation and where to use each. In: 2015 WSC, pp. 1866–1866. IEEE (2015)
20. Ling, J., Feng, Q., Zhang, L.: A business process simulation method supporting resource evolution. In: ICSE, pp. 169–177 (2014)
21. López-Pintado, O., Dumas, M., Berx, J.: Discovery, simulation, and optimization of business processes with differentiated resources. IS **120**, 102289 (2024)
22. López-Pintado, O., Murashko, S., Dumas, M.: Discovery and simulation of data-aware business processes. In: 2024 ICPM, pp. 105–112 (2024)
23. Martin, N., Depaire, B., Caris, A.: The use of process mining in business process simulation model construction: structuring the field. BISE **58** (2016)
24. Martin, N., Depaire, B., Caris, A.: The use of process mining in a business process simulation context: overview and challenges. In: 2014 IEEE CIDM (2014)
25. Măruşter, L., Van Beest, N.R.: Redesigning business processes: a methodology based on simulation and process mining techniques. IS **21** (2009)
26. Meneghello, F., Di Francescomarino, C., Ghidini, C.: Runtime integration of machine learning and simulation for business processes. In: ICPM (2023)
27. Okoli, C., Schabram, K.: A guide to conducting a systematic literature review of information systems research (2015)
28. Owen, C., Albores, P., Greasley, A.: The challenges of representing human behaviour in simulation models: some case examples from supply chain, evacuation modelling and rail disruption (2014)
29. Pourbafrani, M., van der Aalst, W.M.: Hybrid business process simulation: updating detailed process simulation models using high-level simulations. In: RCIS (2022)
30. Pourbafrani, M., van Der Aalst, W.M.: Discovering system dynamics simulation models using process mining. IEEE Access **10**, 78527–78547 (2022)
31. Pourbafrani, M., van Zelst, S.J., van der Aalst, W.: Supporting automatic system dynamics model generation for simulation in the context of process mining. In: Abramowicz, W., Klein, G. (eds.) BIS 2020. LNBIP, vol. 389, pp. 249–263. Springer, Cham (2020). https://doi.org/10.1007/978-3-030-53337-3_19
32. R., A., Singh, S.: Simulation and computer simulations: operations management (2016)
33. Rosenthal, K., Ternes, B., Strecker, S.: Business process simulation: a systematic literature review. In: ECIS, p. 199 (2018)
34. Rozinat, A., Mans, R.S., Song, M., van der Aalst, W.M.: Discovering simulation models. IS **34**, 305–327 (2009)
35. Schuster, D., van Zelst, S.J., van der Aalst, W.M.: Utilizing domain knowledge in data-driven process discovery: a literature review. Comput. Ind. (2022)

36. Siebers, P.O., Macal, C.M., Garnett, J., Buxton, D., Pidd, M.: Discrete-event simulation is dead, long live agent-based simulation! J. Simul. (3) (2010)
37. Sumari, S., Ibrahim, R., Zakaria, N.H., Ab Hamid, A.H.: Comparing three simulation model using taxonomy: system dynamic simulation, discrete event simulation and agent based simulation. IJME (2013)
38. Van Der Aalst, W.M.: Business process simulation survival guide. In: Handbook on Business Process Management 1: Introduction, Methods, and Information Systems (2014)
39. Vasilecas, O., Savickas, T., Normantas, K., Vysockis, T., Kalibatiene, D.: A goal-oriented approach to dynamic business process simulation. In: Databases and Information Systems IX, pp. 143–154 (2016)
40. Wynn, M.T., Dumas, M., Fidge, C.J., Ter Hofstede, A.H., Van Der Aalst, W.M.: Business process simulation for operational decision support. In: BPM Workshops (2008)

Exploring Webcam Eye Tracking Software for Robotic Process Automation: A Pilot Benchmarking Study

Manuel García-Romero(✉), Antonio Martínez-Rojas, José González Enríquez, and Andrés Jiménez-Ramírez

Department of Computer Languages and Systems, University of Seville, Avenida Reina Mercedes, s/n, 41012 Seville, Spain
{mgarcia44,amrojas,jgenriquez,ajramirez}@us.es

Abstract. In the literature, it has been proven that the distinction between relevant and irrelevant information through eye tracking is a technique that optimizes Robotic Process Automation (RPA) processes. To implement this, an eye tracking tool must be installed, both in terms of hardware and software. This paper addresses the need for accessible and cost-effective technologies to enhance productivity and efficiency through user-eye movement tracking. We review current webcam-based eye tracking software, assessing their availability, accessibility, and accuracy in estimating the user's Point of Gaze (POG) on a standard computer setup. Through a systematic review and a pilot benchmarking study, we evaluate *GazeRecorder*, *Webgazer.js*, and *EyeDid* webcam-based eye tracking software on various metrics. Our findings reveal significant performance variations, indicating that while webcam-based eye tracking software may not match the high precision of specialized devices, they offer a viable alternative for RPA applications requiring user interaction analysis. This research improves our understanding of webcam eye tracking capabilities and limitations in RPA, providing insights that could aid the integration of eye tracking technologies into business processes.

Keywords: Robotic Process Automation · Webcam · Eye tracking · Robotic Process Mining · Information filtering

1 Introduction

In the rapidly evolving digital business landscape, where competition is fierce, RPA has emerged as an indispensable tool for organizations undergoing digital transformation. Its primary function is automating repetitive and manual tasks, allowing employees to free themselves from mundane work responsibilities [2].

This research was supported by the EQUAVEL project PID2022-137646OB-C31, funded by MICIU/AEI/10.13039/501100011033 and by ERDF, EU.; the grant FPU20/05984 funded by MICIU/AEI/10.13039/501100011033 and by ESF+.

J. Grabis et al. (Eds.): RCIS 2025, LNBIP 547, pp. 470–485, 2025.
https://doi.org/10.1007/978-3-031-92474-3_28

This liberation empowers them to redirect their efforts toward more strategic and creative endeavors, thus contributing to overall productivity and innovation.

Mining techniques for RPA, such as Robotic Process Mining (RPM) [21] or Task Mining [1,28], play a crucial role in identifying process models that describe human behavior, thus helping automate processes. These techniques rely on User Interface (UI) logs [3,16], which register user actions within a graphical interface, encompassing actions such as mouse clicks and keystrokes. Including screenshots in these UI logs preserves a comprehensive record of user interactions and enables working in environments where the UI can not be processed (e.g., virtual environments such as Citrix) [16].

Nevertheless, a challenge arises due to the huge amount of data contained in the screenshots, i.e., the UI elements displayed on them. Although some of this information may be relevant for the process, others may not [25]. Consequently, extracting meaningful insights from UI logs becomes challenging, especially in terms of efficiency when dealing with complex and information-dense graphical user interfaces. To address this problem, eye tracking technology can be applied to capture the points of the screen where the user focuses her gaze while interacting with the user interface [26]. With additional information from the eye tracking systems, such as the Point of Gaze (POG), gaze fixations, and dispersion [29], it is possible to distinguish between relevant and irrelevant data in screenshots [25].

In the industry, a diverse range of eye tracking tools is available, ranging from those that need specific software and hardware equipment, such as infrared eye tracking tools, to those that use only the built-in webcam or a USB webcam from a standard computer or a smartphone camera [22]. Nowadays, infrared-based eye tracking technology is the most used to measure and record a person's eye movements to understand human visual behavior because of its high accuracy and precision [11]. However, they are still expensive, making this technique unsuitable for many potential users. In contrast, webcam-based eye tracking can offer a more cost-effective tool to the general public, eliminating the need for additional costly equipment. Although it does not offer the same accuracy and precision as dedicated infrared eye tracking tools, webcam-based software still provides valuable information on eye movements and visual behavior [11].

This pilot study seeks to explore and review webcam-based eye tracking software that exists in the industry, supplemented by a benchmark to assess their accuracy and precision in predicting POG on the screen. This work facilitates the use of eye tracking software to incorporate eye tracking data into RPA processes and, in particular, into RPM and Task Mining to improve efficiency in identifying model processes to automate. To meet this objective, we have established two research questions:

- **RQ.1:** *What are the webcam-based eye tracking software accessible by potential users in terms of technical performance, availability of recorded data, and support documentation to use them in RPM?.*

  To analyze and compare obtained eye tracking software, we conducted a preliminary benchmark of the selected ones to answer the following question:

- **RQ.2:** *What are the levels of accuracy and precision of the identified webcam-based eye tracking software in estimating the user's point of gaze (POG) on the screen within the scope of a pilot evaluation?*

The remainder of the paper is organized as follows: Sect. 2 introduces the background concepts. Section 3 describes the related work. Section 4 reports a systematic review of eye tracking software. Section 5 evaluates the identified eye tracking software and analyses the results. Section 6 discusses the results and their implications. Finally, Sect. 7 summarizes this research proposal and describes related future work.

2 Background

2.1 Eye Tracking

Eye tracking is a technique that involves recording and analyzing the movements of a person's eyes as they observe a visual stimulus, such as an image, video, or user interface. [10,15]. This technique provides valuable information on visual attention, cognitive processing, and user preferences. To obtain this information, the points on the screen at which human gazes can be grouped into different eye movements. The two most representative are next:

- **Fixations:** periods of stable gaze when the eyes are relatively still.
- **Saccades:** rapid, jerky eye movements occur when the eyes shift from one fixation point to another.

In addition, precision and accuracy are important concepts for understanding and evaluating the quality of eye tracking data [13]:

- **Accuracy:** offset between the actual gaze position and what the eye tracking software recorded as the gaze position. This systematic error represents the spatial accuracy of the eye tracking data. It is also known as the angular error.
- **Precision:** the measure of variation in the recorded data. It is defined as the ability of an eye tracking software to reliably reproduce the exact gaze point measurement from one sample to the next. Precision is calculated via the sampled points Root Mean Square (RMS).

2.2 Filtering in RPM

Extracting meaningful insights from UI logs in RPA and RPM can be challenging, particularly when dealing with large volumes of complex information from graphical user interfaces. Using the gaze for filtering UI elements, thus reducing the number of features extracted from the screen, can improve computational efficiency and potentially the accuracy of subsequent analysis in RPM [23]. Two gaze filtering algorithms can reduce irrelevant UI elements while retaining meaningful information on the screen [25]. These two alternatives for gaze filtering are pre-filtering and post-filtering:

- **Pre-filtering:** This process entails overlaying a gaze-based mask on the relevant screenshot associated with the UI log [26]. Consequently, the screenshot will preserve the areas of fixation while disregarding the rest. The algorithm iterates through each fixation event and draws an attention area as a circle on the mask. This attention area has the fixation point as the center, and the dispersion determines its radius.
- **Post-filtering:** It excludes irrelevant information based on the percentage determined by a specified threshold where the UI element intersects with the fixation area, defined by a specified threshold percentage.

3 Related Work

In recent years, eye tracking has garnered considerable attention in various scientific domains, notably in data visualization and decision-making. This surge of interest extends to various disciplines, as evidenced by studies in the fields of marketing [31], tourism [30], education [4], or medicine [5].

In an in-depth study of webcam-based eye tracking software, Heck et al. [7] conducted a systematic review to characterize publicly available software for webcam eye tracking that estimates the POG. However, all the work remains theoretical as neither the software's proper functioning nor the POG estimation quality were tested.

No such dedicated study includes a benchmark specifically for webcam-based eye tracking software. However, there are related studies in the field. Burton et al. [11] supervise the work on remote, visible light gaze trackers and analyze the techniques from various perspectives, such as calibration strategies, head pose invariance, and gaze estimation techniques. Kurzhals et al. [19] perform a benchmark of visualization and analysis techniques in the eye tracking of video stimuli. Banitalebi-Dehkordi et al. [6] evaluated the performance of visual attention models generated from eye tracking datasets. Blakey et al. [8]. define gaze tracking metrics, noting a growing disparity between 2D and 3D eye tracking, and highlight how the accuracy reported by different developers of infrared and webcam-based eye tracking tools can be misleading. They provide data on accuracy measurement for both 2D and 3D eye tracking but do not establish any specific benchmark.

Unlike previous studies, this work fills a gap by not only providing a systematic review but also conducting an exploratory pilot benchmark of currently available webcam-based eye tracking software. We provide preliminary insights into the accessibility and usability of webcam-based eye tracking software for users, with a specific focus on their potential suitability for integration into RPA and RPM workflows.

4 Industrial Review

This section conducts a systematic industrial review to identify and select webcam-based eye tracking software that is available and accessible to users.

Table 1. Inclusion and exclusion criteria.

Criteria	Description
C1	The software must be available and documented or supported
C2	The software must work on desktop computers or laptops, excluding mobile devices
C3	The software should be open-source and/or free; if not, a demo version should be available even if it is for a trial period
C4	The software must provide the option to use the built-in/native camera of a PC or webcam as hardware for Eye tracking. It excludes head-mounted devices, virtual reality glasses, and infrared commercial eye tracker hardware
C5	The software must provide Eye tracking as software, not as a feature or extension of a service unrelated about specifically predicting the POG as functionality
C6	The software must be installable in the benchmark test setup (cf. Sect. 5.1)

To achieve this goal, we followed the methodology of Kitchenham et al. [18], as adapted by A. Martínez-Rojas et al. [24], which is specifically tailored for conducting industrial reviews.

4.1 Planning

We plan to conduct the queries using the following web search engines *Bing*[1], *Google*[2] and *DuckDuckGo*[3].

The queries to be introduced in the web search engines were: (*Eye tracking* OR *Gaze tracking* OR *Eye tracker*) AND (*software* OR *tool* OR *github* OR *app*).

Moreover, it was planned to filter the results to include only those from January 2019 to December 2024 and focus on the first page. This approach prioritizes recent and relevant results while minimizing inconsistencies due to outdated information. In fact, these software results must be available and have active support and documentation. We specifically target webcam-based eye tracking software designed for desktop or laptop computers, which can estimate the POG on the screen. It must be available for testing, either as open source, free software, or through a demo trial. Furthermore, they must be compatible when installed and executed in the benchmark setup. These criteria are included in Table 1.

[1] *Microsoft Bing* search engine: https://www.bing.com/.
[2] *Google Search* search engine: https://www.google.com/.
[3] *DuckDuckGo* search engine: https://www.duckduckgo.com/.

4.2 Conducting

After executing all queries in search engines and removing all duplicated or repeated results, we obtained 66 results related to eye tracking software.[4].

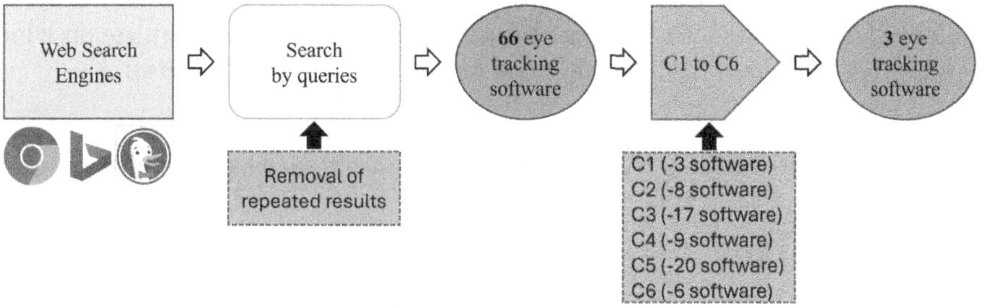

Fig. 1. Search procedure.

According to C1, 3 eye tracking software were removed due to inactive host pages and a lack of support beyond 2024. 8 software did not meet the requirements of C2 as they were exclusively compatible with mobile platforms. Furthermore, C3 excluded 17 eye tracking software due to their inability to undergo free testing (free of cost), lack of open source availability, free software designation, or provision of a trial version. C4 led to the elimination of 9 eye tracking software, which were exclusively tailored for commercial eye tracking tools and lacked compatibility with webcam cameras. According to C5, 20 eye tracking software were dismissed because they did not meet the primary purpose of predicting POG, instead focussing on secondary functions such as generating heatmaps or showing facial expressions. C6 assessed the compatibility of the eye tracking software with the requirements of the benchmark setup, resulting in the exclusion of 6 eye tracking software. Finally, 3 eye tracking software met all inclusion/exclusion criteria, becoming viable candidates for further benchmark evaluation.

A subsequent verification of the number of eye tracking software excluded under each criterion revealed that the selection process was not conducted in the most optimal manner. This is due to the sequential application of the criteria, which follows a progression from more immediate to more detailed assessments. Specifically, C1 allows for an almost instantaneous exclusion, as it verifies whether the host page is inactive. In contrast, C3 requires a more in-depth investigation to determine whether the software offers a free plan or is open-source. Similarly, C5 demands an even more thorough analysis to ascertain whether predicting the POG is among the functionalities provided.

One of the authors conducted this systematic industrial review, while the remaining authors verified the accuracy and validity of the results.

The complete procedure followed to achieve these 3 eye tracking software is illustrated in Fig. 1.

[4] These results are available at: https://doi.org/10.5281/zenodo.11125114.

4.3 Reporting

The webcam-based eye tracking software met the inclusion/exclusion criteria: *Eyedid* (previously known as *Seeso*) [12], *GazeRecorder* [14], and *Webgazer.js* [27]. Table 2 describes the availability of the eye tracking software support and documentation (i.e., Doc. and Support column), platforms on which they are available, usage license, the purpose of use, and programming language on which they are based. These findings allow us to address **RQ.1** (cf. Section 1).

Table 2. Webcam-based eye tracking software obtained from conducting the industrial review.

Software	Ref.	D.&.S*	Platform	License	Purpose	Language
Webgazer.js	[27]	Yes	Multiplatform	Opensource GPLv3	POG-p*	JavaScript
Eyedid	[12]	Yes	Multiplatform	CC-BY-NC-SA (Trial Demo)	POG-p*	JavaScript
GazeRecorder	[14]	Yes	Desktop (Windows)	CC-BY-NC (Trial Demo)	POG-p*, AOI	C#, .NET, JavaScript

* POG-p = POG prediction. D.&.S = Documentation and Support.

For the selected eye tracking software, *GazeRecorder* and *Eyedid* are commercially licensed, although with a free trial. In contrast, *Webgazer.js* is open source.

Both *Webgazer.js* and *Eyedid* are platform-agnostic, operating seamlessly on both PC and mobile platforms as browser-based applications, thus remaining independent of the underlying operating system. In contrast, *GazeRecorder* is only compatible with computers running Windows operating system.

Both *Webgazer.js* and *Eyedid* provide POG during the eye tracking process. Additionally, *GazeRecorder* offers the functionality to display the POG on the screen and heatmaps.

Finally, *Eyedid* and *Webgazer.js* are based on JavaScript and Python programming languages, while *GazeRecorder* is developed by JavaScript and .NET framework.

5 Eye Tracking Software Benchmark

This section focuses on answering **RQ.2**: *What are the levels of accuracy and precision of the identified webcam-based eye tracking software in estimating the user's POG on the screen within the scope of a pilot evaluation?* To this end, a benchmark of the webcam-based eye tracking software obtained from the industrial systematic review (cf. Sect. 4.3) is performed.

5.1 Setup

To benchmark the identified webcam-based eye tracking software, the following setup has been used:

- *Elgato Facecam* webcam 1080p and 60 FPS.
- *Lenovo Legion 5* laptop with AMD Ryzen 5800H processor, Nvidia RTX 3060 graphics card, 16 GB RAM, 1GB SSD, and 15.6-inch screen. Windows 10 is the operating system.
- 27-inch MSI monitor (602 × 338 mm) on a desktop table with the webcam installed at the top center of the monitor.
- Resolution: 1920 × 1080 pixels (1080p) and 81 ppi (pixels per inch), equivalent to 32 pixels per centimeter.
- Eye-to-screen distance: 60 cm and 0° offset.
- The subject is seated on a chair, looking straight ahead at the screen.
- Artificial light is provided by a desk lamp located behind the monitor.

The benchmark evaluation was carried out by a subject, male gender, 26 years old, 181 cm tall, and wearing soft contact lenses. The setup is displayed in Fig. 2.

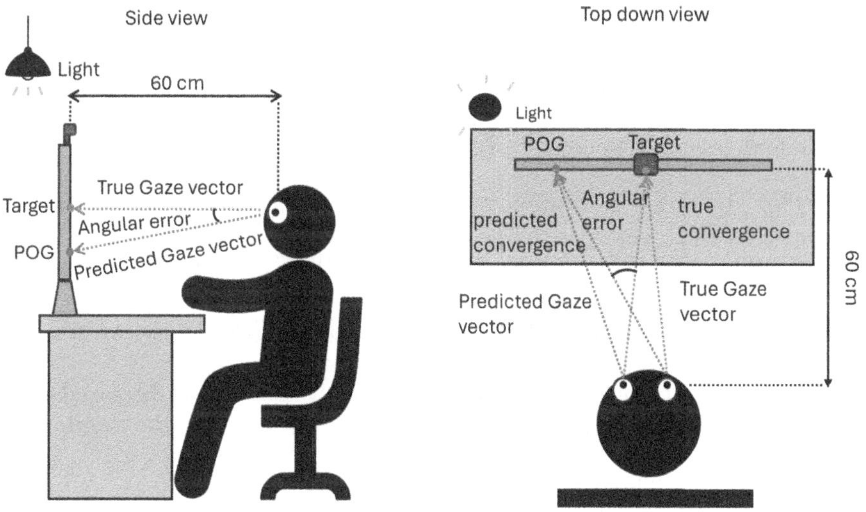

Fig. 2. Experimental Setup. The arrangement of the subject, monitor, and webcam is illustrated from two perspectives: side view and top-down view.

5.2 Design

This section describes the test objects, the benchmark metrics, and how the benchmark is performed.

Four static screens are proposed as test objects (cf. Fig. 3). These test objects differ in their visual elements, increasing complexity from one to the next. Each test object consists of a target with two concentric circles: a 1 cm radius green circle, where the user should focus their gaze, and a 2 cm radius blue circle, specifically used to measure the match on the close target in the POG coincidence category tests as will be discussed in this section.

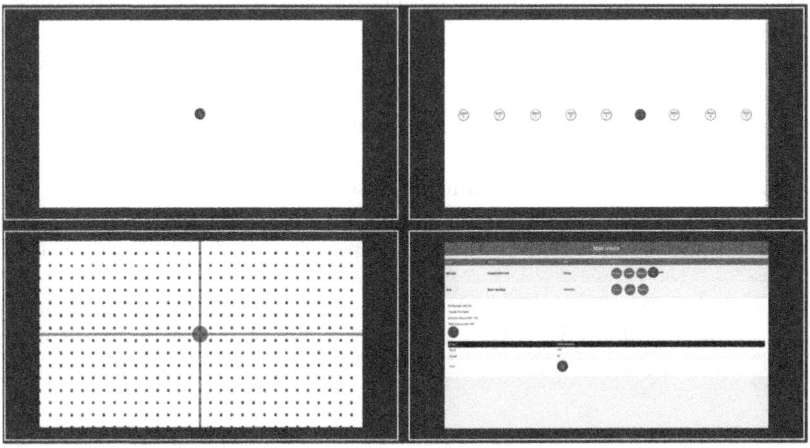

Fig. 3. Tests objects designed for benchmark. Top-left: **Circle**. Top-right: **Buttons**. Bottom-left: **Cartesian system**. Bottom-right: **Invoice email**. (Color figure online)

The four test objects consist of the following:

- **Circle**: The target is located in the center of the screen. The screen's background color is white, with the absence of other elements apart from the target. This test object is the simplest regarding UI elements displayed on the screen.
- **Buttons**: Nine 2 cm circles are displayed horizontally along the white screen. The sixth circle, starting from the left, is the target.
- **Cartesian system**: A coordinates Cartesian system map is displayed on the screen. The target coincides in position with the point at the origin of the Cartesian coordinate system.
- **Invoice email**: The process in which a user downloads an invoice, obtains its value, and records it in a table cell is simulated. Three targets are displayed in (1) the invoice download button, (2) the invoice value, and (3) the empty space in the row where the invoice value must be placed. This test object is the most complex regarding UI elements displayed. In addition, it represents a typical process studied in RPA to be automated, based on transferring data from one UI application to another [20].

Three metrics are defined in total to evaluate the POG estimated by the eye tracking software in the test objects: (1) accuracy, (2) precision, and (3) coincidence. Nonetheless, different aspects are measured in each test object to obtain these metrics. For that reason, these metrics are differentiated into two distinct benchmark categories as follows:

1. **POG accuracy and precision category**. Measures the accuracy and precision of the eye tracking software in the *Cartesian system* test object. This test object is specifically designed for this benchmark category.
   (a) **Accuracy (Acc)**. Quantifies the angular error between the point representing the centroid of the estimated POG from the eye tracking software and the point representing the centroid of the target at which the subject is gazing.
   (b) **Precision**. It is calculated by the RMS of the sampled POGs estimated from the screenshots in one test session.
2. **POG coincidence category**. Classifies where the estimated POG is located in each screenshot by percentages. This category applies to *Circle*, *Buttons*, and *Invoice email* test objects.
   (a) **Match on target**, i.e., the 1cm diameter green circle area. The estimated POG centroid by the eye tracking software on the screenshot is inside the green circle area or touches its circumference or limit.
   (b) **Match on close target**, i.e., the 2 cm diameter blue circle area concentric to the green circle (target). The estimated POG by the eye tracking software on the screenshot is inside the blue circle area or touches its circumference or limit.
   (c) **Off-target**. The estimated POG by the eye tracking software on the screenshot is outside the target limits, i.e., green and blue circles.

Test sessions have to be performed to obtain these defined metrics. Before every test session, the eye tracking calibration needs to be completed. During a test session, the subject must look at the green area of the target. In the *Invoice email* test object specifically, there are three different targets where the subject can look at. The user can observe any of these targets interchangeably during the test session. The screenshots obtained at the end of the test sessions will be generated by *steprecorders*, a built-in Windows application software. The subject will click every two seconds on the screen, and a screenshot will automatically be generated and saved in a UI log. After the test session, the UI log will contain a comprehensive record of all the screenshots captured.

In the POG accuracy and precision category, 2 test sessions will be conducted. A total of 20 POGs will be collected per test session.

In contrast, 3 test sessions will be conducted per test object in the POG coincidence category. A total of 50 POGs will be displayed on the screenshots per test session. Estimating the number of POGs is important for more precise comparisons of the POG coincidence percentage.

5.3 Results

The results obtained from the benchmark of the webcam-based eye tracking software are represented in Table 3 and Table 4.

Table 3. POG accuracy and precision results. Extracted from *cartesian system* test sessions.

Software	Ref.	Session	MEucD* (px)	MEucD* (cm)	Precision	Accuracy
Webgazer.js	[27]	T1	32.24	1.01	**1.13**	**0.96**
		T2	22.51	0.70	**0.75**	**0.67**
Eyedid	[12]	T1	110.30	3.45	3.91	3.29
		T2	74.70	2.33	2.40	2.23
GazeRecorder	[14]	T1	124.82	3.90	4.28	3.72
		T2	109.69	3.43	3.53	3.27

* MEucD= Mean Euclidean Distance.

Table 3 illustrates the accuracy and precision results extracted from each test session on the *Cartesian system* test object for each of the eye tracking software evaluated. The mean Euclidean distance columns in pixels and centimeters represent, respectively, for each unit of measurement, the average distance between the coordinates of the POG centroid estimated by the eye tracking software concerning the target centroid, which corresponds to the center of the screen as the test object is designed. From this information and knowing the distance between the eyes and the screen and the angle from the eyes to the screen established in the evaluation setup (cf. Fig. 2), the angular error and, therefore, the accuracy of each test can be calculated [6]. *Webgazer.js* demonstrates superior performance in terms of precision and accuracy when compared to *Eyedid* and *GazeRecorder*, as lower values closer to zero indicate better results.

Table 4 represents different percentages according to the POG matches estimated by the eye tracking software in each screenshot for each test session of each test object concerning the target displayed. The columns represented are the coincidence on the target (Match), the coincidence on the close radius (Close radius), the sum of the target coincidence (Match + Close Radius) and the off-target (Off-target). It is observed that *Webgazer.js* in the *Invoice email* test object has lower percentages results than other eye tracking software due to the subject focusing his gaze and clicking with the mouse on different targets located in various parts of the screen. Indeed, *Eyedid* and *GazeRecorder* employ a conventional calibration method that involves the observation of points in various regions of the screen, producing accuracy and precision values that are comparatively lower than those of *Webgazer.js*. *Webgazer.js* has the particularity that its way of calibrating itself is by clicking and observing the point where the click is being made. This could explain the better accuracy and precision obtained when fixations are performed on the same target during a certain time. However, it is notable that *Eyedid* exhibits a higher frequency on the close target match coincidence in the *Invoice email* test object. This fact implies that, although *Eyedid* may not reach the same accuracy and precision as *Webgazer.js* in particular locations of the screen, it shows the ability to quickly estimate the POG in proximity to the target during saccadic eye movements and fixations across various targets.

6 Discussion

This section delves into filtering information in RPM according to webcam-based eye tracking and the limitations of the benchmark conducted in this paper.

Table 4. POG Coincidence results. Extracted from *Circle*, *Buttons* and *Invoice email* test sessions.

Software	Scenario	Session Test	Match (%)	Close Radius (%)	Match + Close Radius (%)	Off-target (%)
Webgazer.js	Circle	T1	**38.00**	**46.00**	**84.00**	**16.00**
	Circle	T2	**20.00**	48.00	**68.00**	**32.00**
	Circle	T3	**22.00**	**34.00**	**56.00**	**44.00**
	Buttons	T1	**46.00**	**38.00**	**84.00**	**16.00**
	Buttons	T2	**34.00**	**36.00**	**70.00**	**30.00**
	Buttons	T3	**44.00**	**42.00**	**86.00**	**14.00**
	Email	T1	**24.00**	**32.00**	**56.00**	**44.00**
	Email	T2	14.00	**22.00**	**36.00**	**64.00**
	Email	T3	**26.00**	30.00	**56.00**	**44.00**
Eyedid	Circle	T1	20.00	32.00	52.00	48.00
	Circle	T2	2.00	**54.00**	56.00	44.00
	Circle	T3	4.00	16.00	20.00	80.00
	Buttons	T1	14.00	20.00	34.00	66.00
	Buttons	T2	4.00	16.00	20.00	80.00
	Buttons	T3	8.00	20.00	28.00	72.00
	Email	T1	14.00	16.00	30.00	70.00
	Email	T2	8.00	12.00	20.00	80.00
	Email	T3	10.00	30.00	40.00	60.00
GazeRecorder	Circle	T1	4.00	8.00	12.00	88.00
	Circle	T2	18.00	32.00	50.00	50.00
	Circle	T3	18.00	16.00	34.00	66.00
	Buttons	T1	10.00	10.00	20.00	80.00
	Buttons	T2	0.00	6.00	6.00	94.00
	Buttons	T3	18.00	18.00	36.00	64.00
	Email	T1	4.00	24.00	28.00	72.00
	Email	T2	**16.00**	14.00	30.00	70.00
	Email	T3	8.00	**32.00**	40.00	60.00

6.1 Filtering Information in RPM

The observed differences in accuracy and precision values among the evaluated webcam-based eye tracking software confirm the expected outcome, i.e., they do not exceed their commercial counterparts regarding accuracy and precision levels (cf. Table 3). This aligns with the documented accuracy range of infrared commercial eye tracking tools, which can achieve an angular error within the range of 0.10° to 0.50° [17].

The results obtained from the eye tracking software to answer RQ.2 can be useful for RPM. To perform the filtering task [25], a threshold radius for filtering must be determined. The threshold may vary depending on the case and must be configured accordingly. One approach to determining the appropriate threshold is to use a radius threshold that generally lies between 0.7° and 1.3°. This range corresponds to the acceptable dispersion degree identified by Blignaut et al. [9], which defines the variability of POGs that can be considered part of a fixation.

Therefore, it is feasible to capture regions of interest that have been relevant to the user with a reasonable degree of precision and accuracy. This depends on the specific performance metrics of the eye tracking software in use and the proper adjustment of the threshold radius. In this way, it becomes possible to filter and process information effectively, enabling the extraction of meaningful insights from UI logs and thereby improving the analysis of RPA processes for optimization purposes. All of this can be achieved by employing a standard webcam and open-source eye tracking software, which considerably reduces the cost associated with infrared eye tracking tools. Additionally, webcams are more commonly available in workplace environments compared to infrared commercial eye tracking tools, further enhancing the practicality of this approach.

6.2 Limitations

Obtained Results. These results should be interpreted with caution, as they may fluctuate depending on the specific conditions of the experiment and the characteristics of the subjects involved. Since only one subject participated in the experiment, no robust and generalizable conclusions can be drawn. This pilot study is exploratory in nature, and it would be necessary to replicate the evaluation with a larger number of participants.

Furthermore, the results show variability for the same metrics in the different test sessions for each software. This variability can be explained by the fact that the software requires prior calibration before each test execution. Since it relies on a webcam and general-purpose software-whose accuracy and precision are not comparable to those infrared eye tracking tools designed for this purpose-the calibration performed before each test may lead to variations in terms of accuracy and precision during the tests performing. Therefore, conducting a higher number of tests could yield more consistent average values.

Data Collection. The evaluated webcam-based eye tracking software did not export raw data regarding the exact coordinates of the POG natively. Hence, we took screenshots to calculate the distances of each POG centroid or the matches of the POG concerning the target of the test objects to evaluate them in a benchmark. Although it is not the most precise calculation method, it fairly evaluates the software selected for the experiment.

Test Objects Selection. All test objects were based on statically observing the circular targets on the screen. As previously mentioned, the raw data is obtained based on the distances between the POGs and these targets, derived from screenshots. However, the metrics obtained vary across the test objects, and the methodology for obtaining them also differs. For the **POG accuracy and precision category metric**, we considered 20 POGs to be sufficient to calculate accuracy and precision values. Using more POGs could introduce noise, which would noticeably affect the accuracy and precision of calculations. For instance, POGs that are far from the target due to blinking or that do not even appear in the screenshots could skew the results. Two test sessions were conducted in an initial benchmark evaluation, collecting 40 POGs, during which these issues occurred. Therefore, to make a fair comparison of the eye tracking software, we needed to ensure that each recording contained a minimum number of screenshots with minimal noise for accurate mathematical calculations. Regarding the **POG category metric**, which is expressed as percentage values, a higher level of noise can be tolerated without significantly affecting the results. This is because percentage-based metrics inherently offer greater resilience to small variations or noise in individual data points, as they reflect the overall accuracy relative to the total number of observed points.

7 Conclusions and Future Work

In RPM, it is important to filter the relevant from irrelevant features when extracting data from UI screenshots. These features play a key role in understanding user interaction with the UI and their tasks. An excess of irrelevant data within the UI can lead to noisy features. Recent studies emphasize the effectiveness of eye tracking in filtering relevant information presented in screenshots.

Eye tracking tools require a high budget that is not affordable to the general public. Webcams are a suitable alternative for the general public since most laptops integrate them. Therefore, to perform gaze tracking, only software compatible with the webcam would be necessary.

In this sense, this study has presented an industrial systematic review of the webcam-based eye tracking software industry to analyze the state of the art in this topic. Three eye tracking software were identified: *Webgazer.js*, *Eyedid*, and *GazeRecorder*. They were benchmarked through test sessions to obtain accuracy, precision, and POG coincidence metrics using specific test objects designed (cf. Fig. 3).

The results showed that the *Webgazer.js* software had significantly better accuracy and precision values compared to *Eyedid* and *GazeRecorder*. In the POG coincidence tests, the results of each eye tracking software show different behavior depending on the test object. However, as this is a pilot study, further research is needed to draw robust and generalizable conclusions.

There are several promising directions for future investigation: (1) performing a direct comparison between an infrared eye tracking tool and a webcam-based eye tracker applying filtering techniques on UI Logs to filter relevant information from UI logs; (2) expanding the capabilities of the most prominently positioned open-source eye tracking software to render them particularly effective in scrutinizing the relevance of the UI information within RPM environments; and (3) performing a new benchmark that overcomes the limitations of this evaluation and changing the setup, such as including more participants, different test objects and more test sessions.

References

1. van der Aalst, W.: Process Mining: Data Science in Action. Springer, Heidelberg (2016)
2. Van der Aalst, W.M., Bichler, M., Heinzl, A.: Robotic process automation (2018)
3. Agostinelli, S., Lupia, M., Marrella, A., Mecella, M.: SmartRPA: a tool to reactively synthesize software robots from user interface logs. In: Nurcan, S., Korthaus, A. (eds.) CAiSE 2021. LNBIP, vol. 424, pp. 137–145. Springer, Cham (2021). https://doi.org/10.1007/978-3-030-79108-7_16
4. Alemdag, E., Cagiltay, K.: A systematic review of eye tracking research on multimedia learning. Comput. Educ. **125**, 413–428 (2018)
5. Ashraf, H., Sodergren, M.H., Merali, N., Mylonas, G., Singh, H., Darzi, A.: Eye-tracking technology in medical education: a systematic review. Med. Teach. **40**(1), 62–69 (2018)
6. Banitalebi-Dehkordi, A., Nasiopoulos, E., Pourazad, M.T., Nasiopoulos, P.: Benchmark 3d eye-tracking dataset for visual saliency prediction on stereoscopic 3d video. arXiv preprint arXiv:1803.04845 (2018)
7. Blakey, W.A., Katsigiannis, S., Hajimirza, N., Ramzan, N.: Defining gaze tracking metrics by observing a growing divide between 2d and 3d tracking. In: Proceedings of the 56th Hawaii International Conference on System Sciences (HICSS), pp. 129.1–129.9 (2024)
8. Blakey, W.A., Katsigiannis, S., Hajimirza, N., Ramzan, N.: Defining gaze tracking metrics by observing a growing divide between 2d and 3d tracking. Electron. Imaging **32**(11), 129-1 (2020)
9. Blignaut, P.: Fixation identification: the optimum threshold for a dispersion algorithm. Attent. Percept. Psychophys. **71**, 881–895 (2009)
10. Brunyé, T.T., Drew, T., Weaver, D.L., Elmore, J.G.: A review of eye tracking for understanding and improving diagnostic interpretation. Cogn. Res. Princip. Implicat. **4**(1), 1–16 (2019). https://doi.org/10.1186/s41235-019-0159-2
11. Burton, L., Albert, W., Flynn, M.: A comparison of the performance of webcam vs. infrared eye tracking technology. In: Proceedings of the Human Factors and Ergonomics Society Annual Meeting, vol. 58, pp. 1437–1441. SAGE Publications Sage CA, Los Angeles (2014)
12. eyedid: Eyedid - Eye Tracking Technology. https://eyedid.ai/. Accessed May 2024
13. Feit, A.M., Williams, S., Toledo, A., Paradiso, A., Kulkarni, H., Kane, S., Morris, M.R.: Toward everyday gaze input: accuracy and precision of eye tracking and implications for design. In: Proceedings of the 2017 Chi Conference on Human Factors in Computing Systems, pp. 1118–1130 (2017)
14. GazeRecorder: GazeRecorder - Webcam-Eye-Tracking usability for usability testing. https://gazerecorder.com/gazerecorder/. Accessed May 2024
15. Gwizdka, J.: Characterizing relevance with eye-tracking measures. In: Proceedings of the 5th Information Interaction in Context Symposium, pp. 58–67 (2014)

16. Jimenez-Ramirez, A., Reijers, H.A., Barba, I., Del Valle, C.: A method to improve the early stages of the robotic process automation lifecycle. In: Giorgini, P., Weber, B. (eds.) CAiSE 2019. LNCS, vol. 11483, pp. 446–461. Springer, Cham (2019). https://doi.org/10.1007/978-3-030-21290-2_28
17. Jon, W.: Eyes are the Prize: Evaluating the benefits of eye-tracking equipment. https://uxmag.com/articles/eyes-are-the-prize-evaluating-the-benefits-of-eye-tracking-equipment. Accessed May 2024
18. Kitchenham, B., Brereton, O.P., Budgen, D., Turner, M., Bailey, J., Linkman, S.: Systematic literature reviews in software engineering-a systematic literature review. Inf. Softw. Technol. **51**(1), 7–15 (2009)
19. Kurzhals, K., Bopp, C.F., Bässler, J., Ebinger, F., Weiskopf, D.: Benchmark data for evaluating visualization and analysis techniques for eye tracking for video stimuli. In: Proceedings of the fifth Workshop on Beyond Time and Errors: Novel Evaluation Methods for Visualization, pp. 54–60 (2014)
20. Leno, V., Augusto, A., Dumas, M., La Rosa, M., Maggi, F.M., Polyvyanyy, A.: Discovering data transfer routines from user interaction logs. Inf. Syst. **107**, 101916 (2022)
21. Leno, V., Polyvyanyy, A., Dumas, M., La Rosa, M., Maggi, F.M.: Robotic process mining: vision and challenges. Bus. Inf. Syst. Eng. **63**, 301–314 (2021)
22. Lu, Z., Pesarakli, H.: Seeing is believing: using eye-tracking devices in environmental research. HERD: Health Environ. Res. Des. J. **16**(1), 15–52 (2023)
23. Martínez-Rojas, A., Jiménez-Ramírez, A., Enríquez, J., Reijers, H.: A screenshot-based task mining framework for disclosing the drivers behind variable human actions. Inf. Syst. **121**, 102340 (2024)
24. Martínez-Rojas, A., López-Carnicer, J., González-Enríquez, J., Jiménez-Ramírez, A., Sánchez-Oliva, J.: Intelligent document processing in end-to-end rpa contexts: a systematic literature review. In: Confluence of Artificial Intelligence and Robotic Process Automation, pp. 95–131 (2023)
25. Martínez-Rojas, A., Reijers, H., Jiménez-Ramírez, A., Enríquez, J.: What are you gazing at? an approach to use eye-tracking for robotic process automation. In: International Conference on Business Process Management, pp. 120–134. Springer, Heidelberg (2023). https://doi.org/10.1007/978-3-031-43433-4_8
26. Martínez-Rojas., A., Jiménez-Ramírez., A., Enríquez., J.G., Lizcano-Casas., D.: Incorporating the user attention in user interface logs. In: Proceedings of the 18th International Conference on Web Information Systems and Technologies - WEBIST, pp. 415–421. INSTICC, SciTePress (2022). https://doi.org/10.5220/0011568000003318
27. Papoutsaki, A., Sangkloy, P., Laskey, J., Daskalova, N., Huang, J., Hays, J.: Webgazer: scalable webcam eye tracking using user interactions. In: Proceedings of the 25th International Joint Conference on Artificial Intelligence (IJCAI), pp. 3839–3845. AAAI (2016)
28. Reinkemeyer, L.: Process Mining in Action. Principles, Use Cases and Outlook, Springer, Heidelberg (2020)
29. Salvucci, D.D., Goldberg, J.H.: Identifying fixations and saccades in eye-tracking protocols. In: Proceedings of the 2000 Symposium on Eye Tracking Research & Applications, pp. 71–78 (2000)
30. Scott, N., Zhang, R., Le, D., Moyle, B.: A review of eye-tracking research in tourism. Curr. Issue Tour. **22**(10), 1244–1261 (2019)
31. Wedel, M., Pieters, R.: A review of eye-tracking research in marketing. Rev. Mark. Res., 123–147 (2017)

Author Index

A
Abras, Jordan II-3
Aparicio, Amador II-229
Adamo, Greta I-156
Alfonso, Ivan II-130
Ali, Raian I-331, I-347
Almeida, João Paulo A. I-121
Alonso-Rocha, J. L. II-98
Alshakhsi, Sameha I-347
Al-Thani, Dena I-347
Aly, Adel I-262
Amous, Ikram I-37
Arden-Close, Emily I-331
Asensio, Estefanía Serral I-20

B
Babiker, Areej I-347
Babouchkine, Jean-Marc II-52
Balderas-Díaz, Sara I-52
Bangui, Hind I-87
Baños-González, Miguel II-151
Bártík, Jáchym I-245
Beerepoot, Iris I-71
Bergmann, Ralph I-20
Bertrand, Yannis I-20
Bider, Ilia II-120
Bonino da Silva Santos, Luiz Olavo I-103
Braakman, Mari A. J. I-71
Buchmann, Robert Andrei I-140
Buhnova, Barbora I-87
Burnay, Corentin II-3
Bider, Ilia II-231

C
Cabot, Jordi II-130
Calero, Coral II-151
Calhau, Rodrigo Fernandes I-121
Cancho-Casado, Jorge II-151
Carrión, Emilio I-279
Champagnat, Ronan I-3
Chen, Qian II-140
Cornax, Mario Cortes II-183
Corral-García, Javier II-151
Criado-Lozano, Pablo-Abel II-229

D
Dahache, Amira Ania I-3
Daoudi, Nadia II-130
Delgado-Pérez, Pedro I-52
Di Francescomarino, Chiara I-365
Di Mascolo, Maria II-52
Dias, Mariana II-20
Díaz, Oscar I-192
Dissegna, Sebastiano I-365
dos Reis, Gustavo Rodrigues II-183

E
El-Hajjami, Abdelkarim I-208
Enríquez, J. G. II-98

F
Fast Lappalainen, Katarina I-175
Faulkner, Stéphane II-3

G
García, Félix O. II-151
García-Romero, Manuel I-470
Garmendia, Xabier I-192
Ge, Mouzhi I-87
Gharbi, Mohamed II-52
González Enríquez, José I-470
Granda, Maria Fernanda I-313
Grüger, Joscha I-20
Gruninger, Michael II-109
Guerra, Eduardo I-103
Guerrero-Contreras, Gabriel I-52
Guimarães, Pedro II-218
Guizzardi, Giancarlo I-103, I-121
Guizzardi, Renata I-103
Gurgun, Selin I-331

J. Grabis et al. (Eds.): RCIS 2025, LNBIP 547, pp. 487–489, 2025.
https://doi.org/10.1007/978-3-031-92474-3

H

Haug, Kerstin I-418
Henkel, Martin I-175
Hoffmann, Raquel I-121
Holubová, Irena I-245
Hübner, Karin I-226

J

Jiménez-Ramírez, A. II-98
Jiménez-Ramírez, Andrés I-470
Jmal, Raouf I-37
Joudieh, Noura I-3

K

Kampourakis, Konstantinos E. II-197
Kern, Christopher Julian I-226
Khraiwesh, Samira I-452
Kired, Nour Elhouda II-67
Knies, Eva I-71
Koupil, Pavel I-245
Krauze, Beāte II-162
Kroenung, Julia I-226
Küsters, Aaron I-383

L

Labbé, Cyril II-183
Lee, Suhwan I-401
Lindeberg, Jöran I-175
Logan, James I-121
Lopes Cardoso, Henrique II-89
Lopes, Carla Teixeira II-20
Lu, Xixi I-401

M

Malburg, Lukas I-20
Maldonado, Ricardo I-313
Martínez-Rojas, A. II-98
Martínez-Rojas, Antonio I-470
Masmoudi, Mariam I-37
Medina-Bulo, Inmaculada I-52
Moher, Riley II-109
Moraga, Mł Ángeles II-151
Moreira, João Rebelo I-121
Mos, Adrian II-183
Mughaz, Dror II-38
Martínez-González, M.Mercedes II-229

N

Nuñez, Ana-Gabriela I-313

O

Oliveira, Ítalo I-103

P

Palacio, Ana León I-295
Parra, Otto I-313
Peeters, Maria I-71
Phalp, Keith I-331
Pires, Luís Ferreira I-121
Pivert, Olivier I-262
Poss, Leo I-226
Proper, Henderik I-103
Pufahl, Luise I-452
Pulido, Carlos II-151
Pérez-Fuente, Alejandro II-229

R

Ravat, Franck II-67
Reijers, Hajo A. I-71, I-401
Rinderle-Ma, Stefanie I-418, II-140
Rodrigues, José Frederico II-89
Ronzani, Massimiliano I-365

S

Saarsen, Toomas II-120
Sales, Tiago Prince I-103
Salinesi, Camille I-208
Sarmiento, María-Belén I-313
Schnellmann, Marianne II-207
Schönig, Stefan I-226
Schultheis, Alexander I-20
Schumacher, Pol I-418
Sèdes, Florence I-37
Sidorova, Natalia I-435
Song, Jiefu II-67
Sperotto, Anna I-156
Suire, Cyrille I-3
Supti, Tourjana Islam I-347
Susaiyah, Allmin I-435

T

Teixeira Lopes, Carla II-89
Teste, Olivier II-67

Thion, Virginie I-262
Trabelsi, Marwa I-3

U
Uifălean, Ştefan I-140

V
Valderas, Pedro I-279
Valle, Isadora I-103
van Wingerden, A. B. II-172
van der Aalst, Wil M. P. I-383
Verdier, Christine II-52
Venable, John R. II-228

W
Weigand, H. II-172
Wen, Lijie II-140
Willis, Max I-156
Wittges, Holger I-418

Y
Yankouskaya, Ala I-347

The manufacturer's authorised representative in the EU is Springer Nature Customer Service Centre GmbH, Europaplatz 3, 69115 Heidelberg, Germany. If you have any concerns regarding our products, please contact ProductSafety@springernature.com

Printed and bound by CPI Group (UK) Ltd, Croydon, CR0 4YY

30/09/2026

02225566-0018